History of Science and Technology in China

This is a series of handbooks with high academic values on the general history of Chinese science and technology, with contributions by top-notch scholars in this field. This 5-volume work provides an encyclopedic historical panorama of Chinese scientific and technological development. It unfolds the history of Chinese science and technology through a clarified timeline from as early as the far ancient times to the very present. This work consists of five volumes: Origins of Chinese Sciences, Ancient Chinese Studies of Heaven and Earth, High Tide of Chinese Sciences, Theoretical and Technological Development, and Western Influences.

More information about this series at http://www.springer.com/series/16685

Xiaoyuan Jiang
Editor

The High Tide of Science and Technology Development in China

History of Science and Technology in China
Volume 3

With 65 Figures and 63 Tables

上海交通大學出版社
SHANGHAI JIAO TONG UNIVERSITY PRESS

Springer

Editor
Xiaoyuan Jiang
Science History and Science and Culture
Shanghai Jiao Tong University
Shanghai, China

Translated by
Haoli Zhang
East China University of Science and Technology
Shanghai, China

Jianping Yu
East China University of Science and Technology
Shanghai, China

上海交通大學出版社
SHANGHAI JIAO TONG UNIVERSITY PRESS

ISSN 2730-910X
ISSN 2730-9118 (electronic)
ISBN 978-981-15-7846-5
ISBN 978-981-15-7847-2 (eBook)
ISBN 978-981-15-7848-9 (print and electronic bundle)
https://doi.org/10.1007/978-981-15-7847-2

Jointly published with Shanghai Jiao Tong University Press

The print edition is not for sale in The Mainland of China. Customers from The Mainland of China please order the print book from Shanghai Jiao Tong University Press.

Translation from the language edition: 正午时分 by Xiaoyuan Jiang, © Shanghai Jiao Tong University Press 2016. Published by Shanghai Jiao Tong University Press. All Rights Reserved.

This Springer imprint is published by the registered company Springer Nature Singapore Pte Ltd.
The registered company address is: 152 Beach Road, #21-01/04 Gateway East, Singapore 189721, Singapore

Contents

Contributors

Lisheng Feng The Institution for the History of Science & Technology and Ancient Texts, Tsinghua University, Beijing, China

Zengjian Guan School of History and Culture Science, Shanghai Jiao Tong University, Shanghai, China

Xiaoyuan Jiang School of History and Culture Science, Shanghai Jiao Tong University, Shanghai, China

Yuqun Liao The Institute for the History Natural Sciences, Chinese Academy of Sciences, Beijing, China

Huajie Liu Department of Philosophy and Religious Studies, Peking University, Beijing, China

Shuyong Liu Department of Physics, Capital Normal University, Beijing, China

Yunli Shi Department for the History of Science and Scientific Archaeology, University of Science and Technology of China, Hefei, China

Fei Xu Department of Philosophy of Science and Technology, University of Science and Technology of China, Hefei, China

Zelin Xu Institute of History, School of Humanities, Donghua University, Shanghai, China

Xiongsheng Zeng The Institute for the History of Natural Sciences, Chinese Academy of Sciences, Beijing, China

Exchange and Comparison in Astronomy Between China and Other Countries

1

Xiaoyuan Jiang

Contents

X. Jiang (✉)
School of History and Culture Science, Shanghai Jiao Tong University, Shanghai, China
e-mail: xyjiang@sjtu.edu.cn

X. Jiang (ed.), *The High Tide of Science and Technology Development in China*,
History of Science and Technology in China,
https://doi.org/10.1007/978-981-15-7847-2_1

Abstract

Astronomy in ancient China originated very early in history, so it is impossible that it was introduced from the west later. The origin of astronomy in China was closely tied to the origin of the Chinese civilization. Exchanges in astronomy between China and other countries began from ancient times. The eastward spreading of Buddhism, the expansion of the Mongolian Empire, and the full-scale invasion of Western astronomy from Late Ming Dynasty to Early Qing Dynasty have all influenced the development of astronomy in China.

Keywords

Astronomy · Gai Tian Model · Zhou Bi Suan Jing · Temperate zones · The Theory of the Seven Celestial Bodies · Jiu Zhi Li · Arabic astronomical instruments · Hui Hui Si Tian Tai · The Chongzhen Calendar · "Western Sciences Originated in China" Theory · Emperor Kangxi

1.1 Early Exchanges in Astronomy Between China and Other Countries

1.1.1 Origin of Astronomy in China

The origin of astronomy in China is closely tied to the origin of the Chinese civilization. For centuries, western and Japanese scholars had claimed that civilization and astronomy were introduced to China from the west.

We had discussed the inseparable relationship between astronomy and monarchy in ancient China, based on which we explained the cultural function of astronomy in China. Meanwhile, astronomy in ancient China was known to be inheritance-and-tradition-oriented. Therefore, necessary foundation is laid for discussing the origin of astronomy in China.

According to the majority of those who believe astronomy was introduced to China from the west, standing and nature of astronomy in ancient Chinese culture was not so different from the case of Greece. That's why they transcendentally believe that astronomy in ancient China, like other techniques, could be introduced from somewhere else. For example, the introduction of combat uniforms and skills of nomad tribes into Zhao, a kingdom in the Warring States Period, or Emperor Han Wudi's request of the Ferghana horse. In other words, astronomy could be introduced to China from the west even when the Chinese civilization had been rather advanced.

However, once we understand the relationship between astronomy and monarchy in ancient China, all the similar theories (such as A Iijima Kojima's belief that astronomy was introduced to China during the Warring States Period) would be rendered groundless. And the reason is quite simple. The cultural function of astronomy in ancient China determined that it could only come into being along with the Chinese civilization itself. The role it played in the forming of the Chinese civilization was so important that it was simply impossible that it was introduced to China later.

But another type of theories that claim astronomy was introduced to China from the west in remote ages was often combined with the belief that the Chinese civilization was also from the west. To those who believe in those theories, the cultural function alone is not enough evidence that they are wrong, because, according to those theories, the Chinese civilization itself could have been developed from a culture from the west and the astronomy was already part of that culture (according to Guo Moruo), or that the Chinese civilization was the combination of a western culture, which brought astronomy along, and a native one (according to Vassilyev, a scholar in the former Soviet Union) so that astronomy was introduce to China before or at the same time when the Chinese civilization was established. That's how the contrast between the theory and the cultural function of astronomy was avoided.

The following conclusion, therefore, can be drawn: astronomy in ancient China, as we know it, originated very early in history so it's impossible that it was introduced from the west later. The origin of astronomy in China was closely tied to the origin of the Chinese civilization, and both issues are yet to be discussed.

1.1.2 Relationship Between the Gai Tian Model in Zhou Bi Suan Jing and the Cosmic Model of India

About the origin of astronomy in China, due to ancientness of the history as well as lack of documents, a definite and complete conclusion is hard to come by. Generally speaking, it is a rather illusional scene. On the other hand, in later times (when the system and setup of astronomy were long established in China), various variations of astronomy did spread to China from the west. And some of those might have been absorbed by the Chinese system (merely as a supplementary technical method), but on the whole they didn't have much influence on China's astronomical inquiries. There is a great deal of historical documents supporting that, yet not enough theoretical questions. That's why this is a scene far more factual and detailed comparing to the original one.

In reference to which, the cosmic model in *Zhou Bi Suan Jing* can serve as a typical example and proof.

According to conclusions deemed credible by modern scholars, the *Zhou Bi Suan Jing* was written in 100 BC. Throughout history, it has undoubtedly been regarded as a genuine national treasure of China. It may seem ridiculous to even begin to argue that the part about astronomy in the book might have a foreign origin. However, if we could just move our eyes from ancient China to other old civilizations, we'd be surprised to find that, as we scrutinize the original text of *Zhou Bi Suan Jing*, not only is the question raised above not that ridiculous, but it also has profound meaning in the history of science as well as of philosophy.

According to the original text of *Zhou Bi Suan Jing*, and the expounded demonstration I provided for several crucial questions, we already know that the Gai Tian model in *Zhou Bi Suan Jing* has the following characteristics:

1. The earth and the heaven are two parallel circular planes some 80,000 li apart.
2. There is a high-rise column-shaped object at the center of the earth (known as the Xuanji, standing at a height of 60,000 li with a basal diameter of 23,000 li).

3. Whoever came up with this cosmic model established the position where he was, and it was not at the center but in a southernly location.
4. The column at the center of the earth reaches out to touch the heaven, and that's where the North Pole is.
5. The sun, the moon, and the stars revolve around the North Pole.
6. The sun has multiple concentric orbits, and it migrates to each of them every 6 months (covering all orbits in 1 year).
7. The sun's radiation has a maximum radius, which is 167,000 li.
8. Motion of the sun, as described above, can be used to explain the cause of day and night as well as some astronomical phenomena in the annual apparent motion of the sun.
9. Pi was valued 3 in all calculations.

One amazing fact is that we found the nine characteristics listed above are in complete accordance with the cosmic model of ancient India! This is anything but coincidence, and it's worth noticing and studying, which led to the following findings:

Most accounts of the ancient Indian model were recorded in the *Puranas*, a type of classic literature in Hinduism. The *Puranas* are historical documents as much as they are encyclopedia. Actual time of completion of the books is yet to be determined, but the idea of the cosmic model, according to scholars, can be dated back the time of Vedas, which is about 1000 BC. Suffice to say, it is archaic. The cosmic model in the *Puranas* can be summarized as follows:

The earth is like a round plate with flat bottom, with a towering mountain erecting at the center. The name of the mountain is Meru (or Xumi, as it's known in China, or Sumeru). Surrounding the mountain are circular lands, which are also surrounded by oceans. The same layout goes on to include, in total, seven circles of lands and another seven of oceans.

India is located in the south of the mountain.

There is a series of "sky wheels" above the earth and they revolve around Mount Meru, on the top of which is Dhruva, the north star. The wheels rotate along with other celestial bodies, including the sun, the moon, the stars, and the five planets: Mercury, Venus, Mars, Jupiter, and Saturn (as they are called in English).

Mount Meru is the reason for day and night. There are 180 orbits on the wheel that carries the sun, and it moves to a different orbit every six months, which explains the shifting azimuth of sunrise in the course of a year.

The ancient Indian cosmic model was recorded in the *Shi Jia Fang Zhi*, written by Shi Daoxuan in the Tang dynasty. The book includes some details that are mutual-complementary with the account mentioned above.

> Mount Sumeru, also known as the Xumi, is located in the ocean. According to the chart of the golden wheel, it reaches some 80,000 yojana (ancient Indian length unit) above the ocean surface, with the sun and the moon revolving around it halfway. Surrounded the mountain are seven layers of other mountains, with oceans between every two layers. The ocean water bears eight virtues.

In volume five, titled *the 19th Chapter on the Motion of the Sun and the Moon*, of the Chinese translation of the *Abhidharma Sutra* (number 1644 of the *Tripitaka*), the maximum radius of solar radiation is cited to explain the record of the sun's apparent motion:

> Sunlight has a diameter of 700,021,200 yojana and a perimeter of 2,100,063,600 yojana. When sun rises in Jambudvipa of the south, it sets in Uttarakuru of the north, during which time it's high noon in Furvavideha of the east and midnight in Aparagodaniya of the west, four different stages of a day defined by the sun's movement.

In this case and a great deal of other astronomical data, value of Pi is always set to 3.

According to these accounts, the cosmic model of ancient India and the Gai Tian model in *Zhou Bi Suan Jing* share uncanny resemblances, with details that are almost 100% identical.

1. In both models, the earth and the heaven are both circular planes that are parallel to each other.
2. Both Xuanji and Mount Eru are described as a high-rise erection at the center of the earth.
3. In their respective universes, China and India are both located in the south of the earth.
4. On top of Xuanji and Mount Meru, there is the north star, the axis around which all celestial bodies revolve.
5. The sun, the moon, and the stars revolve around the North Pole at the same level.
6. The seven layers of mountains and seas in the Indian model can be easily associated with the Gai Tian model, which consists of seven concentric circles with six spaces in between, let alone the 180 orbits, which, whether in terms of nature or function, are fully identical to the Gai Tian model (in which the sun moves on to another orbit on a daily basis as well).
7. One detail, in particular, that's worth noticing is the distance between heaven and earth in *Zhou Bi Suan Jing*, which is 80,000 li, while Mount Meru reaches out 80,000 yojana above the ocean surface, with the sky wheels over the top. So, the distance between heaven and earth is also 80,000, only of a different length unit. Could that simply be coincidence?
8. There is a maximum radius of solar radiation, based on which astronomical phenomena related to the sun's apparent motion, such as sunset, sunrise, seasons, day, and night can be explained.
9. In all astronomical calculation, the value of Pi is set to 3.

Throughout the history of human civilization, spontaneous development of different culture was completely possible. That's why many civilizations share similarities and they could be by chance. However, the similarity between the Gai Tian model in *Zhou Bi Suan Jing* and the ancient Indian model is so incredibly high, from the general layout to the smallest details, that coincidence seems to be a farfetched explanation for something like that.

1.1.3 Amazing Knowledge of Temperate Zones in the Zhou Bi Suan Jing

One amazing fact is that the concept of temperate zones, which modern-day people are familiar with but was never mentioned or even believed in Chinese astronomy over the past 2,000 years, can be found in *Zhou Bi Suan Jing*. Relevant accounts in the book are as follows:

> Nothing lives in the polar regions. How do we know that? ...... Ice never melts during summer near the North Pole.
>
> Zhongheng (the central circle) locates at 75,500 li (equivalent of 500 meters) away from the circumference. Near it plants grow in summer don't wither in winter. Zhongheng is a place where Yang is strong and Yin is weak so every living thing thrives and the Five Grains are ripe twice in one year.
>
> Near the North Pole, some crops that sprout in the morning can be harvested in the evening and plants grow in winter.

Here, some explanations are needed.

The so-called Zhongheng as mentioned above, according to Zhao Shuang's annotation, is located between the core and the circumference (as in the Gai Tian model), or the Torrid zone (between 23°30′South and 23°30′North) as we call it today, although the concept of earth being a sphere wasn't mentioned in the *Zhou Bi Suan Jing*.

And to understand the part about some crops that sprout in the morning can be harvest in the evening, we must first talk about deduction and depiction of the polar day/night in the Gai Tian model. The Xuanji (core component of the armillary sphere) area in central earth (a round and flat surface) has a basal diameter of 23,000 li, so the radius is 11,500 li. And according to *Zhou Bi Suan Jing*, the sun's radiation can only reach as far as 167,000 li. Therefore, every year from the spring equinox to the autumn equinox, the midnight sun—when the sun is in the sky during the day as well as the night—can be observed within the Xuanji. And from the autumn equinox to the spring equinox, there comes the polar night—when the sunlight can't reach the Xuanji area at any time. Both phenomena are consistent with Zhao Shang's description because the polar day and the polar night each lasts for half a year.

The accuracy of the knowledge of temperate zones in the Zhou Bi Suan Jing, as mentioned above, is unquestionable. However, it is not part of traditional Chinese astronomy over the past 2000 years.

One reason is that Zhao Shuang, who wrote the annotation for the *Zhou Bi Suan Jing*, clearly expressed his disbelief of the knowledge in the book. For example, to the part about ice never melts in summer near the North Pole, Zhao responded in the annotation that: "If everything is always frozen then it is winter on the Summer Solstice underneath the circumference so everything should die—so on this day winter and summer co-exist from far to near. This is contrary to the law of Yin and Yang so I have my doubts." As for the Torrid zone where plants don't wither in winter with Yang being strong and Yin being weak, and the Five Grains are ripe twice in one year, Zhao wrote: "This is to say that it is always summer between the core and the circumference, yet the distance is too great and it's certainly unheard of." He's never heard of anything like that. From the annotation he wrote for the *Zhou Bi Suan Jing*, we know that he was undoubtedly one of the most qualified astronomers of his time. How come he hadn't heard of the temperate zones? A reasonable explanation would be that this knowledge wasn't part of traditional Chinese astronomy so for most Chinese astronomers it's strange and inconsistent with their learning at that time. That's why they found it unbelievable.

The second reason might be that the Hun Tian school, which was the dominant cosmological theory in ancient China, lacked a proper concept of the earth, so it was impossible for the issue of temperate zones to be raised. That's why even by the Ming Dynasty when Jesuit missionaries introduced the temperate zones in their books, the knowledge still struck the Chinese as a fresh theory that no one had ever heard of. And it was those missionaries' books that urged Chinese scholars to finally accept the idea of temperate zones.

Now we're faced with a series of sharp questions.

Since the Hun Tian school, with no proper concept of the earth, could not have raised the issue of temperate zones, then why was the knowledge recorded in the *Zhou Bi Suan Jing*, which also had no clue what the earth was?

If the author of the *Zhou Bi Suan Jing*, being in the north temperate zone, was able to figure out phenomena such as ice never melts in the North Pole and the Five Grains are ripe twice in one year in the Torrid zone solely based on the fact that it grew colder toward north and hotter south, then why couldn't the followers of the Hun Tian school?

Plus, given the fact that Zhao Shuang annotated the *Zhou Bi Suan Jing*, he was surely a believer of the Gai Tian theory, why, then, didn't he believe the description?

All in all, it is necessary for us to consider the possibility that this knowledge came from outside China.

The earth is a sphere, platitudes and latitudes, temperate zones, etc.—these knowledges, which were already part of astronomy in ancient Greece, are still useful today. The temperate zones were first mentioned in Aristotle's books, and thorough description of the zones can be found in the *Geographika* by Eratosthenes (275–195B.C), according to whom, the Torrid zone is located between 24° South and 24° North; the North and South frigid zones are at 24°, respectively, of the two Poles; the North and South temperate zones are between 24° and 66° South and 24° and 66° North. Timewise, the Greek astronomers' works were written before the *Zhou Bi Suan Jing*. So, the question remains: could the author of the *Zhou Bi Suan Jing*, through direct or indirect channels, have somehow obtained the Greek knowledge?

1.2 Western Astronomy Brought to China by the Eastward Spreading of Buddhism

1.2.1 Traces of Babylon's Zigzag Function in China

Discussion of the relation between two matters often starts with the comparative study of them. And the majority of previous comparative studies (almost all were done by western scholars) of astronomy in ancient Babylon and ancient China had adopted traditional methods of humanities. In other words, they were done by collecting scattered accounts in historical documents as well as from paintings, sculptures, inscriptions, and tools. Based on research methods and findings in linguistics, philology, mythology, ethnography, historiography, and geography, textual research, analysis, and deduction were carried out. The process is pretty much as what Guo Moruo described in *Shi Zhi Gan*. It was till 1955 when the *Astronomical Cuneiform Texts (ACT)* and the *Late Babylonian Astronomical and Related Texts* (*LBART*), edited by O. Neugebauer and A. Sachs, were published that the comparative study of astronomy in ancient Babylon and ancient China was guided to a new path that was more solid and in-depth: the study of mathematical astronomy.

I myself have also published a series of findings of comparative studies on astronomy in ancient Babylon and ancient China using method of mathematical astronomy. My findings involve theories of solar and planetary motion, celestial coordinates, lunar motion, intercalation period, and day length, in the hope of providing new clues for connection between the two civilizations. In summary, my conclusions are as follows:

The Babylonian ephemeris includes a chart of solar motion recording the sun's location on the ecliptic, coordinates of the sun during the monthly syzygy and longitudes the sun covers in the course of a month. This chart already had a very complicated mathematical structure. There were second-order differences and the zigzag function, the most noteworthy mathematical instrument in Babylonian astronomical texts. Almost all problems in mathematical astronomy are described and solved using the zigzag function, which bears multiple mathematical features, with the most important of all being its periodicity. In the issue of solar motion, the zigzag function is manifested in the relation between speed and time.

For a long time, astronomers in ancient China believed that the sun (in its annual apparent motion) moves at a constant speed. It was not until the end of the fifth century that they realized the inconsistent nature of the motion. In the *Huang Ji li*, a calendar edited by Liu Chao in 600 AC, the very first correction table for the inconsistent motion of the sun occurred. Comparing the table with the Babylonian chart, one would be surprised to find that both include the second-order difference and the zigzag function. This "coincidence" lasted for over a century.

In ancient times, both Babylon and China put a premium on the importance of planets in astrology. And in theories of planetary motion, mathematical description of period and movement were emphasized. The Babylonians paid close attention to characteristic phenomena such as opposition, station, reappearance, and disappearance (beginning of lurking). They tried to predict dates and longitudes of said phenomena based on periodicity. During the Seleucid Dynasty and the Western Han Dynasty, approximately in the same time, Babylonian and Chinese astronomers realized, almost simultaneously, a mathematical relationship between the two periods related to planets (synodic period and sidereal period), only in slightly different forms. The two sides also had a rather precise grasp of numbers concerning the periods. Furthermore, they even made the same mistake about the period of planetary motion, believing that the sidereal period of Venus and Mercury was one year.

In terms of mathematical description of planetary motion, the Babylonians were far ahead. For example, back then they were already able to divide the motion of

Mercury, which is considered the most difficult to learn, into six sections, including constant, variable, and variable accelerated motions. There was even the third order difference (ACT, No. 310). On the other hand, up till the *Huang Ji Li*, Chinese astronomers had always believed that all planets moved at a constant speed. Methods similar to those of Babylon only occurred in the *Huang Ji Li* and the *Da Ye Li*, another imperial calendar. Variable motions were mentioned but not variable accelerated motions. When it comes to precision of mathematical description, the Babylonian system was definitely better.

In the ephemeris, the planets' daily position was predicted by calculation instead of observation, same as almost all Babylonian ephemeris, and that requires the interpolation method. By interpolating a series of characteristic phenomena whose time and location (longitudes) of occurrence were given, other relevant numbers can be worked out. With publication of the *Huang Ji Li*, the second-order interpolation method, which was the lowest order of all nonlinear interpolation methods, was adopted in China.

1.2.2 Popularity of the Theory of the Seven Celestial Bodies in China

The seven celestial bodies refer to the sun, the moon, Mercury, Venus, Mars, Jupiter, and Saturn. These celestial bodies had long been the objects of observation, research, and discussion for Chinese astronomers thousands of years ago, but the terms such as "seven celestial bodies," "calendar of the seven celestial bodies," "theory of the seven celestial bodies," or "calendrical theory of the seven celestial bodies" were associated to something else: foreign astronomy, mostly from India, although calendar, astrology, and even numerology from Central Asia might have been added to it during eastward and northward spreading of the theory.

Theory of the seven celestial bodies thrived mainly during the Six Dynasties Period and Tang and Song dynasties. However, its first occurrence in China can be dated back to late Eastern Han dynasty. Then it was shortly popular in a time between Jin and Song dynasties, and the theory spread to North China about the same time. Generally speaking, this foreign theory reached its peak in China during the Northern and Southern dynasties. By Tang dynasty, western astronomy was introduced to China and ascended to eminence, but the old theory was till spreading, all along till Song dynasty. The declining of the theory in China happened at the same time as the fall of Song dynasty. Afterward, the term "theory of the seven celestial bodies" disappeared from all historical documents and other texts.

Here, it is necessary to mention Emperor Liang Wudi's "Lecture at the Changchun Palace" and Indian astronomy.

Connection between the cosmic model in Zhou Bi Suan Jing and India was renewed when Xiao Yan, Emperor of Liang, ascended the throne. It happened during this famous event known as the "Lecture at the Changchun Palace," which is one of the things worth noticing in the cultural history of China. Account of the even can be found in *History of Sui, Astronomy Part One*:

> The emperor called upon a meeting at the Changchun Palace for discussing issues of celestial bodies and their discussion was in full accordance with the *Zhou Bi Suan Jing*. The goal was to establish new rules in order to refute the Hun Tian theory.

Previous studies on the matter, though not meaningless, failed to reveal its background and significance. Two important questions must be answered here: (1) What is the relationship between the cosmological theory advocated at Liang Wudi's lecture and Indian astronomy? And (2) Why did *History of Sui* say the lecture "was in full accordance with the *Zhou Bi Suan Jing*"?

The first question is somehow easier to answer. The main content of the lecture was recorded in volume one of the *Kai Yuan Zhan Jing (Treatise on Astrology of the Kaiyuan Era)*. It seems that Liang Wudi started his argument with quite a pompous statement denying other cosmological theories.

> Many in history have talked about the universe. All of them were ignorant of astronomy and spoke nonsenses. Theories of different schools and individuals varied not little but largely. How could one see the truth with his eyes covered? It's like one trying to comment size of the sky while dwelling in a nook as teeny as a snail's shell or trying to measure depth of the ocean while hiding inside a clam. What blind confidence!

Such argument is nothing but arbitrary. One fact worth noticing is that by that time the Hun Tian theory had risen to dominance; therefore, it was widely accepted among astronomers. In spite of that, Liang Wudi would, without offering any solid evidence, deny it completely. Had he not been the emperor, it would be hardly convincing. Not just that, the cosmic model favored by the emperor himself was put forth, again, without any solid evidence.

> Beyond the four seas stands Mount Jingang, also known as Mount Tiewei. North of Mount Jingang is the Black Mountain, around which the sun and the moon revolve, ever moving day and night. On the southside they show and northside hide. In winter the sun sinks and summer rises. When it rises the day is long and when it sinks it is short. Such is the cause of different seasons of a year as well as different times of a day.

Followers of the Hun Tian theory would find the cosmic model and the cause of seasons unbelievable. However, Liang Wudi's theory did have its ground: the ancient Indian cosmic model as seen in Buddhist classics.

The second question can only be answered on the basis of the first. Since the cosmic model proposed by Liang Wudi came from India, the statement in *History of Sui* was totally understandable, for the cosmic model in *Zhou Bi Suan Jing* was also from India. So, the statement in *History of Sui* was absolutely correct, only with the middle part left out.

1.2.3 The Three Indian Astronomers in Tang Dynasty China

During Tang dynasty, China was a highly open, confident, prosperous, and powerful empire. And it was in that time the Indian astronomy introduced to China reached its peak. Back then talents from all over the world served in the royal court; some even became high-ranking officials. Strictly speaking, the Li family who ruled China then was from a partly non-Chinese bloodline as well. With that backdrop, an Indian family of astronomers serving and leading China's royal institution of astronomy shouldn't be so surprising. And the Indian astronomy introduced by them once received official recognition. Yang Jingfeng wrote in the annotation for volume one of the *Su Yao Jing*, translated by Bu Kong, that:

> Whoever is curious about the position of the five planets can work out the specifics using the Indian method. Kāsyapa, Kumāra and Gautama are the three Indian families of astronomers in charge of the royal institution. Today, Kumāra's calendar is the most widely used, as cross-reference to the Da Shu.

Brief information about the three families is as follows:

The Kāsyapa family: their theory was used as cross-reference to the Da Shu (traditional astronomical method in China) in the royal institution.

The Kumāra family: their name only came up once in the two official history books of Tang dynasty. The account in the *Jiu Tang Shu (Old Book of Tang)* is more thorough. Their theory was also used as cross-reference to the Da Shu.

The Gautama family: the most illustrious among the three. There is a great deal of accounts of the family in historical documents, yet seniority in the family hierarchy had remained a mystery till the inscription on the memorial tablet within the tomb of one of them was discovered in Beitian Village, Chang'an County, Shaanxi, in 1977. Four generations of the Gautama family had served in the royal institution of astronomy, and each of them was either the head or an important official. Accounts of the Gautama family's academic activities in Tang dynasty far exceed those of the other two. The most prominent astronomer from the family was Gautama Siddhartha, whose most outstanding achievements were translating the *Jiu Zhi Li (Calendar of the Nine Celestial Bodies)* and compiling the *Kai Yuan Zhan Jing*.

1.2.4 The Jiu Zhi Li: From India to Astronomy in Ancient Greece

Here is the introduction to the *Jiu Zhi Li* in volume 28, *History of Astronomy*, of the *Xin Tang Shu (New Book of Tang)*:

The *Jiu Zhi Li* was originally introduced from the west, and it was translated into Chinese by Gautama Siddhartha, the imperial astronomer, in the sixth year of Kaiyuan (718 AC). The term "west" had a much extensive meaning in Tang dynasty than it does today, including five regions of India. This can be proved by Xuanzang's *Da Tang Xi Yu Ji (Great Tang Records on the Western Regions)*. The *Jiu Zhi Li* was one hundred percent Indian but at that time it could be labeled as "from the west."

Table 1.1 Origins of the Jiu Zhi Li

中文	英文	中文	英文
古希腊天文学	Ancient Greek Astronomy	婆罗门学派	The Brahman School
夜半学派	The Midnight School	毗坦摩诃历数书	Paitamaha Siddhanta
《历法甘露》	*Li Fa Gan Lu (Essence of Astronomy)*	《五大历数书汇编》	*Treatise on the Five Astronomical Canons*
《太阳历数书》	Tai Yang Li Shu (Astronomical Book of Solar Calendar)	《九执历》	Jiu Zhi Li

The *Jiu Zhi Li* was translated by Gautama Siddhartha on Emperor Xuanzong's command. Text of the translation can be found in volume one of the *Kai Yuan Zhan Jing*, which is the only existing historical document that includes the *Jiu Zhi Li*.

It's likely that there was no original text of *Jiu Zhi Li* in Sanskrit. It's more of a compiled book based on various Indian astronomical documents, which included, at least, the *Treatise on the Five Astronomical Canons* and perhaps the *Li Fa Gan Lu (Essence of Astronomy)*. Involved schools of astronomy included the Midnight School (and perhaps the Brahman School). And all these Indian works and schools were based on theories that originated in ancient Greece (see Table 1.1).

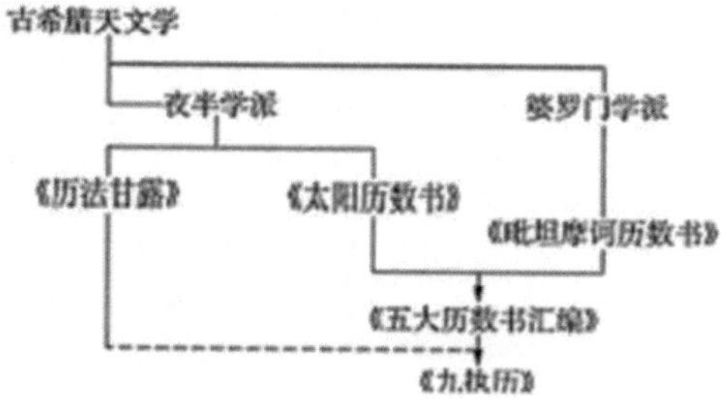

Even after so much translation, the elements of ancient Greek astronomy are still so obvious in the Jiu Zhi Li. The following examples are the most typical:

1. 360° of the circumference
2. The sexagesimal system
3. Ecliptic coordinates
4. Apogee in the annual apparent motion of the sun (set to 10° before the summer solstice, which was in full accordance with actual astronomical phenomena back then)
5. Calculation method for the shifting apparent diameter of the moon
6. Calculation method and table of the sine function

All of the above was unprecedented in traditional Chinese astronomy.

1.3 Exchanges in Astronomy Between China and the Arabic World Brought About by the Mongolian Empire

1.3.1 Arabic Astronomical Instruments and How They Influenced China

In 1267, after Kublai Khan ascended the throne, Jamal al-Din, an Arabic astronomer, brought to China seven pieces of western astronomical instruments as tribute. Phonetically and literally translated names as well as structure and function of the instruments can be found in the chapter on astronomy in *History of Yuan.* Those instruments had aroused tremendous interest in Chinese and foreign scholars. Given that the actual instruments were long gone, scholars now have different opinions pertaining to their nature and function. I hereby list the names, phonetically and literally translated (according to History of Yuan), and the original Arabic pronunciation established by W. Hartner, who also wrote a brief introduction to primary research materials.

1. Dhatu al-halaq-I, or "the Huntianyi seismometer" in Chinese, was, according to Joseph Needham, an equatorial armillary sphere or, as believed by some Chinese scholars, an ecliptic armillary sphere. This was a classic ancient Greek astronomical observation apparatus.
2. Dhatu'sh-shu'batai, or "the instrument for measuring all celestial bodies" in Chinese, was, as believed by both Chinese and foreign scholars, the organon parallacticon mentioned by Ptolemy in his *Almagest.*
3. Rukhamah-i-mu'-wajja, or "spring and autumn dial" in Chinese, was an instrument for measuring the precise time of spring and autumn equinoxes. It was attached to an enclosed house (with only a narrow opening on ridge in the east-west direction).
4. Rukhamah-i-mustawiya, or "winter and summer dial," was an instrument for measuring the precise time of winter and summer solstices. Like the previous instrument, it was also attached to an enclosed house (with a narrow opening on the ridge in the south-north direction).
5. Kura-i-sama, or "Huntiantu chart" in Chinese, was, as agreed by both Chinese and foreign scholars, the celestial globe, which was used in China and the west in ancient times.
6. Kura-i-ard, or "dilizhi" in Chinese, was, as agreed by all scholars, the terrestrial globe.
7. al-Usturlab, or "the instrument for determining hours of day and night", was in fact the astrolabe, which was quite popular in the Arabic world and Europe during the Middle Ages.

Among the seven instruments, (1) (2) (5) and (6) were matured and used in ancient Greece before being inherited by Arabic astronomers. (3) and (4) bear apparent Arabic characteristics. (7) was first used in ancient Greece but then became an iconic Arabic craft. Exquisite astrolabes made by Arabic artisans have long enjoyed a good reputation. It was indeed a significant event that the seven instruments, with such prominent background, should have come to China.

Nine years after Jamal al-Din brought the instruments to China, five years after the Hui Hui Si Tian Tai (an imperial astronomical institution based on the Islamic calendar) in Xanadu was built, and three years after said Islamic institution was put under supervision of the Mi Shu Jian (the imperial library) along with an astronomical institution based on the Chinese calendar, Guo Shoujing, one of the greatest Chinese astronomers in history, was ordered to design and build a batch of astronomical instruments for the said Chinese institution. Three years later (1276–1279), Guo accomplished his mission. The Jianyi, Yangyi and Zhengfang'an, among Guo's other instruments, included many innovative features. However, since Guo's instruments were built after Jamal al-Din's tribute and that most of them were never seen in China before, the question of "whether Guo's instruments were influenced by Arabic astronomy" was raised.

As far as this issue is concerned, most Chinese scholars were inclined to deny the suspicion. They believed that none of Jamal al-Din's instruments "were combined with traditional Chinese astronomy." In direct observation, there are no signs of Arabic influence in any of Guo's instruments. Instead, they bear distinct traits of traditional Chinese astronomy. And the explanation for that is quite convincing:

> In Yuan dynasty, the Islamic and the Chinese astronomical institutions were both under supervision of the Mi Shu Jian, which was a crucial factor because this means Guo and Jamal, as well as the astronomers they led, were rivals. Considering the fact that Guo was ordered to build those instruments, it made perfect sense that he would try his best not to be influenced by his competitors, so as to prove he was equal, if not better. Had he accepted the Arabic influence, he'd have been accused of imitating a superior. How, then, would the Chinese institution hold its ground in the competition?

On the other hand, we should admit that Guo Shoujing was, somehow indirectly, influenced by Arabic astronomy, as seen in the following examples:

One of them is the Jian Yi (see Picture 1.1). The most prominent innovation of this instrument is "Jian" (meaning "simple" in Chinese). Instead of inheriting the

Picture 1.1 The Jian Yi

Picture 1.2 The Gaobiao in Dengfeng, Henan

super-complicated multi-functional structure of the traditional Hun Yi, the Jian Yi simply measures coordinates of one celestial body with a pair of rings perpendicular to each other. It was actually two independent instruments on the same pedestal, respectively, for measuring equatorial and horizontal coordinates. This one-function setup was the traditional style of European astronomical instruments, which can be seen in Jamal al-Din's instruments and the six instruments (which has been well-preserved in the ancient observatory of Beijing to this day) built by F. Verbiest, a Jesuit missionary in emperor Kangxi's time.

Another example is the Gaobiao. Among the seven instruments brought by Jamal al-Din, one was called "winter and summer dial," which bears the same function as the traditional Guibiao in China, with higher precision. Certainly, Guo Shoujing wouldn't want to copy its design. Instead, he decided to improve the Guibiao by building a huge-sized Gaobiao and Liangtianchi, which was a larger version of Guibiao (see Picture 1.2) in Dengfeng, Henan. As is well-known, enlargement was a typical characteristic of Arabic astronomical instruments.

Between the two examples listed above, one was in the European style as introduced through Arabic astronomy; the other took up the Arabic style itself. Both can be regarded as indirect influence of Arabic astronomy on Guo's design.

1.3.2 Chinese Astronomers' Activities in Central Asia

With the rise of the Mongolian Empire, whose territory stretched across Eurasia, nations and cultures were merged, and exchanges between China and other countries in astronomy had reached its peak. However, details and conclusion of contacts made between Chinese and Islamic astronomies are yet to be determined.

First, Yelü Chucai and Qiu Chuji's astronomical activities in central Asia need to be sorted out for the issue is of great importance.

Yelü Chucai (1189–1243) was a native of the Qidan tribe and a lineal descendent of the royal family of Liao dynasty. He once served under Jin, a Manchurian empire, before being summon to Mongolia in 1219 to serve as Genghis Khan's counsel in astrology and medicine. He followed the Mongolian ruler in his expedition to the west, during which time he had an argument with Islamic astronomers over the issue of lunar eclipse. According to the biography of Yelü Chucai in *History of Yuan*:

> Western astronomers reported that on May fifteenth a lunar eclipse would occur but Chucai disagreed. By October next year, Chucai predicted a lunar eclipse and the western astronomers again disagreed. And on that date, a partial lunar eclipse did occur.

This event happened in 1220, the second year of Genghis Khan's expedition to the west, which can be confirmed by the following account in *Part One of Astronomy* in *History of Yuan*: "In the year of Gengchen, the emperor launched the expedition. In May that year, a lunar eclipse did not occur as expected. . .." And the location was in Samarkand, part of Uzbekistan today. It can be inferred from Yelü Chucai's depiction in *Travels to the West*.

Yelü Chucai had a profound knowledge of traditional Chinese astronomy. In early Yuan dynasty the Daming Calendar of Jin was inherited and soon too many errors occurred. One of those was the 1220 eclipse that didn't happen as predicted. In response, Chucai composed the *Xi Zheng Gengwu Yuan Li (Yuan Calendar of the Gengwu Expedition to the West)*, which was included in History of Yuan. In the book, the time difference caused by geographic longitudes was dealt with for the first time, which can be viewed as an example that traditional Chinese astronomy was influenced by western methods because the issue of longitudes and time differences had long been cited in ancient Greek astronomy as well as in Islamic astronomy, which inherited the Greek legacy.

According to other materials, Yelü Chucai was also proficient in Islamic astronomy. For example, Tao Zonyi, a historian in Yuan dynasty, once wrote in his book *Nan Cun Zhui Geng Lu* that:

> Yelü Chucai was an expert in astronomy, astrology, mathematic, music and Confucian classics. He also read plenty of foreign books. It was him who introduced to China the theory of the five planets of the west. He also composed the Madaba Calendar based on Islamic astronomy.

Combining this with the fact that Yelü Chucai was proved right in the two debates against "western astronomers," it can be inferred that he had a good grasp of both Chinese and Arabic astronomical methods, which was why he won the argument every time.

About the same time when Yelü Chucai followed Genghis Khan in his expedition to the west, Qiu Chuji, another important historical figure, was on his way to Central

Asia. He was summoned by Genghis Khan to teach the ruler about Taoist philosophy. Qiu arrived in Samarkand near the end of 1221, almost at the same time as Yelü Chucai's arrival. He then discussed, with local astronomers, about a partial solar eclipse that occurred on May 23rd that year. The following is the account of this event in *Travels to the West of Qiu Chang Chun*, a book written by one of his students:

> He arrived in Samarkand...... and met some astronomers there. Master asked them about the solar eclipse in May. One of them said: It was a partial eclipse that occurred during seven to nine in the morning. Master then asked: I saw the second contact at noon when I was passing by the Luju River. People told me it became a partial eclipse during nine to eleven in the morning. So, it appeared different in three locations.... In retrospect, I saw the second contract when I was underneath the sun and others at a distance had different observations. It's like coving a lamp with a fan. There is no light whatsoever right in the shadow but it gets brighter towards the edge and beyond.

Qiu Chuji was 73 years old at that time, but throughout his long journey he was carrying out field investigations on astronomical issues, which could only be driven by tremendous passion. His explanation and analogy for different types of solar eclipse seen in different locations were absolutely accurate.

Yelü Chucai and Qiu Chuji both made contact and consulted with local astronomers in Samarkand. And this was no coincident. Some 150 years later, the city became the imperial capital of the Timurid dynasty. By the time when Ulugh Beg ascended the throne, an observatory of magnificent scale was built (1420). Ulugh Beg himself conducted the construction. The famous Astronomical Tables of Ulugh Beg, including the first-ever independent sidereal table in the history of western astronomy since Ptolemy's time, were drawn then after observation. Suffice to say that Samarkand was a place with a long astronomical tradition.

During mid-1200s, Hulagu (or Hulegu), the grandson of Genghis Khan, led an expedition to conquer the west, during which his army captured Baghdad in 1258. The khalifa regime of the Abbasid dynasty was overthrown and replaced by the Il-Khanid dynasty. Assisted by prestigious Islamic scholar Nasir al-Din al-Tusi, Hulagu aimed to accomplish in culture what he did in warfare. In Maragha (today's Tabriz, Iran), the imperial capital, a world-class observatory was built (1259) in grand scale and was equipped with sophisticated instruments. The library alone was said to contain over 400,000 books. The Maragha Observatory was once the academic center of the Islamic world, attracting scholars from all over the world to further their studies there.

C.M. D'Ohsson wrote in *History of Mongolia* that there was a Chinese astronomer who followed Hulagu to Persia and had worked in the Maragha Observatory. Later, this subject was mentioned by many western scholars yet name, and the background of this Chinese astronomer was never verified. The only account of him in *History of Mongolia* was the phonetic translation of his name, Fao-moun dji. Due to lack of further information, the actual name in Chinese can only be guessed based on its pronunciation. Joseph Needham had made it out to be Fu Mengji.

D'Ohsson's story was in fact based on a book of chronicles in Persian. The book, which was written in 1317, consisted of nine volumes, and volume eight was *History of China*, which included the following passage:

> It was till the time of Hulagu that their (Chinese) scholars and astronomers had first come here (Iran) with him. Tu Michi, known among his compatriots as "the master", was Nasir al-Din al-Tusi's teacher when the latter was ordered to compile the Astronomical Tables of Ilkhanate and had to learn the calculation methods in Chinese astronomy. Also, when Ghazan Mahmud Khan, the Islamic ruler, ordered the Praised History of Ghazan to be composed, Rashid al-Din, the prime minister, had summoned two Chinese scholars, namely Li Dachi and Ni Kesun. Both were experts in medicine, astronomy and history and they brought with them relevant books from China. According to them, number of years and jiazi (a cycle of sixty years) were uncertain in Chinese chronology.

This passage was the earliest account of the Chinese scholar at the Maragha Observatory. The names "Tu Michi," "Li Dachi," and "Ni Kesun" were reckoned based on their pronunciation in Persian. Their actual names in Chinese were unverifiable. "Tu Michi" could be the "Fu Mengji" mentioned earlier. *Astronomical Tables of Ilkhanate* were the most important accomplishment of Nasir al-Din al-Tusi during his work at the Maragha Observatory. It can also be inferred from which that the *Astronomical Tables of Ilkhanate (or* Zij Il-Khani as in Persian*)* included contribution from Chinese astronomers as well.

However, we can only rely on new discoveries in Chinese historical text to find out the real identities of "Tu Michi" or "Fu Mengji."

1.3.3 Foreign Books in the Hui Hui Si Tian Tai

Given its connection to the Maragha Observatory of the Il-Khanid dynasty and the fact that it was led by Jamal al-Din, an Arabic astronomer, and specialized in studying Arabic astronomy, the Hui Hui Si Tian Tai in Xanadu undoubtfully held a significant position in the history of Arabic astronomy. It can even be seen as the intermediate terminal between the Maragha Observatory and the upcoming Samarkand Observatory of the Timurid dynasty.

For an observatory with such unique status and significance, we know very little about it. And within that limited information, one noteworthy fact is a booklist recorded in volume seven of the *History of the Mi Shu Jian.* The books listed were all included in the Hui Hui Si Tian Tai, among which the 13 works of astronomy are as follows:

1. Si Bo Suan Fa Duan Shu (15 books) by Wuhulie
2. Yun Jie Suan Fa Duan Mu (3 books) by Hanlisuku
3. Zhu Ban Suan Fa Duan Mu Bing Yi Shi (17 books) by Savanna-Hunda-Siya
4. Zao Si Tian Yi Shi (15 books) by Maizhesi
5. Jue Duan Zhu Ban Zai Fu by Akan
6. Zhan Bu Fa Du by Lanmuli
7. Zai Fu Zheng Yi by Mataheli
8. Qiong Li Fa Duan Shu (7 books) by Haiyati
9. Zhu Ban Suan Fa (8 books) by Hexiebiya
10. Ji Chi Zhu Jia Li (48 books)
11. Xing Zuan (4 books) by Suvari-Chova-Chipi

12. Zao Hun Yi Xiang Lou (8 books) by Sanadialate
13. Zhu Ban Fa Du Zuan Yao (12 books) by Safeina

The meaning of "books" here is similar to that of "volumes" in ancient Chinese documents. Number of books for entries 5, 6, and 7 is unknown, but at the beginning of the booklist the author did mention that a total of 195 books are included. So, 195 minus the number of books we do know, those three should contain 58 books totally.

There was no specific account as for in what language those books were written. Although the possibility that they were written in Chinese can't be completely excluded, it's just more likely that they were written in Persian or Arabic.

Titles of those books were translated literally whereas personal names were translated phonetically. And that's why it's almost impossible for us to match them to the original titles and names; therefore so far there has been no major development in verifying those 13 works. Fang Hao believed the first one was in fact the *Elements of Geometry* by Euclides. Given that it's labelled as containing "15 books," same as the *Elements of Geometry*, his theory is somewhat dependable. Someone else thought the fourth one could be the *Almagest* by Ptolemy, but it's a farfetched speculation, because the Zao Si Tian Yi Shi was about making of astronomical instruments while the *Almagest* was not. Also, the *Almagest* has a total of 13 volumes instead of 15 according to the booklist.

1.4 Full-Scale Invasion of Western Astronomy from Late Ming Dynasty to Early Qing Dynasty

1.4.1 The Jesuits' "Shortcut to Heaven"

Jesuits first came to China around the end of the sixteenth century. In 1582, Matteo Ricci (1552~1610) arrived in Macau, becoming the pioneer of Jesuit missionary work. After years of activities and setbacks, as well as extensive association with Chinese people from all walks of life, Matteo Ricci finally found an effective way to preach in China, the so-called academic preaching. In 1601, his request to meet with Emperor Wanli was approved, and he was then allowed to stay in the imperial capital, which was seen as a symbol that Jesuits were accepted by China's upper-class and that the "academic preaching" strategy was working.

Though the credit was often given to Matteo Ricci, "academic preaching" was in fact an invention based on the long tradition of Society of Jesus. The society had always put a premium on education and had built a great many of schools. For example, from 1620s to 1630s, the Jesuits had built 19 schools in Napoli, Italy, alone. And another 18 in Sicily and another 17 in Venice. The Jesuits themselves too had to go through strict education and training. Among them were great scholars, such as Clavius, an eminent mathematician and astronomer who was Matteo Ricci's teacher. His colleagues and friends include Kepler and Galileo, or Johann Adam Schall von Bell (1592~1666), who later became the first president of the Imperial Board of Astronomy in Qing dynasty. Bell's teacher, C. Grinberger, was Clavius' successor at the Collegio Romano. Another outstanding scholar was Johann Terrenz Schreck (1576–1630), who later helped composing the *Chongzhen Calendar*. Schreck was a member of the Accademia dei Lincei, which was the precursor of the National Academy of Sciences of Italy. He was also friends with Kepler and Galileo (who were fellow members of the Accademia dei Lincei). It was the society's tradition of valuing academics and education that gave birth to the "academic preaching."

1.4.2 An Overview of European Astronomy Introduced to China Through the Chongzhen Calendar

Thanks to the great work of Xu Guangqi and Li Tianjing after him, the *Chongzhen Calendar* was presented to Emperor Chongzhen in five batches. The book consists of 44 parts in 137 volumes. The *Chongzhen Calendar* wasn't officially issued in the end of Ming dynasty, but it was published and the first copies were commonly known as the Ming edition. Johann Adam Schall von Bell had possessed a personal collection of the Ming edition by the time the Qing army entered Beijing. The *Xi Yang Xin Fa Li Shu (Calendar Based on New Methods of the West)*, revised by Johann Adam Schall von Bell, was published many times in Qing dynasty, therefore had multipole versions, among which the most wholesome and representative are the one in the Palace Museum of Beijing, which was published in the 2nd year of Shunzhi Period and the one in the Library of Congress in the USA.

The *Chongzhen Calendar* is a magnum opus. *Fa Yuan*, or the introduction to theories, which takes up one third of the entire text, provides a systematical exposition of theories and methods in classical western astronomy, with focuses on the works of Ptolemy, Copernicus, and Tycho. In general, the knowledge introduced was mostly below the level as that of Kepler's three laws of planetary motion with only a few exceptions. The calculations and astronomical tables were all based on the Tycho system. Most of the original materials, along with their authors, from which astronomical theories and works included in the *Chongzhen Calendar* were drawn, have been investigated out by myself, as listed below:

Tycho
- (Astronomiae Instauratae Progymnasmata, 1602)
- (De Mundi, 1588. The *Hui Xing Jie (Explanation for Comets)* mentioned by Jesuits who came to China was actually the same book)
- (Astronomiae Instauratae Mechanica, 1589)
- (De Nova Stella, 1573. Later the full text of this book was republished in the *Astronomiae Instauratae Progymnasmata*)

Ptolemy
- (Almagest)

Copernicus
- (De Revolutionibus, 1543)

Kepler
- (Ad Vitellionem Paralipomena, 1604)
- (Astronomia Nova, 1609)

(Harmonices Mundi, 1619)
(Epitome Astronomiae Copernicanae, 1618~1621)
Galileo
(Sidereus Nuncius, 1610)
(Longomontanus):
(Astronomia Danica. This book, written by Tycho's students, mainly discussed Tycho's theories.)
(Purbach) (Regiomontanus):
(Epitoma Almagesti Ptolemaei, 1496)

Those 13 books, originally written in Latin, were brought to China by the Jesuits after a "relentless trek over mountains and seas" and were cited as reference when compiling the *Chongzhen Calendar.* Ten of them are still part of the collection of Beitang Library in Beijing. And the latest publication was released in 1622, prior to the compiling of the *Chongzhen Calendar.*

1.4.3 Duplicates of the European Astronomical Instruments

In 1673, Ferdinand Verbiest, a Jesuit missionary, built six astronomical instruments for observation by the order of Emperor Kangxi. All of them are still well-kept in the ancient observatory in Beijing's Jianguomen. The six instruments include Tiantiyi (celestial globe), Huangdaojingweiyi (ecliptic armillary sphere), Dipingjingyi (horizon circle), Xiangxianyi (range quadrant), and Jixianyi (sextant). Those instruments were almost direct duplicates (see Picture 1.3) of the European astronomical

Picture 1.3 Duplicate of Tycho's Sextant

instruments built by Tycho nearly one century ago. Ferdinand Verbiest also wrote a book titled *Ling Tai Yi Xiang Zhi (On Astronomical Instruments and Apparatus)*, in which he described in detail mechanics, manufacturing techniques, and auxiliary facilities of the six instruments, making it an extremely valuable manual for European craftsmanship in the sixteenth–seventeenth centuries. The book was also a rare historical document for it's illustrated with numerous exquisite sketches.

There are two other large pieces of astronomical instruments in the ancient observatory in Jianguomen. One is the horizon circle built in 1715, and the other is the Jihengfuchenyi (elaborate equatorial armillary sphere) built in 1744.

1.4.4 Wang Xichan and Mei Wending and Their Connection to European Astronomy

During Ming and Qing dynasties, the most prominent figures among Chinese astronomers (only by then were there genuine astronomers in Chinese society) were Wang Xichan and Mei Wending, who were proficient in both Chinese and western sciences. Wang was a self-claimed Ming adherent who refused to pursue any official title after the downfall of Ming dynasty. He associated with other scholars who were Ming adherent such as Gu Yanwu and lived a simple hermitic life. Mei, on the other hand, was treated as a friend by Emperor Kangxi, although he too refused to work for the Qing government. The emperor openly praised his learning in astronomy and awarded him a plaque for being "devoted and meticulous in academic endeavors." The emperor even gave Mei a book written by himself, asking for his opinion. Though having had such different lives, their learning in astronomy were equally eulogized by later generations. Wang and Mei's studying and improving of the Tycho cosmic model were seen as an outstanding example of Chinese astronomers' work.

Here is the cosmic model put forth by Wang Xichan in his book *On the Angular Motions of the Five Planets*:

> The five planets and their orbits are within the solar system, revolving around the sun, which is located slightly off the center and moves along the solar orbit.

He was not convinced by Tycho's cosmic model, which was the fundamental theory of the *Chongzhen Calendar.* That's why he wanted to replace it with the model in the previous paragraph. However, the definition of orbit, or "bentian" as in the original text, was somewhat altered by Wang in his book. According to the Chongzhen Calendar and other works of western astronomy back then, "bentian" was a widely used term referring to the orbit on which celestial bodies move. It's the equivalent of the deferent in the Ptolemy system. But Wang's "bentian" puts the sun off the center and it didn't play any role in actual calculation. Instead, what did play a crucial role was the solar orbit, same as in Tycho's model. That's why Wang's model was in fact Tycho's model (see Picture 1.4). Qian Xizuo once said, while

Picture 1.4 Tycho's cosmic model

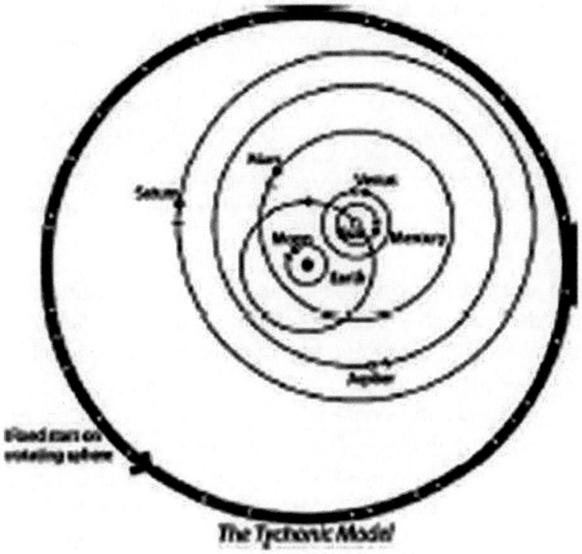

commenting on Wang's model, that it "seemed to be different than the western model but it was the same thing."

Why, then, did Wang Xichan try to argue otherwise? The reason was one of politics and ideology. Wang was a Ming adherent who refused to work for the Qing government. His discontent was twofold. Not only did he oppose the fact that China was ruled by a new regime, but he also opposed the decision to put a western missionary in charge of the imperial institution of astronomy and to officially adapt the western astronomical system. Comparing to the traditional Chinese method, the western astronomy introduced then was far superior in precision of calculation. However, Wang didn't accept that against his best judgment. He strongly believed that the reason why Chinese astronomy was bettered was because people failed to develop the full potential of the traditional method. For that purpose he wrote the *Xiao'an Xin Fa (New Methods of Xiao'an)*, the last calendar written in the classic style, tying to incorporate western methods while preserving the traditional structure. But his attempt was far from achieving what he had hoped. The *Xiao'an Xin Fa* turned out to be an extremely obscure book.

The cosmic model accepted by Mei Wending was nothing different from Ptolemy's model, with the only exception being that he didn't fully agree with Ptolemy on the issue whether celestial bodies move on tangible orbits. Mei did not agree the theory in Tycho's model that planets resolve around the sun. In *Complete Works of Calendrical Learning and Mathematics*, he insisted the five planets revolve around the earth. To avoid openly objecting Tycho's model, which was officially supported by the government, Mei came up with something neutral, the "Solar Belt" theory, which acknowledged Ptolemy's model as being factual while stating that Tycho's model was simply a vision of the former.

> Trace of planetary motion on the year orbit appears to be a belt around the sun. That's why according to the western theory, the five planets revolve around the earth. However, this solar belt is formed by rotation of the year orbit, the center of which moves on an orbit around the earth. Both theories described the three objects in one system so they're not that different.... Or failed to observe the fact, so as to think that the five planets revolve around the sun, which is against the original purpose.

The "year orbit" mentioned by Mei is equivalent to the epicycle in Ptolemy's model. At first, Mei Wending only applied his "solar belt" theory to superior planets. But later, Liu Yungong, one of his students, claimed that the same theory worked on interior planets as well, an idea highly praised by Mei.

As far as self-consistency of the system is concerned, Mei's neutral theory did seem smart in form. He himself believed that his theory was in accordance with Tycho's original ideas. "I had once drawn a chart, trying to determine the ground on which Tycho built his theory, which clearly put the earth at the center of the planetary orbits." Later, Mei's theory was greatly revered by Jiang Yong, who, in his book *Mathematic*, used geometric method to prove how in Mei's model planets are put on the "year orbit" or the "solar belt" when calculating their apparent longitudes. And the results turned out to be all the same, with both interior and superior planets.

However, Jiang Yong was unable to prove that Mei's model bears the same value as Tycho's model introduced in the *Chongzhen Calendar*. Nor could Mei himself provide any observation data to support his model (Mei Wending had nearly no experience in astronomical observation whatsoever). In fact, Mei's cosmic model, despite all its ingenuity, was not how Tycho intended it. And it's even further away from the fact than Tycho's model.

1.4.5 The So-Called "Western Sciences Originated in China" Theory

With western astronomy, mathematics and other sciences being introduced to China by the Jesuits, some in the upper-class of Chinese society, such as Xu Guangqi, Li Zhizao, and Yang Yanyun, were deeply impressed. After the establishment of Qing dynasty, the *Chongzhen Calendar*, compiled by the Jesuits, was renamed *Calendar Based on New Methods of the West* and issued nationwide. Also, Jesuit missionaries were put in charge of the imperial institution of astronomy. Even Emperor Kangxi himself became a student of Jesuit scholars, learning western astronomy, mathematics, and other knowledge.

Things like that had had a strong impact on traditional faith and thoughts of Chinese scholars. One repercussion, among others, was the popularity of the theory that western sciences originated in China within the royal court and the intellectual circle. The theory was mainly referring to astronomy and calendrical methods, and mathematics too, because of its close relationship with the other two disciplines. By the end of Qing dynasty, the theory had made its way into almost all fields in the world of science.

As a matter of fact, the theory itself came from among Ming adherents. The person who put forth the theory in the first place was probably Huang Zongxi. Another pioneer could be Fang Yizhi, a fellow scholar in Huang's time. The two may have invented the theory but they provided no specific evidence. The theory didn't develop any further until Wang Xichan joined the action by giving an elaborate explanation for the theory.

During Qing dynasty, Emperor Kangxi, while being obsessed with western sciences introduced by the Jesuits, openly advocated the theory that western sciences originated in China. In his book *On Triangles*, the emperor wrote: "Ancient knowledge of China spread to the west, where they learned and improved it." And that's just astronomy. His theory concerning mathematics was even more eye-catching. One quote, in particular, that's frequently cited was what the emperor said during a discussion he had with Zhao Hongxie on mathematic in the 50th year of Kangxi period (1711). The following is an account of the quote in *Dong Hua Lu*:

> The western method is good enough but it's originally from China. They call it A'erbazhu'er and the name itself suggests that it has an oriental origin.

The A'erbazhu'er, also known as A'errebada or "A'errebala," is commonly believed to be a phonetical translation of the word "algebra" (or Al-jabr in Arabic). As for how the name suggests "an oriental origin" as the emperor said, we do not know.

As soon as he heard Kangxi's words, Mei Wending responded quite passionately. He decided to supplement and perfect the theory with his being "devoted and meticulous in academic endeavors." In the *Extended Questions of Calendrical Learning*, Mei broke the theory down into three aspects as follows:

One, astrolabes were inspired by the Gaitian model, a cosmic theory in ancient China. Li Zhizao wrote a book titled *Hun Gai Tong Xian Tu Shuo* (*Illustrated Explanation for Principles of Astrolabes*), and Sabbatino de Ursis, a Jesuit scholar in China, wrote another book titled *Jian Ping Yi Shuo (On the Abridged Flat Armilla)*. The former discussed the projection of spheroidal coordinates on planes, while the latter discussed an astronomical instrument called the Jianpingyi (abridged flat armilla), which was similar to an astrolabe. Mei's argument was mainly based on the principles of astrolabes. For example:

> As for the astrolabe, none could've made those except for the saints of yore like Rong Cheng and Li Shou (astronomer and mathematician in ancient China). It fits, in every detail, theories in the *Zhou Bi Suan Jing*, therefore it's a rational conclusion that it was in fact an instrument based on the Gaitian theory.

Obviously, Mei's conclusion was drawn upon a false premise. The Huntian and Gaitian theories in ancient China were in no way like "sculpture" and "portrait" as he put it. Given his expertise in astronomy, it's impossible that Mei didn't know that.

Yet he would go out of his way trying to prove an illusion. Such behavior couldn't simply be explained by a scholar's blind loyalty to his monarch.

And two, by which means and in what forms were Chinese sciences introduced to the west. This was the essential proof for the credibility of the whole theory. Mei started with *Shi Ji (Records of the Grand Historian)*, according to which "Zhou dynasty was weakening after King You and King Li. . .. successors of ancient astronomers scattered across the country. Some went to the principalities and others wound up in barbaric tribes." Based on the account he believed that "they traveled far fleeing from chaos and warfare and some must have brought their books and instruments with them." Another possibility imagined by Mei Wending seemed even more convincing: According to *Shang Shu (Book of Documents)*, King Yao once "put four of his ministers in charge of astronomical observation and they were respectively sent to four corners of the kingdom. And Mei's fabrication, based on this legend, was that because there were oceans on the east and the south and the north was extremely cold so the only one who made it was Zhong He, who was traveling westward. And he went further and further into the west. That's how theories of Chinese astronomy reached Europe. Mei himself went even further with his imagination, saying that he "educated the locals with Chinese culture" along his journey. And here's how spectacular the scene was when he finally arrived in the west:

> People came from afar out of admiration for his virtue. Some were taught in words, others told his stories. All were enlightened. The most intelligent of them followed his teaching and improved it.

Surely, compared to Wang Xichan's accusation that the west "stole" Chinese sciences, Mei Wending's choice of words was a lot softer.

And three, the close relation between western methods and Islamic astronomy. Back then, it was much harder for Mei to recognize that relation than we do today, because Chinese scholars at that time knew very little about the world outside China. Nevertheless, Mei Wending was wrong about the causal relation between the two. In fact, western methods did seem more "accurate" than Islamic astronomy, which originated from western science in the first place. Among the three aspects of Mei Wending's arguments, this is the only one that has some real science in it.

Thanks to Emperor Kangxi's advocation, as well as Mei Wending's vigorous elucidation, the theory was perfected, therefore had a greater influence, even spreading to other fields besides astronomy and mathematics, for example, mechanics. Ruan Yuan, a scholar in Qing dynasty, wrote in an article titled *Zi Ming Zhong Shuo (On Self-chiming Clocks)* from his book *Yan Jing Shi San Ji* (*Collected Essays from the Yanjing Chamber)* that chiming clocks from the west were no different than Kelou, an ancient Chinese timepiece. That's why the design was originally from China. And in medicine, too, Mao Xianglin, another scholar in Qing dynasty, claimed that surgical operations in western medicine were just a tiny proportion of Hua Tuo's knowledge and skills and that surgeries were seldom successful because they didn't learn it thoroughly (volume seven of *Mo Yu Lu*). Such comments were mostly amateur gibberish with no academic value at all.

In fact, rulers of Qing dynasty were faced with a dilemma. On one hand, they were in desperate need of western sciences. For example, they needed western astronomy to formulate the official calendar and the Jesuits' help in dealing with diplomatic affairs (like signing the *Treaty of Nerchinsk* with Russia). They also needed western techniques for making cannons and other instruments, as much as they needed Quinine for curing "royal diseases." On the other hand, they needed to announce themselves as the inheritor of Chinese culture and traditions so that their rule would be deemed legitimate, and they could go on calling China the "Celestial Empire." That's why the issue of "adopting foreign methods" in calendar, the sacred symbol of majesty, became increasing vexing to the Qing government.

Under such circumstances, Kangxi's advocating the "Chinese origin of western sciences" theory was indeed a brilliant scheme, with which, the government could keep benefiting from western techniques while evading the accusation of "adopting foreign methods." Now that western sciences, as much advanced and superior as they may seem, was proved being originally from China, then making use of them was simply an act or reclaiming what's rightfully ours.

The theory was widely popular among Chinese scholars for over 300 years. Another reason was the advocators were trying to lift morale and confidence of the Chinese nation. For thousands of years, the "Celestial Empire" had been admired and worshiped by the rest of the world. Now all of a sudden it was falling behind in almost every way. Of course it brought about a sense of shame, and the theory was merely a means to an end for redeeming our dignity.

1.4.6 How China Missed the Chance of Prosperity and the Role Kangxi Played

In recent years, historians tend to render Kangxi great reviews, comparing his rare gifts and bold strategy to that of King Louis XIV of France. They also praised his generous support for academic inquiries, comparing him to the likes of Cosimo de Medici during the Renaissance. In their letters and reports sent back to Europe, the Jesuits served in Kangxi's court did often use words like merciful, just, magnanimous, brilliant, and great when speaking of the emperor.

Kangxi was very interested in western sciences and technologies, and he was a passionate student of relevant knowledge. It was the truth. Indeed, a monarch like him was a rare case especially in traditional Chinese society. As an individual person, he was undoubtedly an outstanding figure in terms of perspective and intelligence. However, as the ruler of a great nation, his contribution in history is questionable.

First of all, the fact that he invited Jesuits who were familiar with sciences and technologies to work for his government, which was frequently cited by many books as proof of his passion and enthusiasm for science. But, if we take a look at Chinese history before him, we'd see that his deeds were not that different than almost all rulers of China throughout time. Individuals of various expertise had always served in the royal court. The most common ones were Buddhist monks or Taoist priests, who usually provided the emperors with their talents in astrology, fortunetelling, medicine, alchemy, art, music, etc. Generally speaking, they were more like entourage of the emperor except for few who were allowed in the process of important decision-making after they had earned the emperor's trust. The Jesuits' service in Kangxi's court followed the exact same pattern. Although there wasn't any activity of astrology or alchemy, the Jesuits did please the emperor with their talents of medicine, painting, and music. Furthermore, they were responsible for managing western instruments such as chime clocks and designing western-style architecture. Their specific skills and jobs might differ, but the whole setup was just like in all previous dynasties. Having people with fancy tricks from a foreign land served the throne had always been what every ruler of ancient China bragged about, not necessarily because of their passion for science.

A much more serious mistake on Kangxi's part was already pointed out by many. That is, despite his personal interest in western sciences and technologies, the emperor had no intention of promoting them among government officials and the general public whatsoever. He wouldn't even make the smallest effort as to select young aristocrats to be trained by Jesuits, let alone establishing public schools where the Jesuits could teach knowledge of western sciences, or sending young Chinese scholars to study in Europe, which would have been easy tasks with the help of the Jesuits, who'd be more than willing to do so.

In a time when modern science was rising to prosperity, Kangxi was given a once-in-a-millennium opportunity, with which, had he acted on it, China would've caught up with Europe in science and technology. As the ruler of the Chinese empire for over 60 years, he had had plenty of time to make that happen, if he chose to. But his mind, on the whole, was still one of conservatism. And he was only being "visionary" on a rather superficial level, in that he had seen some novelty gadgets than most people in his time.

Meanwhile, the emperor failed, completely, to see the dawn of a new world.

Rod Arithmetic Calculation and Abacuses: Calculation Tools in Traditional Chinese Math

2

Lisheng Feng

Contents

Abstract

There is a close relationship between Chinese traditional mathematics and calculation tools. From counting rods in the spring and Autumn period and the Warring States period (about 500 BC) to the creation of abacuses, a new and efficient calculation tool, in the Tang and Song Dynasties (618–1126), calculation tools always played an important role in the development of traditional math. Abacuses remained in extensive use in China in the twentieth century and are still highly viable. China's counting rods and abacuses subsequently spread to countries of the Chinese culture circle and had longstanding influence on mathematical development in these countries.

Keywords

Counting rods · Counting method · Rod arithmetic calculation · Abacuses · Abacus calculation

L. Feng (✉)
The Institution for the History of Science & Technology and Ancient Texts, Tsinghua University, Beijing, China
e-mail: fenglsh@lib.tsinghua.edu.cn

X. Jiang (ed.), *The High Tide of Science and Technology Development in China*, History of Science and Technology in China,
https://doi.org/10.1007/978-981-15-7847-2_2

Traditional Chinese math had been closely related to calculation tools, which always played an important role in the development of traditional math. In the Spring and Autumn and Warring Kingdoms periods (c.500 B.C.), counting rods were already used for counting and calculations, and in the Western Han Dynasty period (206 B.C.–24 A.D.), a rod arithmetic system was established. Afterwards, rod arithmetic kept evolving and formed a unique math tradition of China. In the Tang and Song Dynasties periods (618–1126), the rod arithmetic method was constantly simplified, while pithy formulae were very popular, giving rise to the need for reforming calculation tools and leading to the creation of a new, efficient calculation tool – abacuses. In the Song and Yuan Dynasties periods (960–1368), abacuses began to run parallel to counting rods, especially in the civil society and replaced counting rods to become the mainstream calculation tool in the Ming Dynasty period (1368–1644). Abacuses remained in extensive use in China in the twentieth century and are still highly viable. China's counting rods and abacuses subsequently spread to countries of the Chinese culture circle, such as Korea, Japan, and Vietnam, and had longstanding influence on mathematical development in these countries.

2.1 Invention and Early Applications of Counting Rods and Counting Method

China's counting method has followed the decimal system since the era when written records emerged. In the inscriptions on tortoise shells during the Shang Dynasty period (pre-1000 B.C.), the decimal numbers and counting method were already used, with 30000 as the maximum number. With development of the production technology and distribution of social wealth in the Western Zhou Dynasty period, a more sophisticated calculation technology was required, and a practical calculation tool was invented. As counting rods had been in long-term use in ancient China, it was possible that the counting rods were used for counting during the Shang and Western Zhou Dynasties periods. However, as the characters "suan" (counting) and "chou" (rod) are absent in the tortoise shell inscriptions, the exact date when counting rods were invented is difficult to determine.

The Spring and Autumn and Warring Kingdoms periods are an important era in the history of scientific and mathematical development in China, and a large number of complicated mathematical calculation problems needed to be addressed. Existing literature and relics indicate counting rods not only existed in the Spring and Autumn and Warring Kingdoms periods but also were widely used. For example, the characters "suan" and "chou" already appeared in many works of the Spring and Autumn and Warring Kingdoms periods, such as *Yi Li* (《仪礼》), *Zuo Zhuan* (《左传》), *Sun Zi* (《孙子》), *Lao Zi* (《老子》), *Guan Zi* (《管子》), *Mo Jing* (《墨经》), and *Xun Zi* (《荀子》).

A counting rod, also known as "suan," "chou," or "ce," is a small rod made of bamboo, wood, or other materials such as tusks or bones. It was a counting or calculation tool based on the decimal system. The counting rods can be placed in a certain way to easily indicate a random number and to carry out the four fundamental arithmetic calculations and more complex calculations. Rod arithmetic calculations are calculations and deductions carried out with the counting rods. According to records, counting rods can indicate nine natural numbers from 1 to 9 in two ways (vertical and horizontal arrangements):

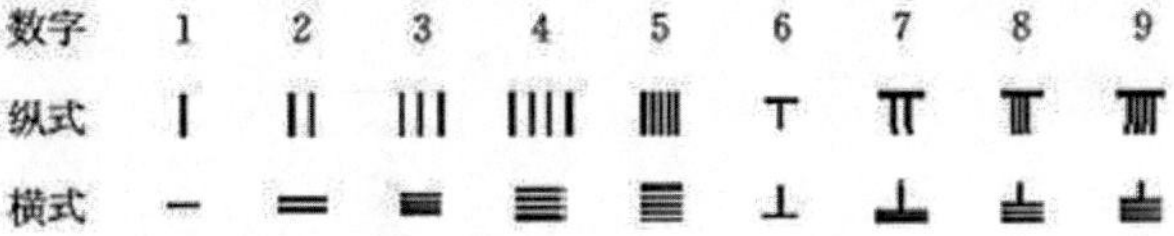

中文	英文	中文	英文
数字	Number	纵式	Vertical arrangement
横式	Horizontal arrangement		

Yi Li – Xiang She Li (《仪礼·乡射礼》) said, "One person followed carrying counting rods," which described how a designated person was responsible for counting and recording the number of hits and misses by an archer during a provincial examination. *Zuo Zhuan – Xiang Gong San Shi Nian* (《左传·襄公三十年》) recorded a word puzzle: An old man's age counted in days is the character "*hai*" (亥). Zhao Shi said: "*Hai* has two heads and six bodies, and two additional 'sixes', which indicated the number of days." And Wenbo Shi said: "So the number of days is 26660." The counting rods were placed to solve this puzzle. The character "亥" can be decomposed into 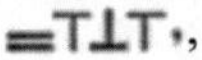, which stands for 26660 days. This story suggested the rod arithmetic method was quite popular during that time. *Lao Zi* (《老子》) said, "Those good at coming up with ideas do not use *chous* or *ces*" (Lao Tse: Chapter 27 of *Tao Te Ching*), which also implied the popularity of counting rods. When describing this counting method, *Mo Jing* (《墨经》) written around the fourth century B.C. said: "If one is at one's place, it is smaller than two; if at ten's place, it is greater than five." That is to say, in addition to its numeric value, the value of a digit also depends on its position in the whole number. The counting rod method already used the decimal system.

As for the shapes and sizes of the counting rods, the earliest record was seen in *Han Shu – Lv Li Zhi* (《汉书·律历志》). According to the record, a counting rod is a round bamboo rod with a diameter of 0.23 cm and a length of 13.69 cm, and every 271 rods form a "handful."

Over 10 batches of counting rods have been unearthed, such as those unearthed in a wooden tomb of the Warring Kingdoms period at Zuojiagong Mountain, Changsha, Hunan Province, in 1954. There were "40 rods, which have the same length, and each is 1–2 cm long" (Hunan Administrative Committee of Cultural Relics: "Wooden Tomb of Warring Kingdoms Period at Zuojiagong Mountain Changsha." Cultural Relics Reference (《文物参考资料》), 1954, issue 12). The excavation brief at that time did not point out these were counting rods. Dunjie Yan later confirmed they were counting rods (Dunjie Yan 1956). In the late 1950s, a bundle of "bamboo

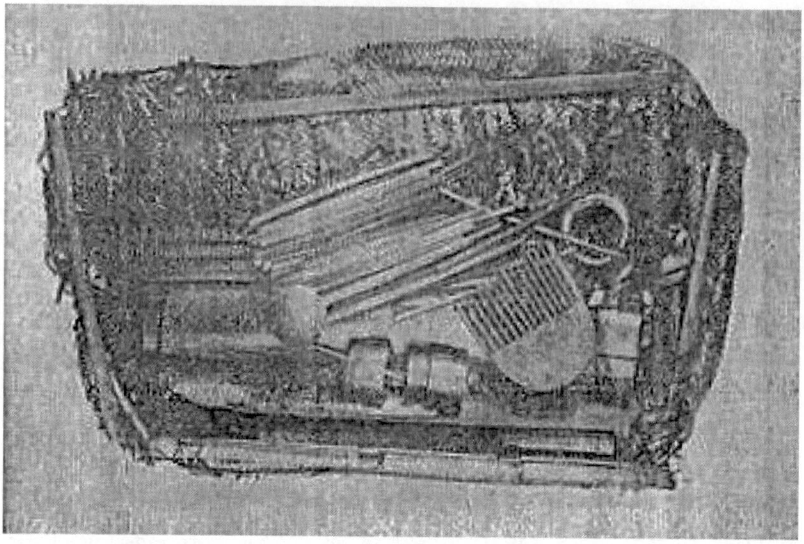

Fig. 2.1 40 counting rods among the bamboo slips unearthed from a Chu Kingdom tomb of the Warring Kingdoms period at Zuojiagong Mountain, Changsha

rods" were dug out from a Chu Kingdom tomb of the Warring Kingdoms period at Deshan Mountain, Changde, Hunan. There were a dozen rods, which were black and mostly rotten. Each was 13 cm long, 0.7 cm wide, and 0.3 cm thick (Provincial Museum of Hunan 1963). These were also counting rods (Pei Zhang: "Study of Unearthed Calculation Rods." *Wen Bo* (《文博》), 1996, issue 4, pp. 53–59). In 1986, 21 round counting rods made of bamboo were unearthed from a tomb of the Warring Kingdoms period at Fangmatan, Gansu, with each 20 cm long and 0.3 cm in diameter (Provincial Relic Archaeological Institute of Gansu: "Excavation of Tombs of Warring Kingdoms, Qin & Han Dynasties periods at Fangmatan, Tianshui, Gansu Province." *Wen Xu* (《文物》), 1989, issue 2). Three groups of bamboo counting rods were mixed in the Qin Dynasty bamboo slips preserved at Peking University, believed to be counting tools used by the owner for calculations (Fig. 2.1) (Wei Han: "Math Literature in Qin Dynasty Bamboo Slips Preserved at Peking University." *Wen Xu* (《文物》), 2012, issue 6).

More counting rods were unearthed from Han Dynasty tombs. For example, over 30 counting rods made of animal bones during Emperor Xuandi's reign of the Western Han Dynasty period (74–49 B.C.) were discovered in Qianyang County, Shaanxi Province (Baoji Prefecture Museum, Qianyang County Cultural Museum and Natural Science History Institute of Chinese Academy of Sciences: "Calculation Rods Unearthed in West Han Dynasty Tomb in Qianyang County." *Archaeology* (《考古》), 1976, issue 2), and their sizes and lengths were basically identical to those recorded in *Han Shu – Lv Li Zhi* (《汉书·律历志》). At a Han Dynasty tomb at the Phoenix Mountain, Jiangling, Hubei, a bundle of bamboo rods made during Emperor Wendi's Reign (179–157 B.C.) were unearthed, and in the bamboo-plated basket

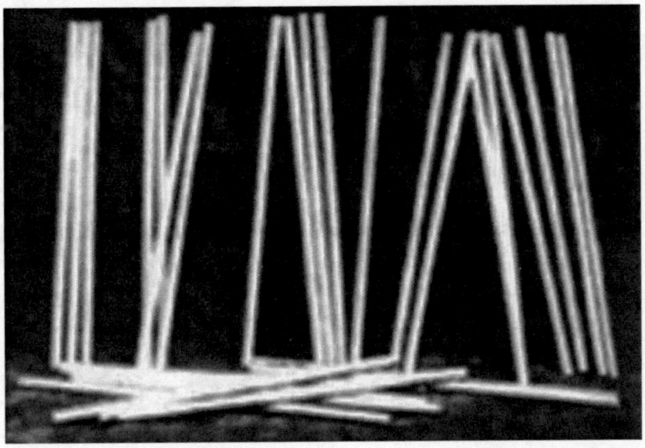

Fig. 2.2 West Han tusk counting rods unearthed in Xunyang, Shaanxi

where these rods were placed, there were also brushes, ink, slips, ink slabs, peelers, balances, measurements, poles, weights, and coins. The counting rod was about 13.5 cm long and 0.3 cm in diameter (Excavation team for Tomb No.168 at Phoenix Mountain: "Excavation Briefing on Tomb No.168 at Phoenix Mountain, Jiangling, Hubei." *Wen Xu* (《文物》), 1975, issue 9), and was a little longer than those found in the Qianyang County. In Nov. 1983, 28 well-made tusk counting rods were unearthed from No.1 Tomb of Yousheng Palace in Xunyang County, Shaanxi Province (see Fig. 2.2), and each rod was 0.4 cm in diameter and 13.5 cm long (Pei Zhang: "Tusk Counting Rods Unearthed at Han Dynasty Tomb in Xunyang, Shaanxi." *China Relics* (《中国文物报》), 6 May 1988, Edition 3).

When indicating numbers, counting rods can be placed in vertical and horizontal lines, and any number can be expressed with the combination of vertical and horizontal numbers. For example, numbers at one's place, hundred's place, and ten thousand's place are expressed with vertically placed rods, numbers at ten's place and thousand's place are expressed with horizontally placed rods, and zero is expressed with space. Once emerging, counting rods strictly followed the decimal system, where carryover is triggered when the number reaches above 9, and a number can indicate several hundred when placed at the hundred's place or tens of thousands when placed at the 10 thousand's place. To denote a number smaller than five, the corresponding number of rods is used, and for six, seven, eight, and nine, a rod is placed above to indicate five, and every other rod indicates one. When denoting a multi-digit number, the counting rods are placed from left to right in the order of the digits and both vertically and horizontally. There are specific rules and space denotes zero. For example, the number 4368 is denoted with the counting rods as follows:

𝍬 𝍢 𝍮 𝍧

Obviously, the rod counting method is essentially the same as the present-day decimal counting system except the form in which numbers are denoted, which is different from the Indian Arabic numbering widely used today. At that time, the rod counting method was the most sophisticated numbering system in the world and had a fundamental role in bringing about the algorithm-centric characteristic of Chinese math.

2.2 Examples of Rod Arithmetic Calculation Method

The calculation procedure of the rod arithmetic method was introduced in works like *Jiu Zhang Suan Shu* (《九章算术》) (100 B.C.), *Shu Ji Yi* (《数术记遗》) (c.200), *Sun Zi Suan Jing* (《孙子算经》) (c.300–400), *Xia Hou Yang Suan Jing* (《夏侯阳算经》) (c.400–500), and *Zhang Qiu Jian Suan Jing* (《张丘建算经》) (500–600). After negative numbers appeared, counting rods were divided into red and black rods, of which the red ones represented positive numbers and the black ones represented negative numbers. The counting rods could also denote all types of algebraic expressions and be used for various algebraic calculations, the method of which was similar to the present-day method of detached coefficients. A few examples are given below to introduce the calculation procedure and method using counting rods.

Ancient literature lacks descriptions of additions and subtractions using counting rods. *Sun Zi Suan Jing* (《孙子算经》) recorded a description of the specific method of multiplication and division using counting rods. A sample problem is to determine the answer of 81 times 81. In what follows, we will give the original text of the problem description and the calculation process and explain it with modern calculation methods.

Original text in *Sun Zi Suan Jing:*

How much is eighty-one times itself
The answer is: six thousand five hundred and sixty-one
The method is: rearrange the positions.

The upper eight multiplies lower eight, and eight multiplies eight equals to sixty-four, so six thousand and four hundred is at the middle position. The upper eight multiplies lower one, and one multiplies eight equals to eight, so eighty is at the middle and shifted to the next position, and the upper eight is removed.

The upper one multiplies lower eight, and one multiplies eight equals to eight, so eighty shifted from the middle. The upper one multiplies the lower one, which equals to one, so one is shifted from the middle, and the middle position is six thousand five hundred and sixty-one.

$$\text{Determine } 81 \times 81 = ?$$

Place the rods at the upper, middle, and lower positions; place the multiplier at the upper position, the multiplicand at the lower position, and the product at the middle position; and the layout in this example is as follows:

Upper position
Middle position
Lower position

This means: the 80 at the upper position multiplies and 81 at the lower position

$$80 \times 81 = 80 \times 80 + 80 \times 1 = 6480,$$

which is placed at the middle position. As 80×81 is complete, the next step is 1×81, and the result is shifted to the next position, and the 80 at the upper position is removed to get the following:

Upper position
Middle position
Lower position

Then, the 1 at the upper position multiplies the 81 at the lower position, namely,

$$1 \times 81 = 80 \times 1 + 1 \times 1 = 81,$$

which is placed at the middle, and the middle position is 6561. After multiplication is complete, the upper 1 and lower 81 are removed to get the following: middle position.

A sample problem of division in *Sun Zi Suan Jing* (《孙子算经》) is to determine $6561 \div 9 = ?$ The original text and modern explanation are given below:

Original text in *Sun Zi Suan Jing*:

Ask people how much is six thousand five hundred and sixty-one divided by nine. The answer is: seven hundred and twenty-nine.

$$\text{How much is } 6561 \div 9 = ?$$
$$\text{Answer : } 729$$

The method is to place six thousand five hundred and sixty-one at the middle as the dividend, and nine is listed below as divisor.

Seven hundred is placed at the upper position, and the upper seven multiplies the lower nine, which equals to sixty-three, namely, the middle position is sixty-three hundred, and the number is shifted to the lower position.

Now twenty is at the upper position, the upper two multiplies the lower nine to get eighteen, namely, one hundred and eighty is at the middle position, and the number is shifted to the lower position.

Nine is placed at the upper position, and the upper nine multiplies the lower nine to get eighty-one, and the eighty-one at the middle position is divided with no remainder, and the number at the upper position is the answer.

The layout is as follows:

Upper position
Middle position
Lower position

The number at the middle is divided by 9 to get the initial quotient of 700, and 6300 is removed from the middle position.

$$6561 - 700 \times 9 = 261$$

Now 261 is divided by 9, and the lower position is shifted to get the following:

Upper position
Middle position
Lower position

The middle position is divided by 9 to get the second quotient of 20.

$$261 - 20 \times 9 = 81$$

Now 81 is divided by 9 and the number is shifted to the lower position as follows:

Upper position
Middle position
Lower position

Then the lower position is divided by 9 to get the third quotient of 9.

$$81 - 9 \times 9 = 0$$

After the number is divided with no remainder, the quotient is obtained at the upper position.

Upper position

Among currently available math works, *Jiu Zhang Suan Shu* (《九章算术》) was the earliest to provide the complete processes of determining square roots and cubic roots using counting rods. In what follows, we will explain the specific algorithmic procedure by using the sample problem of determining the square root of 55225 in *Jiu Zhang Suan Shu* (《九章算术》).

Original text in *Jiu Zhang Suan Shu:*

Now we have fifty-five thousand two-hundred and twenty-five steps. Question: What is the square root?
Answer: two hundred and thirty-five steps.
This is to determine the square root of a number.

The method is:

1. "Place the number as the dividend. Move *Jiesuan* until immovable."
2. "The product is obtained, and one multiplies the number and the result is divided."
3. "When a number is divided by itself, *Beifa* is *Dingfa.*"
4. "The number is divided again, and *Fashu* is shifted."
5. "*Jiesuan* is placed again. The steps are as before and the number multiplies 1 to derive the number for addition and division."
6. "The result acquired is used to determine *Dingfa*. Division is carried out again."

$$\sqrt{55225} = ?$$

This is practically solving the equation $x^2 = 55225$.

Answer: 235.

According to the description of the method, the calculation process is as follows:

As the radicand, 55225 is placed at the upper layer; a rod (*Jiesuan*) moves from the lower digit to the higher digit until the ten-thousand digit, where it can move no more.

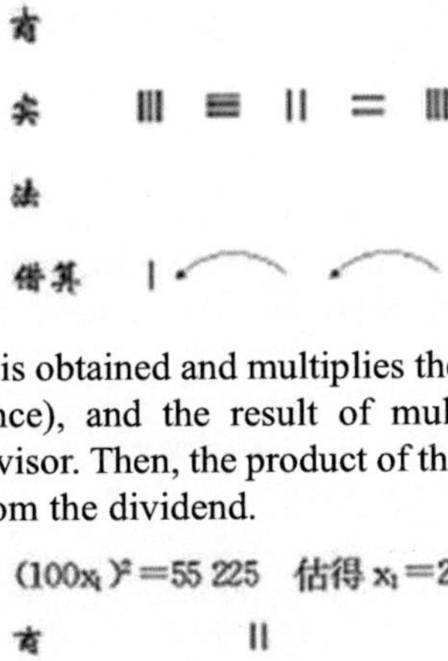

The initial quotient of 2 is obtained and multiplies the number borcolumned from the ten-thousand digit (once), and the result of multiplication is placed at the ten-thousand digit as the divisor. Then, the product of the initial quotient multiplying the divisor is subtracted from the dividend.

$(100x_1)^2 = 55\,225$ 估得 $x_1 = 2$

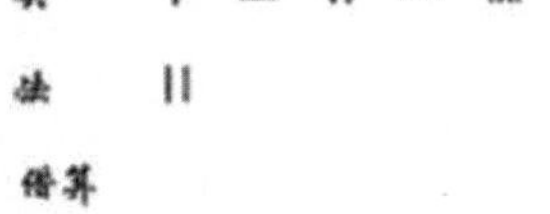

After the product of the initial quotient multiplying the divisor is subtracted from the dividend, the divisor multiplies 2, and the product is called *Dingfa*.

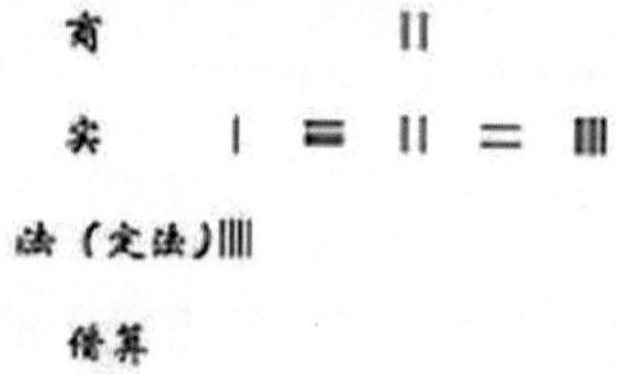

To determine the second quotient in the next step, *Fashu* (namely, *Dingfa* mentioned in 3) is shifted to the lower digit.

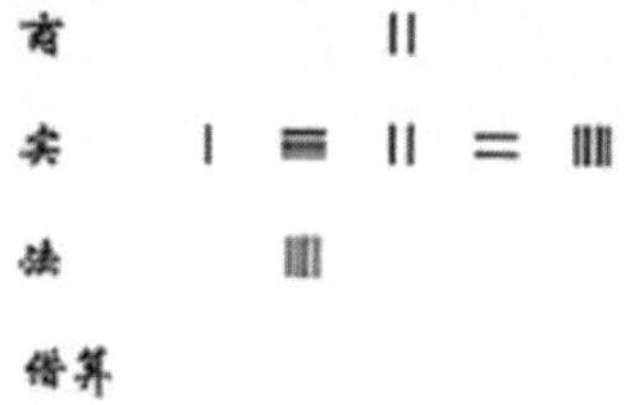

Jiesuan is shifted to the higher digit. (This is to determine the second quotient at the ten's digit. In this case, the shift only occurs once.)

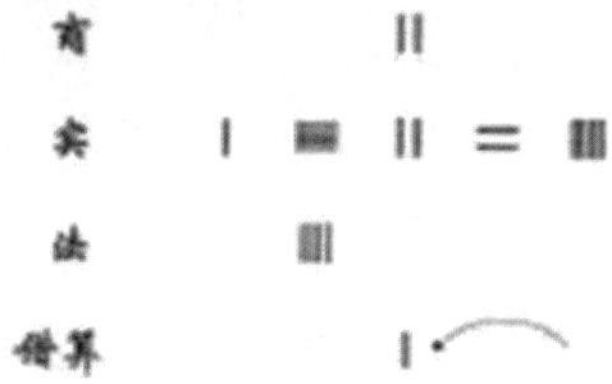

The secondary quotient is determined as 3. The secondary quotient multiplies *Jiesuan* once. After the number derived is recorded in the next column, it is inserted into the divisor position. And the product of the secondary quotient and the divisor is subtracted from the dividend.

$$(10_{x_2})^2 + 400 \cdot (10_{x_2}) = 15$$
$$225$$

Where the squared term in $10x_2 = x - 200$ is omitted and the constant is divided by the coefficient of x_2 to determine $x_2 = 3$.

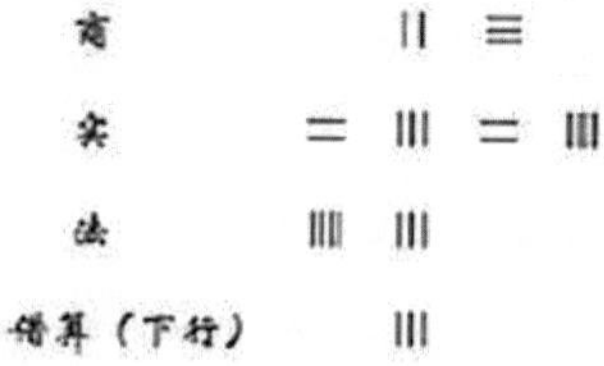

The above columns are integrated into *Dingfa*. The quotient for the next digit is determined in the same way as determining the secondary quotient. *Dingfa* needs to move one place to the lower position and be re-calculated.

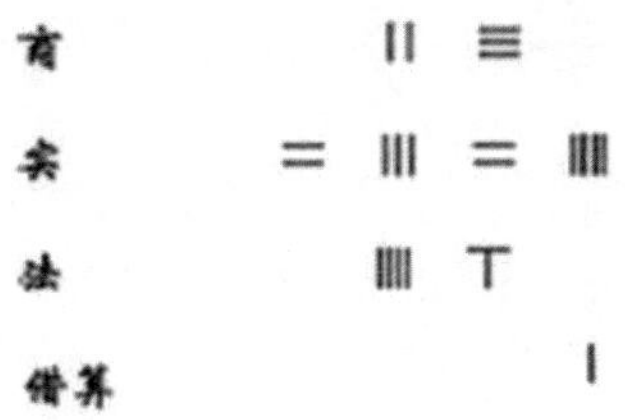

The third quotient is determined as 5.

$$10x_3^2 + 460x_3 = 2325$$

Where $x_3 = x - 230$, for which the squared term is omitted, and the constant is divided by the coefficient of x_3 to determine $x_3 = 5$.

中文	英文	中文	英文
商	Quotient	法	Divisor
实	Dividend	借算	*Jiesuan*

The Equation Chapter of *Jiu Zhang Suan Shu* (《九章算术》) focused on the method of solving linear equations. One of the samples is given below for explanatory purposes:

"Equation Chapter of *Jiu Zhang Suan Shu* (《九章算术》) and Hui Liu's notes":

Now we have three *bings* of upper millet, two *bings* of middle millet and one *bing* of lower millet, which total 39 *dous*; if there are two *bings* of upper millet, three *bings* of middle millet and one *bing* of lower millet, the total quantity is 34 *dous*; and if there are one *bing* of upper millet, two *bings* of middle millet and three *bings* of lower millet, the total quantity is 26 *dous*. How many *dous* of millet does one *bing* of upper, middle and lower millet contain respectively?

Answer: Nine and a quarter dous for upper millet, four and a quarter dous for middle millet and two and three quarters of dous for lower millet.

The equation method is:

Place three bings of upper millet, two bings of middle millet and one bing of lower millet, which total 39 dous to the right. Middle and left millet listed in the same way as the right millet.

So $x = 230 + x_3 = 235$

Thence, the square root of 55225 is determined as 235.

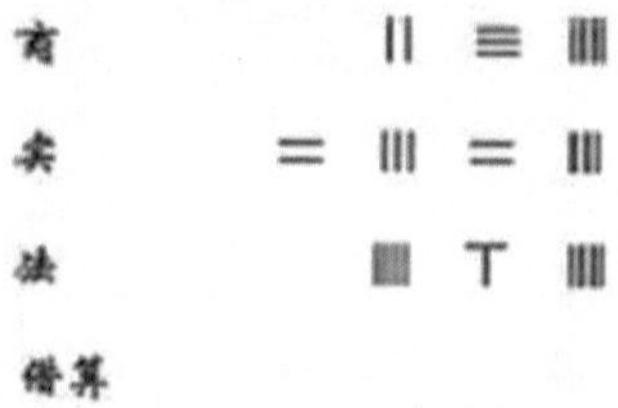

Assume one bing of upper, middle and lower millet contains x dous, y dous and z dous, and solve the linear equations:

$$\begin{cases} 3x + 2y + z = 39 \\ 2x + 3y + z = 34 \\ x + 2y + 3z = 26 \end{cases}$$

According to the equation method, the calculation process is described as follows:

In ancient times, columns were lines, and lines were columns, contrary to today's practice.

中文	英文	中文	英文
商	Quotient	法	Divisor
实	Dividend	借算	*Jiesuan*

The upper millet multiplies the middle column for direct division.

The number multiplies the secondary item and is directly divided. The left column head is removed.

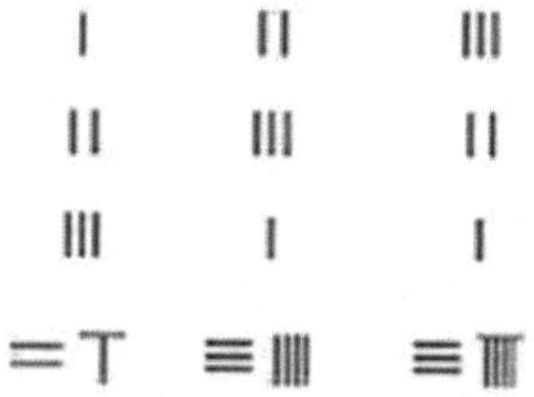

The upper millet coefficient of 3 in the right column multiplies the entire middle column.

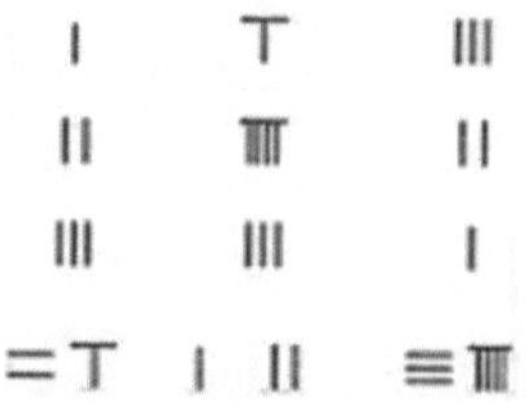

Then, the middle column is subtracted from the right column, and the upper millet coefficient of the middle column becomes 0.

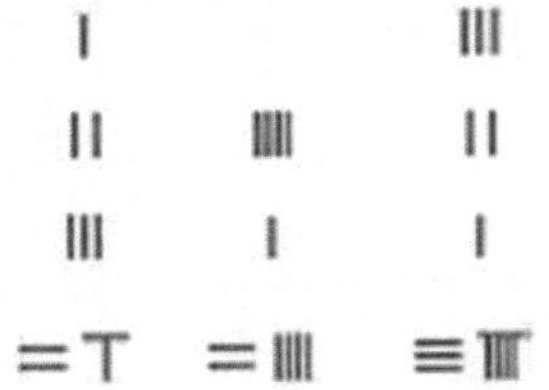

The coefficient of the upper millet in the right column multiplies the entire left column. The left column is subtracted from the right column, and the coefficient of the upper millet in the left column becomes 0.

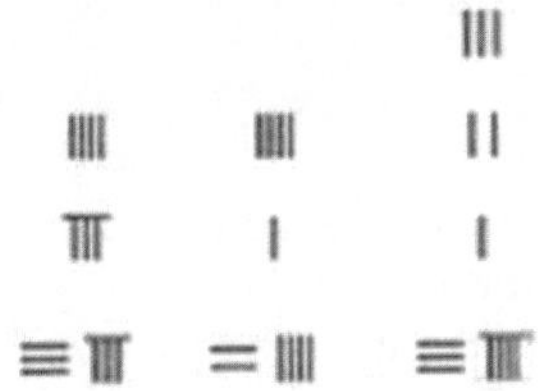

The middle millet coefficient in the middle column (5) multiplies the entire left column, and the left column is subtracted from the middle column four times.

Then, the remaining part of the middle millet in the middle column multiplies the left column and is directly divided. For the remaining part of the lower millet in the left column, the upper one is the dividend, and the lower is the divisor. The divisor is the divisor of the lower millet.

The middle millet coefficient also becomes 0, the lower millet coefficient is 36, and the dividend is 99. The lower millet coefficient and divisor share a common factor of 9, which is simplified. The lower millet coefficient becomes 4 and serves as the divisor, and the dividend is 11 and is only the dividend of the lower millet.

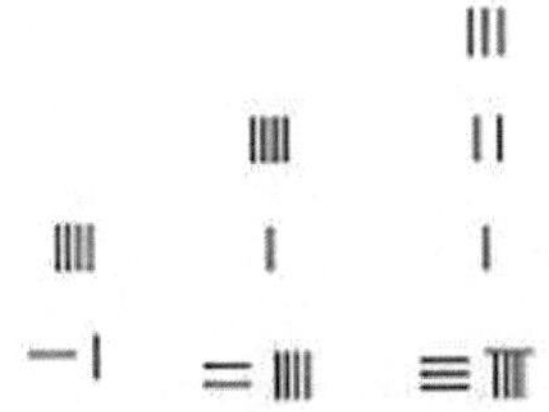

To determine the middle millet, the divisor in the left column multiplies the dividend in the middle column, and the divisor of the lower millet in the left column is subtracted, namely, $24 \times 4 - 11 \times 1$. The remainder of that calculation is divided by the number of middle millet in the middle column, namely, the dividend in the middle column, and the divisor in the left column remains the divisor. In this case, $(25 \times 4 - 11 \times 1)/5 = 17$, with 4 as the divisor.

To determine the middle millet, the divisor multiplies the lower dividend in the middle column, and is divided by the dividend of lower millet. If the number of handfuls of middle millet is one, this is the dividend of middle millet.

To determine the upper millet, the divisor in the left column multiplies the lower dividend, the dividend of the lower millet in the left column is subtracted from the dividend of the lower millet in the right column, and multiplies the number of handfuls of lower millet in the right column, and the dividend of the middle millet in the middle column is subtracted, and multiplies the number of handfuls of middle millet in the right column. In this case, $39 \times 4 - 11 \times 1 - 17 \times 2$. The remainder from this calculation is divided by the number of handfuls of upper millet in the right column to derive the dividend of supper millet. The divisor in the left column remains the divisor.

To determine the upper millet, the divisor multiplies the lower dividend in the right column and is divided by the dividends of the lower and middle millet. If the number of handfuls of upper millet is 1, this is the dividend of the upper millet.

The dividend is the same as the dividend, a dou is derived for each.

In this case, $(39 \times 4 - 11 \times 1 - 17 \times 2)/3 = 27$, and 4 remains the divisor.

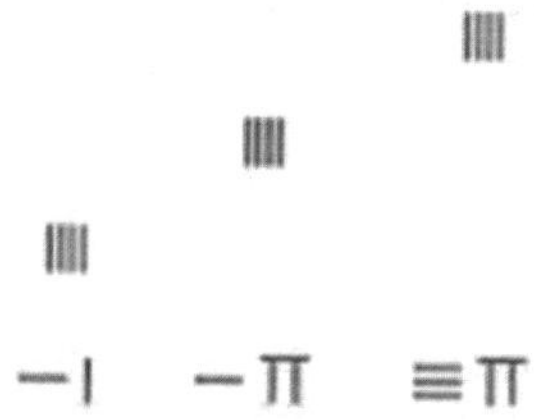

The dividend is divided by the divisor to determine one handful of upper millet is $x = 9\frac{1}{4}$ dou, one handful of middle millet is $y = 4\frac{1}{4}$ dou, and one handful of lower millet is $z = 2\frac{3}{4}$ dou.

In addition to numeric calculations, counting rods are also capable of formula deduction. During the Song and Yuan Dynasties periods, the sophistication of counting rod calculations leapfrogged from square root/cubic root calculations to fourth root calculations. In *Jiu Zhang Suan Fa Zuan Lei* (《九章算法纂类》), Hui Yang recorded Xian Jia's "square root calculation method with increased multiplication" and the "cubic root calculation method with increased multiplication"; the Detailed Explanation to *Jiu Zhang Suan Fa* recorded Xian Jia's illustration of the "square root calculation method," the "sickle grass determination method with increased multiplication," and the method of fourth root calculation with increased multiplication. According to these records, it can be asserted Xian Jia had discovered the table of binomial coefficients and created the method of square root calculation with increased multiplication. Based on this, Jiuzhao Qin created a more generic numerical solution of high-order equations – the positive and negative square roots method.

Math also transformed in the form of calculation during the Song and Yuan Dynasties periods. In order to describe and explain the calculation steps, mathematicians had to record the formulae on paper. Afterwards, some calculation steps could be completed through writing, and symbols such as chips were used, symbolizing calculations to an extent. In order to solve equations, Chinese mathematicians during the Yuan Dynasty period created the *Tianyuan* (translator's note – which means "heavenly center" in Chinese) method, which used *Tianyuan* (equivalent to x) as the symbol of the unknown number and listed equations. This is an issue of introducing some symbols and solving high-order equations by calculating with these symbols. The earliest currently available work on the *Tianyuan* method is *Ce Yuan Hai Jing* (《测圆海镜》) by Ye Li, who marked the character "yuan" to the right of coefficients of the first order (or the character "tai" to the right of constant terms). Coefficients above "yuan" indicate positive powers, and coefficients below "yuan" indicate constant terms and negative powers (which order was reversed in *Yi Gu Yan Duan* (《益古演段》)). The specific method of establishing equations is to list two equal polynomials based on known conditions of the problem and subtract one from the other to derive a numerical high-order equation. The promotion from the *Tianyuan* method to high-order equations with two, three, or four unknowns was another outstanding creation of mathematicians during the Song and Yuan Dynasties periods. Yi Zu said in the postscript of *Si Yuan Yu Jian* (《四元玉鉴》) that *Liang Yi Qun Ying Ji Zhen* (《两仪群英集臻》) by Dezai Li recorded both heavenly and earthly centers, and *Qian Kun Kuo Nang* (《乾坤括囊》) by Dajian Liu mentioned three centers, namely, the centers of heaven, earth, and humanity. Hanqing Zhu "proposed four centers, namely the heaven, earth, humanity and things." The previous two books were lost, and Shijie Zhu's *Si Yuan Yu Jian* (《四元玉鉴》) was the only work still available today and providing a systematic description. Shijie Zhu's expression of high-order equations with four unknowns was undoubtedly based on the Tianyuan method, placing the constant term in the center, the four powers at the upper, lower, left, and right positions, and the rest in the four quadrants. Shijie Zhu's greatest contribution was the method of eliminating unknowns for equations with four unknowns. The method was to select one unknown as x and polynomials consisting of other unknowns as coefficients of x to create several high-order equations with one unknown, and gradually solve the unknown by using multiplication. Repeating this step could eliminate the other unknowns to get a high-order equation with one unknown. Finally, the problem was solved with the square root method with increased multiplication. This was a major development in the solution to linear equations.

2.3 Reform of Rod Arithmetic Calculation and Creation of Abacuses

Counting rods were a primary calculation tool in ancient China, characterized by advantages such as simplicity, intuitiveness, and concreteness as well as disadvantages such as large space required by the rods and difficulty in alignment when calculation speeds up. For this reason, reforms began at an early stage. The currently available work *Shu Shu Ji Yi* (《数术记遗》) (authored by Yue Xu of the Eastern Han Dynasty period and annotated by Ying Zhen of the Northern Zhou Dynasty period) recorded 14 algorithms, which reflected the reforms. According to Ying Zhen's annotations, the "abacus" in *Shu Shu Ji Yi* (《数术记遗》) consisted of three columns, where movable beads were placed in the upper and lower columns and the result was placed in the middle column. The upper bead and the lower four beads were distinguished with colors. For the bead in the upper column, one bead equaled to five, and for the four beads in the lower column, one bead equaled to one. Obviously, this evolved from the expression method using counting rods and shared some similarities with present-day abacuses, while largely different. In the absence of a pithy formula, the abacuses at that time were probably not as easy-to-use as counting rods and therefore were not widely used.

After the mid-Tang Dynasty period, commercial activities burgeoned and the need for numeric calculations increased, and it was imperative that the calculation method be reformed. According to contemporary arithmetic works such as *Xin Tang Shu* (《新唐书》), the reform was about simplifying the multiplication and division algorithms. Three columns were simplified into one column, and multiplication and division were changed into addition and subtraction. In the Song and Yuan Dynasties periods, pithy formulae were created. Meanwhile, people also created the denary scale. The works of Hui Yang during the Southern Song Dynasty period and Shijie Zhu of the Yuan Dynasty period contained a large number of pithy formulae, which were basically consistent with the present-day pithy formulae of abacuses. With the creation of pithy formulae for multiplication and division, the pithy formula method was quick, while manual manipulation was slow and less than handy. As the quick multiplication and division calculations and their pithy formulae became increasingly incongruent with counting rods, it was imperative that the calculation tool be reformed, and as a result, abacuses and abacus calculations emerged.

If the abacuses in *Shu Shu Ji Yi* (《数术记遗》) are improved by changing three columns into two and allowing the movable beads to be shifted to a different column, a modern abacus is created, and if the pithy formulae for counting rods are changed into those for abacuses, the abacus calculation method is created.

It was also during the Song and Yuan Dynasties periods that the reform reached the climax. During this period, practical arithmetic works abounded, far more than in the Tang Dynasty period. Multiplication and division remained the focus of the reform. Shijie Zhu's *Suan Xue Qi Meng* (《算学启蒙》), Kuo Shen's *Meng Xi Bi Tan* (《梦溪笔谈》), Hui Yang's *Cheng Chu Tong Bian Ben Mo* (《乘除通变本末》) (1274), Ju Ding's *Ding Ju Suan Fa* (《丁巨算法》) (1355), Pingyu He's *Xiang Ming Suan Fa* (《详明算法》) (1373), and Heng Jia's *Suan Fa Quan Neng Ji* (《算

法全能集》) all contained specific examples. With emergence of new algorithms, multiplication and division calculations could be carried out in a column without any changes and were identical to the present-day abacus approach.

Abacuses with stringed beads might have already emerged during the Northern Song Dynasty period. In the *Qing Ming Shang He Tu* (《清明上河图》) by Zeduan Zhang during the Northern Song Dynasty period, there are two rectangular plates on the counter in Taichen Zhao's pharmacy, and many abacus historians believe they are abacuses (see Fig. 2.3). Some scholars do not share this view and believe they are coin plates. Anyway, a wooden bead with a diameter of 2.1 cm unearthed at a relic site of the Northern Song Dynasty period in Julu, Hebei, has the same shape and dimensions as an abacus bead. Yin Liu of the Yuan Dynasty period (1248–1293) had a five-character poem titled *Abacus* in *Jingxiu's Collection*. Zhenpeng Wang of the Yuan Dynasty period painted the *Qian Kun Yi Dan Tu* (《乾坤一担图》) (the 3rd year of Emperor Zhida's reign or 1310), where a vendor carries an abacus on the shoulder pole, with the beams, columns, and beads clearly visible. Volume 29 of *Nan Cun Chuo Geng Lu* (《南村辍耕录》) (1366) by Zongyi Tao of the late Yuan Dynasty period contains a metaphor about "abacus beads." A Chinese opera of the Yuan Dynasty period contained such a description "I counted my age with the abacus." These examples are sufficient proof that abacuses were already in use during the Yuan Dynasty period. Evidently, if we regard the modern abacus calculation approach as a system with abacuses with stringed beads and complete algorithms and pithy formulae, it must have emerged no later than the Yuan Dynasty period.

Fig. 2.3 Suspected abacus on the counter of Taichen Zhao's pharmacy in *Qing Ming Shang He Tu* (《清明上河图》)

In *Qing Ming Shang He Tu* (《清明上河图》) painted by painter Zeduan Zhang in the Southern Song Dynasty period (1127), there is an abacus-like instrument in Taichen Zhao's pharmacy. According to the detailed view enlarged by the Beijing Film Studio in 1981, we are certain it must be a 15-column abacus.

Detailed view (Figs. 2.4 and 2.5)

Fig. 2.4 A wooden bead unearthed at the Song Dynasty relic site in Julu

Fig. 2.5 The abacus in *Qian Kun Yi Dan Tu* (《乾坤一担图》)

2.4 Popularization of Abacus Calculation and Loss of Abacus Calculation in China (Fig. 2.6)

From the early to the mid-Ming Dynasty period, commodity economy developed, and abacuses popularized accordingly. Reference to abacuses in *Kui Ben Dui Xiang Si Yan Za Zi* (《魁本对相四言杂字》) (1371) and *Lu Ban Mu Jing* (《鲁班木经》) (sixteenth century) in the early Ming Dynasty period suggests abacuses were already very popular then. The former was a textbook for children to learn to read through pictures, and the latter included abacuses among wooden home appliances as a household necessity. *Kui Ben Dui Xiang Si Yan Za Zi* (《魁本对相四言杂字》) contained pictures of abacuses and counting rods, and the pictures of the abacuses were very clear, with frames, beams, shifts, and beads visible, and with two beads above and five beads below.

The above historical records suggest abacuses were already very popular in the Chinese society from the early fourteenth century to the early fifteenth century. Only when abacuses were widely used could they become commodities sold by peddlers and vendors. Only when abacuses became a household necessity in people's daily lives could low-lives and literati be familiar with them, and could they appear in operas as an ordinary item. The fact that abacuses were included in a children's textbook as an example suggests abacuses were undoubtedly highly prevalent.

In 1987, a wooden abacus was unearthed from a Ming Dynasty tomb in Miaopu Village, Pantuo Township, Zhangpu County, Fujian Province. It is an abacus with one bead in the upper column and five beads in the lower column, and the beads are diamond-shaped (see Fig. 2.7). The abacuses that have been popular since the seventeenth century and are still in use in Japan have exactly the same characteristic. Weizhen Lu (1543–1610), the owner of this Ming tomb, was from Zhangpu, Fujian. During Emperor Wanli's reign, Lu was the Right Assistant Minister of Works and was later promoted to be the Minister of Revenue after his predecessor died. When taking office at the Ministry of Works and the Ministry of Revenue, Lu was

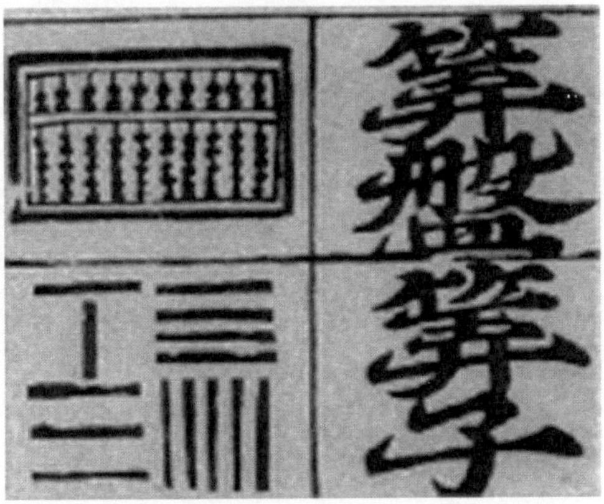

Fig. 2.6 The abacus and rod layout in *Kui Ben Dui Xiang Si Yan Za Zi* (《魁本对相四言杂字》)

Fig. 2.7 Abacus unearthed from Weizhen Lu's tomb (1543–1610)

responsible for overseeing civil engineering, finance, and taxation, which required a lot of calculations, and the symbolic abacus buried alongside him suggests abacuses were used for calculation throughout the Chinese society.

Mathematicians in the early and mid-Ming Dynasty period mostly still used the expression symbols and writing style of counting rods to complete their works. However, they were undoubtedly very familiar with abacuses. Evidence suggests mathematicians Jing Wu, Wensu Wang, Shunzhi Tang, and Yingxiang Gu had all used or studied abacuses. Shunzhi Tang was also an adept user of abacuses. *Yuan Ming Shi Lei Chao* (《元明事类钞》) recorded a story of him using the abacus:

> Shunzhi Tang arrived in Luzhou, when the Prefecture needed to calculate grain quantities. Tang then sent over adept calculators to calculate. Upon completion, the rough number was determined, and Tang operated the abacus by himself and determined the quantity of the grain in no time. All the sophisticated calculators were surprised at his speed. (Zhiyin Yao: *Yuan Ming Shi Lei Chao* (《元明事类钞》) (Volume 18))

After being invented, abacuses were popular alongside counting rods for a long period of time. Abacuses were first popular with the commoners, while the officials and intellectuals and their mathematical works still used counting rods during the Song and Yuan Dynasties periods. No abacuses were used in the mathematical works during the Song and Yuan Dynasties periods. Though abacuses were already popular in the Yuan Dynasty period, counting rods remained the primary calculation tool in mathematical works even in the early Ming Dynasty period, when counting rods and abacuses coexisted. For mathematical works before the mid-sixteenth century, either counting rods were the primary tool, or it cannot be determined whether counting rods or abacuses were used. *Suan Xue Qi Meng* (《算学启蒙》) already contained the *Zhuanggui* (subtracting one from the quotient) operation for division, which became one of the most important pithy formulae in abacus calculations. Mathematical works such as *Suan Fa Quan Neng Ji* 在元末明初的数学著作中《算法全能集》

and *Xiang Ming Suan Fa* (《详明算法》) in the late Yuan Dynasty period and early Ming Dynasty period included the 9 × 9 multiplication table, *Guichu* (long division), and *Zhuanggui* formulae and repeatedly give samples, which reflected the mutual influence between abacuses and counting rods.

After the mid-fifteenth century, works on abacus calculations gradually increased, such as Jing Wu's *Jiu Zhang Xiang Zhu Bi Lei Suan Fa Da Quan* (《九章详注比类算法大全》) (1450); Wensu Wang's *Gu Jin Suan Xue Bao Jian* (《古今算学宝鉴》) (1524) discussed abacus calculations. In the late sixteenth century, works focusing on abacus calculations emerged in great numbers, and abacuses were popularized at an all-out scale. The roles of abacuses and counting rods reversed, with abacuses becoming the mainstream calculation tool and counting rods cast into the dustbin of history. Among works focusing on abacuses, *Pan Zhu Suan Fa* (《盘珠算法》) and *Shu Xue Tong Gui* (《数学通轨》) were earlier and played a demonstrating role for the popularization of abacuses. *Pan Zhu Suan Fa* (《盘珠算法》) provided a comprehensive introduction to the pithy formulae, calculations, and operations of abacuses. The pithy formulae described in the book include *Shangfa*, *Xiawu*, and *Jinshi* for additions, *Xiafa*, *Qiwu*, and *Tuishi* for subtractions, *Guifa*, *Guichu*, and *Zhuanggui* for divisions, *Xiacheng* and the 9 × 9 table for multiplications, and the "Cicada off the Skin" formula and "two-character formula" shared by multiplications and divisions. Besides, there was the "beginner's cumulative algorithm" prepared for beginners. Among these, the 9 × 9 table is a little bit special, as it does not contain the formulae for 1×, but for 10×, such as "10 times 2 equals to 20" and "10 times 3 equals to 30." Second, detailed abacus illustrations were provided. The book listed a total of 54 abacus illustrations to demonstrate the application of the pithy formulae in practical calculations and the operating procedure (Fig. 2.8). For example:

> If we have 914.89 mus of land and 2.9 shengs of grain can be harvested from each mu of land, how much grain can be harvested? The answer is: 26 shis, 5 dous, 3 shengs, one he, eight shaos and one chao.
>
> Nine multiplies nine is eight-one, two multiplies nine is eighteen, and the bead representing eight reverses by two positions, and the bead at the decimal digit advances by one position to indicate ten; eight multiplies nine is seventy-two, two multiplies eight is sixteen, and the bead representing six reverses by four positions to indicate ten; four multiplies nine is thirty-six, the bead representing six moves upward and the bead representing five moves to ten. Two multiples four is eight, and the bead representing

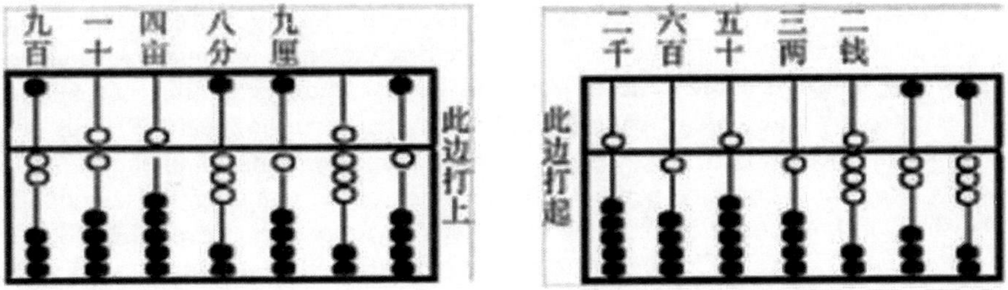

Fig. 2.8 Abacus illustrations in *Pan Zhu Suan Fa* (《盘珠算法》)

eight moves to three and the bead representing five moves to ten. One multiplies nine is nine, and the bead representing nine reverses by one to ten; one multiplies two is two; nine multiplies nine is eighty one, the bead representing one reverses to five; nine multiplies nine is eighty-one, and the bead representing eight reverses by two to indicate ten.

中文	英文	中文	英文
九百	Nine hundred	二千	Two thousand
一十	Ten	六百	Six hundred
四亩	Four mus	五十	Fifty
八分	Eight fens	三两	Three liangs
九里	Nine lis	二钱	Two qians
此边打上	Ending here	此边打起	Starting here

If we have 2653.2 liangs of silver, which is to be split among 515 people. Q: How much silver will each get? A: 5.151823.

The abacus formulae and technique involved in *Shu Xue Tong Gui* (《数学通轨》) are likewise very comprehensive, covering almost all the formulae used by subsequent works. In the late Ming Dynasty period, abacus calculations gradually took shape, and the algorithms were gradually standardized and systematic. One milestone was Dawei Cheng's *Suan Fa Tong Zong* (《算法统宗》), which finally completed the pithy formulae for additions and subtractions, and focused on some types of multiplications and divisions. Few changes occurred in the next five centuries. The previously sporadic pithy formulae were further systematized and improved, and the fingering technique for operating abacuses was covered. For example, "one moves upward and four becomes five" becomes "one reverses and five is divided by four." The order of bead movement for the former is "four is removed before five reverses," and the order for the latter is "five reverses before four is removed," which is much more reasonable.

Abacus calculations followed the same principles as counting rods and particularly had the same counting method. The upper beads of the abacus completely corresponded to the rods. Technically, it wasn't difficult for those proficient with counting rods to shift from beads to rods. One major difference between beads and rods lies in the dependency on and adaptiveness to pithy formulae. Adept users could move beads and determine the number by reciting the pithy formulae. As a calculation tool, abacuses could basically cover all the functions of counting rods, and far outweighed the latter in calculation speed, which happened to meet the needs of a commercial society. Therefore, abacuses thrived in the booming commercial society during the Ming Dynasty period. On the other hand, counting rods gradually vanished, and mathematicians in the early Qing Dynasty period even didn't know what they were when seeing counting rods in Korea. In the seventeenth century, the "four calculations" were popular, namely, abacus calculations, writing calculations, brush pen calculations, and Napier's Rods (See Zhongtong Fang's *Shu Du Yan* (《数

度衍》) for a detailed introduction to the "four calculations."), while traditional counting rods gradually went into oblivion. In 1713, Guozhu He, a prominent calendar scientist during the Qing Dynasty period, implemented surveying projects in Korea and found Korean mathematician Hong Jung Ha (1684-?) was very handy with the counting rods and could calculate square roots and solve equations better than Chinese mathematicians who used western calculation methods. Surprised, he asked: "Since there are no such counting rods in China, what about getting some and taking them to China?" So he asked for 40 counting rods from Hong (Hong Jung Ha: *Jiu Yi Ji* (《九一集》). Rongyun Jin: *Collection of Materials on Science & Technology History of Korea* (《韩国科学技术史资料大系》) (Math 2), Lijiang Press, 1985, p. 493).

Counting rods and abacuses played an important role in the development of mathematics in East Asia. Counting rods spread to Korea and Japan and had positive impact on the development of math in both countries. Japan began to introduce Chinese math in the Tang Dynasty period. Counting rods spread to Japan around the same time and became a primary calculation tool of the Japanese. In the mid-sixteenth century, abacuses spread to Japan and soon became popular as the most important calculation tool for the whole society. Abacus works such as *Pan Zhu Suan Fa* (《盘珠算法》), *Shu Xue Tong Gui* (《数学通轨》), and *Suan Fa Tong Zong* (《算法统宗》) soon spread to Japan after their publication in China and played a major role in facilitating the popularization of abacuses in Japan and the development of *Wasan*. In 1652, Wasan expert Shimada Jinji completed the book *Jiu Shu Suan Fa* based on the format and content of *Suan Fa Tong Zong* (《算法统宗》) and published it in the next year. The version of *Suan Fa Tong Zong* (《算法统宗》) popular in Japan was the privately block-printed copy made by "Zhenhua Wang of Sanguitang" in 1593. This copy was later published after being reviewed by Yuasa in 1675. The reviewed version by Yuasa made *Suan Fa Tong Zong* (《算法统宗》) prevalent in Japan and became one of the most important references for Wasan scholars to study math during the Edo period (Fig. 2.9).

After being spread to Japan, abacuses soon became the most important calculation tool in Japan. However, counting rods were not forgotten as they were in China, but were still used by mathematicians as an auxiliary tool for determining square roots and solving high-order equations.

After spreading to Korea, counting rods had been in use for over 1000 years and remained the primary calculation tool of Korea in the nineteenth century. Though spreading to Korea in the fifteenth and sixteenth centuries and used in civil society, abacuses had failed to play a dominant role. Mathematicians mostly used counting rods.

Given the importance of counting rods and abacuses for math in East Asia, it is appropriate to use the concepts in computational mathematics and computer science for explanatory purposes. If counting rods and abacuses were the hardware, the various calculation methods and pithy formulae were the software, and the pithy formulae could be regarded as the most basic programming language. Counting rods and abacuses provided the hardware and corresponding basic algorithmic programs, so to speak. With popularization of accounting rods and abacuses, the calculation

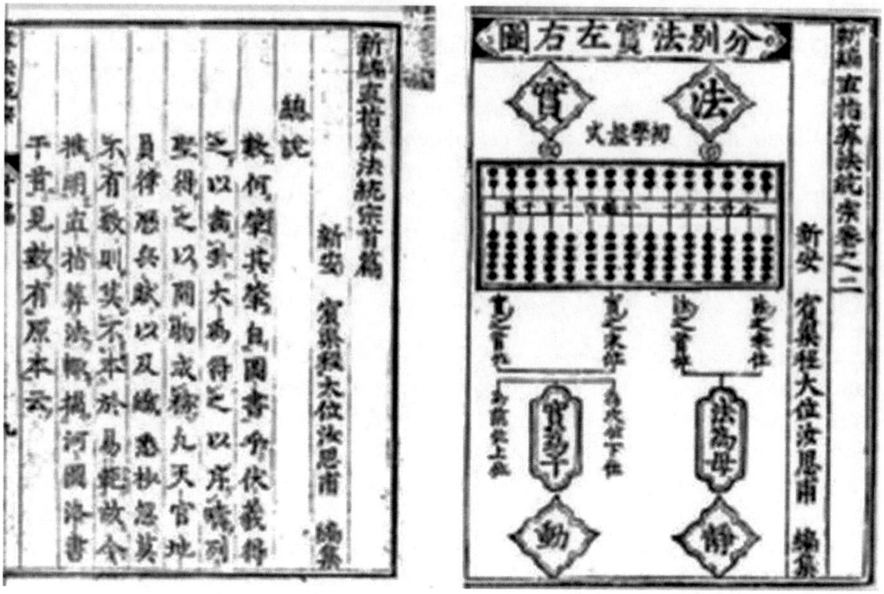

Fig. 2.9 Photographs of reviewed version of *Suan Fa Tong Zong* (《算法统宗》) in 1675

capability of people in the Sinosphere was greatly improved. The programmed algorithms of Chinese mathematics, eastern mathematics, and Wasan were closely related to the use of counting rods and abacuses.

References

1. Dunjie Yan. (1956). *Math achievements in ancient China* (p. 3). China Association for Science and Technology.
2. Provincial Museum of Hunan. (1963). Excavation Report on Chu Kingdom Tomb at Deshan, Changde, Hunan. *Archaeology* (《考古》), 9, 461–473 & 479.

An Overview of Official and Private Agricultural Books Through Chinese History

3

Xiongsheng Zeng

Contents

Abstract

Agricultural works, the books and chapters about agricultural production technology in a broad sense and knowledge directly related to agricultural production,

X. Zeng (✉)
The Institute for the History of Natural Sciences, Chinese Academy of Sciences, Beijing, China
e-mail: zeng@ihns.ac.cn

X. Jiang (ed.), *The High Tide of Science and Technology Development in China*,
History of Science and Technology in China,
https://doi.org/10.1007/978-981-15-7847-2_3

are on the theme of producing grains, vegetables, oil, fiber, special crops, fruit trees, sericulture, animal husbandry and veterinary medicine, forests, flowers, etc. This chapter, in a chronological order, introduces in detail the Chinese agricultural works from the Pre-Qin Period to the Ming and Qing Dynasties.

Keywords

Agronomy · "Shang Nong" in Lu Shi Chun Qiu · Fan Sheng Zhi Shu · Si Min Yue Ling · Qi Min Yao Shu · Si Shi Zuan Yao · Chen Fu Nong Shu · Pulu (谱录Catalogue)

3.1 Overview

Agricultural books refer to works about agricultural production technology in a broad sense and knowledge directly related to agricultural production. In other words, they are books and chapters on the theme of producing grains, vegetables, oil, fiber, some special crops (such as tea, oil, medicinal materials), fruit trees, sericulture, animal husbandry and veterinary medicine, forests, flowers, etc.

There were hundreds of agricultural books in ancient China. Wang Yuhu (王毓瑚)'s *Zhong Guo Nong Xue Shu Lu* (《中国农学书录》) (first edition in 1957, revised edition in 1964) recorded 541 works, including more than 200 lost books. *Zhong Guo Gu Nong Shu Lian He Mu Lu* (《中国古农书联合目录》) edited by Beijing Library in 1959 recorded 643 kinds of existing and lost agricultural books. *Zhong Guo Gu Nong Shu Kao* (《中国古农书考》) written by a Japanese scholar Gennosuke Amano in 1975 evaluated 243 kinds of existing agricultural books. Its attached index listed about 600 kinds of agricultural books and related books. Some scholars, by conducting an investigation into the agricultural books of the Ming and Qing Dynasties in recent years, found that there were more than 830 kinds of agricultural books from the Ming and Qing Dynasties (most of which were from the late Qing Dynasty), and that more than 500 agricultural books had not been included in *Zhong Guo Nong Xue Shu Lu* and *Zhong Guo Gu Nong Shu Kao*, including about 390 extant ones and about 100 whose existence is uncertain (Wang 1989).

According to the style, content, and author's identity of the agricultural books, Wang Yuhu, in his book *Guan Yu Zhong Guo Nong Shu* (《关于中国农书》), classified the ancient agricultural books into nine systems: comprehensive agricultural books, monographs on timing and farming, various special books, sericulture books, veterinary books, potherb books, locust control books, farm monthly orders, and general agricultural books. Later, inspired by Hu Daojing (胡道静), he added 10 categories of systematic books of mountain dwelling (Wang Yuhu: *Zhong Guo Nong Xue Shu Lu* (Appendix)). Shi Shenghan (石声汉) divided ancient agricultural books into three categories: integrated and professional agricultural books, official and private books, national and local agricultural books. In *Zhong Guo Nong Ye Gu Ji Mu Lu* (《中国农业古籍目录》) Zhang Fang (张芳) and Wang Siming (王思明) classified ancient agricultural books in China into 17 categories, namely, comprehensive, seasonal, farmland water conservancy, farm tools, soil cultivation, field crops, horticultural crops, bamboo and wood, plant protection, animal husbandry and veterinary, sericulture, aquatic products, food and processing, products, agricultural politics and economics, disaster relief, and others (Fang and Wang 2002).

In terms of content, ancient agricultural books can basically be divided into two systems: "comprehensive" and "professional." Comprehensive agricultural books cover many aspects of agriculture, while the contents of professional agricultural books only involve one aspect. The two can be further divided into several types. The comprehensive agricultural books include large-scale comprehensive agricultural books, local and seasonal agricultural books, articles encouraging agricultural activities, and maps of farming and weaving. On the other hand, books of the professional type are timing and farming monographs as well as various catalogue monographs, covering tea, flowers, fruit trees, forests, sericulture, agricultural implements, animal husbandry, wild vegetables, locust control, etc.

3.2 Agronomy Knowledge and Works in the Pre-Qin Period

From the Neolithic Age to the Xia, Shang, and Western Zhou Dynasties, the Chinese had accumulated rich agricultural experience and knowledge that was scattered in oracle bone inscriptions, inscriptions on bronze, and ancient books before the Qin Dynasty. The oracle bone inscriptions and inscriptions on bronze both directly reflected people's understanding of farming and animal husbandry production at that time. With the help of this kind of writing, people accumulated and passed on their knowledge of agriculture, hence the further development of agriculture. Writing accelerates the accumulation and spread of agronomic knowledge. The pre-Qin period had already seen the publishing of special agronomic works. At the same time, this knowledge could also be found in the existing classic works, such as *Shi Jing* (《诗经》), *Xia Xiao Zheng* (《夏小正》), *Yu Gong* (《禹贡》), *Guan Zi Di Yuan Pian* (《管子·地员篇》), *Zhou Li* (《周礼》), etc. Although these classics were not specialized agronomic works, the relevant content not only served as the source of agronomic works in the Spring and Autumn Period and Warring States Period, but also influenced the development of agronomy in China later.

3.2.1 Agronomy Knowledge in Pre-Qin Classics

Shi Jing is the earliest collection of poems in China. It includes about 500 years of poetry from the early Western Zhou Dynasty to the middle of the Spring and Autumn Period. Of the 305 poems, more than 20 are related to farming. Among them, "Bin Feng Qi Yue" (《豳风·七月》) is a complete agricultural poem, which describes the monthly work of farming, women workers, gathering, and hunting.

Shang Shu (《尚书》) is a compilation of ancient Chinese historical documents and some works about ancient deeds. Among them, *Yu Shu Yao Dian* (《虞书·尧典》) is the earliest record of farming seasons in traditional agriculture, while *Xia Shu Yu Gong* (《夏书·禹贡》) is the earliest work on soil that divides the soil into 10 kinds according to its color and nature, and clearly records its geographical distribution.

Xia Xiao Zheng is the earliest agricultural almanac. Weather, astronomical phenomena, phenology, and agricultural activities are recorded in month order. Agricultural activities cover farming, fishing and hunting, gathering, sericulture, animal husbandry, etc. The related plants include leek, *yun* (芸), (Or *yuntaicai* (芸苔菜), which refers to rape. See Xia Weiying: *Xia Xiao Zheng: Collated and Annotated*, Agricultural Publishing House, 1981) millet, *fan* (蘩), mulberry, wheat, apricot, plum, orchid, peach, tea, melon, jujube, *luansuan* (卵蒜), etc. The animals concerned are silkworm, chicken, lamb (sheep), colt, horse, etc. It lays the foundation for the later lunar farming books by associating seasons with farming.

Zhou Li is a book compiled according to the official system of the Zhou Dynasty. In the book, the official positions are divided into six ranks: heaven, earth, spring, summer, autumn, and winter. Its basic idea is to realize effective management of land and people through the establishment of all kinds of officials at all levels from the central government to the local government. It is also a book about ancient people's understanding of agriculture. It touches upon grain seed, horticulture, forestry, animal husbandry, sericulture, etc. The book emphasizes adjusting measures to local conditions to develop the various industries of agriculture, forestry, and animal husbandry, and even mentions veterinary medicine and its different branches. Later agricultural books quoted *Zhou Li* as a source as well as a basis of argument.

During the Spring and Autumn Period and the Warring States Period, the development of China's agronomy made a qualitative leap, with the emergence of the agriculturists and agricultural books. *Han Shu Yi Wen Zhi* (《汉书·艺文志》) records 9 agriculturists whose works totaled 114 chapters, of which "Shen Nong Er Shi Pian" (《神农二十篇》) may have been written by Xu Xing (许行) and others in the Warring States Period in the name of Shennong (*Han Shu Yi Wen Zhi*). "Ye Lao Shi Qi Pian" (《野老十七篇》) was written by a hermit in the Warring States Period. In addition, the "Zai Shi Shi Qi Pian" (《宰氏十七篇》) may have been written by Ji Ran (计然), Tao Zhugong Fan Li (陶朱公范蠡)'s master, during the Spring and Autumn Period.

The pre-Qin agricultural books have all been lost. The existing pre-Qin agronomic documents mainly include *Lu Shi Chun Qiu* (《吕氏春秋》), the four articles including "Shang Nong" (上农) and "Shi Er Ji" (十二纪), "Yu Gong Pian" in *Shang Shu*, articles of "Di Yuan Pian" (地员篇) and "Du Di Pian" (度地篇) in *Guan Zi* (《管子》), "Yue Ling" (月令 agricultural calendars) in *Li Ji* (《礼记》) and *Xia Xiao Zheng*, etc. These documents are the main materials for understanding agronomy in the pre-Qin period. Among them, the four articles including "Shang Nong" are the most representative ones.

3.2.2 The Four Articles Including "Shang Nong" in Lu Shi Chun Qiu

Lu Shi Chun Qiu is an eclectic work compiled by Lu Buwei (吕不韦) and his followers who merged the theories of various parties in the pre-Qin period at the end of the Warring States Period. Ma Guohan (马国翰) in the Qing Dynasty believed that the four articles including "Shang Nong" might have been drawn from books such as *Ye Lao* (《野老》) by Lu Buwei's family guests. The article "Shang Nong" discusses the importance of agriculture, whereas the other three articles "Ren Di" (任地), "Bian Tu" (辩土), and "Shen Shi" (审时) form a whole, discussing general theories of agricultural technology. The core issues can be summarized into two aspects: the use of land and the mastery of farming seasons.

After raising ten questions of agricultural production (The ten questions are: Can you turn low-lying land into upland? Can you remove barren soil and replace it with wet soil? Can you make the land suitable and drain it with ridges? Can you sow the seeds in moderate depth and keep them moist in the soil? Can you keep weeds out of the fields? Can you make your fields well ventilated? Can you make the corn in good ear and the stalk strong? Can you make the ears of the crops evenly big? Can you make the grain full and the bran thin? Can you make the rice both nutritious and chewy?), the article "Ren Di" puts forward five principles and specific requirements of soil cultivation, namely, in high fields crops should be planted in ditches while in low-lying areas they should be planted on ridges. The former can contribute to drought resistance and moisture preservation, and the latter drainage and waterlogging prevention. The book also has specific requirements for ridge specifications and standards of matching farm tools.

The article "Bian Tu" sets requirements for the planting density and the thickness of soil coverage. It argues that the seeding rate should be appropriate, neither too dense nor too thin. In fertile lands seeds can be sown denser than in barren lands. Similarly, the covering earth should be neither too thick nor too thin. After the emergence of seedlings, appropriate thinning should be carried out. All these aim to make the plants in the field line up vertically and horizontally in order to ensure ventilation and light transmission in the field so that there will be wind flow even in the center of the field.

The article "Shen Shi" is a monograph on the utilization of farming seasons and the application of weather in agriculture. It is called "Shen Shi"(Timing Consideration) because timing in agriculture should be treated with caution. In the article, "*deshi* (得时good timing)" is compared in detail with "*xianshi* (先时before good-timing)" and "*houshi* (后时after good-timing)" as to the planting of six major food crops in terms of yield (including rice yield) and quality (including taste, hunger resistance, and benefits to consumers). The conclusion is "crops planted at the right time will reap a bumper harvest, while those planted at the wrong time will reduce production."

The four articles including "Shang Nong" in *Lu Shi Chun Qiu* are the foundation works of traditional Chinese agriculture. The agricultural techniques of intensive cultivation described in these articles were inherited and developed by later generations. More importantly, for the first time they scientifically summarize the relationship between heaven, earth, and man in agricultural production, and carry this spirit into all discussions. This spirit and principle has been inherited by later generations of agronomists and has become the most important guiding ideology in China's traditional intensive farming tradition.

3.3 Agricultural Books in the Qin, Han, Wei, Jin, Southern, and Northern Dynasties

3.3.1 Overview

Among the nine agriculturalists whose works totaled 114 chapters in *Han Shu Yi Wen Zhi*, "Shen Nong," "Ye Lao," and "Zai Shi" (宰氏) were written during the Spring and Autumn Period and the Warring States Period. The other chapters may have been written during the Qin and Han Dynasties, which include "Dong An Guo Shi

Liu Pian" (《董安国十六篇》), "Yin Du Wei Shi Si Pian" (《尹都尉十四篇》), "Zhao Shi Wu Pian" (《赵氏五篇》), "Fan Sheng Zhi Shi Ba Pian" (《氾胜之十八篇》), "Wang Shi Liu Pian" (《王氏六篇》), and "Cai Kui Yi Pian" (《蔡癸一篇》). *Sui Shu Jing Ji Zhi* (《隋书·经籍志》) lists five books as agricultural works, namely, *Fan Sheng Zhi Shu* (《氾胜之书》), *Si Ren Yue Ling* (《四人月令》) (or *Si Min Yue Ling* 《四民月令》), *Jin Yuan Shi Lu* (《禁苑实录》), *Qi Min Yao Shu* (《齐民要术》), and *Chun Qiu Ji Shi Liu Chang Ni Yi* (《春秋济世六常拟议》).

The authors of *Han Shu Yi Wen Zhi* and *Sui Shu Jing Ji Zhi*, based on their own understanding of agriculturalists, have excluded many works related to agriculture, for example, the works of agrometeorology, the works of livestock appearance, the monographs on animals and plants, etc. In addition, of all the agricultural books recorded in the Han Annals, only *Fan Sheng Zhi Shu* has been preserved in the Sui Annals, while the rest have been lost and are not included in *Qi Min Yao Shu*. On the contrary, some agricultural books are mentioned in *Qi Min Yao Shu* although they are not recorded in history books.

The representative agricultural books in the Qin, Han, Wei, Jin, Southern, and Northern Dynasties are *Fan Sheng Zhi Shu*, *Si Min Yue Ling*, and *Qi Min Yao Shu*.

3.3.2 Fan Sheng Zhi Shu

Fan Shengzhi (氾胜之) was a parliamentarian during the reign of Emperor Cheng of the Han Dynasty (32 ~ 7 BC) at the end of the Western Han. He taught wheat-planting in the areas of Sanfu (三辅) (Jing Zhao Yin 京兆尹, Zuo Feng Yi 左冯翊, You Fu Feng 右扶风) near the capital Chang'an (now Xi'an, Shaanxi Province). The writing of *Fan Sheng Zhi Shu* is probably closely related to his teaching experience in Guanzhong.

The existing *Fan Sheng Zhi Shu* mainly includes three parts. The first part is the general theory of cultivation. It puts forward the general principles of cultivation, followed by the methods of soil cultivation and those of seed treatment. The former focuses on the timing and methods of soil cultivation and repeatedly illustrates from both positive and negative aspects the importance of correctly grasping the appropriate timing of soil cultivation. The latter includes the selection, preservation, and treatment of crop seeds, emphatically introducing a special method of seed treatment called *squzhong* (溲种). In addition, it also involves the selection of sowing date, etc.

The second part is the theory of crop cultivation, including the cultivation methods of 13 crops: grain, millet, wheat, rice, barnyard grass, soybean, adzuki bean, *xi* (枲), hemp, melon, *hu* (瓠 a kind of edible gourd), taro, mulberry, etc. The contents involve the production process of cultivation, sowing, intertillage, fertilization, irrigation, plant protection, harvest, etc. The third part is about *qutian* (区田 section-field program).

Fan Sheng Zhi Shu puts forward the general principle of cultivation that "the basis of cultivation lies in the selection of time and soil, fertilization and irrigation, as well as early cultivation and harvest." Based on this principle, it introduces cultivation methods of more than 10 crops such as millet, grain, winter wheat, spring wheat, rice, adzuki bean, soybean, *juma* (苴麻 a kind of flax), *cha* (荏) (*yousuzi*) (油苏子),

xima (枲麻 a kind of flax), mulberry, melon, *hu*, taro, etc. Many inventive methods are introduced, for example, a method of controlling the water temperature in the paddy field, of inarching seedlings of *hu*, and of intercropping and interplanting among melons, *xie* (薤), and adzuki beans as well as the theory that "it is inappropriate to take off all the leaves of soybean and adzuki bean." The cultivation techniques of winter wheat are discussed in particular detail.

For the first time, the method of *suixuan* (穗选) and *quzhong* (区种) are recorded, and the method of *quzhong* is described in detail. The method of *suixuan* is a traditional seed selection method while *quzhong* is a kind of seed treatment, or fertilizer application method. First, animal bone juice, *saoyong* (缫蛹) juice, silkworm dung, animal dung, *fuzi* (附子), water, or snow juice are mixed according to a certain proportion into a thick gruel to wash seeds. The washed seeds, looking like *maifanli* (麦饭粒), are then sown in the soil. It is a good method of insect prevention, drought resistance, and fertilization to ensure a bumper harvest.

The method of *quzhong* is a special high-yield cultivation method as well as a method of proper land use. Its invention is related to drought resistance. It is characterized by "making full use of dung in lands not so fertile." Its basic principle is to "make ditches by digging deep into the soil." With a centralized employment of manpower and material resources in the ditches to strengthen management and compact planting so as to ensure sufficient supply of fertilizer and water necessary for crop growth, the maximum production capacity of crops will be exerted and the yield per unit will be increased (According to *Fan Sheng Zhi Shu*, "the method of *quzhong* requires frequent irrigation in dry weather, resulting in the harvest of about 6000 kg per *mu* (about 667 square meters)." The description of the high yield has attracted many people to try *quzhong* method. However, such a high yield is incredible. What was said in the book is probably a propaganda of its effects of increasing crop production). At the same time, the land being cultivated will be expanded to hilly slopes that are difficult to cultivate. Therefore, "even barren mountains ridges, hills and sloping fields can be used as ditches." Furthermore, "wasteland can be used without cultivating." The method of *qutian* embodies the spirit of intensive cultivation and has always been regarded as a way to resist drought and save the poor. It is one of the technologies that can best reflect the characteristics of traditional Chinese agriculture and has a wide and far-reaching influence in history.

Fan Sheng Zhi Shu is the most important agronomy book after the four articles including "Shang Nong" in *Lu Shi Chun Qiu*. It is a new summary of China's agricultural science and technology when the iron plough and cattle ploughing were popularized. As one of the classics of traditional Chinese agriculture, it enjoyed high reputation at that time and had a far-reaching impact on the development of agriculture in the future. What is more, its writing style has become an important model for Chinese traditional comprehensive agricultural books.

3.3.3 Si Min Yue Ling

Cui Shi (崔寔) (103~170), the author of *Si Min Yue Ling*, was born in Anping, Zhuojun (now Hebei Province), in the Eastern Han. Born in a noble family, he served as Court Gentleman for Consultation twice as well as Governer of Wuyuan (now in

Inner Mongolia Autonomous Region) and Liaodong successively and was promoted to Imperial Secretary later. Paying close attention to politics, he wrote *Zheng Lun* (《政论》) and became famous. He was also concerned with the people's livelihoods and wrote *Si Min Yue Ling*, his representative work. The book, written when he lived in Luoyang in his old age, is a manual in the genre of agricultural calendars based on his personal experience of many years working as a scholar, farmer, worker, and businessman. The book may have been written in the Eastern Han from the later years of the reign of Emperor Shun to the early years of the reign of Emperor Huan (143~147) or earlier (Liang 2002).

The contents roughly include nine groups: sacrifice, family ritual, education, maintaining and improving the old and new relationships in family and society; arranging tillage, planting and harvesting of grain, oil and vegetables according to the seasonal climate; doing needle crafts such as raising silkworms, spinning, weaving, dyeing, bleaching, cutting, washing, restructuring, etc.; food processing and brewing; construction projects of housing and farmland water conservancy; collecting wild plants, mainly medicinal materials, and preparing medicines; collecting and preserving household appliances; buying rice and food; other chores, including "health care."

Much of the content related to agriculture originates from *Fan Sheng Zhi Shu*, but there are also many new experiences summed up from practice. For example, although the method of *zhuojuecediqi* (椓橛测地气 measuring the earth power with wooden rods) is derived from *Fan Sheng Zhi Shu*, Cui Shi requires flexible control with the change of time, place, and actual situation. The book first records techniques of *biedao* (别稻) (rice transplanting) and of tree layering for propagation. The addition of many items such as making wine, vinegar, sauce, *qu* (曲 distiller's yeast), *hai* (醢 meat paste), preserved fruits, *ju* (菹 pickles), *ji* (齑 finely chopped ginger, garlic, etc.), *tang* (饧 treacle), *yi* (饴 maltose), *beiqiu* (糒糗 solid food), and even buying, selling, dealing, marketing, etc. has set an example for similar agricultural books such as *Qi Min Yao Shu*. However, its greatest contribution is that it created a new writing style of the agricultural book in terms of "agricultural calendars." Later books such as *Si Shi Zuan Yao* (《四时纂要》), *Nong Sang Yi Shi Cuo Yao* (《农桑衣食撮要》), *Jing Shi Min Shi Lu* (《经世民事录》), and *Nong Pu Bian Lan* (《农圃便览》) all inherited the genre of *Si Min Yue Ling*.

3.3.4 Qi Min Yao Shu

Jia Sixie (贾思勰), the author of *Qi Min Yao Shu*, was from Qijun in the Northern Wei Dynasty (Liang Jiamian: The Author, Annotator and Writing Period of *Qi Min Yao Shu*—A Textural Research on the First Extant Ancient Agricultural Book in China, *Liang Jia Mian Nong Shi Wen Ji,* China Agricultural Press, 2002, pp. 19–26). Living in the late Northern Wei Dynasty, he served as Governor of Gaoyang and also engaged in agricultural business. His book, written between the 30s and 40s in the sixth century, has a total of 10 volumes and 92 articles, covering topics "from farming to the making of *xianhai* (酰醢 meat paste) and recording every occupation he has pursued."

The first, second, and third volumes talk about the cultivation of field crops (including grain crops and economic crops) and vegetables. The content includes technologies of soil cultivation and of seed selection and preservation. The grain crops involved include millet, *shuji* (黍穄), *liangshu* (粱秫) (another variety of millet), soybean, adzuki bean, hemp, *mazi* (麻子), barley and wheat, rice, *handao* (旱稻 upland rice), and *huma* (胡麻 flax). The fiber crops include hemp. The fruits and vegetables mentioned include melon, *hu*, taro, sunflower, *manjing* (蔓菁 turnip), garlic, *cong* (葱 scallion), leek, *shujie* (蜀芥 Sichuan mustard), *yuntai* (芸苔 brassica), *jiezi* (芥子 mustard seed), coriander, *lanxiang* (兰香), *ren* (荏), *liao* (蓼), ginger, *xianghe* (蘘荷 Zingiber mioga), celery, etc. Forage crops involved are mainly *muxu* (苜蓿 alfalfa).

The fourth and fifth volumes deal with fruit trees and timber trees. The first part of the fourth volume is the overview of the two volumes, followed by the topics of fruit trees and economic trees. Fruit trees include jujube, peach, plum, apricot, pear, chestnut, persimmon, pomegranate, papaya, pepper, and *zhuyu* (茱萸 cornel). Timber trees and dye plants include *sangzhe* (桑柘 mulberry trees) (with a silkworm rearing method), elm, poplar, *tang* (棠), *guchu* (谷楮), *qi* (漆 lacquer), *huai* (槐 locust tree), willow, *qiu* (楸 catalpa), *zi* (梓), *wu* (梧), *zuo* (柞 oak), bamboo, *honglanhua* (红蓝花), *zhizi* (栀子 gardenia), *lan* (蓝), and *zicao* (紫草 lithospermum). The last article in volume 5 is connected to the method of planting foxglove.

The sixth volume is about animal husbandry, including poultry and fish. The first article is about raising cattle, horses, donkeys, and mules. Beginning with a general overview of animal husbandry, it continues with the methods of judging horses and cattle by their appearance as well as various ways to cure their diseases. It introduces, for the first time, the method of crossbreeding horses and donkeys to produce mules. Then, the methods of raising sheep, pigs, chickens, geese, ducks, and so on were mentioned in turn. The article on raising sheep is rich in content, accompanied by a method of making crisp cheese and raising colts and foals, lambs and calves, etc. The final article is about fish-farming, mainly citing the method of artificial cultivation of carp from *Tao Zhu Gong Yang Yu Fa* (《陶朱公养鱼法》), and recording for the first time the cultivation methods of aquatic vegetables such as *chun* (莼), *ou* (藕 lotus root), lotus, *qian* (芡), *ji* (芰), etc., in the appendix.

Volumes 7, 8, and 9 deal with brewing, food processing, culinary preparations, and cultural products, which are basically sideline products. It cites reports about the commercial production as well as the sale of agricultural and sideline products from *Shi Ji* (《史记》) and *Han Shu* (《汉书》). It also records the storage, processing, brewing, and cooking technologies of agricultural products. The last two articles are about glue making and preparation of ink, brushes, etc.

Volume 10 is about "plants not indigenous to north China." "China" here refers to the Central Plains, the main territory of the post-Wei Dynasty. There is only one article in this volume, i.e., the 92nd article, which is full of references from predecessors, mainly describing the tropical and subtropical plant resources in the south.

Qi Min Yao Shu systematically summarizes agricultural technology in northern China. Animal-pulling harrows (called iron teeth in the book) are introduced as a method of soil preparation to form a system of land clearing and tilling by combining tilling and leveling. Supplemented by more tillage and more leveling, autumn tillage is emphasized so as to achieve the goal of "guaranteeing sufficient water supply to the land through good tillage even in times of drought." At the same time, specific requirements are put forward as to the times of soil tillage, the timing of preparing land, the depth of ploughing, and the size of plough bars.

The book greatly affirmed the rotation of crops, which is believed to be helpful to increase yield and reduce weeds and pests. It records more than 20 kinds of crop rotation methods, listing the crops that can be rotated one after the other in a better way. It also confirms the sequence of planting crops, for example, leguminous crops are the best for preceding cereal crops and vegetables. Besides that, the book records a variety of inter-cropping, mixed cropping, and intercropping methods as measures to improve land use efficiency, combining cultivation with soil improvement and farming with animal husbandry. The cultivation of green manure crops is also added to the rotation sequence to improve soil fertility by burying grass as green manure. This is also the earliest record of artificial cultivation of green manure in northern China.

In terms of seed technology, *Qi Min Yao Shu* puts forward the idea of "*biezhong* (别种)" on the basis of *suixuan* method, similar to *zhongzitian* (种子田) of later generations. In seed selection, besides high yield and high quality, factors such as early maturity, drought tolerance, insect resistance, taste, wind resistance, bird resistance, easy thrashing, and water resistance are also considered. The book records 107 millet (coarse grain) varieties, 37 rice varieties, and many other varieties of crops. According to the names of millet varieties at that time, the naming rules of crop varieties are summarized, that is, the names of breeders, morphological characteristics of varieties, and things similar to variety characteristics were abstracted respectively to name different varieties of crops. The book also discusses seed treatment techniques before sowing to improve the speed and rate of germination. It also records some special seed treatments to remove weeds.

In *Qi Min Yao Shu*, the sowing period is specifically set to 10 days instead of a month, and there are "*shangshi* (上时)," "*zhongshi* (中时)," and "*xiashi* (下时)." "*Shangshi*" is the most suitable time for sowing, "*zhongshi*" comes next, and "*xiashi*" is the most unsuitable. The sowing date of specific crops should be flexibly set according to phenology, soil fertility, and moisture. Similarly, sowing quantity, density, and depth also need to be controlled accordingly.

Qi Min Yao Shu recognizes that intertillage not only has the function of weeding and preserving soil moisture, but also can mellow the soil and improve the yield and quality of crops. The hoeing methods should also be determined according to crop varieties, growth stages, and climatic conditions.

Qi Min Yao Shu contains many methods related to preventing harm caused by diseases, pests, birds, animals, and grasses, mainly including crop rotation, selection of varieties that are resistant to insects, birds, and animals, burning with fire, water retting, *linxue* (蔺雪), sun exposure, the use of salt and ash, and the use of *hao* (蒿artemisia), *ai* (艾 Chinese mugwort), *zhu* (茿), *shexiang* (麝香 musk), papaya, and lime. It also introduces a method of trapping and killing *guatianyi* (瓜田蚁 a kind of ant) by using cattle and sheep bones with marrow. This is one of the earlier records of killing insects by trapping and killing.

Qi Min Yao Shu emphasizes that harvesting should be carried out rapidly. Besides that, it puts forward the standards of timely harvesting according to the characteristics and maturity of different crops such as *guzi* (谷子 cereal), *shuji*, and beans. *Guzi*, rice, and hemp should be harvested in due course. However, an early harvest is required for crops such as *ji* (穄 millet), *jiaodou* (茭豆), adzuki bean, and *huma* while *shu*, *liangshu* (粱秫), and *chudadou* (春大豆 spring soybean) should be harvested late.

Qi Min Yao Shu summarizes vegetable cultivation techniques systematically, mainly including increasing a multiple cropping index, improving the land utilization rate, fine soil preparation, timely intertillage and harvest, as well as some special cultivation methods. Generally speaking, the vegetable cultivation methods described in the book are similar to field technology, but have higher requirements of fineness. The book also mentions some special vegetable cultivation measures, such as melon sprouting.

Qi Min Yao Shu records a variety of sexual and asexual fruit tree propagation methods such as sowing, cutting, layering, rooting, and grafting, which are respectively used for propagation of different trees (or the same tree under different conditions). Different propagation methods are related to the growth speed of trees, the time of fruiting, and even the yield and quality of fruit. The book also mentions methods to prevent trees from freezing injury and also records the earliest methods to help fruit trees such as jujube and plum to bear fruit.

Qi Min Yao Shu records the techniques of raising cattle, horses, pigs, sheep, donkeys, mules, chickens, geese, ducks, fish, silkworms, and other household animals. It is so rich in content that it can be called the best among ancient agricultural books. This is mainly because *Qi Min Yao Shu* was written in the Northern Wei Dynasty. This regime, established by nomadic people, attaches great importance to the breeding of horses, so there are reports ranging from judging horses to breeding, from the cross breeding of donkeys and horses to the prevention and treatment of horse diseases.

For example, in horse breeding, excessive use of horses and unreasonable feeding will have adverse effects on the *jin* (筋 tendons), bones, skin, *qi* (气), and blood of horses. Besides this, coarse or fine fodder should be fed to horses according to three different situations. Horses should eat less in the morning, eat properly during the day while eating more at night and doing some proper exercise after eating.

The book, for the first time, records in detail the hybridization between donkeys and horses, pointing out that the benefits of cross breeding between male donkeys and mares is not obvious, while hybridizing male horses with female donkeys is better.

As to the prevention and treatment of horse diseases, it lists more than 10 kinds of health problems including abortion, infectious diseases, *zhichuang* (炙疮 post-moxibustion sores), dysuria and constipation, abdominal distension, lying, and dying, and more than 30 kinds of prescription are used. These had not been found in the previous literature.

There is one article about sheep raising in *Qi Min Yao Shu*, discussing in detail topics of selecting stud sheep, grazing or captivity, and the prevention and treatment of infectious diseases. The same is true for pigs, but the focus is on the rapid fattening of pigs. Poultry, such as chickens, geese, ducks, and the like used for meat, are also required to be fat and tender. Therefore, there are similarities with pig raising in cages, feed, etc. For the purpose of laying eggs, the chickens, geese, and ducks that lay more eggs should be selected as breeding birds.

There is not much about sericulture. However, it explains in detail the management of silkworm rearing, the rational use of mulberry, the temperature, humidity and lighting of the silkworm room, and even the building materials used in the construction of the silkworm nursery. Each measure is put forward according to the physiological characteristics of the silkworm in different growth periods.

Qi Min Yao Shu has a great deal of discussion about the storage and processing of agricultural products, which is almost unprecedented in traditional agricultural works. When discussing the storage of cereal such millet and wheat, diseases and insect pests that may occur in the storage process are treated by physical means such as fire and steaming, so that the cereals are easy to pound and the grains are firm. Cereal is mainly used for processing into a staple food. In addition, it can also be used for brewing wine and vinegar, and processed into a non-staple food such as sauce, *chi* (豉), and sugar. There are more parts in the book to discuss brewing, vinegar making, sauce making, and other techniques, which are not found in other ancient literature.

In terms of the storage of fruits and vegetables, salt, *cao* (漕), honey, *qu* (曲), and rice soup are added so that the storage time is prolonged and the edible quality is also improved. *Youchaomian* (油炒面 fried flour with oil) is mixed with processed apricot and plum to turn them into instant food. The book also has methods for retaining the freshness of fruits and vegetables to ensure that they can be available in winter.

The processing of animal products such as livestock, poultry, and fish involves meat, dairy products, fur, and eggs. Meat can be treated by means of *jiang* (酱), *zha* (鲊), *fu* (脯), *la* (腊), steaming, and frying for longer preservation and better eating quality. Wool is mainly processed into felt. Animal skins are used for making glue. The discussion of dairy products is mainly about goat milk, including various cheese-making methods. There is a method of making salted duck eggs. In addition to salt, the bark of a southern tree, *yuanzi* (杬子), is added.

What Jia Sixie was concerned with in his book was people's affluence. He is opposed to profit-making activities conducted away from agricultural production, but he is not opposed to trading activities of agricultural and sideline products based on agricultural production. In the book, it can be clearly judged that all or part of the items produced for profit include vegetables (melons, *hu*, sunflowers, *manjing*, *song* 菘, *lufu*芦菔, *xie*, and coriander), fruit trees and economic plants (*honglanhua*, *lan*, and lithospermum), timber trees (*zhe*, *yu*, *baiyang* 白杨, *tang*, *chu* 楮, willow, *jiliu* 箕柳, *qiu*, *zuo*), and livestock (donkey, horse, cattle, sheep, chicken, and fish). Even grain production has been commercialized to a certain extent.

For this reason, the article "Shou Zhong (收种 collection of seed grain)" emphasizes the importance of selecting grain seeds. In the book, the income of grain fields is used as the comparative price to calculate the income of other products, which clearly shows that he attaches great importance to grain production. On the other hand, although grain production is discussed in the first part and occupies the largest space, he does not attribute agriculture to grain production alone. Instead, he values the development of a diversified economy. He lays stress on increasing the added value of agricultural products through processing and comprehensive utilization of agricultural and sideline products. Likewise, he attaches importance to reducing production costs and expanding profits so as to get the biggest return with the smallest investment. The use of certain agricultural technologies and crop varieties is also based on commercial profit considerations.

Qi Min Yao Shu is a book with world influence. It reached Japan in the Tang Dynasty and had a wide and lasting influence. Some of the contents of the book transferred to Europe and influenced academic development there. It is well reflected in the works of Charles Robert Darwin (1809~1882). In 1859 Darwin published the book *Origin of Species* and put forward a complete theory on biological evolution. Later he published two books successively: *The Variation of Animals and Plants under Domestication* (1868) and *The Descent of Man, and Selection in Relation to Sex* (1871). According to the statistics, there are more than 100 references to Chinese materials in these three books (Shiran et al. 1982). Most of them come from "an encyclopedia of ancient China," which, some scholars infer, refers to *Qi Min Yao Shu* (Pan Jixing: Darwin and *Qi Min Yao Shu*, *Nong Ye Kao Gu* (《农业考古》), 1990, Issue 2).

3.4 Agricultural Books in the Sui, Tang, Song, and Yuan Dynasties

3.4.1 Overview

In the Sui, Tang, Song, and Yuan Dynasties, the number of agricultural books increased at an unprecedented speed. *Song Shi Yi Wen Zhi* (《宋史·艺文志》) alone recorded 107 agricultural books with 423 volumes. This does not include the works on animal husbandry and veterinary medicine included in "Wu Xing Zhi" ("五行志") and "Yi Jia Lei" ("医家类"). Most of these agricultural books recorded in *Song Shi* appeared in the Tang and Song Dynasties, and agricultural books continued to come out at a fast speed later on. The Yuan Dynasty ruled China for less than a hundred years, but produced many agricultural books. Among them, there were three large agricultural books, namely, *Nong Sang Ji Yao* (《农桑辑要》), *Wang Zhen Nong Shu* (《王祯农书》), and *Nong Sang Yi Shi Cuo Yao* (《农桑衣食撮要》). In addition, at the turn of the Song and Yuan Dynasties, there were many small agricultural books. For example, books cited in *Nong Sang Ji Yao* and *Wang Zhen Nong Shu* were *Zhong Shi Zhi Shuo* (《种莳直说》), *Han Shi Zhi Shuo* (《韩氏直说》), *Nong Sang Zhi Shuo* (《农桑直说》), *Nong Sang Yao Zhi* (《农桑要旨》), *Shi Nong Bi Yong* (《士农必用》), and *Wu Ben Xin Shu* (《务本新书》). According to the description of

Wang Yuhu's *Zhong Guo Nong Xue Shu Lu*, there were more than 30 kinds of agricultural book in the 1400 years from the Spring and Autumn period and Warring States to the Tang Dynasty, while there were more than 170 kinds of agricultural book in about 800 years of the Sui, Tang, Song, and Yuan Dynasties.

In the Sui, Tang, Song, and Yuan Dynasties, there were some remarkable study directions reflected in agricultural books, for example, the first appearance of the southern agricultural books, the emergence of professional agricultural books "Pu Lu" ("谱录"), the publication of official agricultural books, articles encouraging agriculture and *Geng Zhi Tu* (《耕织图》), the appearance of hermit agricultural books for mountain dwellers, the publication of agricultural books for the exchange and comparison of agricultural technologies between the north and the south, and the emerging popularity of agricultural books with illustrations and pictures. The contents of agricultural books were also innovative: sericulture, farm tools, herbs, flowers, and trees became the main topics in agricultural books.

During this period, important representative or pioneering works appeared in almost every area. For example, *Chen Fu Nong Shu* (《陈旉农书》) is the earliest existing local private agricultural book that reflects the knowledge of agricultural production in the Yangtze River basin and areas south of it. *Nong Sang Ji Yao* is the earliest extant official agricultural book. *Wang Zhen Nong Shu* is the first agronomy book covering agronomy in both the north and south. Others, such as *Cha Jing* (《茶经》), *Lei Si Jing* (《耒耜经》), *Can Shu* (《蚕书》), *Si Mu An Ji Ji* (《司牧安骥集》), *Ju Lu* (《橘录》), *Li Zhi Pu* (《荔枝谱》), *Luo Yang Mu Dan Ji* (《洛阳牡丹记》), *Jun Pu* (《菌谱》), and *Tang Shuang Pu* (《糖霜谱》), are all representative and pioneering works in various professional fields.

3.4.2 Private Agronomy Works

After the imperial examination system was established in the Sui and Tang Dynasties, some scholars who failed the examination and were forced to do farm work to earn their own living began to write agricultural books to sum up the production experience. Besides these, some scholars tired of officialdom also diverted their attention to agronomy, or became obsessed with flowers, plants, insects, fish, birds, and animals. There were other people who, because of their religious beliefs, grew medicinal herbs in order to keep in good health. They pushed the tradition of private agriculture to a new stage. Some retired scholar-officials or monks worked on their own in the mountains or fields and wrote farming books after gaining some experience in agronomy. These agricultural books sum up direct and indirect experience in agricultural technology, and also talk a lot about how to keep fit. Therefore, these kinds of farming book are also called "systematic farming books of mountain dwelling."

Different from the books about agricultural technology, the systematic farming books of mountain dwelling are a mixture of arts, health care, and leisure (Hu Daojing: *Meng Xi Wang Huai Lu,* an Agricural Masterpiece by Shen Kuo, *Nong Shu Nong Shi Lun Ji* (《农书·农史论集》), pp. 30–31). In fact, "Cha Jing,"

"Sun Pu" (笋谱), "Zhu Pu" (竹谱), *Chen Fu Nong Shu*, as well as "Tong Pu" written by Chen Zhu (陈翥) can also be regarded as farming books of the system. The contents of these books reflect the characteristics of mountain areas, or take mountain plants such as tea, *tong* (桐), and bamboo as the main subjects. Moreover, most of the authors were reclusive monks in mountain areas. The recorded agricultural books of this kind are: *Shan Ju Yao Shu* (《山居要术》), *Shan Ju Za Yao* (《山居杂要》), and *Shan Ju Zhong Shi Yao Shu* (《山居种莳要术》) by Wang Min (王旻) during the Tang Dynasty, *Bu Shan Jing* (《补山经》) by Zhou Jiang (周绛) (*Song Shi Yi Wen Zhi Nong Jia Lei* (《宋史·艺文志·农家类》)), *Ping Quan Shan Ju Cao Mu Ji* (《平泉山居草木记》) by Li Deyu (李德裕) (Pingquan is where Li Deyu's villa located, 30 miles away from Luoyang City. The book records exotic flowers and plants), *Meng Xi Wang Huai Lu* (《梦溪忘怀录》) and *Cha Lun* (《茶论》) by Shen Kuo (沈括) during the Song Dynasty, *Shan Jia Qing Gong* (《山家清供》) by Lin Hong (林洪), *Bei Shan Jiu Jing* (《北山酒经》) by Zhu Gong (朱肱), as well as *Shan Ju Si Yao* (《山居四要》) by Wang Rumao (汪汝懋) during the Yuan Dynasty. Among them, *Shan Ju Yao Shu*, *Tong Pu*, and *Shan Ju Si Yao* are representative ones.

Shan Ju Yao Shu, also known as *Shan Ju Lu* (《山居录》), was written by Wang Min, a Taoist priest from the Tang Dynasty, who had another name, Tai He Xian Sheng (太和先生). The *Shan Ju Lu* recorded in *Ju Jia Bi Yong Shi Lei Quan Ji* (《居家必用事类全集》) edited in the Yuan Dynasty has the following content: the general description of mountain dwelling, the method of making garden fences, planting medicinal herbs, vegetables, fruit and timber, flowers, bamboo and wood, bamboo ware, etc. However, the complete edition of *Shan Ju Lu* contains writings by authors of later generations. The book incorporates the previous agricultural books such as *Qi Min Yao Shu*, but there are also many things that were not found in the previous agricultural books. The two records of "planting tea" and "collecting tea seeds" are the earliest known and most detailed records of tea cultivation methods and management. The records of tea cultivation in some later agricultural books or tea books do not go beyond the contents of this book. There are also earlier records about biological control in ancient literature. The book is an influential agricultural book from the Tang and Song Dynasties. Much of the content in *Si Shi Zuan Yao* may come from it. The part about "herbs" is even used and referenced by the later *Nong Sang Ji Yao* and *Ben Cao Gang Mu* (《本草纲目》).

Chen Zhu (1009~1061), also named Zixiang (子翔) and known as Tong Zhu Jun (桐竹君), is the author of *Tong Pu*. Born in Tongling, Anhui province, he once lived in seclusion in the south of Mount Xishan (now Mount Fenghuang in Tongling city), so he called himself "Tongling Yi Min" (铜陵逸民 an escaped person in Tongling). At the age of 40, feeling that his official career was hopeless, he retired to plant *tong* trees and bamboo south of Mount Xishan and wrote a book entitled *Tong Pu*. The book systematically and comprehensively summarizes the experience of planting and utilizing *tong* in the Northern Song Dynasty and before.

Among the articles, the article "Xu Yuan" (叙源) is a textual research piece on the name of *tong*. It also discusses and introduces the morphological and biological characteristics of the tree, the material quality of it, and the comprehensive utilization of its flowers, leaves, etc.

The article "Lei Shu" (类属) discusses the species and classification of *tong*. It divides the varieties of the tree into seven species and three categories, which clarifies both individual differences and similarities.

The article "Zhong Zhi"(种植) introduces the technologies concerning seedling breeding, afforestation, and the tending of *tong* growth.

"Suo Yi" (所宜) is an article about the suitable growth environment for *tong*, including topography, soil fertility, light, temperature, moisture, etc., and puts forward some corresponding technical measures, such as intertillage, weeding, fertilization, leaf thinning, etc.

The article "Suo Chu" (所出) records the original distribution of *tong*. It shows that the natural distribution and artificial cultivation of *tong* were very common in the middle and lower reaches of the Yangtze River at that time, especially in the south of the Yangtze River, of which the central part of Shu was the most famous place.

The article "Cai Zhuo" (采斫) records the experience of pruning and thinning *tong* and timber harvesting while the article "Qi Yong"(器用) summarizes the experience of utilizing *tong*, indicating that the ancients had a deep understanding of its material quality.

"Za Shuo" (杂说) is a selection of anecdotes about *tong*, and "Ji Zhi" (记志) includes two articles, describing the author's experience in planting *tong* and bamboo in the south of Mount Xishan. The article "Shi Fu" (诗赋) includes the author's poems and odes about *tong*. In a nutshell, *Tong Pu* is the earliest book dedicated to the *tong* tree in both China and the world.

Wang Rumao, with another name Yijing (以敬), was born in Jingdezhen, Jiangxi Province. Once a Junior Compiler of the State History Museum in the years of Zhizheng of the Yuan Dynasty, he abandoned his post to give lectures. By 1360 he had written a book called *Shan Ju Si Yao*. The so-called "Si Yao" refers to the four essentials of "She Sheng" (摄生), "Yang Sheng" (养生), "Wei Sheng" (卫生), and "Zhi Sheng" (治生). Among them, the most important part of "Zhi Sheng" is about farming. Following the genre of agricultural calendars, it lists methods of seeding, cutting, planting, transplanting, collection, and other miscellaneous activities. However, it only generally records the names of crops, flowers, and fruit and seldom involves specific operating methods. It has an appendix of methods of planting flowers, fruit, vegetables, etc. In addition, several prescriptions for livestock diseases are attached to the part of "Wei Sheng."

3.4.3 Si Shi Zuan Yao

Si Shi Zuan Yao is an influential agricultural book. The writer is Han E, probably from the late Tang Dynasty to the early Five Dynasties (Qiyu 1981).

Si Shi Zuan Yao is an agricultural calendar book written on the basis of previous relevant materials, such as *Fan Sheng Zhi Shu*, *Qi Min Yao Shu*, and *Bao Sheng Yue Lu* (《保生月录》). Compared with *Qi Min Yao Shu*, one of its most significant contributions lies in the addition of new contents not recorded in *Qi Min Yao Shu*, such as tea, *yiyi* (薏苡 coix seed), shuyu (薯蓣 yam), buckwheat, *zhongjunzi* (种菌子), beekeeping, etc., although it seems that most of these contents may come from Wang Min's *Shan Ju Yao Shu*. In addition, *Si Shi Zuan Yao* develops some thoughts in *Qi Min Yao Shu*. For example, the book proposes a new method of fruit tree grafting, which is an important development in grafting theory. In addition, the book has a lot of content about predicting natural disasters and weather changes according to astronomical phenomena. This is also the result of the shift of China's agricultural center from the north to the south. The climate in the south is relatively changeable so this has become one of the important points that people pay attention to.

In history, *Si Shi Zuan Yao* was equally as important as *Qi Min Yao Shu* and *Nong Sang Ji Yao*. Although *Si Shi Zuan Yao* was lost in China after the Ming Dynasty, it has been preserved in North Korea and has become an important reference book for the study of the 600-year agricultural history from *Qi Min Yao Shu* to *Chen Fu Nong Shu*. *Si Shi Zuan Yao* appeared in the Tang and Five Dynasties, which was the time when China's economic center shifted from north to south. It also appeared in the history of Chinese agriculture, and *Si Shi Zuan Yao* can be regarded as a turning point. Although the book focuses on agricultural production in the north, it also includes some information about agriculture in southern China. Therefore, this book has important reference value for studying the social and economic history and agricultural technological history in the transitional period.

3.4.4 Chen Fu Nong Shu

The author of *Chen Fu Nong Shu* is Chen Fu, who was born in the 9th year of Xining (1076) in the Northern Song Dynasty. He was a reclusive scholar who did not seek official advancement, but grew medicinal herbs and lived a self-sufficient life. He believed that agricultural books such as *Qi Min Yao Shu* and *Si Shi Zuan Yao* were out of date and impractical, and hoped that the agricultural book he wrote would "be helpful to the people in later generations." The book was written in the 19th year of Shaoxing in the Southern Song Dynasty (1149).

The book is divided into three volumes with more than 10,000 words. The first volume deals with soil cultivation and crop cultivation, the middle volume the management of farm animals, and the third volume the techniques of mulberry planting and silkworm rearing. The integration of the three volumes forms an organic whole, of which the first volume is the main body and occupies two thirds of the whole book. The middle volume is "Niu Shuo" (牛说 about cattle) as farm cattle are the main driving force of farming. This book is the first existing ancient agricultural book that used a special article to discuss the problem of cattle farming systematically. Likewise, the third volume discusses sericulture because sericulture is an important part of farming. However, it is again the first time an agricultural book treated sericulture as a key issue.

The first volume consists of 12 articles of "Shi Yi" (适宜 suitability) as well as two articles of "Qi Bao" (祈报 praying) and "Shan Qi Gen Miao" (善其根苗篇 proper treatment of roots and shoots) to form an organic whole. "Cai Li Zhi Yi" (财力之宜

suitability for financial resources) emphasizes that the scale of production (especially the area of cultivated land) should be commensurate with financial resources and manpower. The article of "Di Shi Zhi Yi" (地势之宜 suitability for land use) discusses the proper use of land. According to the article "Geng Nou Zhi Yi" (耕耨之宜 suitability of tillage), the terrain not only affects the utilization of the land, but also affects the sequence and speed of cultivation and the depth of plowing. Then there is the article of "Tian Shi Zhi Yi" (天时之宜 the suitability of time). It emphasizes that agricultural activities must be arranged according to the changes of the sun. It also pays attention to the troubles brought by abnormal climate to agriculture and closely correlates the heaven, the earth, and time, reflecting the idea of choosing the proper time according to different situations. Chen Fu also links weather with crops, hence the article "Liu Zhong Zhi Yi" (六种之宜 six kinds of suitable crops), mainly discussing the sequence of cultivating several dryland crops, which is also the first summary of dryland crop cultivation techniques in southern China. For the first time, the article "Ju Chu Zhi Yi" (居处之宜 living in a proper place) discusses the relationship between farmland and residence, proposing that living in a place close to farmland is convenient for taking care of the land. In the following article of "Fen Tian Zhi Yi" (粪田之宜 appropriate field manure), soil and fertilizer become the focus, and famous judgments such as "paying attention to soil improvement" were put forward. *Chen Fu Nong Shu* also includes "thrift," "rewarding diligence and punish laziness," "material preparation," "spiritual preparation," and "respect for ghosts and gods" in its 12 articles of "Shi Yi," which are all key factors affecting agricultural production in the author's view.

The middle volume systematically discusses the problem of cattle farming. Besides demonstrating the work of cattle, it emphasizes the husbandry, use, and treatment of cattle. When herding, one must "adapt to the times," combine herding with raising, keep the stockyards clean, and at the same time prevent cattle from getting sick. The straw should be clean, "finely chopped" and "moistened with wheat bran, chaff or beans," and placed in a trough to feed the cattle. When grazing in spring and summer, cattle must first drink water before eating grass to avoid abdominal distension. Fresh feed should also be provided to cattle in winter (*Ling Wao Dai Da Ta Li* (《岭外代答·踏犁》), Volume 4). When having cattle do its work, one "must not exhaust the cattle's strength," "prevent it from catching cold," and "not make it too tired." The work that cattle do should be adjusted according to the temperature in different seasons. Medical treatment requires the differentiation of symptoms and signs, the right remedy for the case, and isolation measures for epidemic diseases. This is also a remarkable moment of progress in the history of veterinary medicine.

The last volume discusses sericulture and includes five articles. The article "Zhong Sang" (种桑 planting mulberry) introduces a method of mulberry seed propagation as well as asexual propagation methods such as layering and grafting. Mulberry trees should be pruned many times after planting to improve the yield and quality of mulberry leaves. The article "Shou Can Zhong" (收蚕种 collection of silkworm eggs) introduces the preservation of silkworm eggs, *yucan* (浴蚕 washing of silkworm eggs), silkworm nursery, and techniques for feeding young silkworms.

It mentions the use of low temperature to select fine silkworm eggs and eliminate inferior species. The article "Yu Can" (育蚕 silkworm rearing) emphasizes self-picking to ensure the unanimous emergence of seedlings. The article "Yong Huo Cai Sang" (用火采桑 collecting mulberry with fire) proposes a method to control the humidity and temperature of silkworm chambers by using fire when feeding leaves to silkworms. The article "Cu Bo He Cang Jian" (簇箔和藏茧 *cubo* and cocoon storage) introduces the making of a tool called *cubo* (簇箔) and the methods of cocoon collection and storage.

Chen Fu Nong Shu is the first book systematically discussing agronomy in southern China. The topography and terrain in the south are relatively complex. The article "Di Shi Zhi Yi" (地势之宜 the appropriate terrain) first discusses the uses of land systematically. Among them, the planning and utilization of fields in high lands is the most detailed, which fully reflects the idea of adjusting measures to local conditions and rational planning and utilization. Southern Chinese agriculture is mainly composed of rice, sericulture, cattle, and livestock as well as a certain number of dryland crops. Most of the "twelve suitabilities" are related to rice cultivation, while the articles of "Hao Yun Zhi Yi" (薅耘之宜 suitability of weeding) and "Shan Qi Gen Miao" are devoted to rice field management and rice seedling raising techniques. Chen Fu believes that "when planting, appropriate land should be chosen, and the manure should be used properly." In addition, careful management is the key to cultivating good seedlings, and weeding is an important link in rice field management, because it can eliminate grass damage, improve soil fertility, and improve soil structure.

In the article of "Fen Tian Zhi Yi" focusing on soil and fertilizer, Chen Fu puts forward two outstanding theories on soil. One is the theory of "*fenyao*" (粪药 manure as medicine), which holds that soil of different properties can achieve good harvest as long as it is properly treated, and the key to treatment lies in the use of manure. At that time, people used "*fenyao*" ("using dung is like taking medicine") to mean treating soil with dung according to different properties of soil. The second theory is that for land that has been cultivated all year round where the soil fertility has declined, "if the soil of new fertile land can be added and treated with manure, then it will be more fertile, and never become barren."

The core of the two theories is manure (fertilizer). Not only does the book use special articles to talk about fertilizers, other articles also have specific and detailed discussions on it, and there are many innovations and developments in the accumulation and application methods of fertilizers. In the Yuan Dynasty, Wang Zhen (王祯) inherited and carried forward Chen Fu's theory of soil and fertilizer, and put forward the idea that "to cherish dung is the same as cherishing gold" and "a dung field is better than a newly bought field." He divided fertilizers into five types; besides stable manure, there are also seedling manure, grass manure, fire manure, and mud manure, further expanding the sources of fertilizers and proposing the principle of adequate and sufficient fertilization (*Wang Zhen Nong Shu Fen Rang Pian* (《王祯农书·粪壤篇》)), which is an amendment to the theory of "fertilizing the field with as much dung as possible" held since the Warring States Period.

3.4.5 Official Agricultural Books

During the Sui, Tang, Song, and Yuan Dynasties, governments at all levels promoted the development of agriculture by compiling and publishing agricultural books, which led to the appearance of official agricultural books. On April 7, 686, the second year of Chuigong during the Tang Dynasty, the Imperial Mother edited *Yue Liao Xin Jie* (《月寮新诫》) and *Zhao Ren Ben Ye* (《兆人本业》), and distributed them to the Territorial Representatives (*Tang Hui Yao* (《唐会要》), Volume 36). *Zhao Ren Ben Ye* is about "agricultural customs and the seasonal methods of cultivation," with a total of 80 events. It is the earliest official agricultural book known (*Kun Xue Ji Wen* (《困学纪闻》), Volume 5).

In the fourth year of Tianxi during the Song Dynasty (1020), Emporer Zhenzong ordered the books of *Si Shi Zuan Yao* and *Qi Min Yao Shu* to be inscribed and encouraged farmers to use them. This is probably the earliest edition of these two agricultural books. Emporer Zhenzong also instructed officials to compile a 12-volume book, *Shou Shi Yao Lu* (《授时要录》), which is an official agricultural book in the genre of agricultural calendars. In the Song Dynasty, similar official books include *Da Nong Xiao Jing* (《大农孝经》) and *Ben Shu* (《本书》). Emporer Zhenzong once issued a planting method in the name of the government, introducing the cultivation method of *zhancheng* rice (占城稻) in detail, which has strong operability, and it also pays attention to the adjustment of measures to local conditions. In the 23rd year of the Yuan Dynasty (1286), the government issued the officially edited book *Nong Sang Ji Yao* to its subordinate states and counties.

The farming exhortation is a publication issued by the local government to encourage farming. During the Song and Yuan Dynasties, local officials went to the countryside at certain times of the year (usually at the beginning of spring ploughing) to encourage farming. In this process, they often issued proclamations. During the Northern Song Dynasty, there were many poems and articles on farming encouragement. However, most of them were official articles and engaged in formalism. During the Southern Song Dynasty, this style of work changed slightly, and added in the technical content. For example, Zhu Xi (朱熹)'s *Nan Kang Jun Quan Nong Wen* (《南康军劝农文》), Gao Side (高斯得)'s *Ning Guo Fu Quan Nong Wen* (《宁国府劝农文》), Zhen Dexiu (真德秀)'s *Quan Zhou Quan Nong Wen* (《泉州劝农文》), *Fu Zhou Quan Nong Wen* (《福州劝农文》), and so on, some of which were aimed at technical problems in agricultural production. However, the effect was limited. Although there were officials doing the same thing in the Yuan, Ming, and Qing Dynasties, the number of farming exhortations was far smaller than that of the Song Dynasty.

A ploughing and weaving map is actually a kind of farming exhortation in the form of pictures and illustrations. It expresses some key links in agricultural production in the form of images, accompanied by poems and ballads. Its purpose is also to emphasize agriculture and encourage farmers. The expression of farming and sericulture in the form of fine arts appeared in the Five Dynasties Period. Ploughing and weaving pictures appeared in the court during the first year of Baoyuan under the reign of Emperor Renzong in the Song Dynasty (1038~1040). At that time, the farming and weaving situation of peasant families was painted on the wall of Yanchun Pavilion (Wang Yinglin: *Kun Xue Ji Wen*, Volume 15). These works of art, while being appreciated by emperors, remind them not to forget the hardship of farming. Emperor Gaozong also mentioned this matter (*Jian Yan Yi Lai Xi Nian Yao Lu* (《建炎以来系年要录》), Volume 87). Under his reign in the Southern Song Dynasty, Lou Shu (楼璹), the District Magistrate of Qianxian County, compiled a set of *Geng Zhi Tu* (《耕织图》) based on a visit to sericulture farmers. The Ploughing Drawings include 21 pictures of soaking seeds, ploughing, *banuobayang* (耙耨拔秧), transplanting seedlings, first weeding, second weeding, third weeding, irrigation, screening, warehousing, etc. The Weaving Pictures include 24 pictures of silkworm feeding, mulberry picking, cocoon selection, flower picking, and silk cutting. Each picture is accompanied by a five-word-and-eight-sentence verse.

Lou Shu's *Geng Zhi Tu* is a very influential work. Cheng Qi (程棨)'s *Ploughing and Weaving Map* of the Yuan Dynasty may be a reproduction of it. Wang Zhen also quoted "Tu Shi" (图诗 drawings and verses) and the *Ploughing and Weaving Map* in many places in his "Nong Qi Tu Pu" (农器图谱 Drawings of Agricultural Devices). The addition of sericulture and textiles may be influenced by the *Ploughing and Weaving Map*, because this part of the content was not included in Zeng Zhijin (曾之谨)'s *Nong Qi Pu* (《农器谱》) in the Southern Song Dynasty.

The development of the "Ploughing and Weaving Map" reached its peak in the Qing Dynasty. The Emperors Kangxi, Yongzheng, Ganlong, Jiaqing, and Guangxu all had an imperial "Ploughing and Weaving Map" published. Kangxi's *Ploughing and Weaving Map* was painted by court painter Jiao Bingzhen (焦秉贞) in the 35th year of Kangxi's reign (1696). It is an edited version of Lou Shu's *Ploughing and Weaving Drawings* (or a copy), of which 23 are "Ploughing" and 23 are "Weaving," totaling 46. The drawings feature the law of western drawing that "the object is big when near and small when far," or the perspective method. Therefore, the village scenery in the paintings and the hardship of the farmers can be best presented, making it an excellent work of art.

The drawings and contents of Yongzheng's *Geng Zhi Tu* are basically the same as Kangxi's. There are 23 drawings of ploughing and weaving respectively, totaling 46 pieces, but the arrangement is slightly changed. Ganlong's stone inscription "Geng Zhi Tu" (1769) was carved by the Painting Institute based on copies of Cheng Qi in the Yuan Dynasty under the instruction of Emperor Ganlong. It is divided into 21 drawings of ploughing and 24 of weaving, totaling 45 pieces. Each picture is 53 cm long and 34 cm high. It was partially destroyed in 1860 during the Second Opium War when British and French troops invaded Beijing.

Guangxu's stone inscription "Geng Zhi Tu" was found on the wall of a farmhouse gate building in Bo'ai County, Henan Province. There are 10 pieces of ploughing and weaving respectively, totaling 20 pieces, each engraved on 4 bluestones that are 200 cm long and 30 cm wide. The content is very different from the traditional ploughing and weaving maps. The 10 ploughing maps depict the process from planting to harvesting of rice and the 10 weaving maps the process of cotton from planting to processing into cloth. In addition, the original silk weaving pattern was replaced by cotton planting, seed removal, and textile. "Geng Zhi Tu" was introduced to Europe in large number in the form of export paintings during the Qing Dynasty and became the object of people's collections and displays (Watabe Takeshi: China Craze in Europe and *Geng Zhi Tu*, a speech at the Institute for the History of Natural Science, Sep. 5, 2006).

3.4.6 Three Agricultural Books in the Yuan Dynasty

The Yuan Dynasty ruled for less than 100 years, but it left behind three remarkable agricultural books in the history of Chinese agriculture: *Nong Sang Ji Yao*, *Wang Zhen Nong Shu*, and *Nong Sang Yi Shi Cuo Yao*.

3.4.6.1 *Nong Sang Ji Yao*

Nong Sang Ji Yao was compiled by the central authorities in charge of agriculture and water conservancy in the Yuan Dynasty. The specific writers are Meng Qi (孟祺), Zhang Wenqian (张文谦), Chang Shiwen (畅师文), and Miao Haoqian (苗好谦). The book was written before the 10th year of Zhiyuan (1273). The content of agricultural technology contained in the book was tried out around the Houzai Gate (present-day Di'an Gate) in the city of Dadu of the Yuan Dynasty with satisfactory results (Mengxiang 1983). After the book was written, it was issued to officials in charge of farming encouragement at all levels to guide agricultural production. It is the earliest extant official agricultural book.

Nong Sang Ji Yao basically inherits the content of *Qi Min Yao Shu*, but also adds some new material, for example, ramie, kapok, watermelon, carrot, chrysanthemum, amaranth, piper longum, sugarcane, beekeeping, etc. This content is the first-hand material written by summarizing the experience at that time. For example, the summary of China's cotton and ramie cultivation techniques is the earliest summary of these two important cash crops. In order to promote ramie and cotton cultivation to the north, the book also discusses the local conditions and customs and thinks that different local conditions and customs are suitable for different crops. The book introduces the human factor into the old concept of local conditions and customs, emphasizing the exertion of human subjective initiative and human intelligence, and became a milestone in the history of agricultural thought.

Nong Sang Ji Yao places sericulture production at the same important position as agriculture, reflecting the characteristics of agriculture books in the Song and Yuan Dynasties. Judging from the length, although planting mulberry and raising silkworm each occupy one volume, the two volumes occupy nearly one third of the whole book. This is not so in *Qi Min Yao Shu*. The book also collects agricultural books such as *Shi Nong Bi Yong*, *Wu Ben Xin Shu*, *Si Shi Lei Yao* (《四时类要》), *Bo Wen Lu* (《博闻录》), *Han Shi Zhi Shuo*, *Nong Sang Yao Zhi*, and *Zhong Shi Zhi Shuo* (《种莳直说》). Since most of these agricultural books are now lost, *Nong Sang Ji Yao* objectively plays a role in preserving and disseminating ancient agricultural science and technology.

From the newly added and edited contents, we can see the progress of agricultural technology since *Qi Min Yao Shu*. For example, in field crop production technology, attention is paid to autumn tillage and the role of *pa* (耙 raking) is highlighted. The method of *yunmiao* (芸苗 improving seedlings) is mentioned, and the animal farm tool "*louchu*" (耧锄 hoe) is used for intertillage. When this farm tool passed through Europe, it brought about the beginning of modern agriculture after the improvement of the Englishman Jethro Tull (1674~1741). The book also proposes the harvest of wheat. In the aspect of mulberry planting and silkworm rearing technology, the book quotes *Shi Nong Bi Yong* and *Wu Ben Xin Shu* to make detailed records on mulberry grafting, layering, pruning (branch cutting), pest control, and intercropping of mulberry and other crops. There are many discussions on cocoon selection, *yuluan*, feeding, and prevention of silkworm diseases, and it also effectively summarized the sericulture techniques.

3.4.6.2 *Wang Zhen Nong Shu*

Wang Zhen, also named Boshan (伯善), was a native of Dongping, Shandong province (now Dongping County). He lived between the thirteenth and fourteenth centuries. In the first year of Yuanzhen during the reign of Emperor Chengzong of the Yuan Dynasty (1295), he served as District Magistrate of Jingde, Xuanzhou (now in Anhui), and later transferred to Yongfeng, Xinzhou (now Guangfeng in Jiangxi province). The book was written during his tenure in these two places and was written around 1300 years ago.

The book is composed of three parts. The first part, "Nong Sang Tong Jue" (农桑通诀), discusses the origin of agriculture, cattle farming, and sericulture as well as the relationship between agriculture and weather, geographical location, and manpower. Then according to the basic sequence of agricultural production such as sowing in spring, ploughing in summer, harvesting in autumn, and storage in winter, it records some common basic measures that should be taken in each stage of the field crop production process. Finally, there are three articles entitled "Planting," "Livestock Raising," and "Silkworm Reeling," which record the techniques related to tree planting such as mulberry trees, livestock raising, and cocoon processing. This part is also interspersed with some content that has little to do with agricultural production technology, such as "Qi Bao." The second part, "Bai Gu Pu" (百谷谱), describes the cultivation techniques and methods of more than 80 kinds of plants, followed by a passage of "Bei Huang Lun" (备荒论), which started the discussion of "Huang Zheng" (荒政 famine policy) by Xu Guangqi (徐光启) and others. The third part, "Nong Qi Tu Pu" (农器图谱), introduces agricultural implements by means of pictures and texts, and some of them do not belong to the category of agricultural implements, such as farmland system, land registry, *taishe* (太社), etc.

There are mainly two characteristics of *Wang Zhen Nong Shu*. First, it is the first time that agricultural technology in both the north and south are included in the same book. When Wang Zhen wrote the book, he was already "a famous senior scholar in eastern Shandong province, a traveling official in the north and south, and an experienced veteran" (Copy of Imperial Edicts in *Wang Zhen Nong Shu* Carved during the Reign of Emperor Yuan, *Wang Zhen Nong Shu*, collated by Wang Yuhu, Agricultural Publishing House, 1981, p. 446). This enabled him to make a comparison between the north and the south during his writing, with the purpose of "making the north and the south notify each other so that there is no waste, and then the methods of farming can be regarded as a total achievement." For example, he puts forward a solution to the problem that the dry land surface in the northern plain is generally large and there are some difficulties in leveling it. The previous generation of agricultural books rarely discuss the cultivation of paddy fields in the south while *Wang Zhen Nong Shu* classifies those fields into three types: early maturity in high fields, late maturity in lower fields, and water slurry fields. The first two types had already been mentioned in *Chen Fu Nong Shu*, but the third had not.

The second feature of *Wang Zhen Nong Shu* is the writing of "Nong Qi Tu Pu." Agricultural implements are divided into 20 categories, each of which is further divided into several subgroups. Each item is accompanied by a painting, with a total of more than 300 paintings. The structure, source, and usage of the agricultural implements are described in texts. Rhymes and poems are attached to most of the pictures and texts as a summary. This pioneering work became a model for agricultural tools recorded in later agricultural books and encyclopedias. Judging from the agricultural implements contained, the tools used had the characteristics of high efficiency, labor saving, exclusive use, perfection, and matching. Its contribution is not confined to farming implements in a general sense, but has extended to land use. Furthermore, it discusses land use and gives the concept and building method of terraced fields for the first time. There are also some comments on *tutian* (涂田 tidal-flat fields) and *shatian* (沙田 sand fields) which are not covered in *Chen Fu Nong Shu*. Drawn according to the astronomical phenomena, solar terms, climate change, and their relationship with farming, the paintings are convenient and direct for farmers to grasp farming time. The method of printing attached to the book is also a great contribution to printing.

3.4.6.3 *Nong Sang Yi Shi Cuo Yao*

Shortly after the publication of *Wang Zhen Nong Shu*, in the year 1314, *Nong Sang Yi Shi Cuo Yao* was published. The writer Lu Mingshan (鲁明善) is a Uighur. He lived in the mainland with his father for a long time and was deeply influenced by Han culture. He once did some literature work for the emperor in the imperial court. Later he worked in Jiangxi, Anhui, and other places. The book was written and published during his tenure at Anfeng Circuit, and was published again in 1330.

From the title and content, *Nong Sang Yi Shi Cuo Yao* is similar to *Nong Sang Ji Yao* by the Department of Agriculture, but it also has its own characteristics. First of all, in terms of genre, it adopts the traditional type of agricultural calendars. This genre is "concise and easy to understand and read" and is a development of *Nong Sang Ji Yao*. Secondly, the content is richer than *Nong Sang Ji Yao*. Some southern products, such as *jitou* (鸡头) (i.e., gordon euryale seed), *ling* (菱), lotus shoot, *jiaosun* (茭笋), *cigu* (茈菰), bamboo shoot, mandarin fish, etc., are all introduced in the book. It mentions rice growing six times. It is worth noting that as an agronomist of the Uigur nationality, Lu Mingshan also introduces the production techniques and experiences of some ethnic minorities, such as the collection of sheep breeds, the prevention and treatment of scabies, oronasal sores, cocoon hoofs and other diseases of sheep, the cultivation of grape, and the methods of making cheese, ghee, etc.

Nong Sang Yi Shi Cuo Yao is a relatively complete agricultural calendar book after *Si Shi Zuan Yao* that has been preserved till today.

3.4.7 Agricultural Books in the Type of Pulu (谱录Catalogue)

Pulu was originally a category in ancient book classification, originated by You Mao (尤袤) in the Southern Song Dynasty. He set up the type of "pulu" in his book *Sui Chu Tang Shu Mu* (《遂初堂书目》) to "*Xiang Pu* (《香谱》), *Shi Pu* (《石谱》) and *Xie Lu* (《蟹录》) that cannot be attached to any other books." Among the numerous works, there are many related to agronomy, covering agricultural implements, field crops, horticultural crops, animal husbandry, even meteorology and disasters, etc. Among them, horticultural works are the most prominent and have the highest agronomic achievements.

3.4.7.1 Agricultural machinery and field crops

This kind of *pulu* is represented by *Lei Si Jing* (《耒耜经》) written by Lu Guimeng (陆龟蒙) in the Tang Dynasty. This article, including the preface, has a total of only 600 words, but it records in detail a kind of paddy field farming tool, *quyuanli* (曲辕犁), commonly used in the Jiangdong area at that time. It also mentions three kinds of farming tools such as *pa* (raking), *liuzhou* (碌碡), and *lize* (礰礋). *Jiangdongli* (江东犁) is a farm implement composed of 11 parts, namely, *lichan* (犁镵), *libi* (犁壁), *lidi* (犁底), *yachan* (压镵), *ce'e* (策额), *liyuan* (犁辕), *lijian* (犁箭), *liping* (犁评), *lijian* (犁建), *lishao* (犁梢), and *lipan* (犁盘). *Lichan* and *libi* are made of metal, and the rest are made of wood. In this paper, the raw materials used in the manufacture of ploughs as well as the names, shapes, sizes, and dimensions of each part are described in detail, which is convenient for imitation and circulation. The paper also mentions *pa*, which is mainly used for soil preparation and can break up clods and remove weeds, just like *lize* and *liuzhou*. The appearance of *Jiangdongli* indicates that the traditional Chinese plough had become fixed. Ploughing and harrowing with the matching use of *lize* and *liuzhou* also indicate that the rice planting system of paddy fields in the south had also formed. Despite its short length, *Lei Si Jing* is regarded as the earliest book on agricultural tools in the history of Chinese agronomy.

After *Lei Si Jing*, agricultural works on farming tools include *Nong Qi Pu* written by Zeng Zhijin (曾之谨) in the Southern Song Dynasty, which is related to *He Pu* by Zeng Anzhi (曾安止) in the Northern Song Dynasty. *He Pu* is the first monograph on rice cultivation in Chinese history. The book includes articles on rice names, rice varieties, rice planting, rice cultivation, fertilizer, and soil as well as *qibao*. Like other *pulu* works, *He Pu* also attaches great importance to the classification of rice. Rice is divided first according to the different sowing and harvesting periods, second, the quality and use, and third, the comprehensive characteristics of the growth period and soil adaptability. In order to accurately record varieties of *taihe* (泰和) rice, Zeng Anzhi has made some meaningful discussions on the names of different varieties. The first is the distinction between "zongming" (总名 general name), which distinguishes rice from other crops, pointing out that "gu" (谷grain) in ancient books includes not only millet but also rice. *Dao* (稻) is the general name of rice. Under the general name, rice is divided into early rice and late rice, *jingdao* (秔稻) and *nuodao* (糯稻). Early rice and late rice are determined according to the sowing and harvest dates. The second is the distinction between "fuming" (复名 other names). *Dao* is also called *tu* (稌). The third is the distinction between "sanming" (散名 different names). Zeng Anzhi lists the records of a variety of names in ancient books and mentions the phenomenon of different names for the same things.

Su Shi (苏轼), the literary leader at that time, commented that *He Pu* is elegant and full of facts, but lacks information about agricultural implements. Zeng Anzhi accepted this with an open mind, but because of his blindness, he could not write again. After more than 100 years, Zeng Anzhi's grand-nephew Zeng Zhijin made up for this shortcoming. Zhou Bida (周必大), a fellow countryman of Zeng Zhijin, mentioned in the preface to *Nong Qi Pu* that the book recorded 10 kinds of agricultural devices with "miscellaneous notes." From the title of the book to the names of the items listed in the book, it is not difficult to see that Wang Zhen's *Nong Qi Tu Pu* is to some extent identical or similar to Zeng Zhijin's *Nong Qi Pu*. There are many reasons to believe that Wang Zhen's writing in the Yuan Dynasty was completed on the basis of Zeng Zhijin (Zeng Xiongsheng: A Probe into the Relationship between *Nong Qi Tu Pu* and *Nong Qi Pu*, *Nong Ye Kao Gu* (《农业考古》), 2003, Issue 1).

3.4.7.2 Gardening

Horticultural books are the most popular in *pulu* works, which can be divided into tea, flowers, fruit trees, timbers, vegetables, hobbies, etc.

Lu Yu (陆羽)'s *Cha Jing* (《茶经》) is the earliest tea work. Lu Yu (733~804) was born in Jingling, Fuzhou, during the Tang Dynasty (now Tianmen, Hubei Province). Originally a temple monk, he lived in seclusion in Tiaoxi (now Huzhou, Zhejiang province) and planted, drank, and discussed tea in Mount Guzhu. *Cha Jing* is divided into three volumes and ten chapters. The first chapter describes the production and characteristics of tea. The second is about the utensils used to pick tea leaves. The third records the processing of tea leaves. The fourth introduces the appliances used in tea processing. Section five is about the brewing methods of tea leaves and section six, the procedures in drinking tea. The seventh section collects material about tea from ancient books and narrates its history, occupying a long section. Section eight describes the origin of tea and ranks the quality of tea in different places into the top, secondary, and third grades. Section nine talks about the related matters of field tea processing. Section ten shows the above nine aspects in the form of pictures for easy reference, so there are actually nine aspects. The book systematically summarizes the tea planting experience before the Tang Dynasty and his own experience, covering the topics of the origin, types, characteristics, preparation methods, cooking, tea sets, water quality, tea drinking customs, famous tea producing areas, allusions and medicinal values of tea, etc. It is the world's first monograph on tea.

Cha Jing systematically records the history of tea-related activities in ancient China, discusses the efficacy of tea drinking, and attributes the functions of tea drinking to antipyretic thirst, dispelling congealing tightness, relieving brain pain, eye-brightening, relieving dysphoria, relaxing joints, drowsiness, etc. It also summarizes the tea-making process up to the middle of the Tang Dynasty, and lists seven steps in the tea-making process, namely, picking, steaming, pounding, patting, baking, *chuan* (穿), and sealing. It also proposes the method of judging tea quality and records a whole set of tea-making and drinking methods, called the "tea-frying method." All this has improved the world's understanding of tea drinking, and Lu Yu has been enshrined as a "tea god." With the popularity of tea drinking, the tea and horse exchanges between the mainland and the neighboring ethnic groups as well as other countries in the world also started (*Xin Tang Shu Lu Yu Zhuan* (《新唐书·陆羽传》)).

After *Cha Jing*, various works related to tea appeared one after another. According to the records of *Song Shi Yi Wen Zhi*, there were as many as 12. Most of the tea works are about Jian'an Tea (建安茶). Jian'an is now located in Fujian. Beiyuan is a tea producing area under Jian'an. The tea it produces is also called *Jiancha* (建茶). The first book on *Jiancha* was *Bei Yuan Cha Lu* (《北苑茶录》) by Ding Wei (丁谓). Also entitled *Jian An Cha Lu* (《建安茶录》), it took the form of drawings. Therefore, some people call this book "Cha Tu" (茶图 drawing on tea).

The second book on *Jiancha* is *Cha Lu* (《茶录》) by Cai Xiang (蔡襄). Cai Xiang believes that Lu Yu's *Cha Jing* did not contain tea produced in Fujian province while Ding Wei's *Cha Tu* only discusses the making of tea without introducing the cooking methods. Therefore, the author wrote two volumes of *Cha Lu* and carved them on stone in the first year of Zhiping (1064). The first part of the book deals with tea and the second with tea ware.

After Cai Xiang's *Cha Lu*, Song Zi'an (宋子安) wrote one volume of *Dong Xi Shi Cha Lu* (《东溪试茶录》). Dongxi is the place in Jian'an. Song Zi'an wrote this book as a supplement to *Cha Tu* and *Cha Lu*. The book is divided into eight items, and its basic idea is to distinguish between the places where tea is produced. Some places of production are close to each other, but the quality of tea differs greatly.

After that, Huang Ru (黄儒) wrote two volumes of *Pin Cha Yao Lu* (《品茶要录》) about *Jiancha*. Huang Ru was from Jian'an. This book deals specifically with the collection and processing of *Jiancha* and exposes fake and inferior products. After Huang Ru, Xiong Fan (熊蕃) wrote one volume of *Xuan He Bei Yuan Gong Cha Lu* (《宣和北苑贡茶录》). Beiyuan is located at the foot of Mount Fenghuang in the east of Jian'an and is famous for its abundant *gongcha* (贡茶). Xiong Fan (from Jianyang) personally observed and recorded the tea production. His exposition includes the evolution of Jian'an tea, the development of *gongcha*, and the grade of tea buds. He also lists the names of more than 40 kinds of *gongcha* and their years of manufacture. Attached to the book are 38 pictures, supplemented by Xiong Ke (熊克), the son of Xiong Fan, showing the shape and size of *gongcha* of each variety. After that, Zhao Ruli (赵汝砺), Xiong Fan's protege, wrote one volume of *Beiyuan Bie Lu* (《北苑别录》) to supplement *pulu* works on *gongcha*.

Flower books are the most popular in the *pulu* works on gardening. Such books aim at appreciating flowers, so the varieties and ranks of flowers constitute the main content. Each book records dozens or hundreds of varieties. For example, Shi Zhu (史铸)'s *Bai Ju Ji Pu* (《百菊集谱》) records more than 160 kinds of chrysanthemum. Zhou Shihou (周师厚)'s *Luo Yang Hua Mu Ji* (《洛阳花木记》) records the names and varieties of many flowers and trees in Luoyang, including 109 kinds of peony, 41 kinds of *shaoyao* (芍药herbaceous peony), 82 kinds of miscellaneous flowers, 147 kinds of fruit flowers, 37 kinds of *cihua* (刺花 thorn flowers), 89 kinds of *caohua* (草花 grass flowers), 19 kinds of *shuihua* (水花 water flowers), and 6 kinds of *manhua* (蔓花 tendril flowers). It is not difficult to imagine the grand occasion of flower cultivation at that time. Each book also describes the cultivation methods of flowers as well as flower varieties. Sometimes flower cultivation is also recorded as a custom.

Luo Yang Mu Dan Ji (《洛阳牡丹记》) by Ouyang Xiu (欧阳修) in the Song Dynasty is the earliest extant *pulu* work of the peony. Luoyang is famous for its peony cultivation. Local people say that all other flowers have names, but peony is just called flower. Ouyang Xiu lived in Luoyang for 4 years. The book is divided into three parts. The first part describes the names and characteristics of 24 kinds of top-quality peony such as "yaohuang" (姚黄) and "weihua" (魏花). The second part talks about the origin of these names, including the place of origin, color, and surname of the cultivator. The third part focuses on the description of the love of flowers of the Luoyang people and also introduces the methods of grafting excellent varieties of peony and the key points of planting peony. To grow flowers, one must first select the right place, remove the old soil, and plant in new soil mixed with 500 grams of the powdered drug *bailian* (白敛), because peony roots are often eaten by insects but *bailian* can kill them. There are also certain rules for watering flowers, which must be before sunrise or after sunset, and the amount of water must be controlled according to seasonal changes. When buds appear on peony flowers, cut off the small ones and leave only one or two large ones. If there are insect pests, insect holes must be found and insects be killed with sulfur so that the flowers can grow well.

After Ouyang Xiu, a series of works on peony appeared one after another. Among them are Zhang Bangji (张邦基)'s *Chen Zhou Mu Dan Ji* (《陈州牡丹记》) and Lu You (陆游)'s *Tian Peng Mu Dan Pu* (《天彭牡丹谱》). Both Chen Zhou and Tianpeng (in Chengdu) are famous for peony. Lu You went to Tianpeng to admire the flowers while he was an official in Shu and wrote a volume of *Mu Dan Pu* (《牡丹谱》). His book is divided into three chapters as Ouyang Xiu did in his book *Luo Yang Mu Dan Ji*. The first chapter evaluates and ranks peony according to color. The second records the names of the flowers, excluding all that can be found in Ouyang Xiu's book but those famous in Tianpeng, and describes the form of the flowers. The third chapter is "Feng Su Ji" (风俗记 Custom Record), which contains anecdotes and stories about people enjoying flowers in Shu.

Just as people from Luoyang like peony, people from Yangzhou like *shaoyao* (芍药). The *Shaoyao* in Yangzhou has a reputation as the best in the world. The earliest *pulu* work on *shaoyao* was written by Liu Ban (刘攽), a native of Qingjiang, Jiangxi (now Zhangshu City). He came to Yangzhou in Guangling, the sixth year of Xining in the Song Dynasty (1073). Since it was April and the *shaoyao* were blooming, he invited friends to appreciate the flowers. He wrote the book and asked painters to draw all kinds of *shaoyao*. It records 31 species of *shaoyao* in Yangzhou, rated into 7 grades. In the third year after he wrote the book, or the eighth year of Xining (1075), Wang Guan (王观), a Rugao native, when working in Jiangdu County, Yangzhou, wrote another book entitled *Yang Zhou Shao Yao Pu* (《扬州芍药谱》) on the basis of Liu's book, describing 39 kinds of *shaoyao*, of which 31 were originally contained in Liu's book and 8 were newly added. Other authors of the same period include Kong Wuzhong (孔武仲), who also wrote 1 volume of *Shao Yao Pu* (《芍药谱》). The full text is collected in *Neng Gai Zhai Man Lu* (《能改斋漫录》) by Wu Zeng (吴曾) in the Song Dynasty. It records 33 species of *shaoyao*, all named after the shape of flowers.

A *pulu* work on chrysanthemum was first written by Liu Meng (刘蒙), a Pengcheng (now Xuzhou City, Jiangsu Province) native. In September, 1104, he went to Longmen and visited Liu Yuansun (刘元孙), a hermit who planted chrysanthemum and recited poems, on the bank of River Yishui. He then discussed varieties and cultivation techniques of chrysanthemum with him. According to the color classification of chrysanthemum, Liu Meng focused on describing their flower shape and leaf appearance, and compiled the book. After Liu Meng, Shi Zhengzhi (史正志), Fan Chengda (范成大), Shen Jing (沈竞), Hu Rong (胡融), Ma Ji (马楫), and others also wrote *pulu* books on chrysanthemum. The most comprehensive work is the six volumes of *Bai Ju Ji Pu* (《百菊集谱》) by Shi Zhu (史铸). More than 100 varieties of chrysanthemum are listed. While creating the new catalogue, special catalogues by the abovementioned authors are also collected. Besides this, it includes the chrysanthemum names contained in Zhou Shihou (周师厚)'s *Luo Yang Hua Mu Ji* (《洛阳花木记》) as well as related arts, stories, miscellaneous sayings, poems, chapters, songs, etc. Some of the *pulu* works on chrysanthemum in the Song Dynasty, such as Shen Jing's *Ju Ming Pian* (《菊名篇》) and Hu Rong's *Tu Xing Ju Pu* (《图形菊谱》), have been preserved to this day precisely because of *Bai Ju Ji Pu*.

The first Orchid work was *Jin Zhang Lan Pu* (《金漳兰谱》) by Zhao Shigeng (赵时庚). The author wrote the preface in 1233, and so the book might also have been written that year. The book has three volumes and is divided into five chapters. The cultivation methods recorded in the book vary from variety to variety and are very detailed. Although his book is the earliest orchid book, it is not the best. The one wrote by Wang Jinshu (王进叔), who is from Linjiang (now Zhangshu City, Jiangxi Province), is the best. He wrote the book in 1247, more than 10 years later than Zhao's. The book is divided into six parts.

Peony flourished in the Tang Dynasty and plum blossom had a tendency to replace it in the Song Dynasty. Fan Chengda wrote another flower monograph, *Fan Cun Mei Pu* (《范村梅谱》), in addition to *Ju Pu*. It is about the 12 kinds of plum blossom in Fan's private garden. This is the first book about plum blossom. The value of plum lies in eating and appreciating. Color, fragrance, taste, and appearance have become the aesthetic standards of the plum blossom.

Quan Fang Bei Zu (《全芳备祖》) is the epitome of *pulu* works on flower in the Song Dynasty. The author is Chen Yong (陈泳), a native of Tiantai (now Tiantai

County, Zhejiang Province). The book was written in his thirties, so it claimed to be "the book of a young man." It was printed from about 1253 to 1256.

Litchi and citrus are the main fruit trees in tropical and subtropical regions. *Li Zhi Pu* (《荔枝谱》) usually refers to the book written by Cai Xiang (蔡襄). In fact, several monographs on litchi appeared in the Tang and Song Dynasties, such as *Guang Zhong Li Zhi Pu* (《广中荔枝谱》), *Zeng Cheng Li Zhi Pu* (《增城荔枝谱》), and *Pu Tian Li Zhi Pu* (《莆田荔枝谱》). But all of them are now lost. Cai Xiang, a native of Xianyou, Xinghua Prefecture (now Xianyou County, Fujian Province), wrote the book in the fourth year of Jiayou in the Song Dynasty (1059) when he served as the prefect of Quanzhou. The book is divided into seven chapters: the first one mainly talks about the history and distribution of litchi and its biological characteristics such as "its vulnerability to frigidity and intolerance to transplantation." The second chapter focuses on the characteristics of the famous litchi product "chenzi" (陈紫), and uses it as the standard to classify all kinds of litchi into upper, middle, and lower grades according to their quality. The third chapter describes the production and marketing of litchi in Fujian. At that time, exports were already made by sea as far away as Japan, Arabia, and other places, and production and sales were booming. The fourth chapter talks about physiological characteristics of litchi and the effect of eating it. The fifth is about the cultivation and management of litchi. The sixth chapter discusses its processing methods. According to the characteristics that fresh litchi fruit deteriorate easily and rots after being picked, three processing methods, namely, *hongyan* (红盐 salted), *baishai* (白晒 dried in the sun), and *mijian* (蜜煎 preserved with honey), were put forward. The seventh chapter records 32 litchi varieties. This book is both the earliest extant litchi monograph and the earliest extant fruit tree cultivation monograph in China.

The earliest extant monograph on citrus is *Yong Jia Ju Lu* (《永嘉橘录》) by Han Yanzhi (韩彦直) in the Southern Song Dynasty. About 100 years before this, Chen Shunyu (陈舜俞) of the Northern Song Dynasty wrote a poem "Shan Zhong Yong Ju Chang Yong" (山中咏橘长咏) and introduced in detail, by adding notes to the poem, the citrus production and marketing in his hometown of Mount Dongting on Taihu Lake. The contents include the geographical environment, yield and quality, harvest date, seedling technology, seed source, variety classification, orange orchard plantation, pest control, pruning and cultivation, irrigation and fertilization, folk customs, storage and processing, transportation and distribution, and so on. Some of the contents are similar to those of *Yong Jia Ju Lu*, but its historical influence is far less.

Yong Jia Ju Lu is divided into three volumes and one "Preface" by the author. The upper and middle volumes mainly record the citrus varieties (including some species) in Wenzhou at that time, with a total of 27 varieties or species. Among them, there are 8 kinds of *gan* (柑), 14 kinds of *ju* (橘), and 5 kinds of *cheng* (橙). It also introduces the basis for naming each variety and its specific area. The last volume is divided into 9 sections: seed treatment, initial planting, cultivation, disease removal, watering, picking, collection, treatment, and medicine application. It summarizes the experience of local orange farmers in great detail. It is the earliest monograph on citrus in China and the world's first complete work on citrus cultivation, having a far-reaching influence in the world.

The *pulu* books on vegetables mainly include *Sun Pu* (《笋谱》) and *Jun Pu* (《菌谱》). Besides these, Dai Kaizhi (戴凯之), a native of Jin, mentioned various uses of bamboo in his book *Zhu Pu* (《竹谱》). At the beginning of the Song Dynasty, the monk Zanning (赞宁) wrote *Sun Pu*. He imitates the writing style of *Cha Jing* and divides the book into 5 sections: name (10), origin (98), eating (13), anecdotes (60), and miscellaneous theories (8). Among them the "name" section lists not only the alias of bamboo shoots but also the cultivation methods.

The author of *Jun Pu* is Chen Renyu (陈仁玉) from Xianju, Taizhou, in the Southern Song Dynasty. The book records 11 kinds of *jun* (菌) which are special local products. Its purpose is to "make the best use of them." The book describes the origin, collection period, shape, color, and taste of them.

The *pulu* on hobbies include *Tang Shuang Pu* (《糖霜谱》). "*Tangshuang*" (糖霜) is sugar processed from sugar cane. In the Song Dynasty, "sugar cane is planted everywhere, but *tangbing* (糖冰) exists only in Futang, Siming, Panyu, Guanghan and Suining, while Suining has the best" (Hong Mai: *Rong Zhai Wu Bi* (《容斋五笔》), Volume 6). The author of *Tang Shuang Pu*, which records sugar cane planting and sugar frost production, is Wang Zhuo (王灼) from Suining. The book was written around the 24th year of Shaoxing (1154). The chapter "First" describes the origin of frosting. "Second" narrates the history of sugar cane consumption and celebrity allusions. "Third" is about sugarcane planting, including variety, seed storage, soil preparation, sowing, fertilization, weeding, intertillage, ridging and harvesting, sugarcane field planting systems, the distribution of frost processing areas, etc. "Fourth" discusses the basic process of sugar making equipment and sugar cane processing. "Fifth" focuses on the disposal and collection of sugar water after it is put into the urn. "Sixth" is about the economic chain reaction caused by the manufacture of *tangshuang* (糖霜). "Seventh" talks about the function and possible side effects of finished icing products. The book is the earliest extant monograph on planting sugar cane and sugar production.

Si Mu An Ji Ji and *Can Shu* are specialized agricultural books that appeared in the Sui, Tang, Song, and Yuan Dynasties and are preserved to this day and related to animals.

It is generally believed that the author of *Si Mu An Ji Ji*, Li Shi (李石), was from the Tang Dynasty. However, the existing *Si Mu An Ji Ji* has been supplemented in the process of repeated publication after its completion. The length and content of the book are very different. Although the book is mainly based on veterinary prescriptions, the rich experience of horse-judging compiled in the first volume also partly reflects the progress of livestock topography in this period. It is the oldest comprehensive veterinary work in existence in China. From the Tang Dynasty to the Ming Dynasty, it was always a must-read book for veterinarians. Yu Benyuan (喻本元) and Yu Benheng (喻本亨), veterinarians from the Ming Dynasty, compiled the *Yuan Heng Liao Ma Ji* (《元亨疗马集》) on the basis of reference and absorption of ancient books such as *Si Mu An Ji Ji*.

The author of *Can Shu* is Qin Guan (秦观), who wrote it in or before the 7th year of Yuanfeng of the Song Dynasty (1084) (Huang Shirui: A Study of Qin Guan's *Can Shu*, *Nong Shi Yan Jiu* (《农史研究》), Volume 5, 1985, pp. 251–252). While visiting

Yanzhou, Shandong province, he found that sericulture technology there was different from that of Wuzhong, so he made records. The book has a total of more than 1000 words, with concise and practical records of every stage from *yuluan* to silk reeling. Its main achievements include quantitative descriptions of silkworm physiology, a systematic record of sericulture technology and improvements of reeling machines (Wei Dong: On Qin Guan's *Can Shu*, *Zhong Guo Nong Shi* (《中国农史》), 1987, Volume 1). This book is the earliest special book on sericulture.

China's agricultural policies traditionally take solving the problem of food and clothing as its own duty. Therefore, preparing for and rescuing people from famine and even agrometeorology became one of the contents of China's traditional agronomy. The previous *Fan Sheng Zhi Shu* and *Qi Min Yao Shu* contained related content. However, as a monograph to discuss the issue of famine preparedness and relief, Dong Wei (董煟)'s book *Jiu Huang Huo Min Shu* (《救荒活民书》) from the Southern Song Dynasty is the first one. Dong Wei was born in Poyang (now Poyang County, Jiangxi Province). He was a Metropolitan Graduate in the 5th year of Shaoxi of the Southern Song Dynasty (1196) and served as a District Magistrate in Rui'an, Zhejiang Province. The book is divided into three volumes. The first volume is the "selection of ancient materials as reference for today." It is a selection of documents and materials on famine management and famine relief from ancient times to the ninth year of Chunxi in the Southern Song Dynasty (1182). It discusses the previous arguments and practices and puts forward its own opinions and propositions. For example, taking Emperor Yang of the Sui Dynasty as an example, he points out the consequences of tax collection without distribution. The middle volume puts forward some specific and auxiliary measures to relieve the famine. The last volume records the words and deeds of different authors in the Song Dynasty regarding famine policy. There is also 1 volume of *Shi Yi* (《拾遗》), which is close to volumes 1 and 2 in writing style, i.e., while quoting some documents, it includes some important documents from history.

Tian Jia Wu Xing (《田家五行》) is a book published at the end of the Yuan Dynasty and the beginning of the Ming Dynasty, which collects folk experiences on agrometeorology and climate change. Before the book was officially written, the relevant knowledge was already widely popular among the people south of the Yangtze River. The book is divided into three volumes, each further divided into several categories. The first volume, from the "first month of lunar calendar" to "the twelfth month," records methods of weather forecasting in daily order. The middle volume deals with astronomy, geography, vegetation, birds and animals, scaly fish and other categories, most of which are phenological in nature. The last volume is about three *xun* (旬), *liujia* (六甲), climate, *juanji* (涓吉), *xiangfu* (祥符), etc. In the book, there are more than 140 proverbs that use astronomical phenomena and phenology to predict the weather, more than 100 proverbs on medium and long-term forecasts, and nearly 40 proverbs on agrometeorology. These proverbs reveal some laws of weather and climate change from different aspects and were widely circulated after tests of time and practice. Other works similar to *Tian Jia Wu Xing* include 12 volumes of *Tian Jia Li* (《田家历》), which are contained in the category of *wuxing* (五行) in *Sui Shu Jing Ji Zhi* and are lost. *Shuo Fu* (《说郛》) contains one volume of *Xiang Yu Shu* (《相雨书》), written by Huang Zifa (黄子发) in the Tang Dynasty. There are only 11 items in the book, which are all about the situation of clouds before precipitation, and also the experience of farmers.

3.5 Agricultural Books in the Ming and Qing Dynasties

3.5.1 Overview

The Ming and Qing Dynasties were the period with the most publications of agricultural books. Wang Yuhu's *Zhong Guo Nong Xue Shu Lu* contains 541 agricultural books from all dynasties, of which 329 were written in the Ming and Qing Dynasties (128 and 201 in the Ming and Qing Dynasties, respectively), accounting for about 60% of the total. The agricultural books written and published in more than 500 years in the Ming and Qing Dynasties were more than those published in the previous 2000 years combined. According to statistics by Wang Da (王达), there are more than 830 agricultural books from the Ming and Qing Dynasties, including more than 500 that were not included in *Zhong Guo Nong Xue Shu Lu*. There are 204 agricultural books on sericulture (Wang Da: Collection and Introduction of Sericultural Books during the Ming and Qing Dynasties, *Zhong Guo Nong Shi*, 1986, Volume 4; 1989, Bound Volume). In addition, 698 books were collected from *Zhong Guo Gu Nong Shu Cun Mu* (《中国古农书存目》) attached to *Zhong Guo Nong Ye Bai Ke Quan Shu Nong Ye Li Shi Juan* (《中国农业百科全书·农业历史卷》), 601 of which belong to the Ming and Qing Dynasties, accounting for 86% of the total number of existing agricultural books. Among the numerous agricultural books, there are not only national, integrated or comprehensive agricultural books, but also local and professional ones. There are both official works and private writings.

Comprehensive or integrated farming books include agricultural calendars such as *Nong Pu Bian Lan* (《农圃便览》) by Ding Yizeng (丁宜曾) and *Yang Yu Yue Ling* (《养余月令》) by Dai Xi (戴羲). There are also comprehensive agricultural books such as *Nong Zheng Quan Shu* (《农政全书》) and *Shou Shi Tong Kao* (《授时通考》) as well as general agricultural books such as *Bian Min Tu Zuan* (《便民图纂》).

The most striking are various local agricultural books, such as *Shen Shi Nong Shu* (《沈氏农书》) (written in 1640 or a little before) by a person with the surname of Shen in the late Ming and early Qing Dynasties and *Bu Nong Shu* (《补农书》) (1658) by Zhang Luxiang (张履祥) of Tongxiang, both of which reflect agricultural production in the Jiahu region. Books reflecting the agriculture in the Guanzhong region include *Bin Feng Guang Yi* (《豳风广义》) and *Zhi Ben Ti Gang Nong Ze* (《知本提纲·农则》) by Yang Shen (杨屾) and *Nong Yan Zhu Shi* (《农言著实》) by Yang Xiuyuan (Yichen) (杨秀元(一臣)) (1856). Works reflecting agriculture in the Shandong area are *Nong Pu Bian Lan* by Ding Yizeng (1755) and *Nong Sang Jing* (《农桑经》) by Pu Songling (蒲松龄). *San Nong Ji* (《三农纪》) (1760) written by Zhang Zongfa (张宗法) is about agriculture in Sichuan. Qi Junzao (祁寯藻)'s *Ma Shou Nong Yan* (《马首农言》) (1836) is about agriculture in Shanxi. Liu Yingtang (刘应

棠)'s *Suo Shan Nong Pu* (《梭山农谱》) (1674) and He Gangde (何刚德)'s *Fu Jun Nong Chan Kao Lue* (《抚郡农产考略》) (1903) are about agriculture in Jiangxi. Jiang Gao (姜皋)'s *Pu Mao Nong Zi* (《浦泖农咨》) (1834) is about agriculture in Shanghai.

The largest quantity is professional agricultural books. The writing style of "*pulu*" formed in the Tang and Song Dynasties continued to flourish in the Ming and Qing Dynasties. In the Ming Dynasty, Huang Shengzeng (黄省曾) (in Wuxian, now Suzhou) alone was the author of one volume of *Dao Pin* (《稻品》), one volume of *Can Jing* (《蚕经》), one volume of *Yang Yu Jing* (《养鱼经》), and one volume of *Yi Ju Shu* (《艺菊书》). These four books are collectively called "Nong Pu Si Shu" (农圃四书 Four Books of Agriculture and Gardening). In addition, there is one volume of *Yu Jing* (《芋经》) (also called *Zhong Yu Fa* 《种芋法》) and one volume of *Shou Jing* (《兽经》). These are all *pulu* books. *Zhong Guo Nong Xue Shu Lu* stated that there were 36 *pulu* works on chrysanthemum, 15 on orchid, and 10 on peony as well as 12 fruit monographs in the Qing Dynasty, of which 9 were on litchi. There are more tea books. According to statistics 188 tea books have been initially confirmed from the Tang Dynasty to the Qing Dynasty, including 96 complete ones, 28 collected books, and 64 lost books. In terms of dynasties, there are 16 from the Tang and Five Dynasties, 47 from the Song and Yuan Dynasties, 79 from the Ming Dynasty, 42 from the Qing Dynasty, and 4 from the undetermined Ming and Qing Dynasties (Zhang Chuanzheng, Zhu Zizhen, Li Xinghui: Tea Books in the Ming and Qing Dynasties and Their Historical Value, *Gu Jin Nong Ye* (《古今农业》), 2006, Volume 3). In terms of field crops, Huang Shengzeng 's *Li Sheng Yu Jing Dao Pin* (《理生玉镜稻品》) (also known as *Dao Pin*) and Li Yanzhang (李彦章)'s *Jiang Nan Cui Geng Ke Dao Bian* (《江南催耕课稻编》) (1834) are two examples. During the Ming and Qing Dynasties, some cultivated crops and agricultural organisms were first added as writing subjects into the catalogue of agricultural books, such as rose and lotus to ornamental plants, peach, plum, and *longyan* (龙眼) to fruit trees, sweet potato, and cotton to field crops, turnip, and taro to vegetables. During the Ming and Qing Dynasties, a variety of wild vegetables were added to *pulu*.

Other professional agricultural books include those on locust control. The locust is one of the most important pests in Chinese agriculture. Records of locust damage and locust control have been published ever since the fifth year of Lu Huangong in *Chun Qiu* (《春秋》) (707 BC). However, the discussion of locusts as a special issue was first found in the book *Zhi Huang Shu* (《治蝗疏》) collected by Xu Guangqi (徐光启)'s *Nong Zheng Quan Shu* (《农政全书》). But almost all the monographs on locust capture were written and published in the Qing Dynasty, including *Bu Huang Kao* (《捕蝗考》) by Chen Fangsheng (陈芳生) (about 1684), *Bu Huang Hui Bian* (《捕蝗汇编》) by Chen Jin (陈仅) (about 1837–1845), *Zhi Huang Quan Fa* (《治蝗全法》) by Gu Yan (顾彦) (1857), *Liu Yun Ge Bu Huag Ji* (《留云阁捕蝗记》) by Peng Shoushan (彭寿山) (1836), *Bu Huang Yao Shuo* (《捕蝗要说》) (or *Bu Huang Yao Jue* 《捕蝗要诀》)) by an unknown author (1856), *Chu Nan Ba Yao* (《除蝻八要》) by Li Xingfu (李惺甫) (1850), and *Zhi Huang Shu* (《治蝗书》) by Chen Chongdi (陈崇砥) (1874).

The most distinctive professional agricultural books are those on *quzhong* (*qutian*) that appeared during this period. However, the most professional agricultural books in the Ming and Qing Dynasties were about sericulture. About 90% of the sericulture works in the Ming and Qing Dynasties appeared in the late Qing Dynasty. The purpose of compiling and publishing such books was to teach farmers from different areas how to plant mulberry and raise silkworms. The contents are mostly based on the advanced experience of the western Zhejiang region, and then amended according to the actual situation in various places. The books mainly include *Yang Can Cheng Fa* (《养蚕成法》) by Han Mengzhou (韩梦周) (1766), *Can Sang Shuo* (《蚕桑说》) by Shen Lian (沈练) (1840), *Can Sang Ji Yao* (《蚕桑辑要》) by Shen Bingcheng (沈秉成) (1871), *Hu Can Shu* (《湖蚕述》) by Wang Rizhen (汪日祯) (1874), *Bi Nong Zui Yao* (《裨农最要》) by Chen Kaizhi (陈开沚) (1897), and *Can Sang Cui Bian* (《蚕桑萃编》) by Wei Jie (卫杰) (1892).

As a professional agricultural book, animal husbandry and veterinary works include *Yuan Heng Liao Ma Ji* (1608), co-authored by the Yu Ren and Yu Jie brothers, which is a summative classic of Chinese veterinary medicine. There are also *Yang Geng Ji* (《养耕集》), *Bao Du Ji* (《抱犊集》), *Xiang Niu Xin Jing Yao Lan* (《相牛心经要览》), *Zhu Jing Da Quan* (《猪经大全》), and so on. The Qing Dynasty also saw the only monograph on beekeeping in Chinese history, *Feng Ya Xiao Ji* (《蜂衙小记》) by Hao Yixing (郝懿行).

During the Ming and Qing Dynasties, there were many works on Pentsaology, influential ones of which include *Jiu Huang Ben Cao* (《救荒本草》), *Ben Cao Gang Mu* (《本草纲目》), *Qun Fang Pu* (《群芳谱》) and *Guang Qun Fang Pu* (《广群芳谱》), *Huang Jing* (《花镜》), etc. Among traditional botanical books, *Zhi Wu Ming Shi Tu Kao* (《植物名实图考》), written in the latter part of the Qing Dynasty, is a book of the highest level and also the last one. These works mainly focus on botany and are also related to farming.

During the years of Ganlong and Jiaqing in the Qing Dynasty, there was a strong trend of textual research. The discussion of "*nongdao*"(农道 agronomy theories) based on Neo-Confucianism and books about the naming of plants by the use of exegetical methods also appeared one after another. *Jiu Gu Kao* (《九谷考》) (1803) by Cheng Yaotian (程瑶田) (1725~1814) examines nine kinds of grain crops, including *liang* (粱), *su* (黍), *ji* (稷), rice, wheat, soybean, adzuki bean, hemp, etc. It is attached to four papers analyzing the names of grains, and two letters discussing crops with friends. *Bao Xun* (《宝训》) (written in 1790) by Hao Yixing (1755~1823), compiled from excerpts of ancient books, is a collection of folk proverbs based on agricultural language and various books. The book consists of eight volumes, which contain miscellaneous sayings, crops, sericulture, vegetables, fruits, timbers, herbs, animals, etc. In addition, there is 1 volume of *Ji Hai Cuo* (《记海错》) (1867) in *Hao Shi Yi Shu* (《郝氏遗书》), which records 49 species of marine products. One volume of *Feng Ya Xiao Ji* describes the morphology, ecology, habits of bees, and methods of honey collection, which is the only ancient monograph on bee keeping as previously mentioned. *Shi Gu* (《释谷》) (1840) by Liu Baonan (刘宝楠) (1791~1855) is a further study based on *Jiu Gu Kao*. The book consists of four volumes, with a detailed analysis of wheat, beans, and hemp. *Zhi Wu Ming Shi Tu Kao* (1848) by Wu Qixun (吴其濬) (1789~1847) has 38 volumes, collecting 1714 kinds of plants, more than 1800 pictures and 700,000 words, and is regarded as a monumental work in Chinese botany. The book also contains many achievements in agronomy. For example, 53 kinds of edible plants, 176 kinds of vegetables, and 156 kinds of fruit trees are incorporated into cereals, and more than 100 kinds of medicinal plants have been added to the ancient herbal medicines. The plants contained in the book have detailed records on their morphological characteristics, environment of origin, and various uses. The book has made special textual research on different plants with the same name and the plant with different names, which shows practical characteristics in classification.

3.5.2 Official Agricultural Books

After the formation of the official agronomy tradition in the Tang and Song Dynasties, it further developed in the Ming and Qing Dynasties, especially in the Qing Dynasty, when there were many agricultural books with official backgrounds. One is the traditional ploughing and weaving paintings, cotton paintings, as well as silkworm and mulberry paintings inherited from *Geng Zhi Tu* in the Song Dynasty. The second is a large-scale agricultural encyclopedia *Shou Shi Tong Kao* (《授时通考》) compiled by the government. The third are agriculture books written by local officials for the purpose of encouraging farming.

Geng Zhi Tu focuses on planting rice, mulberry, and rearing silkworms in the south, which obviously does not conform to the reality that cotton has become the material for the masses in the north and south since the Song and Yuan Dynasties. Moreover, *Geng Zhi Tu* itself has a trend of specialization. So, in the 30th year of Ganlong of the Qing Dynasty (1765), Fang Guancheng (方观承), Governor-general of Metropolitan Area, presided over the drawing of a set of *Mian Hua Tu* (《棉花图》 Cotton Paintings). There are 16 pictures in total, including paintings depicting the whole production process from cotton planting to dyeing and weaving. Each painting has a written explanation. After the completion, Fang Guancheng presented it to the imperial court. Emperor Ganlong liked it very much. He specially wrote a poem on each painting. It was combined with Emperor Kangxi's *Mu Mian Fu* (《木棉赋》) and made into *Yu Zhi Mian Hua Tu* (《御制棉花图》). Later, Fang Guancheng carved *Yu Zhi Mian Hua Tu* on stones, whose rubbings were very popular. In the 13th year of Jiaqing's reign (1808), the office of the Palace Treasury also issued a woodcut block-printed *Mian Hua Tu* and renamed it *Shou Yi Guang Xun* (《授衣广训》). The content is mainly taken from *Mian Hua Tu* by Fang Guancheng. In addition, it also contains Emperor Kangxi's *Mu Mian Fu* and poems by Emperors Ganlong and Jiaqing. In fact, it is just another version of *Mian Hua Tu* rather than a new one.

In the 15th year of Guangxu's reign (1889), the woodcut picture album *Sang Zhi Tu* (《桑织图》) was published by the Silkworm and Mulberry Bureau of Sanyuan County. There are 24 original pictures, mainly including three parts: mulberry planting, silkworm rearing, and weaving.

In the 16th year of Guangxu's reign, Zong Chenglie (宗承烈), a Qiantang native, wrote *Cang Sang Tu Shuo* (《蚕桑图说》), which was based on the book *Cang Sang Shuo Lue* (《蚕桑说略》) written by Zong Jingfan (宗景藩), a District Magistrate of

Puyin County in Hubei Province during the years of Tongzhi. He asked Wu Jiayou (吴嘉猷), a famous painter at that time, to draw pictures as illustrations for the book. There are 5 mulberry pictures and 10 silkworm pictures. At the top of each picture, there are more detailed written instructions to introduce people's experience in mulberry and silkworm rearing in Zhejiang.

In the second year of Ganlong of the Qing Dynasty (1737), E Ertai (鄂尔泰) and Zhang Tingyu (张廷玉), in accordance with the emperor's will, led more than 40 grand ministers to compile *Shou Shi Tong Kao*, a large-scale agricultural book. It collected the previous records on farm work and was finished 5 years later, in the seventh year of Ganlong (1742). The 78-volume book, with 980,000 words, is 15 times longer than *Nong Sang Ji Yao* from the Yuan Dynasty. The content centered around field production and takes into account all trades of forestry, animal husbandry, sidelines, and fishing. It consists of eight parts, namely, Tian Shi (天时 timing), Tu Yi (土宜 soil suitability), Gu Zhong (谷种 rice seed), Gong Zuo (功作 cultivation), Quan Ke (劝课 encouragement), Xu Ju (蓄聚 accumulation), Nong Yu (农余 surplus farming), and Can Sang (蚕桑 sericulture). The beginning of each part collects and researches relevant documents from past dynasties. Then the relevant production experience, imperial edicts, policies, etc. from previous literature are cited.

"Tian Shi" includes six volumes, including spring, summer, autumn, and winter, which are divided into four seasons of farming activities. "Tu Yi" includes the contents of identification, material and soil, land system, land mapping theory, water conservancy, and so on, totaling 12 volumes. "Gu Zhong" includes the source and names of food crop varieties (rice, *ji*, *shu*, *su*, wheat, beans, and hemp), among which the collection of rice variety resources from different regions is the most detailed, but cultivation techniques are not described.

The most technical part of the book is about the cultivation process of crops, which is divided into eight parts, namely, cultivation, *palao* (耙耢 harrowing), sowing, fertilization, *yunzi* (耘耔 weeding), irrigation, harvest, storage and processing, and so on, which are described in eight volumes. Attached to the "irrigation volume" is 1 volume of "Tai Xi Shui Fa" (泰西水法), which introduces the western irrigation tools introduced at that time. Finally, there is also one volume on animal husbandry, describing the raising of farm animals.

"Quan Ke" includes 12 volumes of *Zhao Ling* (诏令 imperial orders), *Zhang Zou* (章奏 memorials), *Guan Si* (官司), *Qi Bao* (祈报), *Chi Yu* (敕谕 imperial edicts), *Qi Gu* (祈谷), imperial poems (2 volumes), and ploughing and weaving maps (2 volumes), of which the ploughing and weaving maps are more valuable. "Xu Ju" has four volumes, specifically containing "Ping Cang" (平仓), "She Cang" (社仓), "Yi Cang" (义仓), and related schemata, describing the system and decrees of accumulating grain and preparing for famine. "Nong Yu," the largest part with 14 volumes and complex content, includes four volumes on vegetables, four volumes on fruit, two volumes on timber, one volume on various plants, and two volumes on livestock. "Cang Sang" has seven volumes, the first five of which deal with silkworm rearing, *fenbo* (分箔), cocooning, cocoon selection, silk reeling, weaving, and dyeing as well as mulberry administration. The second two volumes of "Sang Yu" describe cotton planting and other fiber crops.

Shou Shi Tong Kao is the last large-scale integrated agricultural book in ancient China. As the compilers are a group of "imperial officials" who do not understand agricultural production, their understanding of agriculture is far less than the authors of *Nong Sang Ji Yao* in the Yuan Dynasty. Therefore, this book is not innovative except for its length, but its collected literature is extremely detailed, and it is the one with the biggest collection of literature. It quotes 3575 documents from 553 ancient books (Another view is there are 427 kinds. Shi Shenghan: *Zhong Guo Gu Nong Shu Ping Jie* (《中国古农书评介》), Agricultural Publishing House, 1980, p. 77). There are 521 illustrations in the book. As the last large agricultural book to collect agricultural literature, its position in philology is undeniable. What is more important is that the book has a relatively strict system for the arrangement of agronomy contents. Compared with previous large-scale agricultural books, it shows the perspectives of the Qing government on agriculture.

Shou Shi Tong Kao has a rich content with a wide range of materials, but its core is still "agriculture and sericulture." "Gu Zhong" and "Gong Zuo" mainly refer to grain crops and their cultivation techniques. Among them, the collection of rice varieties from all over the country is the most detailed, with a total of 3429 (including duplicates) counted. These varieties come from the local records of 233 prefectures, departments, and counties in 16 provinces in the Qing Dynasty. In contrast, contents about vegetables, fruit trees, timber, and animal husbandry were reduced relatively. Food processing and cooking, as well as commercial activities of buying at low prices while selling at high prices, are all deleted. Major adjustments were made to the sections about water conservancy, wasteland reclamation, and wasteland management, which are the most distinctive in *Nong Sang Ji Yao*. Water conservancy is attached to "Tu Yi" as a subsidiary. *Tai Xi Shui Fa* is about irrigation and so is included in "Gong Zuo." The cultivation of wasteland and the preparation for famine are totally excluded. The contents related to plants for relieving famine were also deleted or omitted.

Shou Shi Tong Kao is another large-scale official agricultural book after the Yuan Dynasty's *Nong Sang Ji Yao*. It is another important agricultural book after *Nong Zheng Quan Shu*. Although the evaluation of later generations on *Shou Shi Tong Kao* is not as high as that of *Nong Sang Ji Yao* and *Nong Zheng Quan Shu*, its influence has been growing. Since it was an official book compiled by order of the emperor, which was re-engraved in the provinces, it was widely circulated. In the early years, western scholars often came into contact with it, so the international reputation was very high (Shenghan 1980). In 1837, Stanislas Julien (1797~1873), a Frenchman, translated volumes 72 to 76 of *Shou Shi Tong Kao, Can Shi,* into French, and included the sericulture part of *Tian Gong Kai Wu Nai Fu* (《天工开物·乃服》). The book was entitled "Resume des principaux traites Chinois sur la culture des muries et l'education des vers a soie" ("Summary of Main Chinese Works on Mulberry Planting and Silkworm Raising"), with 224 pages in total and another 24 pages of introduction. On the title page the Chinese title *Can Sang Ji Yao* was printed. It was then translated into Italian (1837), German (1838), English (1838), and Russian (1840) and other European languages and was widely circulated.

During the Ming and Qing Dynasties, there were also agricultural books written by local officials, whose nature was similar to that of the articles of farming encouragement in the Song Dynasty. However, they were not notices but real books. Among them, the famous ones are *Bian Min Tu Zuan*, *Bao Di Quan Nong Shu* (《宝坻劝农书》), *Jiang Nan Cui Geng Ke Dao Pian*, and *Ying Tian Ji Yao* (《营田辑要》).

Bian Min Tu Zuan was carved by Kuang Fan (邝璠) (1465~1505) from Renqiu County, Hebei Province, during his tenure in Wuxian County, Jiangsu Province. The book compiles some information from past agricultural books and also collects experiences in the region of Taihu Lake. The book consists of 16 volumes. Volume 1, *Nong Wu Zhi Tu* (《农务之图》), depicts the whole process from rice sowing to harvest in 15 paintings, while Volume 2, *Nu Gong Zhi Tu* (《女红之图》), depicts 16 images of silkworm, textile, and garment production. These two volumes of paintings were re-engraved based on Lou Shu's *Geng Zhi Tu* in the Southern Song Dynasty. In the book, the original ancient poems are replaced by songs from the Wu area, which were easy to understand for the people of Jiangsu and Zhejiang. The 14 volumes from volume 3 to 16 are text parts. Volume 3, *Geng Huo Lei* (《耕获类》), introduces the cultivation, processing, and collection techniques of grain, oil, and fiber crops with an emphasis on rice. Volume 4, *Sang Can Lei* (《桑蚕类》), introduces the techniques of mulberry planting and silkworm rearing. Volumes 5 and 6 are *Shu Yi Lei* (《树艺类》), recording the practical experience of growing fruit trees, flowers, and vegetables. Volume 7, *Za Zhan Lei* (《杂占类》), contains mostly agricultural proverbs for weather prediction, mainly from *Tian Jia Wu Xing*. Volume 14, *Mu Yang Lei* (《牧养类》), describes the identification, breeding, and disease prevention of livestock and poultry. Volume 15 and volume 16 are *Zhi Zao Lei* (《制造类》), which are recorded from *Duo Neng Bi Shi* (《多能鄙事》). As for volumes 12 and 13, which deal with medicine and hygiene, most of the prescriptions contained are taken from the medical books of the Song, Yuan, and Ming Dynasties. There is also Volume 8, *Yue Zhan Lei* (《月占类》), Volume 9, *Qi Rang Lei* (《祈禳类》), and Volume 10, *Juan Ji Lei* (《涓吉类》). The greatest contribution of the book is its summary of rice farming techniques in the Wu area.

Bian Min Tu Zuan was written by a northerner when he was serving as an official in the south, while *Bao Di Quan Nong Shu* was written by a southerner when serving as an official in the north. The author is Yuan Huang (袁黄), a native of Wujiang, Jiangsu Province, and a Metropolitan Graduate in the year of Bingchen in Ming Emperor Wanli's reign (1585). In the 19th year of Wanli (1591), during his tenure as a district magistrate of Baodi county, he wrote a book to encourage farming and mulberry. There are more than 10,000 words in the book, including 8 articles on *tianshi*, *dili*, field system, sowing, tillage, irrigation, manure, and *zhanyan* (占验). Although the contents of each article are also collected from previous works, they are all combined with the actual situation of Baodi. Yuan Huang hoped to transfer the rice planting experience from his hometown in the south to the north. Therefore, much of content of the book comes from south of the Yangtze River. On the other hand, Yuan Huang also tried to combine the reality in the north and put forward some improvement measures. For example, he argues that "it is a pity that the manure is not collected because the pigs and sheep in the north are not reared in pens," trying to extend the fertilizer technology from the south to the north.

Jiang Nan Cui Geng Ke Dao Bian was written by Li Yanzhang (李彦章), a Houguan native in Fujian Province, in the Qing Dynasty. When he served as an official in the Tonghai Circuit, Changzhen, he hoped that the Suzhou region could also accomplish "harvesting rice twice a year" like in central Fujian, or Jiangyou and Jingxiang, thus vigorously promoting early rice. The contents of the book are divided into 10 chapters, mainly composed of contents from various agricultural books, annals, and other relevant records, but the author makes detailed comments at the end of each chapter.

Ying Tian Ji Yao was written by Huang Fuchen (黄辅辰) (1798~1866) at the request of Liu Rong (刘蓉), the Provincial Governor of Shaanxi Province, in 1864, the third year of Tongzhi in the Qing Dynasty. It was a monograph on wasteland cultivation by troops in all dynasties. The book consists of 4 chapters with more than 40,000 words. The frontispiece "General Theory" shows the differences between farming and wasteland cultivation by troops. The former is for ordinary people and the other is for soldiers, but both are conducive to the defense of the border and the country. The first two articles describe the experiences of farming and water conservancy in past dynasties. The third article recounts the malpractice in the cultivation by troops of past dynasties. The fourth article is mainly about agricultural techniques related to tillage and reclamation. They are subdivided into 12 parts. As a monograph on reclamation, the book pays equal attention to both policy and technology, and dares to expose malpractices. Although it basically compiles the old theories of its predecessors, it still reflects the author's own original ideas. He advocates for the cultivation of wasteland by farmers instead of soldiers. Tax should be light, for the benefit of the people, rather than harming the people. The varieties of plants compiled in the book are also far greater than those in the general agricultural books.

3.5.3 Local Agricultural Books

Local agricultural books are books about agriculture in a specific area. The titles of such books often bear the names of the places. This is a tradition formed after the farming encouragement texts in the Song Dynasty, when a local official issued a text of exhortation to farmers in a certain place and entitled it a farming encouragement text in a certain place, such as *E Zhou Quan Nong Wen* (《鄂州劝农文》), *Long Xing Fu Quan Nong Wen* (《隆兴府劝农文》), or *Fu Zhou Quan Nong Wen* (《福州劝农文》). During the Ming and Qing Dynasties, there were some local agricultural books, such as *Bao Di Quan Nong Shu*, *Chang Shu Xian Shui Li Quan Shu* (《常熟县水利全书》), *Suo Shan Nong Pu* (《梭山农谱》), and *Fu Jun Nong Chan Kao Lue* (《抚郡农产考略》). Unlike national agricultural books, local agricultural books are more based on the actual needs of local agricultural production, adjust measures to local conditions, and emphasize practicability and feasibility.

3.5.3.1 Agricultural Books in the South of Yangtze River

Since ancient times, the water towns south of the Yangtze River have been dominated by rice cultivation, followed by sericulture and a certain amount of animal breeding. Therefore, rice and sericulture have also become the main contents of agricultural books there. In the late Ming Dynasty, *Nong Shu* by an author surnamed

Shen was written about “Yun Tian Di Fa” (运田地法). *Tian* refers to paddy fields while *di* is the dry land of mulberry fields. Paddy fields require deep tillage and corresponding measures. The fertilizing of paddy fields should be “seedling-based,” that is, when the young panicle of rice is differentiated, if “the seedling color is yellow,” further fertilization is urgently needed, but if the leaf color is still black and not yellow, no further fertilization is needed; otherwise it will only lead to good seedlings without good rice. The fertilizer used depends not only on pig and sheep manure but also on market mechanisms to solve the problem of sourcing fertilizer. The book also applies the technology of *kaotian* (烤田) to seedling management to improve the survival rate of transplanted seedlings. For rice and wheat fields that harvest twice a year, Shen proposes that “only dry fields are the best, and if the fields are muddy, they must be reclaimed until the surface is dry a few days later... Don’t tread it tight when it is muddy.” In order to solve the contradiction between the late ripening of the second rice and early planting of wheat, and to prolong the growth period of wheat, the book also mentions the technology of transplanting wheat seedlings.

Bu Nong Shu, a supplement to *She Shi Nong Shu,* is written by Zhang Luxiang, a neo-Confucian in Tongxiang, Zhejiang Province, in the early Qing Dynasty. He copied Shen’s *Nong Shu* precisely because Lianchuan and Tongxiang are “not far away,” so the book can be used as reference. His supplements precisely take into account that Shen’s ideas “are not fully suitable” to Tongxiang. He said, “the soils and capabilities of people are different,” and then “farming should be based on tradition.” Zhang Luxiang supplements the techniques of soil preparation of rice and wheat fields with double harvests and puts forward the requirements of “*di (lin)zao, tushu, goushen*” (地(疄)燥、土疏、沟深dry land, sparse soil, and deep furrow) with the emphasis on the requirement of “early.” He develops the technology of “*kailinzuogou*” (开疄作沟) after the Song and Yuan Dynasties to a very complete level.

Zhang Luxiang and Shen have one thing in common in agricultural management, that is, they stress the economic benefits. According to the calculation of input and output, Zhang Luxiang puts forward that “it is better to raise more geese than chickens,” “the profit of raising hens is slightly higher than that of raising roosters,” “the profit of raising silkworms is high,” and “more land is better than more field.” Under the ideological guidance of adjusting measures to local conditions, Zhang Luxiang advocated diversified management. In addition to planting rice and mulberry, he also planted vegetables, medicinal herbs, and raised chickens, geese, sheep, pigs, etc. on his own fields and lands. The livelihood planning he undertook for his friend Wu Xingsu (邬行素)’s family after Wu died also reflects this. The management of human resources is also one aspect of Zhang Luxiang’s thoughts on agricultural management. He is very particular about the way to use people and only uses them according to their ability. He distinguishes between busyness and idleness, diligence and laziness, as well as difficult work and easy work. He gives different wages and meals, rewards diligence, and punishes laziness, so as to arouse people’s enthusiasm for work.

Compared with *Shen Shi Nong Shu*, sericulture technology in *Bu Nong Shu* is more prominent. Zhang Luxiang invented the method of rearing silkworms with a wooden frame in a type of pine shed. This method not only saves labor but can also prevent silkworm diseases, which is of great significance for improving the quality and yield of cocoon filaments. He also summarized the propagation method of mulberry layering and said that the propagation of mulberry layering is a must in the process of cultivating land.

South of the Yangtze River is where agronomy developed the best in the Ming and Qing Dynasties. Apart from *Shen Shi Nong Shu* and *Bu Nong Shu*, there were other agricultural books there during the Ming and Qing Dynasties, of which *Pu Mao Nong Zi* (《浦泖农咨》) by Jiang Gao (姜皋) was representative. *Pu* refers to the Huangpu River while *Mao* is Mao Lake, which is located in Songjiang Prefecture of Shanghai. Jiang Gao is a native of Pumao. The book was written in the 14th year of Daoguang in the Qing Dynasty (1834). The book consists of only one volume, containing about 7000 words and 40 items. The book talks about the local farming situation in a farmer's tone. The contents include water conservancy, *tianshi*, sowing, *yangtian* (秧田), *yuntang* (耘耥), *yihuo* (刈获), fertilization, farm cattle, farm tools, as well as farmers' tax burdens and economic life. It also summarizes in great detail the situation of rice production in the areas of Songjiang and Shanghai, especially the land system, farming methods, and technical experience of rice production areas.

There were also some "water conservancy books" in the area south of the Yangtze River during the Ming and Qing Dynasties. For example, *Changshu Xian Shui Li Quan Shu* by Geng Ju (耿桔), *Wu Zhong Shui Li Shu* (《吴中水利书》) by Zhang Guowei (张国维), *Zhu Wei Shuo* (《筑围说》) by Chen Hu (陈瑚), and *Zhu Wei Tu Shuo* (《筑圩图说》) by Sun Jun (孙峻). *Changshu Xian Shui Li Quan Shu* is a monograph that systematically summarizes water conservancy of *weitian* (圩田) in light of Changshu's local conditions. The writer Geng Ju was a District Magistrate of Changshu for 3 years. This book is based on his plan and the construction summary of his water control implementation. Xu Guangqi comments on the book as being "extraordinary in both water conservancy and famine relief," which is not accidental.

Wu Zhong Shui Li Shu was written by Zhang Guowei (1595~1646) in the late Ming Dynasty when he was the Provincial Governor of Jiangnan and Senior Vice Minister of the Ministry of Works. The book is written according to his own experience and based on previous literature. The 28-volume book first lists 52 general paintings on water conservancy in the seven prefectures in southeast China, including water sources, the run of the water vein, water names, etc. It also compiles the imperial edicts, orders, notes, ballads, and classifications.

3.5.3.2 Shandong Agricultural Books

Agronomy in Shandong has always been developed. Fan Shengzhi of the Western Han Dynasty, Jia Sixie, author of *Qi Min Yao Shu*, and Wang Zhen of the Yuan Dynasty are all from Shandong and are great agronomists. The earliest local agricultural book appearing in Shandong in the Ming and Qing Dynasties is *Guo Mai Min Tian* (《国脉民天》). Other famous books include *Nong Sang Jing* and *Nong Pu Bian Lan*.

Nong Sang Jing is an agricultural book from Shandong during the Ming and Qing Dynasties. The author is Pu Songling (蒲松龄) (1640~1715), also called Liuxian (留仙), a native of Zichuan, Shandong. He is also the author of the collection of classical Chinese novels *Liao Zhai Zhi Yi* (《聊斋志异》). *Nong Sang Jing* is an unprinted manuscript, which was completed around the year of Yiyou (1705) under the reign of Qing Emperor Kangxi. The book is divided into two parts: "Nong Jing" (农经) and "Can Jing" (蚕经). The status of sericulture has been significantly improved in Shandong's local agricultural books. The content of planting cotton and sorghum is also first seen in Shandong's local agricultural books. The book provides new contents such as the rotation of winter wheat and sorghum for soil preparation, the prevention and control of soybean pests, the selection of insect-resistant crops, soil and water conservation, etc.

About half a century after *Nong Sang Jing*, another agricultural book, *Nong Pu Bian Lan*, was published in the Shandong Peninsula. The author Ding Yizeng was from Rizhao, Shandong Province. The book represents the author's long farming experience after returning to his hometown because of his failure in the imperial examinations in his thirties. The book is in the form of agricultural calendar. It covers fields, gardening, climate forecasts, the processing of agricultural products, maternal and child health care, medicine and hygiene, and even poetry and spring festival couplets. The author once lived in the village of Xishiliang, and the contents of the book closely follow the local reality. For example, he proposes the measure of "*gengdi*" (耕地 land tillage) according to the fact that "all low-lying land with accumulated water will definitely harden after it dries up and when land is wet, the ground must be muddy, so it is not only difficult to cultivate, but also full of growing grass." Even when quoting from previous agricultural books, he strives to combine them with the local situation. Because of the prominent local characteristics, the book also reflects some new developments in agriculture in Shandong in the Qing Dynasty. The contents about Chinese cabbage and tobacco are examples as these two crops began to emerge in Shandong only after the Qing Dynasty had begun. The book also introduces seafood such as white shrimp, flounder, whitebait, saury, mackerel, pomfret, sole, squid, jellyfish, sea cucumber, abalone, etc. These contents are closely related to Rizhao's special geographical location near the sea.

3.5.3.3 Hebei Agricultural Books

Hebei uses mainly dry land agriculture, but paddy fields also exist in some low-lying areas. During the Ming and Qing Dynasties, some people of insight were committed to the development of rice cultivation in the Hebei region. This is reflected in Hebei's regional agricultural books. Among them, there are *Bao Di Quan Nong Shu* and *Ze Nong Yao Lu* (《泽农要录》). Wu Bangqing (吴邦庆) (1766~1848), the author of *Ze Nong Yao Lu*, was from Bazhou, Hebei Province. Bazhou is a low-lying area. He Chengju (何承矩) and others used to grow rice in this area in order to consolidate national defense in the Song Dynasty. During the Ming and Qing Dynasties, Bazhou was a major place near the capital city. Wu Bangqing devoted himself to the development of paddy field agriculture in the capital city and its environs. He compiled documents from previous dynasties and his own works into a set of

8 books called *Ji Fu He Dao Shui Li Cong Shu* (《畿辅河道水利丛书》) (compiled in 1824). *Ze Nong Yao Lu* is one of the books, which compiles discussions on paddy field agriculture in ancient agricultural books. The six-volume book is divided into 10 sections, each being preceded by an introduction written by the author, which explains the author's practical experience. In view of the lack of rice planting experience among northerners, he emphasizes methods of cultivation, irrigation, and rice planting.

3.5.3.4 Jiangxi Agricultural Books

Jiangxi is a province with relatively developed agriculture, and also has a tradition of agronomy writing. This tradition appeared at least in the Song Dynasty, when Zeng Anzhi and his grand-nephew Zeng Zhijin wrote two agricultural books, *He Pu* and *Nong Qi Pu*, respectively. In the Song Dynasty, there was also a famous author Sheng Si (胜私) from Jiangxi who wrote the three volumes of *Nong Shu* (《农书》) (Zhu Xi: To My Old Friend Sheng Si, *Hui An Ji* (《晦庵集》), Volume 7). If we take Zhu Xi's work *Quan Nong Wen* into account, Jiangxi's contribution to agriculture in the Song Dynasty is already considerable. During the Ming and Qing Dynasties, Jiangxi's agricultural tradition was still developing, especially with three people from Fengxin: Song Yingxing (宋应星), Liu Yingtang (刘应棠), and Shuai Nianzu (帅念祖). They respectively wrote *Tian Gong Kai Wu* (《天工开物》), *Suo Shan Nong Pu* (《梭山农谱》), and *Qu Tian Bian* (《区田编》). The books with the most local characteristics of Jiangxi's agriculture are *Suo Shan Nong Pu* and *Fu Jun Nong Chan Kao Lue* (《抚郡农产考略》).

Tian Gong Kai Wu is an encyclopaedia of agricultural and handicraft technology written by Song Yingxing, a Jiangxi native in the Ming Dynasty. It was published in the tenth year of Chongzhen during the Ming Dynasty (1637). Among them, *Nai Li* (《乃粒》) and *Nai Fu* (《乃服》) are related to farming, while *Cui Jing* (《粹精》) (grain), *Gan Shi* (《甘嗜》) (sugar making), and *Gao Ye* (《膏液》) (oil squeezing) are related to processing, which covers about 1/3 of the total space. Most of the content about agriculture in the book is drawn from agricultural practices in Jiangxi. For example, the method of *shuidui* (水碓) in Xinjun, Jiangnan (now Shangrao, Jiangxi), has "three usages (grinding flour, rice hulling and irrigation)." There is also the "planting method in Jijun Jiangxi" in which soybeans are planted in the stubble of *daogao* (稻藁) to rotate rice and beans. The fertilizing technique of dropping ashes onto the seedling roots used in *lengjiang* fields (冷浆田) is also from Jiangxi Province. Xu Guangqi also mentioned, "Jiangxi people use lime or ashes of cattle and pigs and store them in baskets, then transplant rice seedlings with the roots dipped in them" (*Scripts of Xu Guangqi, Personal Letters in Nong Zheng Quan Shu*, Zhong Hua Book Company, 1962). The method farmers in Jiangxi use to cultivate rice fields with feet is also different from the manual cultivation popular in Jiangsu and Zhejiang provinces as recorded in *Chen Fu Nong Shu*. Other issues such as the ratio of *yangtian* (秧田 seedling fields) to *bentian* (本田 rice fields), the water requirements for early and late rice, multiple cropping systems in rice fields, including *shuangjilian* (双季连), and paddy-upland rotation of rice and beans, rice and wheat, etc. may all come from agricultural practices in Jiangxi.

Suo Shan Nong Pu (1674) was written by Liu Yingtang when he was in Suoshan, located in Fengxin County, Jiangxi Province. The book is divided into three volumes, i.e., ploughing, weeding, and harvesting, each of which has a short preface and is divided into two parts, work and equipment, with a total of over 20,000 words. The book describes the whole process of rice production from planting to harvesting, and is characterized by putting weeding, harvest, and cultivation on an equal footing. The idea was emphasized that every link of agricultural production cannot be ignored. Apart from farming, it is especially valuable to have a fairly accurate description of the farming tools used in paddy field production. There is no quotation from agricultural books of the previous generations. Instead, most of the book is about the practical experience of the farmers in the author's hometown, which has many unique local practices. For example, *chengchan* (塍铲) and *chengdao* (塍刀) are two kinds of farm tools for paddy field operations in the hilly areas of the south, which are specially used to clean up the ridge. The book also records for the first time a tool *chongshu* (虫梳) used to remove insects from rice fields. It also mentions two major agricultural disasters in the area of Suoshan. One is "*qingfeng*" (青风) (or *hanlufeng* 寒露风) caused by a sudden cold when rice is blooming. The second is the insect pest that occurs during the mid-summer heat fumigation. Both are natural disasters and do more harm than floods and droughts, especially *qingfeng*.

Fu Jun Nong Chan Kao Lue is an agricultural work written by He Gangde at the end of the Qing Dynasty on local agriculture in Fuzhou, Jiangxi province. In the 27th year of Guangxu during the Qing Dynasty (1901), He Gangde was transferred to Fuzhou as prefect. He visited squires, consulted old farmers, investigated agricultural products, and engaged in writing in his spare time. The purpose was to make "farmers follow the rules and learn deeply about the reason why they are declining and flourishing in farming." It took a year and was finally completed in 1903, the 29th year of Guangxu, and was printed by the Fu Jun School. In the 33rd year of Guangxu (1907), it was reprinted by the Jiangsu Provincial Printing Bureau. Most of the contents recorded in the book are "what rural farmers say and what the author has tested" (preface by Huang Shenfu 黄申甫). The two volumes of the book record 143 kinds of common cultivated plants in local villages, such as grain, grass, and wood. Each entry generally describes the biological characteristics of agricultural products and their varieties. The products recorded in the book are described separately using the traditional four aspects of *tianshi*, *dili*, *renshi* (人事), and *wuyong* (物用). At the same time, some new contents are added, especially related to the modern agricultural chemistry introduced from the west. For example, when talking about soil fertilizer, the concept of nitrogen was used. One third of the book is about rice, especially the cultivation techniques and experiences of 56 rice varieties. These contents both inherit from and develop upon the previous generation of Jiangxi agricultural books such as *Tian Gong Kai Wu*. Attached to the book is an article "Zhong Tian Za Shuo" (种田杂说) by Jiang Zhaotang (江召棠), a district magistrate in Linchuan under Fuzhou's control, which introduces farming experience on fertilization and pest control in Linchuan (Xu Xin: *Zhong Tian Za Shuo* by Jiang Shaotang, *Nong Ye Kao Gu* (《农业考古》), 1987, Volume 2, p. 69).

3.5.3.5 Hunan Agricultural Books

Jia Pu Chu Xue Ji (《稼圃初学记》) was written by Li Jinxing (李晋兴), a Linwu native of Hunan Province in the early Qing Dynasty. Linwu County is located in the southernmost part of Hunan Province, at the northern foot of the eastern section of the Nanling Mountains. The mountain area is prone to drought. While hiring people to work, Li had to make up his mind to observe the water in the field and ask the old farmers for advice on how to cultivate the field. What the book records is the results of his consultation. The book was written in the fiftieth year of Kangxi in the Qing Dynasty (1711), and is only 1600 words long. The content is about farming and vegetable growing, with the emphasis on farming. The book records in detail the rice production technology in the Linwu area. It mainly includes rice varieties, soil preparation of *yangtian* and *bentian*, fertilization, *chashi* (插莳), weeding, irrigation, etc. Some contents are not found in other agricultural books. For example, fish ponds are used to breed seedlings, and the density of sowing, the growth of seedlings, and the size of seedling stems (tubes) can be determined according to the fertility of the seedling bed, which can be divided into *shanguyang* (扇骨秧) and *simaoyang* (丝毛秧). From the aspect of soil preparation, winter tillage is emphasized. During spring ploughing, it was mentioned to trim the ridge of the field, first for planting beans and second for stabilizing the water in the field. The discussion of planting beans on the ridge of fields is the earliest record about the utilization of field ridges in agricultural books. The book also mentions "putting a suitable amount of fish into the rice field," which is also the earliest documentary record of fish farming in paddy fields in the true modern sense in agricultural books.

3.5.3.6 Sichuan Agricultural Books

Sichuan's local agricultural books in the Qing Dynasty are represented by *San Nong Ji* (《三农纪》). The book (completed in 1760) was written by Zhang Zongfa (张宗法) (1714~1803), a native of Shifang County, Sichuan Province. He wrote the book in the style of conversations between a farmer, an old man, and a shepherd boy to urge and persuade each other. In addition to quoting the old farmer's comments, the book also has its own viewpoints. It quotes nearly 220 documents and has a wide range of contents, covering many aspects of production and life. Agriculture alone includes fields of grain, economic trees, fruits and vegetables, animal husbandry, household sideline production, etc., but it does not mention processing and trade. The first five volumes of this book discuss the conditions related to agricultural production such as climate, soil, and environment. There are 185 species of cultivated plants and 18 species of animals recorded in the book, which is quite comprehensive and not found in other local farming books. As a local agricultural book, *San Nong Ji* has a detailed description of the special agricultural products in the local and neighboring areas, such as *pianniu* (犏牛 a kind of crossbred cattle), and some newly introduced local crops, such as corn. There are also many innovations in the cultivation techniques of sorghum, soybean, and spinach, as well as the techniques of raising cattle and feeding pigs.

3.5.3.7 Guanzhong Agricultural Books

The Guanzhong region of Shaanxi has always been famous for its agriculture. There were a large number of agricultural talents here. Legend has it that Hou Ji (后稷), one of the ancestors of Chinese agriculture, was born here. In the Qing Dynasty, there were three agricultural books, two of which were *Bin Feng Guang Yi* and *Zhi Ben Ti Gang* (《知本提纲》) by Yang Shen from Xingping. The third one is *Nong Yan Zhu Shi* (《农言著实》) by Yang Xiuyuan (杨秀元).

Bin Feng Guang Yi was written in the 5th year of Ganlong (1740) and was engraved in the 7th year (1742) during the Qing Dynasty. The book is divided into three volumes. The first volume focuses on mulberry land suitability and methods used in the process of planting mulberry trees. The middle volume describes various silkworm rearing equipment and the making of them, as well as the steps from the selection of silkworm species to silk reeling. The first part of the last volume talks about silk weaving and machinery. In addition, there is also a method of raising *hucan* (槲蚕). The original intention of this book is to revive sericulture. The foreword includes Yang Shen's suggestions to the local government, listing four reasons why the north can plant mulberry and raise silkworms, and explaining his enthusiasm for developing the local folk production. After the book was completed, he associated what Mencius referred to as "clothes and silks" with "meat eating." Therefore, the second part of the last volume is about animal husbandry, mainly the raising of livestock and poultry and the treatment of diseases. There are also some discussions on gardening.

The book is characterized by its focus on the conditions and practicality in the Guanzhong region of Shaanxi province. For example, it specifically introduces an excellent mulberry variety discovered by Yang Shen in the region, which is better than southern mulberry. The book also makes an analysis and comparison between ancient and present China, north and south, and concludes that the suitable time for raising silkworms in Shaanxi is three to four days before *guyu* (谷雨). In order to facilitate extensive publicity, the book has many drawings and the text is simple and easy to understand.

The volume of *Nong Yan Ji Shi* is a lecture on farming given by the author Yang Xiuyuan, a native of Sanyuan, Shaanxi province, to his family. The book is divided into two parts. The first part narrates the farming activities that should be done in each month. The second part is "miscellaneous notes," with a total of 10 articles, emphasizing matters that should be given special attention. The farm run by the Yang family is located near the Xianling tomb of Emperor Tang Gaozu on the dry tableland, with piles of rubble everywhere. Therefore, it had become one of the farming activities to "pick up tiles and bricks in the wheat fields when there is nothing to do in the first month of lunar calendar." In June, there were always *huangshu* (黄鼠 suslik) that cause damage to the wheat fields. If there is a nest of *huangshu* in the valley, they can lure them out with dozens of bamboo fishing rods. The plague of *huangshu* and its control mentioned here is rarely discussed in general agricultural books, but it is common in agricultural practice. Yang Xiuyuan also introduces in detail the cultivation techniques of winter wheat, millet, pea, alfalfa, and other crops on the dry tableland according to local conditions and planting seasons. From manure to soil preparation and sowing, then to intertillage and harvest, there is a unique set of methods. Take intertillage, for example. The book mentions an intertillage weeding tool called *louchu* (漏锄), a small hoe with a square gap in the middle. It is characterized by not turning over the soil when hoeing. After hoeing, the land is flat, which is favorable for preserving soil moisture, and it is easy to use. This is also the reason why it has this name as opposed to *benchu* (笨锄). Another example is his emphasis on courtyard economy. It is rarely recorded in other agricultural books to plant radish, cabbage, and other crops at the gates, around hamlets, or under the walls to expand food sources.

3.5.3.8 Shanxi Agricultural Books

Ma Shou Nong Yan (《马首农言》) is the representative of Shanxi agriculture books. The author, Qi Junzao, was born in Shouyang, Shanxi, and held important positions in the imperial court. This book was written in the years of Daoguang during the Qing Dynasty by the author when he came home to mourn. The book records the number of coarse grains per spike of different varieties of local cereal in the 14th year of Daoguang (1834). The harvest that year was only ninety percent, and the number of coarse grains per spike ranged from 7892 to 9835. As a result of this, he thought, "if it's a year with a good harvest, the number of coarse grains per spike may be over ten thousand." His careful observation is obvious.

The book was written in the 16th year of Daoguang (1836). Mashou is the ancient name of Shouyang. The book narrates the local agricultural situation and so is entitled *Ma Shou Nong Yan*. The book includes 14 articles on topography and climate, planting, agricultural implements, agricultural proverbs, *zhanyan*, local proverbs, grain diseases, food prices, water conservancy, animal husbandry, famine preparedness, shrine worship, weaving, miscellaneous sayings, etc. In describing the specific technology, it considers the local climate, soil, terrain, and crop characteristics and puts forward many cultivation measures. The book also collects many local proverbs, and also set up a section of "Fang Yan" (方言) to record local proverbs related to agricultural production and rural life. After that, there are proverbs of various crop diseases, "wugubing" (五谷病), which were not found in previous agricultural books. However, the book does not mention the prevention and treatment of these diseases. In the book, a special section entitled "Liang Jia Wu Jia" (粮价物价 price of rice and commodities) is included, which also reflects the characteristics of Shanxi people who are good at doing business.

3.5.3.9 Jianghuai Agricultural Books

In the 24th year of Daoguang in the Qing Dynasty (1844), Bao Shichen (包世臣) compiled the major works in his life into *An Wu Si Zhong* (《安吴四种》), including 12 volumes of *Qi Min Si Shu* (《齐民四术》), one of which is *Nong Zheng Pian* (《农政篇》), which was also entitled *Jun Xian Nong Zheng* (《郡县农政》). It is an agricultural book summarizing and describing agricultural production techniques in the Jianghuai region during the Qing Dynasty. *Jun Xian Nong Zheng* borrows some concepts from previous agricultural books, such as "rentu" (任土) and "lao" (劳). At the same time, it is also discussed in combination with the actual production in the Jianghuai region. For example, the explanation on the key points of rice cultivation in winter fallow fields and the characteristics of paddy soil is different from *Qi Min Yao Shu* and *Chen Fu Nong Shu*. The book mentions a variety of farming systems in the Jianghuai region, such as double cropping rice, multiple cropping of rice and wheat, multiple cropping of rice and beans, interplanting of wheat and cotton, etc.

3.5.4 Professional Agricultural Books

During the Ming and Qing Dynasties, professional agricultural books developed and increased in number, and the subjects of writing also increased. Some special agricultural content was discussed for the first time. The examples are rose, lotus, and other ornamental plants; peach, plum, *longyan*, and other fruit trees; sweet potato, cotton, and other field crops; and turnips, taros, and other vegetable crops. In terms of animal husbandry and veterinary medicine, besides the original horses and cattle, there are pigs, chickens, and bees. During the Ming and Qing Dynasties, a variety of wild vegetables, locust control, and *qutian* were added as subjects for professional writing. Even in the professional fields previously established, the content of this period also increased. For example, the varieties of litchi produced in the areas listed in *Li Zhi Pu* surpass the previous generation. However, most of the professional agricultural books in the Ming and Qing Dynasties were sericulture books. In the previous dynasties, only the one by Qin Guan was handed down. In contrast, there were 204 sericulture books that could be counted in this period, of which only one was written during the Ming Dynasty, and the rest were written and engraved in the Qing Dynasty, about 90% of which appeared in the late Qing period.

3.5.4.1 Type of *Quzhong*

Works on *quzhong* are one of the most common among professional agricultural books in the Ming and Qing Dynasties. In the 22nd year of Daoguang in the Qing Dynasty (1842), Zhao Mengling (赵梦龄) compiled Song Baochun (宋葆淳)'s collection of scattered *Fan Sheng Zhi Shu*, Sun Zhaikui (孙宅揆)'s *Jiao Jia Shu* (《教稼书》), Shuai Nianzu's *Qu Tian Bian*, Zhuo Zheng Lao Ren (拙政老人)'s *Jia Shu Bian* (《加庶编》), and Pan Zengyi (潘曾沂)'s *Feng Yu Zhuang Ben Shu* (《丰豫庄本书》), and integrated them into *Qu Zhong Wu Zhong* (《区种五种》). Fan Liang (范梁), his disciple, published it and also included Geng Yinlou (耿荫楼)'s *Guo Mai Min Tian* (《国脉民天》) as an appendix. Therefore, there are actually six works. In the 1950s, Wang Yuhu compiled another 5, which were combined into *Qu Zhong Shi Zhong* (《区种十种》) (Yuhu 1955), namely, *Guo Mai Min Tian, Lun Qu Tian* (《论区田》), *Jiao Jia Shu* (illustrations on *qutian*), *Qu Tian Fa* (《区田法》), *Qu Tian Bian*, *Xiu Qi Zhi Zhi* (《修齐直指》), *Zeng Ding Jiao Jia Shu* (《增订教稼书》), *Jia Shu Bian*, *Qu Zhong Fa* (《区种法》), and *Duo Jia Ji* (《多稼集》). These 10 books were all written by scholars in the Ming and Qing Dynasties.

Guo Mai Min Tian is written by Geng Yinlou (?~1638) during the Ming Dynasty. There is only one volume with over 3000 words, and the contents are divided into

7 items: *qutian*, *qintian* (亲田), *yangzhong* (养种), *shaizhong* (晒种), dung storage, drought control, and famine preparedness. Among them, the method of "*qintian*" is the most distinctive, which is consistent in principle with the method of *qutian* of "not tilling more lands."

Lun Qu Tian was written by Lu Shiyi (陆世仪) (1611~1672) in the late Ming and early Qing Dynasties. It was originally part of his book *Si Bian Lu* (《思辨录》). It is notable in that the method of *qutian* is applied to paddy fields. The book gives quite a practical account of farming, and the author often puts forward his own opinions, such as the method of determining how good the harvest year is by counting the number of "*pu*" (铺).

Jiao Jia Shu (1721) was written by Sun Zhaikui in the Qing Dynasty. Before him, Zhu Longyao (朱龙耀) was successful trialling *qutian* in the 53rd year of Kangxi (1714) during the Qing Dynasty in Pingding County, Shanxi Province, and compiled *Qu Tian Shuo* (《区田说》). Sun got it and expanded it into *Jiao Jia Shu*. He added the following items: the theory of *zhenmu* (甽亩), the method of *fenzhong* (粪种), the method of *zhifen* (制粪), the method of *zhengfen* (蒸粪), the method of *zaofen* (造粪), and the theory of *zhiyi* (制宜). It shows the importance attached to fertilizer.

Qu Tian Fa was written by Wang Xinjing (王心敬) (1656~1738) during the Qing Dynasty. It advocates replacing a small square area with *tongqu*an (通畎).

Qu Tian Bian (1742) was written by Shuai Nianzu during the Qing Dynasty and annotated by Xu Ruji (许汝济). In the 12 paragraphs of the full text, the beginning is a painting of *qutian*. The following sections describe the methods of planting, using dung, watering, and accumulating water. In the process of spreading this book, several books with "annotations" appeared. In addition to Xu's annotation, the engraver also added notes and annotations. What is more, methods such as "Dong Yue Zhong Gu Fang" (冬月种谷方 sowing wheat in winter months), "Fang Lao Wei Tian Fa" (防涝围田法 reclamation of field to prevent floods), "Chuan Jing Fa" (穿井法), "Fen Tian Fa" (粪田法 dung use), and "Bian Neng Qu Tian Sheng Gong Fa" (变能区田省工法 increasing efficiency) were used as appendices. Among them, the method of sowing wheat in winter months is a remedial method adopted in the northern regions when wheat planting is missed for some reason in autumn.

Xiu Qi Zhi Zhi (1776) was written by Yang Shen and annotated by his disciple Qi Zhuo (齐倬) from Lintong. While expounding the principles of farming, raising livestock, and rearing silkworms, the specific techniques of farming, planting mulberries, planting trees, and raising livestock are further discussed in light of the actual situation in the Guanzhong region.

Zeng Ding Jiao Jia Shu was written by Sheng Baier (盛百二) during the Qing Dynasty. It was written in the forty-third year of Ganlong (1778). It is based on Sun's *Jiao Jia Shu* and adds several articles. The original book is the first volume and the second volume is a supplement. Among the material, the method of *jiagu* (架谷) is very special.

Jia Shu Bian was written by Xu Jiayou (许嘉猷) during the Qing Dynasty, or "Zhuo Zheng Lao Ren" as he signed himself. The book mainly talks about the method of dividing *qutian*, which is based on *Qu Tian Tu Kan Wu* (《区田图勘误》) by mathematician Mei Wending (1633~1721) and adds some comments.

Qu Zhong Fa was written by Pan Zengyi (1792~1852) during the Qing Dynasty. The author lived in his hometown of Wuxian, Jiangsu, for a long time and enthusiastically advocated the *qutian* method and used it in paddy field rice production. In the 8th year of Daoguang in the Qing Dynasty (1828), after trying out the method on the farmland of his clan, he wrote *Ke Nong Qu Zhong Fa Zhi Jiang San Shi Er Tiao* (《课农区种法直讲三十二条》). It explains in detail the methods of *quzhi,* sowing, cultivation, and manure application. It advocates deep tillage and early sowing, thin planting and more harvesting, harvesting once a year, and no planting of wheat. The experiment was carried out for 2 years and yielded good harvests. In the 14th year of Daoguang's reign (1834), *Feng Yu Zhuang You Zhong Liang Ge* (《丰豫庄诱种粮歌》) and *Ke Nong Qu Zhong Fa Tu* (《课农区种法图》) were also engraved. Another book engraved in the same period is *Bian Nong Yao Fang* (《便农药方》), which includes the recipe for curing various livestock diseases. Later, the author's nephew, Zu Yin (祖荫), inscribed the book *Feng Yu Zhuang Ben Shu* (《丰豫庄本书》) in 1877, the third year of Qing emperor Guangxu's reign, by compiling these words together with petitions to the Suzhou office, government instructions, and some essays, including an excerpt from Geng Yinlou's book *Guo Mai Min Tian* (entitled *Zhong Tian Shuo* 《种田说》).

Duo Jia Ji, also known as *Geng Xin Nong Hua* (《耕心农话》), was written by Xi Cheng (奚诚), a native of Wuxian County, Jiangsu Province. The book is divided into two volumes. The first volume "Zhong Tian Xin Fa" (种田新法 New Farming Law) introduces 13 simple new laws, which are designed on the basic methods of *qutian* and *daitian* (代田) in combination with farmers' practices. It emphasizes planting tall crops and short ones alternatively, planting between ridges, and using different land each year, so as to accumulate soil fertility and make full use of every piece of land. The next volume is "Nong Zheng Fa Ming" (农政发明 agricultural inventions), a total of six items, with a focus on the *qutian* method. The book mentions a method called "jiaofen" (窖粪) to accumulate manure. It shows further understanding of the effect of irrigation and winter ploughing implemented since the Song Dynasty. It also applies the soil preparation technology of "ditching and making ridges" originally used for rice and wheat cropping in the Jiangnan region to cotton planting. It is suggested that early rice replace late rice and the rice double cropping system replace the rice-wheat double cropping system.

In addition to what was included in *Qu Zhong Shi Zhong*, there are also several influential monographs on *qutian*.

Qu Zhong Zu Shi Yue Yan (《区种足食约言》) was compiled by Shou Zhuo Ju Shi (守拙居士) during the Qing Dynasty. This book is also divided into 12 items like *Qu Tian Bian*, with the addition on *fenrang* (粪壤 dung soil).

Lun Qu Tian Shu (《论区田书》) was written by Chen Pu (陈溥) during the Qing Dynasty. The manuscript contains letters he wrote to his friend "Liu Xiong" (六兄). Therefore, it is certain that these letters were written around 1857.

Qu Tian Fa (1877) was compiled by Deng Chen (邓琛) as District Magistrate of Pu County, Shanxi, during the reign of Qing Emperor Guangxu. The book is compiled from excerpts from *Nong Sang Ji Yao*, Sun Zhai's *Qu Tian Shuo* (that is, *Jiao Jia Shu*), part of Lu Shiyi's *Lun Qu Tian*, and the full text of Pan Zengyi's *Qu Tian Fa*.

Qu Tian Tu Shuo was compiled and published by Yang Baoyi (杨葆彝) in Hangzhou in the 10th year of Guangxu (1884) and was included in his series, *Da Ting Shan Guan Cong Shu* (《大亭山馆丛书》). This book is very short and is made up of excerpts from various books.

Qu Tian Shi Zhong Shi Yan Tu Shuo (《区田试种实验图说》) was written by Feng Xiu (冯绣) (1860~1909) during the Qing Dynasty (1908). The book is divided into 12 chapters on field reclamation, dung accumulation, seeds, soil suitability, timing, seedling, cultivation, watering, weeding, miscellaneous planting, interest on labor costs, and historical situations. The following five sections attached are the theory of management based on time, of prevention of pests, of prevention of mold (referring to plant diseases) infection, a list of miscellaneous notes, and a list of budget expenditures and losses in rented fields. Finally, five maps are attached, including a *qutian* map, a type of *zhongzipan* (种子盘), adapting *qutian* for planting cereals, adapting *qutian* for planting sorghum, and adapting *qutian* for planting *dian* (靛). The *qutian* method mentioned is very practical and detailed, and *zhongzipan* used in sowing as well as methods of *qutian* adaptation are created. Among them, there is a long section about management based on time. According to the order of solar terms, the work to be done and matters needing attention in each season are described in detail. It not only explains agricultural technology, but also discusses the gain and loss of operation and management. At the end of the section, there is a detailed list of budget expenditures and losses in rented fields, which is the most detailed among trials of *qutian* methods.

3.5.4.2 Field Crops

During the Ming and Qing Dynasties, specialized agricultural books related to field crops mainly involved rice, taro, sweet potato, tobacco, cotton, etc.

Huang Shengzeng's *Dao Pin* (also known as *Li Sheng Yu Jing Dao Pin* 《理生玉镜稻品》) is a record of rice varieties. In the book, the concepts of rice (*tu* 稌, *nuo* 稬), *nuo* (糯) (*shu* 秫), *jing* (秔) (*jing* 粳), and *xian* (籼) are explained first, which is followed by the characteristics, sowing period, maturity period, economic value and alias of 34 rice varieties, etc. The book contains rice varieties in Suzhou as well as those in surrounding areas, most of which are also grown in Suzhou but have different names. It notes factors such as seed, texture, shape, *fumang* (稃芒), stem, stress resistance, yield, and quality when recording the traits of rice varieties. The sowing and maturation time of each variety is also recorded. *Dao Pin* is the earliest extant complete special record on rice varieties.

Yu Jing (《芋经》) by Huang Shengzeng is a monograph on taro cultivation. The book consists of 4 chapters. The chapter "Ming" (名) quotes the records of taro names and varieties from *Shuo Wen* (《说文》), *Guang Ya* (《广雅》), *Guang Zhi* (《广志》), and *Tang Ben Cao* (《唐本草》). In the chapter, "Shi Ji" (食忌 contraindication), the matters needing attention in eating taro and preventing taro poisoning are discussed. The chapter "Shi" (事) quotes some historical anecdotes of eating taro to satisfy hunger. In the chapter "Yi Fa" (艺法), books such as *Fan Sheng Zhi Shu*, *Qi Min Yao Shu*, *Si Min Yue Ling*, *Jia Zheng Fa* (《家政法》), *Wu Ben Xin Shu*, and *Wu Xing Xiang Gan Zhi* (《物类相感志》) as well as methods of taro planting at that time are collected, mainly including seed selection, soil preparation and seedling raising, planting and *tangtu* (塘土), etc. The methods are innovative. For example, in order to prevent taro from freezing, a cellar storage method in winter was proposed. Another example is *tangtu*, i.e., digging between taro trees and covering the earth over the taro roots to keep the soil loose to produce large and round taro. *Yu Jing* is the only monograph on taro in history.

Since the introduction of the sweet potato in the Ming Dynasty, books on it have appeared continuously. First, Xu Guangqi wrote *Gan Shu Shu* (《甘薯疏》), and then Wang Xiangjin (王象晋) extracted some including its preface and added it into *Er Ru Ting Qun Fang Pu* (《二如亭群芳谱》) (1621). After the book was introduced into North Korea, it was compiled into the book *Zhong Shu Pu* (《种薯谱》) in 1834 by *Xu Youju* (徐有榘) with full quotations together with his comments and excerpts from *Gan Shu Pu* (《甘薯谱》) by Kim and Jiang of North Korea. Chen Zhenlong (陈振龙), a Fujian native, was one of the first to introduce sweet potatoes, and his descendants also actively promoted sweet potatoes. Chen Shiyuan (陈世元), Chen Zhenlong's descendant, compiled *Jin Shu Chuan Xi Lu* (《金薯传习录》) by collecting and editing the sweet potato literature publicized and promoted by his ancestors (including the records in various books and relevant files in various places). The book was written in the 33rd year of Ganlong (1768) and edited in the year of Bingshen (1776). The new edition is based on the editing and supplementation of the original book and *Zhi Huang Chuan Xi Lu* (《治蝗传习录》) originally attached to the book. The extant *Jin Shu Chuan Xi Lu* consists of two volumes, while the new version consists of three volumes. Volume 3 is *Zhi Huang Chuan Xi Lu* (Xiongsheng 1992), which includes *Zhi Huang Bi Ji* (《治蝗笔记》) written by Chen Jinglun (陈经纶) in the 25th year of Wanli (1597) during the Ming Dynasty, in which there are the earliest records on locust control by breeding ducks. The book also records an artificial method of barrel incubation. The extant *Jin Shu Chuan Xi Lu* consists of two volumes. The first volume introduces the methods of planting, eating, preserving, and processing, and attached to it is *Jin Shu Lun* (《金薯论》) by the author's son Chen Yun (陈云). The next volume contains poems and verses about sweet potato. In the year 1776 when *Jin Shu Chuan Xi Lu* was edited and supplemented, Lu Yaozhuan (陆耀撰) compiled *Gan Shu Lu* (《甘薯录》), which was made to teach farmers how to plant sweet potatoes when he was an official in Shandong. The contents of the book are all compiled from previous discussions. The book has 6 items and is very practical and should have been helpful in the process of growing sweet potatoes. Besides these, Lu Yao (陆耀) also has one volume of *Yan Pu* (《烟谱》), which is the earliest book on tobacco. It has 5 sections: "Sheng Chan" (生产) (the introduction and distribution of tobacco), "Zhi Zao"(制造) (tobacco processing), "Qi Ju"(器具) (smoking utensils), "Hao Shang" (好尚) (social fashion), and "Yi Ji" (宜忌) (smoking appropriate and taboo). Attached to the book are "Yan Cao Ge" (烟草歌 tobacco songs) and "Hou Yan Cao Ge" (后烟草歌 post-tobacco songs).

Cotton cultivation techniques were first found in Han E's *Si Shi Zuan Yao*, followed by *Nong Sang Ji Yao* in the Yuan Dynasty. Cotton monographs started in the Ming Dynasty. Xu Guangqi is the author of *Ji Bei Shu* (《吉贝疏》) (another name of it may be *Zhong Mian Hua Fa*《种棉花法》), and he once said in the sericulture part of volume 35 of *Nong Zheng Quan Shu* that "I have had a detailed discussion on cotton in my book *Ji Bei Shu*." This is the real origin of the part about cotton in the complete book of agriculture. Chu Hua (褚华) wrote *Mu Mian Pu* (《木棉谱》) in the Qing Dynasty, which, in addition to quoting previous records and textual research, summarizes and records the methods of planting and processing cotton and the tools used. When Ren Shusen (任树森) was an official in Guizhou, in order to publicize and explain the cotton planting method, he reprinted Chu Hua's *Mu Mian Pu*. Feeling Chu's language was profound and difficult, he wrote the easy-to-understand *Zhong Mian Fa* (《种棉法》), briefly introducing his native cotton planting method in combination with the local conditions of Guizhou. The content is more practical and pertinent. Also included in the cotton monograph is *Mian Hua Tu* (《棉花图》), which was drawn under the charge of Fang Guancheng, the Governor-general of the Metropolitan Area, in the 30th year of Ganlong during the Qing Dynasty (1765).

3.5.4.3 Horticultural Crops

The Ming Dynasty saw the earliest appearance of agricultural books with the title of "Pu" (圃 garden). The first book on horticulture under the name of "Pu" is *Lao Pu Shu* (《老圃书》). The original book has a preface by Gu Cheng Shan Ren (古城山人) written in the 15th year of Zhengde during the Ming Dynasty (1520). Later, there were *Xue Pu Za Shu* (《学圃杂疏》), *Zhi Pu Xu Zhi* (《治圃须知》), *Lao Pu Yi De* (《老圃一得》), *Ru Nan Pu Shi* (《汝南圃史》), *Lao Pu Liang Yan* (《老圃良言》), *Lao Pu Zhi* (《老圃志》), *Lao Pu Za Shuo* (《老圃杂说》), *Zai Pei Pu Shi* (《栽培圃史》), and so on. Related to these are *Guan Yuan Shi* (《灌园史》) and *Guan Yuan Cao Mu Shi* (《灌园草木识》). These are mainly general works on horticulture. In addition, many of them are professional agricultural books related to appreciation, fruit trees, and vegetables.

Zhong Shu Shu (《种树书》) (1379) was probably written by Yu Zongben (俞宗本) (?~1401). The author gave a brief explanation of the title and content in the preface "Introduction to *Zhong Shu Shu*," from which it can be seen that this is an agricultural book mainly focusing on farming, with an emphasis on gardening, and its farming activities such as *qipu* (畦圃) and *yuanchi* (园池) account for a large proportion. The book has about 10,000 words and is divided into 8 items, namely, 12 months of planting activities, beans and wheat, mulberry, bamboo, wood, flowers, fruits, and vegetables. It reflects the level and achievements of agricultural production practice at the end of the Yuan Dynasty and the beginning of the Ming Dynasty. Later agricultural books, such as *Bian Min Zuan Tu*, *Nong Zheng Quan Shu*, and *Shou Shi Tong Kao*, all quoted a lot of from it.

Xue Pu Za Shu (1587) was written by Wang Shimao. The book is divided into three volumes. The first volume is the focus of the book, with more than 30 kinds of flowers recorded. The second volume is about fruits, vegetables, melons, beans, and bamboo. The third volume makes good additions, with several pieces of peony cultivation methods transcribed from other books.

Ru Nan Pu Shi (1620), written by Zhou Wenhua (周文华) in the late Ming Dynasty, is divided into 12 categories: planting on an agricultural calendar, *huaguo*

(花果), *muguo* (木果), *shuiguo* (水果), woody flowers, thorn flowers, herb flowers, bamboo, wood, grass, vegetables, melons, and beans. The agricultural calendar introduces gardening activities suitable for each month and involves weather forecasts. The planting part includes 12 steps in the process of cultivation from sowing to harvest, which is followed by the planting techniques of 32 kinds of fruit, 91 kinds of flowers, 22 kinds of bamboo and wood, as well as 40 kinds of vegetable, mostly based on the author's own experience.

Qun Fang Pu (1621), formerly known as *Er Ru Ting Qun Fang Pu*, was compiled by Wang Xiangjin. The book consists of 28 volumes and is divided into four parts: yuan (元), heng (亨), li (利), and zhen (贞). The 4 parts are further divided into 14 subsections: tian (天 heaven), sui (岁 year), grain, vegetable, fruit, tea, bamboo, mulberry, hemp and *ge* (葛), cotton, medicine, wood, flower, grass, and *heyu* (鹤鱼). Although the style of the book is still the same as that of *Quan Fang Bei Zu*, it contains more plant varieties and is more detailed in content. Except for the three categories at the beginning and end of the book, the rest belong to the range of botany and contain nearly 400 species. The content is "a little bit about planting, but more about the treatment of diseases and allusions."

Guang Qun Fang Pu (1708) was adapted from Wang Xiangjin's *Qun Fang Pu* under the instruction of Emperor Kangxi during the Qing Dynasty. The 100-volume book is divided into 11 parts, namely, *tianshi*, grains, mulberry and hemp, vegetables, tea, flowers, fruits, wood, bamboo, grass, and medicine. The book deletes some contents unrelated to farming and makes corrections and additions. The contents are more substantial and the style tends to be more complete, thus enjoying a higher practical and academic level.

Hua Jing (《花镜》) (1688) by Chen Haozi (陈淏子) is famous for its profound, detailed, and systematic contents. There are more than 110,000 words in 6 volumes. Volume 1 is a new edition of the flower calendar, including 10 items such as planting, transplanting, cutting, grafting, layering, sowing, harvesting, watering, cultivation, and rectification. Volume 2 deals with the 18 methods of *kehua* (课花), which is equivalent to the general theory of cultivation. It is full of creative ideas and is the essence of the book. Volumes 3 to 5 are about the categories of flowers, vines, and grasses, respectively, each with cultivation techniques, totaling 352 species, which is actually the theory of cultivation. Volume 6 contains notes on the methods of raising fowls, scaly aquatic animals, and insects, outlining the methods of raising and managing 45 kinds of ornamental animals. The discussion on ornamental plant classification, grafting mechanisms and plant variability in the book is based on the original ideas obtained from ornamental practice, which is indeed valuable. The summary on the plan of garden layouts shows its extraordinary talent with its elegant conception and exquisite collocation. The book has been reprinted many times since its publication.

The catalogue of ornamental plants in the Ming and Qing Dynasties was still based on a limited number of flowers such as orchid, chrysanthemum, peony, and *shaoyao*, among which 36 were chrysanthemum, 15 were orchid, and 10 were peony, accounting for the vast majority. The rest were *shaoyao*, camellia, and Chinese flowering crabapple, while special books on rose, lotus, and impatiens were first published in the Qing Dynasty.

Pulu works on orchids during the Ming and Qing Dynasties are as follows.

Lan Pu Ao Fa (《兰谱奥法》), written by Zhou Lujing (周履靖) during the Ming Dynasty (or Zhao Shigeng (赵时庚) in the Song Dynasty), includes seven items about the method of orchid planting. It is an excellent work despite its short length.

Lan Pu (《兰谱》) was written by Gao Lian (高濂), a famous scholar during the reign of Wanli in the Ming Dynasty. The book is mainly copied from *Jin Zhang Lan Pu* (《金漳兰谱》), with an appendix of *Zhong Lan Ao Jue* (《种兰奥诀》). The book proposes tips on planting orchids in all four seasons.

Di Yi Xiang Bi Ji (《第一香笔记》) (1796) was written by Zhu Kerou (朱克柔), which was originally entitled *Zu Xiang Xiao Pu* (《祖香小谱》). It not only compiles the relevant documents of its predecessors, but also narrates them according to his personal experience. The book includes articles on flower quality, nature, appearance, cultivation, protection, miscellany, citation, and so on. It discusses the biological characteristics of orchids in great detail. Some terms pioneered by the book are still in use today. The description of the orchid trading market and orchid production base is also rare historical material with reference value.

Lan Yan Lue Shu (《兰言略述》) written by Yuang Shijun (袁世俊) at the end of the Qing Dynasty includes 97 varieties of *lanhui* (兰蕙), reviewing their shapes, habits, and districts, respectively.

Most of the *pulu* works on chrysanthemum in the Ming Dynasty record the excellent achievements of chrysanthemum cultivation in the Taihu Lake region, of which *Yi Ju Shu* (《艺菊书》) by Huang Shengzeng was the earliest one. The book is divided into six parts. Different from books during the Song Dynasty, which focused on the description of flower qualities, it focuses on the discussion of cultivation techniques and has certain practicability. Besides this, *Ju Pu* by Zhou Lujing also attaches great importance to cultivation methods. The first volume of the book, entitled "Yi Ju Fa" (艺菊法), includes 15 items, in which flower qualities are kept but no longer occupy the main position. The next volume includes Huang's *Yi Ju Shu*. Furthermore, *Ju Pu* (1746), written by Shu Qiu Ming Zhu Ren (署秋明主人) during the Qing Dynasty, records hundreds of varieties purchased from the south. The book records the experience of successfully making the flowers prosper after a lot of study to overcome the unfavorable conditions of climate, water, and soil.

Peony production areas expanded from Luoyang, Chenzhou, and Pengzhou in the Song Dynasty to Bozhou and Caozhou in the Ming and Qing Dynasties, resulting in the emergence of peony works such as *Bo Zhou Mu Dan Shi* (《亳州牡丹史》) and *Cao Zhou Mu Dan Pu* (《曹州牡丹谱》).

Bo Zhou Mu Dan Shi (1617) was compiled by Xue Fengxiang (薛凤翔) in the Ming Dynasty. He was a Bozhou native fond of peonies and planted many in his own garden. He summarized cultivation and management techniques, recounted anecdotes, collected famous poems from the Tang and Song Dynasties, and compiled them into a book. According to the book, there were as many as 160 peony varieties in Bozhou at that time.

Other works related to the peony in Bozhou include *Bo Zhou Mu Dan Zhi* (《亳州牡丹志》) written in the Ming Dynasty and *Bo Zhou Mu Dan Shu* (《亳州牡丹述》) by Niu Xiu (钮琇) during the Qing Dynasty. Besides these, *Cao Zhou Mu Dan Pu* (《曹州牡丹谱》) (1792), written by Yu Pengnian (余鹏年) during Qing Dynasty, records 56 species of peony in Caozhou (now Heze of Shandong Province) and 7 cultivation techniques commonly used in the area. Other works related to Cao Zhou peonies include *Cao Nan Mu Dan Pu* (《曹南牡丹谱》) written by Su Yumei (苏毓眉) and *Zhong Mu Dan Pu* (《种牡丹谱》) by Guo Ruyi (郭如仪) in the Qing Dynasty. What is more, *Mu Dan Pu* (1809) by Ji Nan (计楠) in the Qing Dynasty contains 103 species, of which 24 are of Bozhou origin, 19 are of Caozhou origin, 47 are of Songjiang origin, 8 are of Mount Dongting origin, and 5 are of Pingwang origin. There are short explanations under the names of the flowers, and the planting methods are described in great detail.

One volume of *Yue Ji Hua Pu* (《月季花谱》) signed by "Ping Hua Guan Zhu" (评花馆主) is included in *Nong Xue Cong Shu*. The book is divided into 9 parts, namely, watering, cultivation, breeding, pruning, cold shelter, cutting, seeding, pest removal, and categorization.

There are 33 species of lotus recorded in *Xiang He Pu* (《瓨荷谱》) written by Yang Zhongbao (杨钟宝), including 10 categories of single petal, 1 category of *chongtai* (重台), 9 categories of *qianye* (千叶), 7 species of single petal, and 6 species of *qianye*, with descriptions of each one. The following are "Yi Fa Liu Tiao" (艺法六条 six techniques) entitled "Chu Yang" (出秧), "Shi Ou" (莳藕), "Wei Zhi" (位置), "Pei Yang" (培养), "Xi Ji" (喜忌), and "Cang Yang"(藏秧).

Zhao Xuemin (赵学敏) (1719~1805) wrote *Feng Xian Pu* (《凤仙谱》) (1790) in 2 volumes. The frontispiece is "Pu Li" (谱例), stating that the main theme of the compilation is "having comprehensive content about categorization, with necessary information on planting and irrigation, methods to treat diseases as well as their medicinal values." The first volume deals with names and categories, and extensively quotes from the literature of past dynasties to study the naming of impatiens. The flower is divided into 11 categories and the shape, color, characteristics, and appearance of about 180 kinds are narrated. The second volume includes 8 parts: planting, irrigation, harvesting, *yihua* (医花), pest control, medicine preparation, general discussion, and miscellaneous remarks. The book contains more than 30,000 words. It must be the earliest and only book on impatiens.

Since the appearance of Cai Xiang's *Li Zhi Pu* (1059), monographs on fruit trees became a popular topic of writing. According to *Zhong Guo Nong Xue Shu Lu*, there are 12 monographs on fruit trees in the Ming and Qing Dynasties, of which 9 are devoted to litchi. The authors' native place and the main recorded places of origin are mostly in Fujian. For example, in the late Ming Dynasty, Tu Benjun (屠本畯) compiled *Min Zhong Li Zhi Pu* (《闽中荔枝谱》), which was written when he was an official in Fujian. No other books appeared until *Ling Nan Li Zhi Pu* (《岭南荔枝谱》) by Wu Yingkui (吴应逵) under the reign of Ganlong in the Qing Dynasty.

Shui Mi Tao Pu (《水蜜桃谱》) (1813) written by Chu Hua (褚华) and *Ju Li Pu* (《槜李谱》) by Wang Fengchen (王逢辰) (1857) record peach and plum produced in Shanghai and Jiaxing, respectively. Both of the two books describe the methods of planting, transplanting, deinsectization, and harvesting in detail. As the authors of the two books are both locals, the content is full and accurate.

The one volume of *Long Yan Pu* (《龙眼谱》) written by Zhao Gunong (赵古农) during the Qing Dynasty is the earliest book on *longyan*.

3.5.4.4 Wild Vegetables and Herbs

The emergence of a large number of wild vegetable works is another major feature of agricultural works in the Ming and Qing Dynasties. Among them, *Jiu Huang Ben Cao* (《救荒本草》) (1406), compiled by Zhu Su (朱橚), had the greatest impact. Zhu Su was fond of learning and was good at writing verse. He collected and cultivated more than 400 kinds of plant seedlings in his garden. He personally observed and recorded them to identify the nature and taste. For edible ones, he employed painters to draw maps according to their actual appearance. A total of 414 species of plants were collected in the book, of which 138 species have been found in herbal medicine of past dynasties and 276 species have been added, which is exactly 2/3 of the book. They are divided into 245 species of grass, 80 species of wood, 20 species of rice and cereals, 23 species of fruit, and 46 species of vegetables. It records the name, alias, origin, and characteristics of plants accurately. This book obviously surpasses the cursive script of past dynasties in terms of the accuracy of description, the richness of terms, and the fineness of drawings. Xu Guangqi quoted the full text of the famine policy in *Nong Zheng Quan Shu*. Another influential book is *Ye Cai Pu* (《野菜谱》) written by Wang Pan (王磐). The book is only one volume long and contains more than 60 kinds of wild vegetables, each with pictures and poems. Xu Guangqi once included it in *Nong Zheng Quan Shu*. Later, during the Ming Dynasty, Hua Hao (滑浩) deleted the drawings, inscribed poems in turn, changed the order, and printed it under the original title *Ye Cai Pu*.

Bao Shan (鲍山) wrote *Ye Cai Bo Lu* (《野菜博录》) during the Ming Dynasty (1622), recording 435 kinds of potherb. The Book is divided into two parts of grass and wood, and then subdivided into groups according to their edible parts. Each kind of potherb is illustrated and its characteristics and eating method are briefly recorded.

Jiu Huang Ye Pu (《救荒野谱》) (1642) written by Yao Kecheng (姚可成) during the Ming Dynasty was compiled from *Shi Wu Ben Cao* (《食物本草》) by the famous doctor Li Dongyuan (李东垣) (1620), which included 60 species of edible grass, supplemented by 45 species of grass and 15 species of wood. Besides the illustrative drawings, there were also songs and detailed notes on the eating method, making it easy for people to understand and remember.

Zhou Lujing wrote *Ru Cao Bian* (《茹草编》) in the late Ming Dynasty. The book is divided into four volumes. The first two volumes record 105 kinds of wild vegetables with attached pictures. The second two volumes are a compilation of relevant anecdotes and old proverbs, and the eating method is extremely well-described.

Ye Su Pin (《野蔌品》) (1591) by Gao Lian (高濂) is a single-line compilation of a part of the 12th volume of the book *Zun Sheng Ba Jian* (《遵生八笺》). It records nearly 100 kinds of wild vegetables, "all known as edible."

Tu Benjun wrote *Ye Cai Jian* (《野菜笺》) during the Ming Dynasty and collected only 22 kinds of wild vegetables from Siming Mountain.

Ye Cai Zan (《野菜赞》) (1652) by Yan Jingxing (颜景星) recorded the 44 kinds of wild vegetables he had eaten, and indicated their characteristics and eating methods. After the description of each kind of wild vegetable, they were praised for their contribution to life.

From the Tang Dynasty and Qing Dynasty, 188 tea books have been initially confirmed, including 96 complete ones, 28 collected, and 64 lost books. In terms of dynasty, there are 16 from the Tang and Five Dynasties, 47 from the Song and Yuan Dynasties, 79 from the Ming Dynasty, 42 from the Qing Dynasty, and 4 from unknown years during the Ming and Qing Dynasties. Tea books from the Ming and Qing Dynasties accounted for more than 72% of the ancient tea books. The Ming Dynasty was the dynasty with the largest number of tea books written and published in ancient China (Chuanzheng and Zhu 2006).

Among these numerous tea books, the following ones are outstanding and detailed. *Cha Jie* (《茶解》) (completed in 1609) by Luo Lin (罗廪) is a more specific and systematic one. The collection, cultivation, and processing of tea and the location and management of tea gardens are all discussed in detail. It surpasses books of the previous generation to some extent. For example, the summary of the main technical points for *chaoqing* (炒青) is still the technological principle followed in the processing and manufacturing of high-grade green tea today.

Under the reign of Wanli in the Ming Dynasty, Tu Long (屠隆) and Wen Long (闻龙), respectively, wrote and printed two books with same name, *Cha Jian* (《茶笺》). The former book records famous products produced in various places, including items on picking, sun drying, baking, and storing of tea. The latter is mainly about baking, which reflects the situation of drinking, production, and processing of tea in the Ming Dynasty.

Of the 11 tea books in the Qing Dynasty, 7 belong to the early Qing Dynasty, 1 belongs to the late Qing Dynasty, and the remaining 3 cannot be verified. Lu Tingcan (陆廷灿) wrote *Xu Cha Jing* (《续茶经》) when he was the District Magistrate of Chong'an county in Fujian province. The book, as a supplement to *Cha Jing* by Lu Yu during the Tang Dynasty, compiles the previous works and adds his own experiences. It is divided into 10 items, and there is 1 volume appendix, describing the tea methods of past dynasties.

During the Sui, Tang, Song, and Yuan Dynasties, herbs became an important part of agricultural books, but the writing of a special *pulu* book for certain drugs began in the Qing Dynasty. Wang Yuyang (王渔洋) aspired to write *Ren Shen Pu* (《人参谱》) but did not finish it. In the thirty-first year of Ganlong (1766), Lu Xuan (陆烜) wrote *Ren Shen Pu* in the process of curing his own disease. It compiles hundreds of documents of past dynasties. In the frontispiece there is a full picture of ginseng, and the main text is divided into six categories: explanation of name, origin, nature and taste, prescription, source, allusion, and poetry. It is probably the earliest monograph on ginseng published in China.

3.5.4.5 Sericulture

From the middle of the Qing Dynasty, silk exports increased, prompting the development of sericulture in the Zhujiang River Delta and the Yangtze River Delta, and the introduction of mulberry trees to try out sericulture was quite popular throughout the country. Some officials also tried to write and distribute silkworm books to popularize knowledge about the silkworm and mulberry. *Zhong Guo Nong Xue Shu Lu* included 30 books on sericulture from the Qing Dynasty. What is more, according to statistics from Wang Da (王达), there are still 204 silkworm and mulberry farming books from the Ming and Qing Dynasties not included, among which 197 are from the Qing Dynasty (Wang Da: Collection and Introduction of Sericultural Books during the Ming and Qing Dynasties, *Zhong Guo Nong Shi*, 1986, Volume 4; 1989, Bound Edition 2).

There is only one sericulture work from the Ming Dynasty, *Can Jing* by Huang Shengzeng. It is a special book on mulberry cultivation and silkworm rearing in the south of Yangtze River. Huang Shengzeng summarizes the experience of mulberry planting and silkworm rearing in Suzhou and Hangzhou. The book is divided into nine parts, eight of which are mainly about silkworm rearing, hence the title *Can Jing*.

There are five silkworm books from the early Qing Dynasty. *Bin Feng Guang Yi* (1740), a book written by Yang Shen, aims at promoting mulberry planting and silkworm rearing in Shaanxi. The book is divided into three volumes, which explain and analyze in turn the key points of operation in each link of the mulberry planting, silkworm rearing, and silk reeling process. There are more than 50 attached drawings. After the book was completed, it was rewritten to make it easier to understand by using folk sayings as much as possible, resulting in the new book *Can Zheng Zhai Yao* (《蚕政摘要》) (1756). According to the order of the operating rules, it first talks about mulberry planting, then about utensils, and finally silkworm reeling.

There are two books entitled *Can Sang Shuo* (《蚕桑说》). One was written by Shen Qian (沈潜) when he was the District Magistrate in Luojiang County in Sichuan and is included in *Luo Jiang Xian Zhi* (《罗江县志》). The other was written by Li Ba (李拔), an official in Fujian, and is included in *Huang Chao Jing Shi Wen Bian* (《皇朝经世文编》).

Yang Can Cheng Fa (《养蚕成法》) (1766) was compiled by Han Mengzhou (韩梦周), the District Magistrate of Lai'an County in Anhui province in order to promote the rearing of tussah. The book was slightly modified from *Yang Shan Can Cheng Fa* (《养山蚕成法》) (1743) published by the Shandong Provincial Governor's Office in the early years of Ganlong. It describes the method of rearing tussah in Shandong and was widely circulated. Another book is *Yang Shan Can Shuo* (《养山蚕说》) (1771), which has been lost. It was published by Hao Jingxiu (郝敬修), then District Magistrate of Hanyin County in Shaanxi, to advocate rearing tussah.

The number of silkworm books from the middle of the Qing Dynasty is still small, but their level obviously improved.

Wu Xing Can Shu (《吴兴蚕书》), written by Gao Quan (高铨), a native of Guian, Zhejiang Province, recounts in detail the methods of planting mulberry and raising silkworms in the west of Zhejiang Province. It also extracts relevant information from *Shen Shi Nong Shu* and other works, which were printed by Shen Xizhou (沈锡周) in the 16th year of Guangxu during the Qing Dynasty (1890). Since the author was a local, the content was very accurate.

Can Sang Shuo (《蚕桑说》) (1840) was compiled by Shen Lian (沈练) in Liyang, Jiangsu province, when he was an assistant instructor in Jixi County, Anhui province, to promote mulberry planting and sericulture and teach local people about sericulture methods in Liyang, his hometown. After settling in Xiuning in his later years, he supplemented the book and re-titled it *Guang Can Sang Shuo* (《广蚕桑说》) (1855), based on the newly published *Can Sang Ji Yao* at that time. In the early years of Guangxu, Zong Yuanhan (宗源瀚), the prefect of Yanzhou, Zhejiang Province, set up a Silkworm Bureau to promote the planting of mulberry and sericulture. He asked Zhong Xuelu (仲学辂) of Chun'an County to supplement it, and entitled it *Guang Can Sang Shuo Ji Bu* (《广蚕桑说辑补》) (1875). The book is divided into two volumes, including 19 items on mulberry cultivation, 66 on silkworm rearing, 16 miscellaneous essays, and a newly added general introduction of silkworm and mulberry. The book was printed in several places and enjoyed high popularity due to its good reasoning and organization as well as its simple content.

The late Qing Dynasty silkworm books were compiled and published after the 1880s, with a total of more than 100 books, including the following books.

Can Sang Ji Yao (1871) was written by Shen Bingcheng, a native of Zhejiang Province and a Metropolitan Graduate in the year of Bingchen under the reign of Xianfeng (1856). When he was a Circuit Intendant at the Tonghai Circuit, Changzhen, Jiangsu, he wrote the book to advocate the promotion of sericulture. It is divided into four categories: norms, miscellaneous remarks, illustrations, and *yuefu* (乐府). Among them, the miscellaneous remarks contain a collection of *Can Sang Qian Shuo* (《蚕桑浅说》) by He Shi'an (何石安) under the reign of Daoguang. It systematically and briefly narrates the cultivation of silkworm and mulberry in different sections. The illustrations depict 36 silkworm and mulberry tools, each with explanations to make them easy to copy. It is often included in later silkworm books and was widely circulated and engraved many times.

Hu Can Shu (1874) was compiled by Wang Rizhen, a native of Wucheng, Zhejiang Province, who participated in the reconstruction of *Hu Zhou Fu Zhi* (《湖州府志》) and was exclusively in charge of the section on sericulture. Later on, it was published separately as a book with a few additions and deletions. What was cited in the book was the recent sericulture data recorded in the documents from the Huzhou area, and was only for practical purposes. The four-volume book, which talks in turn about silkworm tools and mulberry planting, silkworm rearing techniques, *shangshan* (上山) and silk reeling, silk selling and weaving, is a comprehensive silkworm book on silkworm rearing techniques in Huzhou during the Qing Dynasty.

Can Shi Yao Lue (《蚕事要略》) was written by Zhang Xingfu (张行孚), a native of Anji, Zhejiang. In view of the fact that the sericultural techniques used in Huzhou were quite different from those in ancient books, he argued that comparison should be made so as to choose good ones. The ancient methods and theories quoted in the book were mostly based on *Nong Sang Ji Yao*, so in addition to the original inscriptions, the later *Nong Sang Ji Yao* from the series *Jian Xi Cun She* (渐西村舍) and *Si Bu Bei Yao* (四部备要) both included and attached the book for reference.

Bi Nong Zui Yao (《裨农最要》) (1897) was written by Chen Kaizhi (陈开沚), a native of Santai, Sichuan. Mostly quoting from previous theories, the book also includes some remarks on his own experience, making it a systematic and detailed monograph highlighting local characteristics.

Can Sang Cui Bian (《蚕桑萃编》) (1892) was written by Wei Jie (卫杰), a native of Sichuan. When Li Hongzhang (李鸿章) (1823~1901) created the Silkworm Bureau for the establishment of industry as Governor-general of the Metropolitan Area and Minister of Commerce of Beiyang, he conferred on Wei Jie the title of Circuit Intendant and instructed him to be in charge of the compilation. The 15-volume book is divided into seven parts, namely, imperial edicts, mulberry policies, silkworm policies, silk reeling policies, weaving policies, paintings, and external records, starting with the imperial edicts of past dynasties and ending with the collection of sericulture from western and oriental countries. It is the largest volume of all ancient silkworm books, which comprehensively includes mulberry, silkworm, reeling, spinning, dying, and weaving.

3.5.4.6 Animal Husbandry and Veterinary

The animal husbandry and veterinary works of the Ming Dynasty are still mainly concentrated on raising horses. There were many veterinary works such as *Lei Fang Ma Jing* (《类方马经》), *Ma Shu* (《马书》), and *Yuan Heng Liao Ma Ji* (《元亨疗马集》). Of the more than 30 veterinary monographs cited in *Yuan Heng Liao Ma Ji*, only *Si Mu An Ji Ji* and *Quan Ji Tong Xuan Lun* (《痊骥通玄论》) are works of the previous generation, while the remaining 30 or more are works from the Ming Dynasty.

Lei Fang Ma Jing (《类方马经》) is a kind of official horse medical book written in the period of Chenghua (1465~1485) under the reign of Xianzong during the Ming Dynasty. It was originally carved in 6 volumes and is a large special veterinary book.

Ma Shu (《马书》) (completed in 1594) was written by Yang Shiqiao (杨时乔) (?~1609) when he was in charge of raising war horses. The book covers horse breeding, horse judging, and the therapy of horse diseases, but mainly focuses on diagnosis and treatment. The book consists of 14 volumes, and nearly 4/5 of the book focuses on diagnosis, treatment, and disease symptoms. The book gives information on the breeding season, breeding age, breeding of stallions, and methods of judging whether mares are pregnant. In the book, the advisable breeding season for horses is spring, but breeding in summer is also mentioned. The main requirement is to breed in the early morning of a sunny day or in the evening when it is cool. The stallion has to be three years old. As a stallion, a male horse "must be fed to make it fat and strong" when grazing in spring. Mares "must eat grass before drinking water, are not allowed to feed on buckwheat straw, millet, coarse cereals or rice swill, and if dirty water is fed to them, it would be difficult to bear a foal." The book also records the method of using a male horse to test and determine whether the mare is pregnant. If the mare is found to "kick" and no longer mate with the male horse, it means she is pregnant. *Ma Shu* inherits the experience of *Qi Min Yao Shu* and sums up the "three rules concerning drinking and feeding." First, let horses drink less when thirsty, weak, or pregnant and don't feed them too much when they are hungry, going out or coming from far away. Second, stop them drinking dirty water and keep their fodder clean. Third, horses are not allowed to drink when riding, sweating, or after eating fodder, and excessive fodder is not appropriate when horses are getting fat, riding less, or during a hot summer. *Ma Shu* includes documents from *Shi Huang Wen Dui Mai Se Lun* (《师皇问对脉色论》), *Ba Yao Tu Lun* (《八要图论》), and "Qi Shi Er Da Zheng" ("七十二大症"). It makes a preliminary summary of the diagnosis theory and treatment based on the syndrome differentiation of Chinese veterinarians and begins the relevant discussion in *Yuan Heng Liao Ma Ji*.

Yuan Heng Liao Ma Ji was compiled by Yu Ren and Yu Jie, veterinarians from the Ming Dynasty. It was first engraved in the 36th year of Wanli during the Ming Dynasty (1608) and has been edited and amended many times. The main content includes three parts: horse therapy collection, cattle therapy collection, and camel theories. The horse therapy collection is the central part of the book and is divided into four volumes: spring, summer, autumn, and winter. The book has 113 pictures, 3 *fu*s, 150 songs, and more than 300 prescriptions. For most of the veterinary diseases described in the book, the part of "Lun" (论 theory) explains the cause of disease, "Yin" (因 cause) indicates symptoms, and "Fang" (方 prescription) indicates treatment. They are expressed in the form of "song" or "ode" and are easy to recite and remember.

As a summary veterinary classic, the main achievements of the book include the following aspects. First, the development of *semai* (色脉) diagnosis theory. For the tips on detecting diseases, *semai* is the best. *Yuan Heng Liao Ma Niu Tuo Jing Quan Ji* (《元亨疗马牛驼经全集》) contains several representative documents on "Mai Se Lun" (脉色论), which develops and improves the theory and makes it an important means to help Chinese veterinarians to diagnose diseases.

The second is the systematic summary of the "Ba Zheng Lun" (八证论). *Bazheng* (八证 eight symptoms) are the outline of syndrome differentiation for Chinese veterinarians and the basis for treatment. They refer to eight syndromes of "zhengxie" (正邪), "hanre" (寒热), "xushi" (虚实), and "biaoli" (表里). They are developed from "bayao" (八要) in human medicine. The proposal of the "Ba Zheng Lun" set a clear standard for veterinary treatment and promoted the further development of syndrome differentiation and treatment.

The third is the summary of "seventy-two major symptoms," which are common and refractory diseases in the treatment of equine diseases. In the book, the etiology and pathogenesis of each disease is indicated, and the characteristics of each one are described in detail. It especially introduces the key points of difference when the symptoms are the same. It is an important achievement to sum up the experience that when symptoms are the same but the etiology and pathogenesis are different, the curative effect is only achieved by adopting different therapeutic methods.

At the beginning of the Qing Dynasty, the rural areas in the mainland restricted the Han people from raising horses, and the horse medicine industry was in depression. Instead, the development of animal husbandry and veterinary science related to cattle and pigs prospered. During this time books such as *Niu Jing Da Quan* (《牛经大全》), *Yang Geng Ji* (《养耕集》), *Bao Du Ji* (《抱犊集》), *Xiang Niu Xin Jing Yao Lan* (《相牛心经要览》), *Zhu Jing Da Quan* (《猪经大全》), *Chuan Ya Shou Yi Fang* (《串雅兽医方》), and *Huo Shou Ci Zhou* (《活兽慈舟》) appeared.

Niu Jing Da Quan, also known as *Shui Huang Niu Jing He Bing Da Quan* (《水黄牛经合并大全》), written by an unknown author, is divided into two volumes. It was later included in the re-edited *Yuan Heng Liao Ma Ji*.

Yang Geng Ji (1800) was dictated by veterinarian Fu Shufeng (傅述凤), and compiled into a book by his son Fu Shanchang (傅善苌). It is divided into two parts. The first part focuses on acupuncture. Before this, only one picture of "Niu Ti Xue Fa Ming Tu" (牛体穴法名图) was published in the world, lacking written description. However, *Yang Geng Ji* describes the correct position of more than 40 acupuncture points, the depth and technique of needle insertion, and lists more than 20 corresponding special acupuncture methods, thus forming a complete system of cattle acupuncture. The second part contains a variety of prescriptions, of which the supplementary prescriptions are mainly for digestive diseases, which highlights the therapeutic principle of combining acupuncture and medicine to bring out the best in each other.

Bao Du Ji is a book on cattle medicine written in the late Qing Dynasty. Its contents mainly include: introduction to medical treatment, acupuncture, symptoms of cow diseases, drug properties and prescriptions, etc. The introductory chapter systematically discusses the basic theory and emphasizes that we should not use drugs without investigating the feelings or knowing the symptoms.

Xiang Niu Xin Jing Yao Lan (1822), written by an unknown author, is divided into 31 sections. It describes the identification criteria for each part of the cow's body in a very detailed way, mainly from the aspects of servitude and temperament. The main target group is buffaloes, and there is another section entitled "Huang Niu Zong Lun" (黄牛总论).

Zhu Jing Da Quan (1891) is a special book on swine diseases popular in Sichuan, Guizhou, and other southwest regions in the late Qing Dynasty, written by an unknown author. It has a short preface and describes 50 kinds of common diseases, many of which are illustrated with disease images and all of which are listed with treatment methods. The prescriptions include single prescription, simple prescription, and classic prescription.

Huo Shou Ci Zhou (1873) was written by Li Nanhui (李南晖) during the Qing Dynasty. It is divided into 7 parts: cattle, buffalo, horse, pig, sheep, dog, and cat. The book has more than 200,000 words, with cattle and buffalo each accounting for half. In the book, the feeding and management of each kind of livestock is discussed first, and the appearance is identified. Then the syndrome differentiation and treatment are carried out according to *wuzang wujing* (五脏五经). The book contains 240 symptoms and more than 700 prescriptions. One symptom is often combined with two prescriptions, one for traditional Chinese medicine and the other for herbal medicine, reflecting the characteristics of the combination of prescription and medicine.

Chuan Ya Shou Yi Fang is a monograph on veterinary prescription compiled from *Chuan Ya Wai Pian* (《串雅外篇》), which was co-compiled by Zhao Xuemin (赵学敏) (about 1719~1805), author of *Ben Cao Gang Mu Shi Yi* (《本草纲目拾遗》) (1875), and Zong Boyun (宗伯云), a traditional Chinese medicine practitioner. Some prescriptions in the book had not been recorded in ancient veterinary books in the past.

3.5.4.7 Poultry and Fish

During the Ming and Qing Dynasties, monographs on fish and chicken raising also appeared. *Bu Ji* (《哺记》) was written by Huang Baijia (黄百家), son of Huang

Zongxi (黄宗羲), a famous scholar in the Qing Dynasty. When Huang Baijia was living on Mount Ya, he investigated the local feeding workshop and wrote a thousand words of *Bu Ji*, which included the identification of male and female eggs, incubation techniques, and the experience of identifying embryos by observing illuminated eggs. Another book is *Ji Pu* (《鸡谱》), a book on raising cockfighting chickens. The book was written in the Qing Dynasty under the reign of Ganlong. It consists of 51 articles with more than 14,000 words. The main contents include description and judgment of the appearance of fighting cocks, selection and breeding of improved varieties of fighting cocks, hatching of eggs, feeding and management of chickens, various diseases and their prevention and control measures, and the selection of fighting cocks and their treatment after fighting. It is the only ancient monograph on chicken raising that has ever been discovered. Besides this, there is *An Chun Pu* (《鹌鹑谱》) written by Cheng Shilin (程石麟) during the reign of Qing Emperor Kangxi. It was a monograph on raising and training quails. In the book, the methods of raising fighting quails are introduced in different categories, and the experience of raising quails is summarized.

Yu Jing (《鱼经》) written by Huang Shengceng is a special book on fish culture and fishery resources. The book is divided into three parts. The first part introduces the reproduction methods of several fish, including carp, trout, grass carp, silver carp, mullet, etc. The reproduction methods can be summed up in two ways: spawning and hatching, and taking seedlings for pond breeding. The feed used is the egg yolk of chickens and ducks, barley bran, or fried soybean flour. The second part introduces the method of fish culture, focusing on two aspects: chiseling the pond and feeding. The third part introduces 19 major fish species in rivers, lakes and seas, most of which are valuable ones, including sturgeon, bambusa, perch, sole, pomfret, bighead, whitefish, bream, whitefish, shad, tuna, saury, roe, crucian carp, tiger shrimp, anchovy, and puffer fish (spotted fish).

Zhu Sha Yu Pu (《朱砂鱼谱》) (1596) written by Zhang Chou (张丑) during the Ming Dynasty contains two chapters. The first chapter describes the morphological variety, genetic variation, and artificial selection of goldfish from 10 aspects. The other chapter records the ecological habits, reproduction, and rearing methods of goldfish also from 10 aspects. The book correctly points out that a goldfish's tail, color, pattern, trunk, and various parts are different from ordinary fish, with its trunk especially thick, short, and fat. It also describes for the first time the application of a mixed selection method to cultivate new varieties of goldfish. The book is an early monograph on ornamental goldfish in ancient China.

Tu Benjun, the author of *Ye Cai Jian*, wrote *Min Zhong Hai Cuo Shu* (《闽中海错疏》) (1596) when he was Vice Magistrate of Fujian Salt Distribution Commission during the Wanli period. The book has three volumes as well as one "self-preface" and one "appendix." It records more than 200 species of aquatic animals in the Fujian coastal areas, of which marine animals account for 3/4 of the whole book (including a few freshwater species). Many aquatic animals mentioned in the book have a relatively high economic value, such as the large yellow croaker, yellow croaker, hairtail and cuttlefish, as well as mackerel, herring, sardines, sole, pomfret, and prawn. Later, the author also wrote the 1 volume *Hai Wei Suo Yin* (《海味索引》).

3.5.4.8 Locust Control

In the Ming and Qing Dynasties, locust control continued to be a concern and led to the emergence of special books on the topic. Most of their compilers are local administrative officials. Because of their practical use, local authorities often reprint them and spread them widely. There are more than 20 preserved so far. The more important ones are as follows.

Bu Huang Kao (《捕蝗考》) (about 1684), written by Chen Fangsheng (陈芳生) during the Qing Dynasty, is the earliest monograph on locust hunting that can be seen now. The first part of the book is about the prevention of locusts, with a total of 10 items. The latter part is about locust trapping methods of the previous generation.

The six articles including "Hu Tian" (护田) can be found in *Hui Zheng Zhi* (《惠政志》), volume 5 of *Xing Tang Xian Xin Zhi* (《行唐县新志》) in Hebei province. The article introduces "hutianfu" (护田夫), a basic-level organization established in the 29th year of Ganlong during the Qing Dynasty (1764), which "specializes in searching for locust nymphs." The article "Bian Lei" (辨类) gives a detailed description of the life history, species, morphology, and ecology of locusts (including gnats and slime worms). The article "Bu Fa" (捕法) explains in detail the methods of killing locusts. The article "Sou Zhong" (搜种) details the method of searching for nymphs to eradicate locust plagues. The article "Qi Dao" (祈祷) introduces some popular folk methods of using prayer to eliminate locusts, and the article "Quan Yu" (劝谕) mentions incentives to catch locusts and ways to turn harm into profit.

Zhi Huang Chuan Xi Lu (《治蝗传习录》) is a compilation of locust control materials compiled and published by Chen Shiyuan in the year of Bingshen during the reign of Ganlong in the Qing Dynasty (1776). It is composed of eight articles such as "Bu Ban Tong Shang Ge Sheng Pu Bu Huang Nan Fa Ze " (《部颁通伤各省扑捕蝗蝻法则》) and "Shu Zu Chuan Zhi Huang Yi Fa Shi Mo Gen You " (《述祖传治蝗遗法始末根由》) as well as verses on duck breeding and locust control. It contains *Zhi Huang Bi Ji* written by Chen Jinglun in the 25th year of Wanli during the Ming Dynasty (1597), which contains the earliest records on duck breeding and locust control.

Bu Huang Hui Bian was written by Chen Jin (陈仅) when he was a District Magistrate of Ziyang County in Shaanxi Province in the 16th year of Daoguang during the Qing Dynasty (1836) (Ni Genjin: Investigation into the Life and Works of Chen Jin, the Author of *Bu Huang Hui Bian*, *Gu Jin Nong Ye*, 2005, Volume 3). The book consists of four volumes, following Emperor Kangxi's "Bu Huang Shuo" (捕蝗说) in the front. The four volumes are in turn "Bu Huang Ba Lun" (捕蝗八论 eight theories on locust hunting), "Bu Huang Shi Yi" (捕蝗十宜 ten appropriate conditions for locust hunting), "Bu Huang Shi Fa" (捕蝗十法 ten methods for locust hunting), "Shi Shi Si Zheng" (史事四证 four pieces of historical evidence), and "Cheng Fa Si Zheng" (成法四证 methodological proofs for four established methods of locusts hunting). The contents of the book are basically compiled from previous works. The four established methods cited are from *Bu Huang Ji* (《捕蝗记》) by Ma Yuan (马源), *Chu Huang Ji* (《除蝗记》) by Lu Shiyi (陆世仪), *Bu Huang Fa* (《捕蝗法》) by Li Zhonfen (李钟份), and *Bu Qiang Bu Nan Fa* (《布墙捕蝻法》) by Ren Hongye (任宏业). The book is also interspersed with comments from the authors.

Zhi Huang Quan Fa (《治蝗全法》) was written by Gu Yan (顾彦) from Wuxi in the 7th year of Xianfeng during the Qing Dynasty (1857). A year before, a plague of locusts had occurred in the Wuxi area. Gu Yan edited *Jian Ming Bu Huang Fa* (《简明捕蝗法》) and issued it to farmers. The following year, it was expanded to four volumes, adding official methods of the locust control, previous theories and records on famine relief, and was entitled *Zhi Huang Quan Fa*. In the book, the breeding and areas of locust spread, the life history and habits of locusts, and the methods of mobilizing people to catch and kill locusts are all explained in detail.

Liu Yun Ge Bu Huang Ji (《留云阁捕蝗记》) (1836) was written by Peng Shoushan when he was the District Magistrate in Leping County, Jiangxi Province, by compiling various official documents and people's experiences of locust hunting. The book gives a detailed description of the common location of the eggs and larvae of locusts and how frogs, as natural enemies, can be used to catch and eat locusts and reduce the harm.

Bu Huang Yao Shuo (1856), also called *Bu Huang Yao Jue,* was written by an unknown author. The book is preceded by the preface of Qian Xinhe (钱炘和) when he was the Provincial Administration Commissioner of the Metropolitan Area. It may have been written by previous local officials based on folk experience. There are clear records on locust's habit of fearing dampness and preferring fire, and that the number of generations common to locusts is usually one per year, but two generations can occur in individual years, "even three times after a long drought."

Chu Nan Ba Yao (1850) was written by Li Xingfu, then the District Magistrate of Changan County in Shaanxi Province. It gives a detailed account of cankerworm's social and phototactic nature and points out clearly that when trapping and killing cankerworms by digging trenches, the shapes of trenches should be adjusted to topography. This book, together with two other related locust control books written by him, *Zhi Fei Huang Jie Fa* (《治飞蝗捷法》) and *Sou Wa Huang Zi Zhang Cheng* (《搜挖蝗子章程》), were jointly compiled and engraved by Xi'an Prefect Shen Shousong (沈寿嵩), entitled *Xian Xing Bu Chu Huang Nan Yao Fa* (《现行捕除蝗蝻要法》).

Zhi Huang Shu (1874) was written by Chen Chongdi (陈崇砥), a native of Houguan, Fujian Province, when he was an official in Hebei Province. The book details methods of controlling locusts and is accompanied by pictures. It is especially novel in suggesting "setting up an organization" to control locusts. The book also introduces a kind of tackle specially used to treat stick insects.

3.5.4.9 Theory

The development in the Ming and Qing Dynasties applied the concepts of *yin* (阴) and *yang* (阳), five elements and *qi* (气) from traditional philosophy to explain soil fertility, and problems related to agricultural production. Among them, Ma Yilong (马一龙) and Yang Shen contributed the most.

Nong Shuo (《农说》) by Ma Yilong inherits the traditional thought of emphasizing agriculture and the theory of three talents. While emphasizing manpower (physical strength), he attaches great importance to the role of knowledge (i.e., intelligence). He believes that a good farmer must have both knowledge and physical strength. Intelligence is mainly manifested in a deep understanding of agricultural principles while physical strength refers to "being diligent when farming." The function of intelligence is mainly manifested in three aspects: knowing the time (i.e., the weather), knowing the soil (i.e., the soil nature), and knowing what is appropriate to use (i.e., crops). He said, "it is necessary to harmonize the time, the soil, and the physical properties of crops and if there is no mismatch, you will get twice the result with half the effort." Ma Yilong expounded in detail the relationship between weather, soil, and agricultural production with the theory of *yin*, *yang*, and *qi*. According to the principle of "*yang* governing the occurrence and *yin* the convergence of *qi*," the theory of "*xuyang*" (畜阳 accumulating *yang*) is put forward, and it is believed that "the way of reproduction is only to combine *yang* with earth." For the sake of raising *yang*, he suggests that "winter ploughing should be done earlier and spring ploughing later." "High-lying fields should be dug deep and low-lying fields shallow." It is advisable to "till several times" to prevent "shrinkage" and in order to "intensify cultivation and smoothness," "harrow while tilling and plant while harrowing."

According to the dialectical principle of *yin* and *yang*, Ma Yilong also puts forward a method to prevent the "overgrowth" of crops. By "breaking their floating roots, cutting their attached leaves, and removing the dirt in the field to dry their skin," etc., you can inhibit the growth of roots and leaves and prevent crops from overgrowing. But the more important way is to use a base fertilizer and he attaches great importance to intertillage and ridging.

Nong Shuo puts forward the prevention and control methods of various crop diseases and insect pests, such as irrigation, *changqian* (长牵), and use of lime, tung, or *buye* (布叶). In order to prevent and control disease fundamentally, he proposes starting with seed selection, believing that fully mature millet must be used as seed. He puts forward two methods of raising seedlings. First, after the winter solstice, select a nursery in a high-lying place, cultivate it, spread seeds, cover with sparse grass to prevent birds, cover with plant ash, irrigate with water, and then during the Qingming Festival, irrigate with fertilizer water to promote germination, and weed and fertilize to promote growth. The second is to wrap the seeds with grass and hang them under eaves exposed to wind, then after spring put them in deep water pools but not close to the soil and plant them half a month later so that they will sprout.

Ma Yilong explains the significance of rice transplanting with the theory of *yin* and *yang* and believes that transplanting makes the seedling field and the rice field more viable. Transplanting requires vertical and horizontal rows to facilitate weeding. The density should be determined according to how fertile the soil is. The density of planting in fertile fields should be appropriate while in barren fields dense planting is not allowed. Weeding should be done earlier to prevent trouble later. Through weeding, the growth of transverse roots is inhibited, and the top roots are promoted to enter the soil to absorb more nutrients and increase the number of ears and grains per plant.

The article "Xiu Ye Zhang Nong Ze" (修业章·农则) in *Zhi Ben Ti Gang* (《知本提纲》) (1738), written by Yang Shen and his disciple Zheng Shiduo (郑世铎), explore the "cultivation way" and "farming way" through cultivation and by promoting techniques of sericulture, animal husbandry according to the theory of *yin*, *yang*, and the five elements. It adapts the ancient "gold, wood, water, fire and soil" to "heaven, earth, water, fire and gas." Heaven and fire belong to *yang* while earth and water belong to *yin*. *Yin* and *yang* are equally divided and through the interaction of *qi*, they meet each other, circulate harmoniously, and produce everything. It is used as the basic principle to discuss the growth and development of crops and agricultural technology.

The book explains the meaning, depth, and time of ploughing and farming that are often involved in traditional agriculture with two terms, "yindi" (因地 land-based) and "chentian" (乘天 weather-based). Being weather-based also means early autumn ploughing and late spring ploughing. Only in this way can we "avoid frost and gather *yang*" and "prevent weeds growing and lead to harmony." By land-based, he means that since mountain soil is dry and lacks *yin*, it is advisable to plough deep to connect to the *yin* of the land, but in damp places which are full of water and lack *yang*, hoe shallow so that it can be exposed to the sun. Ma Yilong said, "It is better to plough deep in high mountain but shallow in marshy land."

The book abandons the traditional practices of emphasizing deep ploughing alone and points out that "the first ploughing should be shallow just to turn over the soil and prevent weeds growing, while the second ploughing should be deeper until mud is seen to get rid of the roots." It is proposed to deepen the cultivation step by step through *zhuangeng* (转耕) and *taogeng* (套耕). While preserving the merits of the *qutian* method, a "short-wall cultivation" is proposed to solve the problem of having a large population but little land. The suggested method is to build several short-walls in the field in summer months and level them to the fields after autumn. In this way the soil becomes self-fertilizing and the roots of the crop grow deep so that the yield will be many times higher than before.

For the first time, the book mentions the transplanting density of wheat, millet, and so on. "The wheat seedlings are clustered in tens, which are four inches apart. Millet seedlings are in clumps of 2 or 3, three inches apart." For the first time, the theory of the five elements of *qi* is used to explain the relationship between crop yield, quality, human health, and even intelligence with "good timing" and "bad timing" put forward in *Lu Shi Chun Qiu Shen Shi*. In terms of dung, it is pointed out that a new method of soil fertility is "to make *yuqi* (余气) *xiangpei* (相培)." "*Yuqi*" refers to the residual nutrients not absorbed in food; "*Xiangpei*" means that it returns to the field after being excreted and can be used to cultivate crops for their growth. This idea is actually close to the concept of nutrient elements in modern science (Xiuling 1990).

In the book, the interplanting method of combining grain crops with cash crops and vegetables is mentioned in order to achieve as much as three harvests a year or 13 harvests in 2 years. Fertilization is regarded as the key to achieve more harvests. Based on predecessors' ideas of "cherishing manure as gold" and "a manure field is better than a newly-bought field," the author puts forward the ideas that "a manure field is better than a reclaimed field" and "accumulating manure is similar to accumulating gold." The author also summarizes the traditional methods of fertilizer accumulation and puts forward that "there are ten methods for making manure," including human manure, livestock manure, grass manure, fire manure, mud manure, bone clam ash manure, seedling manure, and slag manure. In order to improve the effect of fertilizer, it is also proposed that "both fermented and unfermented dung have their appropriate uses." "Unfermented dung can only be used on *muguo* (木果), while others, especially snake melons, are unsuited. However, fermented dung is suitable for everything, but should be used according to time, place and plant." What he means is that different fertilizers are used according to weather, soil, and crops. He also applies the same principle to animal production and farmland irrigation.

3.5.5 Journal of Agriculture and Agronomy Series

In 1894, China was defeated in the Sino-Japanese War. In 1895 Kang Youwei (康有为) and others, in the Gongche Shangshu Movement, proposed to revitalize agriculture by translating foreign agricultural books to imitate foreign countries. They also organized agricultural societies everywhere. In 1896, Luo Zhenyu (罗振玉) and others established the "Agriculture Association" in Shanghai. In May 1897, the Journal of Agriculture (《农学报》) was founded with the aim of "giving priority to revitalizing farming as well as sericulture and animal husbandry, excluding other matters" and "detailing the agricultural policies of various provinces, introducing the work of the Association, and translating eastern and western agricultural books and newspapers as materials to study" (*Nong Xue Bao Lue Li* 《农学报略例》). It was a semi-monthly journal in the beginning and changed to a periodical appearing once every ten days after 1898. As a semi-monthly, the number of words in each issue was about 8000, but it was reduced to 4000–5000 after being published every ten days. Its publication had lasted for 10 years and was stopped in December of the 32nd year of Guangxu in the Qing Dynasty (January 1907), with 315 issues published altogether.

A large number of agronomy articles were published in Journal of Agriculture. The contents can be divided into monographs, advanced experiences and new technologies, information and intelligence, experimental research and investigation reports, biographies of people (*Se Zhe Zhuan* 《穑者传》), scientific sketches, etc. The fields involved included general theory, crops, gardening, plant protection, soil fertilizers, animal husbandry and veterinary medicine, silkworm, bee and tea medicine, meteorology, farm tools, manufacturing, forestry, and aquatic products, as well as local products and agricultural trade. Most of these contents were translated from abroad. There were also agricultural documents written by the Chinese, such as *Yin Du Wei Shu* (《尹都尉书》), *Fan Sheng Zhi Shu*, *Ju Lu*, and other ancient agricultural books, as well as some agricultural books written and published in the late Qing Dynasty, such as *Xiang He Pu*, *Mu Mian Pu, Shui Mi Tao Pu* (《水蜜桃谱》), *Chu Jian Pu* (《樗茧谱》), *Zui Li Pu*, *Yi Ju Fa*, and *Yue Ji Hua Pu* (《月季花谱》), and the newly written *Zhong Yan Ye Fa* (《种烟叶法》), *Pu Kui Zai Pei Fa* (《蒲葵栽培法》), *Yang Can Cheng Fa*, *Xin Bian Ji Cheng Niu Yi Fang* (《新编集成牛医方》), and so on. Luo Zhenyu himself also published many articles on agriculture in the Journal of Agriculture, such as "Ken Huang Si Yi" (垦荒私议 Private Discussion on

Reclamation), "Lun Nong Ye Yi Zhi Ji Gai Liang" (论农业移殖及改良 On Agricultural Migration and Improvement), "Ri Ben Nong Zheng Wei Xin Ji" (日本农政维新记The Reform of Japanese Agriculture), "Pi Di Fei Tian Shuo" (僻地肥田说 The Theory of Reclaiming Farmlands and Fertile Fields), and "Zhen Xing Lin Ye Ce" (振兴林业策 Strategies for Revitalizing Forestry) (Zhu Xianli: *Nong Xue Bao*—the First Professional Scientific and Technological Journal in China, Index of *Nong Xue Bao*, *Zhong Guo Ke Ji Shi Liao* (《中国科技史料》), 1986, Volume 2). Most of them were later compiled in the book *Nong Shi Si Yi* (《农事私议》). The published articles were sometimes annotated with editor's comments, which were mostly written by Luo.

Nong Xue Cong Shu (《农学丛书》A Series on Agronomy) is mainly compiled from articles published in the "Journal of Agriculture." However, new translations and monographs were also included in the compilation. From 1899 to 1906, there were 7 episodes, 82 volumes (20 volumes in the first episode, 12 volumes in the fourth episode, and 10 volumes in the remaining episodes), and 235 species. Most are translations from Japanese agricultural books, and some are Japanese translations of European and American agricultural books. In addition, ancient Chinese agricultural books also account for a certain proportion in the series. Its purpose is "to use Chinese methods, combined with new theories from the east and west, to verify each other and to excel in the long run."

References

1. Wang, D. (1989). On the Characteristics and Achievements of Agricultural Books During the Ming and Qing Dynasties, *Nong Shi Yan Jiu* (《农史研究》) (the 8th issue), Agricultural Publishing House.
2. Fang, Z., & Wang, S. (2002). *Zhong Guo Nong Ye Gu Ji Mu Lu*. Beijing Library Press.
3. Liang, J. (2002). A Probe into Several Earliest Recorded Ancient Agricural Literatures in China, *Liang Jia Mian Nong Shi Wen Ji* (《梁家勉农史文集》) (p. 16). China Agricultural Press.
4. Shiran, D., et al. (1982). *History of science and technology in China* (p. 272). Science Press.
5. Qiyu, M. (1981). *Praface to Si Shi Zuan Yao: Collated and Annotated* (Vol. 1). Agricultural Publishing House.
6. Mengxiang, X. (1983). *Xi Jin Zhi Ji Yi (《析津志辑佚》)* (p. 2 & p. 114). Beijing Ancient Books Publishing House.
7. Shenghan, S. (1980). *Zhong Guo Gu Nong Shu Ping Jie* (p. 75). Agricultural Publishing House.
8. Yuhu, W. (1955). *Qu Zhong Shi Zhong*. Financial and Economic Press.
9. Xiongsheng, Z. (1992). Reading Notes on *Jin Shu Chuan Xi Lu*. *Gu Jin Nong Ye, 4*, 39–40.
10. Chuanzheng, Z., & Zhu, Z. (2006). Li Xinghui: Tea Books in the Ming and Qing Dynasties and Their Historical Value. *Gu Jin Nong Ye, 3*.
11. Xiuling, Y. (1990). The achievements and problems in agronomy in the Qing Dynasty. *Nong Ye Kao Gu, 1*.

Medical Schools and Theories

4

Yuqun Liao

Contents

Abstract

The ancient "scientific knowledge" is less evidence-based than the modern science but more constructive. Under the guidance of constructivism, people apply traditional Chinese medicine in medical activities. This chapter introduces the four great medical schools in the Jin and Yuan dynasties as well as the medical schools of "cold damage" and "warm disease," which were founded early but flourished late.

Y. Liao (✉)
The Institute for the History Natural Sciences, Chinese Academy of Sciences, Beijing, China
e-mail: liaogg@ihns.ac.cn

X. Jiang (ed.), *The High Tide of Science and Technology Development in China*,
History of Science and Technology in China,
https://doi.org/10.1007/978-981-15-7847-2_4

Keywords

Home school · Fire theory · Theory of attacking pathogenic factors · Theory of spleen and stomach · Theory of nourishing Yin · Cold damage · Warm disease

The sociology of science represented by Merton's *Science, Technology and Society in Seventeenth Century England* (1938) witnessed an important change in the post 1970s. Along with the emergence of the Western European constructivism, the research focus of sociology of science shifted – no longer centering on the "science" but on the "scientific knowledge" itself. It advocated that scientific knowledge is constructed socially by mankind; during the construction process, nature doesn't play any role. Undoubtedly, the viewpoints of scientific sociology of the constructivism seem extreme, but they could be referred to observe and analyze the traditional medicine, since the ancient "scientific knowledge" is less "evidence-based" than the modern science but more "constructive." At the same time, it must be seen that traditional Chinese medicine is applied in medical activities under the guidance of such "constructivism" theory. Hence, it's not an easy task to evaluate the values of those theories.

4.1 The Era of Home School

The English word "school" has two meanings when it's used as a noun. First, it means "study place" that we are all familiar with; second, it means "school of thought." According to researchers, "this word has already been used during the philosophers' era in ancient Greece. The Greek etymology is σχολη(skholê), which has a broad definition mainly indicating the use of leisure time, including study and other knowledge exploration activities. Afterwards, it refers to a place for study, which is quite similar with the meaning of 'school' in modern language. Until 1612, the word school started to have the connotation of 'school of thought' in modern terms, meaning people united by a general similarity of principles and methods" (Luo Xingbo and Liu Wei: "Scientific Schools and Scientific Progress." Edited by the Academic Department of the Chinese Association for Science and Technology: *Disciplinary Development and Technological Innovation: Proceedings of the Fifth Academic Exchange Theory Symposium*, Tsinghua University Press, 2010, p. 11).

4.1.1 Western "Academy," Chinese "Home," and "School"

Speaking of the origin of western science or the earliest philosopher, most of people will think of Plato from ancient Greece (B.C. 427–B.C. 347) and Aristotle (B.C. 384–B.C. 322), as well as their academies.

At the age of 17, Aristotle went to Athens and studied in Plato's academy. He's been staying there until the death of Plato. During this period, on one hand he learnt philosophy from the master, and on the other hand he developed strong interests in natural science that was ignored by Plato and insisted on independent research. After he travelled around the world, he came back to Athens in around B.C. 344 and established his own academy. Due to his unique method of teaching while strolling, he was crowned with an elegant title of "Peripatetic School." The academic center in the late ancient Greece was the "Alexandria" built in B.C. 332. As the largest city and capital of Greece at that time, there were also many academies, among which the largest one was called "Museum" where scholars were gathered (Written by Whitfield, translated by Fan Yizu: *Colorful Graphic: World History of Science and Technology*, Science Popularization Press, 2006, p. 46).

There are too many similar cases to be mentioned here one by one. All in all, in that era, those who can establish "academies" in their own names must be famous "sages"; the curriculum taught in the academy is the theories of the sages; if there are plenty of disciples who come to study due to the "reputation" and "professionalism" of the academy, then this becomes "school." On one hand, those who can found a certain academic system could be called spiritual and academic leaders; their academic systems must have their unique characteristics; if people admire their "reputation" and "professionalism," then they could come to the academy and communicate with the teachers; after learning and acknowledging their theories, the disciples could pass them on to others; if opposing the theories or pioneering new theories, they could establish a new school of thought and become masters of their own. Such situations are basically the same as our "school" in modern world. Based on the past "academies," it's not hard to understand why the word "school" is generally translated as "a place for study" and then derived to have the meaning of "school of thought."

It's generally believed that the earlier form of knowledge sharing in ancient Chinese society is "official school" – study in official academy (The transition from "official school" to "private school" is not a topic covered by this article. Those who are interested can refer to Luo Genze's masterpiece: "There was no private book before the Warring States period." The article is edited by Gu Jiegang: *Ancient History* (Volume 4), Shanghai Ancient Books Publishing House, 1982, page 8). As the Zhou Dynasty's imperial power waned and social etiquettes collapsed, the "private school" emerged. Since the inception of private schooling, it has led to the co-existence of various theories and academic prosperity, which should not be neglected during the research on the origins of academic schools in Chinese history. An individual teaching is "a school of thought"; those with the same or similar standpoints and propositions belong to the "one school of thought." However, either of these actually, to some extent, include the meanings of "genre," "sect," or "school."

The earliest comprehensive summary and evaluation of various schools was Sima Tan's *Lun Liu Jia Yao Zhi* (On the Six Schools of Thought) quoted in Sima Qian's *Shi Ji* (the Historical Records). The so-called Six Schools of Thought are the Yin-Yang School, the Confucian School, the Mohist School, the Logician School, the Legalist School, and the Taoist School. By the end of the Western Han Dynasty, Liu Xiang and Liu Xin, the father and son, received the imperial order to compile the first bibliography works *Qi Lue* (七略) in China. Basically, one person with one book is regarded as "one school." Although this concept is slightly different from that of "School" by Sima Tan, of course different sects have varying propositions.

Actually, it has both the meanings of "genre" and "school." What's more, the father and son Liu Xiang and Liu Xin, and Ban Gu, the author of the *Han Shu·Yi Wen Zhi* (Book of Han: Art and Literature), didn't reject the theory of Sima Tan, but placed it in the "Zhu Zi Lue" (诸子略) after the most respected "Six Classical Arts" (六艺). In addition to the "Six Schools of Thought" of the Yin-Yang, Confucian, Mohist, Logician, Legalist, and Taoist, other four schools are added including the Diplomacy School, the Eclectic School, the Agriculturist School, and the Fictionist School. Each catalogue of the "Ten Schools" is followed by a paragraph of comment which explains its origination and purpose.

In terms of knowledge inheritance, the most common expression method is "Qin Zhi" (亲炙, the student is taught by the teacher personally) and "Si Shu" (私淑, although the student is not taught by the teacher, he still admires and regards the teacher as his model). The former means the "disciple" that is taught by the master, such as the form of western "academy" and Chinese "home school." The positive side is that such disciples could get more chances to learn from the masters; at the same time, when staying together with the master day and night, the disciples can cultivate spiritual aspirations, improve morality and academic achievement, and get more likely to be promoted in career. In opposite, the negative side is that no matter how accomplished the masters are, it's after all their individual achievements. Some of the teachers are even very arrogant and narrow-minded, who don't allow their disciples to learn other theories. On top of that, disciples strictly abide by the teachers' principles, look up to them as "idols," and indiscriminately believe in their theories like "superstition," but never realize and criticize the mistakes made by them; or when confronted with conflict of interest, disciples would help each other and unite with those of the same views but alienate those with different views. In this way, the bond between teacher and student and the scientific research team would always evolve to be "sect" instead of "school."

The "Si Shu" emerged along with the "qualitative" change of the way of sharing and spreading knowledge. When scholars obtain certain knowledge through written records, if they show favorable attitudes of "buy-in" or even respect that author as teacher, then they could be called "Si Shu Disciple." On one hand, the number, scope, and time of inheriting certain school of theory could be increased; on the other hand, the characteristics of the so-called school of thought is more towards "sharing the same academic views, propositions and methods," whereas the "direct inheritance relationship" is weaken. Meanwhile, the method of knowledge sharing and inheritance brings about another change: it creates the opposite sides with opposing attitudes. When the opponents are majority, it will form or create one or more than one opposing "schools." As long as the opponents hold the same opposing reasons, no matter which era they live in, although they don't have any connections or even don't know each other, historians and researchers still categorize them into the same "school of thought." This is another representation of how the change of ways of sharing and spreading knowledge promotes the development of school. For sure, we should also see that the change of ways of sharing and spreading knowledge has another function, except for "facilitating the disappearance of home school" and "weakening schools of thought." Simply speaking, if we look at all sorts of "science biography," we can find one basic rule: the introduction of outstanding figures in earlier years normally include their masters/teachers, or brief education background; their research focus and achievements might not be the same as the teachers, but their relationship is relatively closer. As the times evolve, those characteristics become not that striking. From entering the "school" not the "academy," people will be influenced by teachers and professors of different subjects. The change of research direction and remarkable achievements might be inspired by an article that one reads accidentally or by attending a relevant or non-relevant academic conference. Of course, most of them should be attributed to absorbing (maximum) ancestors' thoughts and theories, or even "interdisciplinary" and "transdisciplinary."

4.1.2 Various Schools of Thought in Early Years

The "Fang Ji Lue" (experience and technology) in the *Han Shu·Yi Wen Zhi* mentioned above is divided into "Medical Classics" (medical theories), "Experience and Prescription," "Immortal" (longevity and life extension), and "Intercourse" (sexual medicine and technique), which records four types of books related to "all life tools" and "the science of life." Once some scholars inappropriately regarded this classification as different "schools," but from various "book titles," they could get a glimpse of the problems of "home school" and "school of thought" in early medicine field.

4.1.2.1 Medical Classics

The "Medical Classics" recorded in the *Han Shu·Yi Wen Zhi* is shown in Table 4.1. Through this limited material, it could be known that there were at least three independent systems of medical theories circulating at that time. Although the "Medical Classics" is similar to the author's "explanatory" words – the contents involved have commonalities – but they are titled with different masters like Yellow Emperor, Bian Que, and Bai Shi, which have different inheritance systems or the so-called different "sects"; their theoretic systems must also vary or the so-called different "schools."

类 别	书名与卷数	说 明
医经 (7家,175卷)	黄帝内经……18卷 外经……37卷 扁鹊内经……9卷 外经……12卷 白氏内经……38卷 外经……36卷 旁篇……25卷	医经者,原人血脉、经络、骨髓、阴阳、表里,以起百病之本,死生之分,而用度针石汤火所施,调百药齐和之所宜。至齐(剂)之得,犹慈石取铁,以物相使。拙者失理,以瘉为剧,以生为死。

4.1.2.2 Experience and Prescription

The "Experience and Prescription" in the *Han Shu·Yi Wen Zhi* does not mean the so-called classic prescription by the late generations, but the "accumulation of knowledge and methodologies" (see Table 4.2). The normal way to compile the

Table 4.1 The "Medical Classics" recorded in the *Han Shu·Yi Wen Zhi*

中文	英文
类别	Category
医经(7家, 175卷)	Medical Classics (7 Sects, 175 Volumes)
书名与卷数	Book Title and Volume No.
黄帝内经……18卷	*Huang Di Nei Jing* (The Inner Cannon of Yellow Emperor) ———— 18 Volumes
外经……37卷	*Wai Jing* (The External Cannon of Medicine) ———— 37 Volumes
扁鹊内经……9卷	*Bian Que Nei Jing* (The Inner Cannon of Bian Que) ———— 9 Volumes
外经……12卷	*Wai Jing* (The External Cannon of Medicine) ———— 12 Volumes
白氏内经……38	*Bai Shi Nei Jing* (The Inner Cannon of Bai Shi) ———— 38 Volumes
外经……36卷	*Wai Jing* (The External Cannon of Medicine) ———— 36 Volumes
旁篇……25	*Pang Pian* (The Supplement Chapter) ———— 25 Volumes
说明	Explanation
医经者, 原人血脉, 经络, 骨髓, 阴阳, 表里, 以起百病之本, 死生之分, 而用度针石汤火所施, 调百药齐和之所宜。至齐(剂)之得, 犹慈石取铁, 以物相使。拙者失理, 以瘉为剧, 以生为死。	The medical classics are used to explore the physiological characteristics of human blood, meridians, bone marrow, yin and yang, and exterior and interior, to clarify the root causes of various diseases, to distinguish the boundary between death and life, to apply the treatment methods of acupuncture, stone, decoction, moxibustion, etc., and to use various medicines suitable for clinical needs. The efficacy of the most appropriate medicine, like a magnet for iron, depends on the interaction of the medicine. If poorly skilled doctors violate medical principles, they would make illnesses to deteriorate and cause the death of those who should have been saved

"experiences and prescriptions" is through officials' or other scholars' conscious collection in case of disease treatment or recommendation to colleagues, friends, and people in need. Therefore, regarding its origination, it's more like a "family," starting from "small family" (owner, user, and inheritor of one or multiple methods) and then to "large family" (compilation of different sources), such as someone or some doctor's "home secret prescription" normally seen in medical books. Another one is the large-size prescription books compiled by individuals, such as Sun Simiao's *Qian Jin Fang* (千金方) and Wang Tao's *Wai Tai Mi Yao*

Table 4.2 "Experience and Prescription" recorded in the *Han Shu·Yi Wen Zhi*

中文	英文
类别	Category
经方 (11家, 295卷)	Prescription (11 Schools, 295 Volumes)
书名与卷数	Book Title and Volume No.
五脏六腑痺十二病方......30卷	12 Prescriptions for Viscera Diseases —————— 30 Volumes
五脏六腑疝十六病方......40卷	16 Prescriptions for Viscera Hernia —————— 40 Volumes
五脏六腑瘅十二病方......40卷	12 Prescriptions for Viscera Pyretic Symptoms —————— 40 Volumes
风寒热十六病方......26卷	16 Prescriptions for Cold and Fever —————— 26 Volumes
泰始黄帝扁鹊俞拊方......23卷	Prescriptions of Doctor Yu Fu and Bian Que during the Yellow Emperor Era —————— 23 Volumes
五藏伤中十一病方......31卷	11 Prescriptions for Viscera Injuries —————— 31 Volumes
客疾五藏狂癫病方......17卷	Prescriptions for Epilepsy Diseases —————— 17 Volumes
金创瘲瘛方......30卷	Prescriptions for Cramp Diseases —————— 30 Volumes
妇人婴儿方......19卷	Prescriptions for Women and Child —————— 19 Volumes
汤液经法......32卷	Traditional Chinese Medicine Prescriptions —————— 32 Volumes
神农皇帝食禁......7卷	Shen Nong Emperor's Diet —————— 7 Volumes
说明	Explanation
中文	英文
经方者, 本草石之寒温, 量疾病之浅深, 假药味之滋, 因气感之宜, 辨五苦六辛, 致水火之齐, 以通闭解结, 反之于平。及失其宜者, 以热益热, 以寒增寒, 精气内伤, 不见于外, 是所独失也。故谚曰: "有病不治, 常得中医。"	The classical prescriptions mean the medicine is divided into warm and cold two types. In order to judge whether the disease is mild or serious, it should be based on the characteristics of these medicines, following the movement of qi, and categorize those medicines according to their nature and taste. There are also two types of prescriptions, water-natured and fire-natured. Through purgation, vomiting, and sweating, the body can be restored to a state of balance. What will happen if the physician cannot tell the cold from warm and use the wrong methods, like he should have used the cold treatment but ended up with the warm treatment? It would make the cold disease colder, and the warm disease warmer. Physicians should not only observe the cold and warm, but also the essence and qi in the body. Or otherwise, he could not make a correct judgment. That's why there is a saying: "If you get sick and don't see a doctor, the result would be the same as you see a Chinese medicine doctor."

Fang (外台秘要方), or the collective works like the Tai Ping Sheng Hui Fang (太平盛惠方), the Pu Ji Fang (普济方), and others. This determined the nature of "school" of this kind of works.

类别	书名与卷数	说明
经方 (11家,295卷)	五脏六腑痹十二病方 ……30卷 五脏六腑疝十六病方 ……40卷 五脏六腑瘅十二病方 ……40卷 风寒热十六病方 ……26卷 泰始黄帝扁鹊俞拊方 ……23卷 五藏伤中十一病方 ……31卷 客疾五藏狂癫病方 ……17卷 金创瘲瘛方 ……30卷 妇人婴儿方 ……19卷 汤液经法 ……32卷 神农皇帝食禁 ……7卷	经方者,本草石之寒温,量疾病之浅深,假药味之滋,因气感之宜,辨五苦六辛,致水火之齐,以通闭解结,反之于平。及失其宜者,以热益热,以寒增寒,精气内伤,不见于外,是所独失也。故谚曰:"有病不治,常得中医。"

4.1.2.3 Intercourse

Thanks to the unearthed medical records of Mawangdui bamboo slips and silks, we have a better understanding of the specific content of such works, but the name "Fang Zhong" (intercourse) was first seen in the *Han Shu·Yi Wen Zhi*, and its basic information is shown in Table 4.3. Because human sexual behavior is not simply for the reproduction of offspring, but as a part of cultural and spiritual life and constitutes one of the elements of interpersonal relationship, the emergence of "intercourse" monograph also marks the progress of human civilization. Also noteworthy is that in the feudal society where the ethics and morality of men are superior to women, there is no such abnormal consciousness in these "intercourse" works. On the contrary, women are placed in an extremely important position, which is mainly manifested in two aspects: First, when sexual behavior is regarded as a spiritual and cultural life, women's willingness and joy are fully emphasized, and they believe that only by this can they achieve harmony in the interpersonal relationship between the two genders and between their families. Secondly, when sexual behavior appears as a physiological activity, it not only emphasizes the importance of foreplay and the whole process, so that female orgasm can be achieved, but also believes that men's semen has a tonic effect on women. In the "intercourse"-related books, correct and incorrect sexual behaviors are summarized as "eight benefits" and "seven losses," respectively. Generally speaking, "intercourse" can be defined as focusing on sexual skills and art, with the nature of sexual medicine.

Table 4.3 The "Intercourse" works recorded in the *Han Shu·Yi Wen Zhi*

中文	英文
类别	Category
房中 (8家, 191卷)	Intercourse (8 Schools, 191 Volumes)
书名与卷数	Book Title and Volume No.
容成阴道......26卷	Rong Cheng Vagina ——— 26 Volumes
务成子阴道......36卷	Wu Cheng Zi Vagina ——— 36 Volumes
汤盘庚阴道......20卷	Tang Pan Geng Vagina ——— 20 Volumes
天老杂子阴道......25卷	Tian Lao Za Zi Vagina ——— 25 Volumes
天一阴道......24卷	Tian Yi Vagina ——— 24 Volumes
黄帝三王养阳方......20卷	The Talk of Health Preservation between Yellow Emperor and Three Nobles ——— 20 Volumes
三家内房有子方......17卷	San Jia Nei Fang You Zi Fang (Fertility Prescription) ——— 17 Volumes
说明	Explanation
房中者, 情性之极, 至道之际, 是以圣王制外乐以禁内情, 而为之节文。《传》曰: "先王之作乐, 所以节百事也。" 乐而有节, 则和平寿考。及迷者弗顾, 以生疾而陨性命。	Intercourse represents the extreme of feelings and emotions and a way of cultivation in Taoism. As the Biography mentioned: "if the emperor indulged in pursuing intercourse pleasure, then he could not govern the country well." If people are not addicted to intercourse, then they would live a longer life.

类　别	书名与卷数	说　明
房中 (8家,191卷)	容成阴道　……26卷 务成子阴道　……36卷 尧舜阴道　……23卷 汤盘庚阴道　……20卷 天老杂子阴道　……25卷 天一阴道　……24卷 黄帝三王养阳方……20卷 三家内房有子方……17卷	房中者,情性之极,至道之际,是以圣王制外乐以禁内情,而为之节文。《传》曰:"先王之作乐,所以节百事也。"乐而有节,则和平寿考。及迷者弗顾,以生疾而陨性命。

In terms of its specific method, different schools "express their own opinions"; but in general, there is no conflict of views between those main themes. But most importantly, its essence is not a theory, but mainly an empirical sex "skill," so that each other will be happy to get something useful and new from other people or other works without any obstacles. Therefore, it is better to say that they build a "discipline" together rather than a "school."

4.1.2.4 Immortal

类 别	书名与卷数	说 明
神仙 (10家,201卷)	宓戏杂子道 ……20篇 上圣杂子道 ……26卷 道要杂子 ……18卷 黄帝杂子步引 ……12卷 黄帝岐伯按摩 ……10卷 黄帝岐伯芝菌 ……18卷 黄帝杂子十九家方 ……21卷 泰壹杂子十五家方 ……22卷 神农杂子技道 ……23卷 泰壹杂子黄冶 ……31卷	神仙者,所以保性命之真,而游求于其外者也。聊以荡意平心,同死生之域,而无怵惕于胸中。然而或者专以为务,则诞欺怪迂之文弥以益多,非圣王之所以教也。孔子曰:"索隐行怪,后世有述焉,吾不为之矣。"

According to the "Immortal" works in Table 4.4, from the book titles we could learn that there are at least methods of "Bu Yin" (Guidance), "Massage," "Lucid Ganoderma," etc. Also as the name suggests, the eight "Za Zi" represent the

Table 4.4 The "Immortal"-related works recorded in the *Han Shu·Yi Wen Zhi*

中文	英文
类别	Category
神仙 (10家, 201卷)	Immortal (10 Schools, 201 Volumes)
书名与卷数	Book Title and Volume No.
宓戏杂子道……20篇	Mi Xi Za Zi Dao ———— 20 Chapters
上圣杂子道……26卷	Shang Sheng Za Zi Dao ———— 26 Volumes
道要杂子……18卷	Dao Yao Za Zi ———— 18 Volumes
黄帝杂子步引……12卷	Yellow Emperor Za Zi Bu Yin ———— 12 Volumes
黄帝岐伯按摩……10卷	Yellow Emperor and Qi Bo Massage ———— 10 Volumes
黄帝岐伯芝菌……18卷	Yellow Emperor Qi Bo Lucid Ganoderma ———— 18 Volumes
黄帝杂子十九家方……21卷	Yellow Emperor Za Zi Shi Jiu Jia Fang ———— 21 Volumes
泰壹杂子十五家方……22卷	Tai Yi Za Zi Shi Wu Jia Fang ———— 22 Volumes
神农杂子技道……23卷	Shen Nong Za Zi Ji Dao ———— 23 Volumes
泰壹杂子黄冶……31卷	Tao Yi Za Zi Huang Zhi ———— 31 Volumes
说明	Explanation
神仙者, 所以保性命之真, 而游求于其外者也。聊以盪意平心, 同死生之域, 而无怵惕于胸中。然而或者专以为务, 则诞欺怪迂之文弥以益多, 非圣王之所以教也。孔子曰: "索隐行怪, 后世有述焉, 吾不为之矣。"	The essence of immortals lies in keeping in good health, keeping one's mind calm, and not taking life and death too seriously. Confucius said: "I will not hoodwink the public by doing some absurd things to win popularity among others."

Table 4.5 The specific methods in each chapter of the *Shi Wen*

中文	英文	中文	英文
序号	No.	大成	Da Cheng
问者	Question By	曹熬	Cao Ao
黄帝	Yellow Emperor	容成	Rong Cheng
黄帝	Yellow Emperor	舜	Shun
黄帝	Yellow Emperor	彭祖	Peng Zu
黄帝	Yellow Emperor	耆老	Qi Lao
尧	Yao	癸	Gui
王子巧父	Wang Zi Qiao's Father	文挚	Wen Zhi
磐庚	Pan Geng	王期	Wang Qi
禹	Yu	内容	Content
齐威王	King Wei of Qi	接阴法	Intercourse Method
秦昭王	King Zhao of Qin	服食法	Fu Shi (Diet Method)
答者	Answer By	食气法	Shi Qi (Breathing Method)
天师	Tian Shi	禁欲法	Abstinence Method
中文	**英文**	**中文**	**英文**
导引法	Dao Yin (Exercise Method)	卧法	Sleeping Method
自我按摩法	Self-Massage Method	续表	Continued Table

schools of Yellow Emperor, Tai Yi, and Shen Nong, but the contents and methods are miscellaneous. Here the sample analysis is the *Shi Wen* (Ten Questions) named by archaeologists, one of the unearthed medical records of Mawangdui bamboo slips and silks (see Table 4.5).

序号	问者	答者	内容						
			接阴法	服食法	食气法	禁欲法	导引法	卧法	自我按摩法
1	黄帝	天师	▲						
2	黄帝	大成		▲					
3	黄帝	曹熬	▲						
4	黄帝	容成			▲				
5	尧	舜	▲						▲
6	王子巧父	彭祖				▲			▲

（续表）

序号	问者	答者	内容						
			接阴法	服食法	食气法	禁欲法	导引法	卧法	自我按摩法
7	磐庚	耆老	▲						
8	禹	癸					▲		
9	齐威王	文挚		▲				▲	
10	秦昭王	王期	▲	▲	▲				

Referring to other texts unearthed in Mawangdui, it can be roughly seen that the so-called "Immortal" (longevity and life extension) method has the following types:

"Pi Gu" (Fasting), also known as "Qu (or Que) Gu," means not eating grains, but only eating certain specific vegetables to maintain life and achieve the purpose of curing diseases and longevity. The Shi Ji·Liu Hou Shi Jia (Historical Records: The Family of the Nobles) mentions that Zhang Liang, after helping Liu Bang ascend the throne, was in frail health, so he decided to "learn Pi Gu to recover his health."

"Shi Qi" refers to a certain way of breathing for health preservation. In the fourth chapter of the *Shi Wen* (Ten Questions), "Yellow Emperor Asks Rong Cheng," the explanation for the theoretic basis is: "The essence of heaven and earth is invisible and intangible. Those who obtain the essence of heaven and earth will live a long wife, while those who lose it will die soon."

"Fu Shi" (Oral Intake of Food) includes many specific methods, such as pine and cypress, egg, ripe wine, and the like. "Yellow Emperor Qi Bo Lucid Ganoderma" recorded in the *Han Shu·Yi Wen Zhi* belongs to this category.

"Dao Yin" (Exercise Methods) means curing disease through the movements of limbs, or called "medical sports," which is in accordance with the record quoted from the *Zhuang Zi·Ke Yi* that "People engaged in self-cultivation" likes "mimicking the movements of bear and bird."

"Jie Yin" (Sexual Intercourse) indicates that men could achieve longevity through certain kind of sexual intercourse. In the *Shi Wen* that records the longevity methods of various schools, five of them have mentioned this, which reflects the importance of life preservation at that time. Accordingly, only the school of "Peng Zu" advocates complete abstinence. The fundamental difference between "Jie Yin" and "Fang Zhong" is that the center is men while women are only tools – not only men must not "ejaculate," but also obtain "benefits" from the upper body of women. From this point, it could be seen that female is actually regarded as the "food" of "Fu Shi" method and the "breathing" of the "Shi Qi" method, like an objective material that provides certain benefits.

In short, regarding this category, there are not only various styles of schools because of differences in methods and inheritance pedigrees, and "theories." Therefore, there must exist a problem of mutual recognition – which gives birth to the definition of "school."

Through the four types of "Fang Ji" (skills), we could probably develop a simple but generalized impression: Different theories are easy to lead to different schools, including "science" in modern times; practical skills, because of their empirical nature, are easy to cause conservation, but also very easy to be updated. The difference between "science" in modern times and ancient theories is that if it can be verified – whether it is right or wrong – then the "school" that exists before verification will disappear.

4.1.3 Different Sublation Forms

4.1.3.1 The Received Version of the *Huang Di Nei Jing*: "Compatibility and Transformation"

Since the articles in the *Huang Di Nei Jing* are written independently, there is no coherent and progressive relationship between each other, and the angles and rationale for the same issue are often different, so its nature as "Collected Papers" is widely recognized by the academic circles without any doubt. However, perhaps because of this, scholars have ignored some of the important articles with academic pursuit of "unifying different opinions." To illustrate that, let's take the example of "Viscera Theory" that involves core basic theories.

At the beginning of the *Su Wen·Wu Zang Bie Lun Pian* (Five Zangs Chapter), a question is raised:

> Emperor Huang asked: I heard some alchemists regarded brain marrow as Cang, while some regarded intestines and stomach as Fu; Their opinions were opposite but they both argued that they were right. I'm not sure which one is correct. Can you tell me more about it?

From this, it's learnt that the "Fu and Zang (Viscera) Theory" in traditional Chinese medicine has no unified model in early times. Instead, its development is very diversified. The following is the answer given by Yellow Emperor's medical teacher Qi Bo:

> The extraordinary fu-organs are characterized by hollowness, similar to the six fu-organs in morphology, and storage of essence, similar to the five zang-organs in function. That is why this group of tissues and organs are called the extraordinary fu-organs, including the brain, the marrow, the bones, the vessels, the gallbladder and the uterus. Among the extraordinary fu-organs, the gallbladder is a special one. It is hollow inside and excretes bile to promote digestion, similar to the functions of the six fu-organs, but it also stores bile which is part of the essence, similar to the functions of the five zang-organs. That is why the gallbladder belongs both to the six fu-organs and the extraordinary fu-organs.

Table 4.6 The comparison between "The Extraordinary Fu-organs" and "The Fu-Organs for Transmission and Digestion"

中文	英文	中文	英文
分类	Category	比类	Manifestation
奇恒之腑	The Extraordinary Fu-organs	象地	Manifest earth
传化之腑	The Fu-organs for transmission and digestion	象天	Manifest heaven
数字	Number	藏而不泻	Storage without excretion
器官	Organ	泻而不藏	Excretion without storage
脑, 髓, 骨, 脉, 胆, 女子胞	Brain, marrow, bone, vessel, gallbladder, and uterus	胃, 大肠, 小肠, 三焦, 膀胱	Stomach, large intestine, small intestine, San Jiao (Triple Burner, the sixth organ of Fu in TCM), bladder
功能	Function		

In order to facilitate understanding and comparison, the "key points" are listed in Table 4.6. Among them, the nature of "The Extraordinary Fu-organs (奇恒之腑)" is "Yang," so it "manifests earth (象地)," while the nature of "The Fu-Organs for Transmission and Digestion (传化之腑)" is "Yang," so it "manifests heaven." The number of 6 or 5 also conforms with the rule of "Yang Odd and Yin Even." In terms of the function, "Yin" means "storage" (static) and "Yang" means "excretion" (dynamic), which constitutes a complete and self-consistent theoretical system of the Yin-Yang theory.

分类	数字	器　官	比类	功能
奇恒之腑	6	脑、髓、骨、脉、胆、女子胞	象地	藏而不泻
传化之腑	5	胃、大肠、小肠、三焦、膀胱	象天	泻而不藏

Then he continued to talk about:

> The so-called Five Zang-organs (viscera) store the essence Qi without excretion, so it is full but not excess. The Six Fu-organs (viscera) transmit and digest essence without storage, so it is excess but not full.

The theory of "The Five Zang-organs and Six Fu-organs" is seen in many chapters of the received version of the *Huang Di Nei Jing*, which is based on the basic rules of the Yin-Yang theory. Except for not adhering to the "Yang Odd and Yin Even" paradigm in the *Yi Jing* (Book of Changes), there are no other problems (Table 4.7).

Table 4.7 The basic content of "The Five Zang-organs and Six Fu-organs"

中文	英文	中文	英文
分类	Category	象地	Manifest earth
五脏	Five Zang-organs	象天	Manifest heaven
中文	英文	中文	英文
六腑	Six Fu-organs	功能	Function
数字	Number	藏而不泻	Storage without excretion
器官	Organ	胆泻而不藏	Gallbladder has excretion without storage
比类	Manifestation	心, 肝, 脾, 肺, 肾	Heart, liver, spleen, lung, kidney
胃, 大肠, 小肠, 三焦, 膀胱, 胆	Stomach, large intestine, small intestine, San Jiao (Triple Burner, the sixth organ of Fu in TCM), bladder, gallbladder		

分类	数字	器官	比类	功能
五脏	5	心、肝、脾、肺、肾	象地	藏而不泻
六腑	6	胃、大肠、小肠、三焦、膀胱、胆	象天	泻而不藏

As shown in Tables 4.6 and 4.7, different "schools" established their respective "Chinese medical theories of organs" based on the Yin-Yang theory; the author of "Su Wen · Wu Zang Bie Lun Pian," unified those theories into one. Since then, the so-called Zang-Fu Organ Theory in the "Fundamental Theory" of Chinese traditional medicine has been used until now (Table 4.8).

分类	器官	功能
五脏	心、肝、脾、肺、肾	藏精气而不泻，满而不能实
六腑	胃、大肠、小肠、三焦、膀胱、胆	传化物而不藏，故实而不能满
奇恒之腑	脑、髓、骨、脉、胆、女子胞	藏而不泻

However, there are some parts that are not self-consistent in the systematic and classic "Zang-Fu Organ Theory" consisting of the three types of organs. For instance, why "gallbladder" is one of the Six Fu-organs and one of the Extraordinary Organs at the same time? Every time students raise that question, those who only repeat what the book says cannot give answers.

Unifying different theories in a comprehensive and compromised manner is regarded as a unique way to deal with the non-empirical "various opinions of

Table 4.8 The categorization and function of Zang-Fu Organs until now

中文	英文	中文	英文
分类	Category	奇恒之腑	The Extraordinary Fu-organs
五脏	Five Zang-organs	器官	Organ
六腑	Six Fu-organs	功能	Function
心, 肝, 脾, 肺, 肾	Heart, liver, spleen, lung, kidney	藏而不泻	It stores essence Qi without excretion
胃, 大肠, 小肠, 三焦, 膀胱, 胆	Stomach, large intestine, small intestine, San Jiao (Triple Burner, the sixth organ of Fu in TCM), bladder, gallbladder	脑, 髓, 骨, 脉, 胆, 女子胞	Brain, marrow, bone, vessel, gallbladder, and uterus
藏精气而不泻, 满而不能实	It stores the essence Qi without excretion, so it is full but not excess	传化物而不藏, 故实而不能满	It transmits and digests the essence Qi without storage, so it is excess not full

schools." In the era when the received version of the *Huang Di Nei Jing* was produced, Wang Mang who owned the real power of the Han Dynasty once built a large number of houses and courtyards to "search the great talents in the world." Thousands of scholars came to his imperial court and were ordered to compiled Wang Mang's theories" (the *Han Shu·Wang Mang Zhuan*, Book of Han·Biography of Wang Mang). During the Eastern Han Dynasty, Emperor Zhang of the Han Dynasty was keen on studying knowledge. He visited the Bai Hu Guan and discussed with other Confucian scholars about "the similarities and differences of the Five Classics." He also made judgments concerning how to resolve the conflictions arising from the discussion (the *Hou Han Shu · Zhang Wu Ji*, Book of the Later Han Dynasty·Biography of Emperor Zhang). But such "Theory of Yun Qi (Transportation of Qi)" was established later, so there still existed multiple schools of thoughts in the era when the received version of the Huang Di Nei Jing was produced and several hundred years afterwards – which are called the "Seven Classic Monographs" in "The School of Yun Qi" by the Chinese traditional medicine. By the Song Dynasty, different methods of leveraging Qi in such theories were integrated together to form "A Unified Theory." Since those theories are incorporated in a "big basket," among the two concepts of "Yun" and "Qi," there emerge a series of sub-concepts like "Zhu Yun," "Da Yun," "Zhu Qi," "Ke Qi," "Si Tian Zhi Qi," and "Zai Quan Zhi Qi" (Yun: The Five Elements, metal, wood, water, fire, earth; Qi: The Six Climates, wind, heat, fire, damp, dry, cold).

Zhu Qi: according to the sequence of "deficient yin-wind-wood, less yin-heart-fire, less yang-damp-earth, yang ming-dry-gold, tai yang-cold-water," the twenty-four solar terms in the lunar calendar are called the Zhu Qi. Each "Zhu Qi" influences four solar terms, respectively. This sequence is permanent and always starts from deficient yin-wind-wood and ends with tai yang-cold-water.

Ke Qi: according to the sequence of "deficient yin-wind-wood, less yin-heart-fire, tai yin-damp-earth, shao yin-damp-gold, tai yang-cold-water," the twenty-four solar terms in the lunar calendar are called the Ke Qi. The start of Ke Qi is not fixed and varies with the changes of earthly branches in each year. For example, the Year of Wuxu, taiyang-cold-water "Si Tian," taiyang-damp-earth "Zai Quan."

Si Tian: The Qi of Si Tian in the "Ke Qi" that influences the whole year. For example, the whole Year Chenxu will be influenced by taiyang-cold-water.

Zai Quan: The Qi of Zai Quan in the "Ke Qi" that influences only the second half of the year. For example, only the second half of the Year Chenxu will be influenced by tai yin-damp-earth.

4.1.3.2 Drug System: "Focus on Practical Values"

What's listed in Table 4.9 is the milestone works in the development of Chinese Materia Medica recognized by the academic circles. Their common features or writing mode are all later works are based on previous works. Under the premise that the basic content is copied (differentiate with different fonts), supplement is added. Therefore, although there are some sublations, the overall number of drugs is constantly increasing. This is a typical representative of continuously growing knowledge system through "experience accumulation." In the development of this discipline from the Han Dynasty to the Ming Dynasty, there is basically no "school of thought." The most important thing is acquiring new knowledge. The criterion for selection is based on whether it has practical value (including foreign objects) – basically in line with the development law of the "technical" knowledge system mentioned above.

Table 4.9 The representatives of Materia Medica

中文	英文	中文	英文
书名	Book Title	朝代	Dynasty
《神农本草经》	The *Shen Nong Ben Cao Jing* (Sheng Nong's herbal classic)	东汉	Eastern Han Dynasty
《新修本草》	The *Xin Xiu Ben Cao* (Tang Materia Medica)	唐	Tang Dynasty
中文	**英文**	**中文**	**英文**
《本草纲目》	The *Ben Cao Gang Mu* (Compendium of Materia Medica)	明	Ming Dynasty
药物数	Number of Medicine	书名	Book Title
《本草经集注》	The *Ben Cao Jing Ji Zhu* (Collected Notes to Canon of Materia Medica)	《证类本草》	The *Zheng Lei Ben Cao* (Classified Materia Medica)
朝代	Dynasty	南朝·梁	The Southern Dynasty ·Liang
宋	Song Dynasty	药物数	Number of Medicine

书名	朝代	药物数	书名	朝代	药物数
《神农本草经》	东汉	365	《本草经集注》	南朝・梁	720
《新修本草》	唐	850	《证类本草》	宋	1 748
《本草纲目》	明	1892			

4.1.3.3 Pulse Diagnosis and Acupuncture: "New System"

The pulse diagnosis technology "that the patient does not need to speak and the doctor knows the source of the disease" adds some magic charm to Chinese medicine. "Acupuncture" is a unique treatment technique of Chinese medicine. They belong to two independent categories, but because their systems were formed in the Wei and Jin Dynasties, and together constituted the characteristics of the medical development at this stage, so they are described together.

Most of the works on medical history are based on Wang Shuhe and Huangfu Mi who are the representative figures of the famous physicians in the Wei and Jin Dynasties. However, the reason why they are crowned with eternal glory is not because they have superb medical skills, but because they each completed an epoch-making important book. The former compiled the 10 volumes of the *Mai Jing* (Pulse Diagnosis Classics) which includes the "formulas and gist" from famous physicians like "Qi Bo to Hua Tuo," known as the first pulse diagnosis monograph. The latter compiled the first Acupuncture works "*The Zhen Jiu Jia Yi Jing*" (A–B Classic of Acupuncture and Moxibustion). Both works have been passed down to today as must-read classics for later learners.

The method of pulse diagnosis originated from the fact that doctors knew that there are many beating "pulses" in the human body. Therefore, the most primitive method of pulse diagnosis is to examine the pulses of the body and analyze the condition of the disease based on this. Later, there emerged various pulse diagnosis methods and related theories. For example, when palpating the pulse beating, the fingers are divided into several levels from light to heavy to diagnose the diseases of each organ separately, which is called the "light and heavy pulse method." The floating pulse manifests the image of spring (liver and wood), and the full pulse manifests the image of summer (heart, fire), etc., so it is called the "Four Season Pulse-taking Method"; 3 arteries are taken from the head, feet, and hands (heaven, earth, and human) to diagnose the upper, middle, and lower Qi and Blood, which is called the "Three Portions and Nine Pulses" method. And the method most related to the pulse-taking nowadays is to take people's carotid artery for the pulse of "Yang" and radial artery of the wrist for the pulse of "Yin," to compare the changes of pulse sizes at the two places to check whether Yin and Yang are balanced. This is called the "Ren Ying Cun Kou (Carotid Artery and Radial Artery Pulse-taking Method)." During the Eastern Han Dynasty, this method was further simplified as the Cunkou (radial artery of the wrist) which was divided into Yin and Yang by the middle finger

Table 4.10 Classic pulse-taking methods

中文	英文	中文	英文
左手 (阴, 血)	Left hand (Yin, Blood)	心	Heart
右手 (阳, 气)	Right hand (Yang, Qi)	肺	Lung
寸 (食指)——上	Cun (Index Finger)	肝	Liver
关 (中指)——中	Guan (Middle Finger)	脾	Spleen
尺 (无名指)——下	Chi (Ring Finger)	肾	Kidney
命门	Ming Men (The Gate of Vitality)		

as the boundary (called "Guan"), so that the index finger diagnoses "Yang" and the ring finger diagnosed "Yin."

Until the Mai Jing compiled by the imperial physician Wang Shuhe in the Western Jin Dynasty, could we see the pulse diagnosis method used by modern Chinese medicine doctors. In other words, the middle finger "Guan" as boundary also becomes a pulse-taking part – "Cun Kou" (radial artery) is divided into three parts Cun (index finger), Guan (middle finger), and Chi (ring finger), which are matched with the Five Zang-organs and Six Fu-organs (Table 4.10).

	寸(食指)——上	关(中指)——中	尺(无名指)——下
左手(阴、血)	心	肝	肾
右手(阳、气)	肺	脾	命门

According to this, "Chi" and "Cun" are no longer matched with Yin and Yang, but correspond to the upper and lower parts of the human body; the Yin and Yang are changed to match with the left and right. The corresponding relationship of "Left hand: heart, liver, kidney; Right hand: lung, spleen, Ming Men (the gate of vitality)" reflects the division of the human body into three parts of "upper, middle, and lower," which gradually evolved the concept of "San Jiao." In conjunction with this pulse diagnosis method, the number of "pulse manifestation" has also increased to as many as 24; and in the current version of the *Huang Di Nei Jing*, there are still fewer than 10 types of "pulse manifestations." This classic pulse diagnosis method is still in use today.

Another important figure is Huang Fumi (215–282), who was ranked among the list of "Recluse" in the history due to his scholarly and official family background, his refusal to be recruited as a high-ranking official repeatedly, and his literary and historical attainments known as "Book Obsession." He selected the received version of the *Huang Di Nei Jing* that focuses on meridians and the *Ming Tang Jing* that focuses on acupuncture and moxibustion to compile them into the *Zhen Jiu Jia Yi Jing*. Since then, the basic framework of acupuncture and moxibustion that integrates "meridians" and "acupoints" has been established.

The work of Wang Shuhe and Huang Fumi is similar. Since their works, the previous "schools" of pulse diagnosis and acupuncture therapy were gone. Because their work is basically "collection" and "selection" of predecessors' works, medical historians often say that the characteristics of medical development during this historical period are "literature documentation." However, they neglected the significance of such works during the development of medical knowledge system.

4.2 The Most Important Medical School: The Four Great Medical Schools in the Jin and Yuan Dynasties

According to the *Si Ku Quan Shu Zong Mu Ti Yao* (The General Catalogue Summary of Imperial Book Collections), "The division of Confucian School is in the Song Dynasty. The division of Medical School is in the Jin and Yuan Dynasties." Because during that time, "The four great medical schools representatives in the Jin and Yuan Dynasties," Liu Xiansu, Zhang Congzheng, Li Gao, and Zhu Zhenheng, had their own understandings and theories of various aspects from "pathogenesis" to "clinical treatment" in the medical circles.

Their theories not only directly influence their own "disciples," but also attracted some physicians in different eras and regions through the circulation of their works. Even their theories were spread overseas, while the so-called Later Generation School (Hou Shi Pai, 后世派) was formed in Japan following the doctrines of "Li Gao and Zhu Zhenheng." Even nowadays, their theories are still used as theoretical basis for TCM practitioners to study and inherit and therapeutic methods that are widely applied in clinics. Therefore, they totally satisfy any of the standards in the definition of "school."

4.2.1 Liu Wansu and His "Fire Theory" (火热论)

Liu Wansu, courtesy name Shou Zhen and art name Tong Xuan Chu Shi (通玄处士). Since he lived in He Jian for a long time, he was also called "Mr. He Jian" or "Liu Hejian." Liu's date of birth and death is unknown, but according to his preface of "Su Wen Bing Ji Qi Yi Bao Ming Ji" (素问病机气宜保命集) written in 1186 in the Jin Dynasty: "When I was 25 years old, I focused on the study of the *Nei Jing* (The Internal Cannon of Medicine) . . .. . . until I was over 60 years old . . .." It could be estimated that he was born during 1120 to 1130. Since Zhang Congzheng was crowned as "Chang Sha, He Jian Fu Sheng" (长沙, 河间复生) in the early thirteenth century, it could be learnt that he died at around the end of the twelfth century.

Liu Wansu despises those who rely on their well-respected medical family background but do not seek to make progress. "I see many physicians who only rely on their ancestors' reputation and use old prescriptions, but do not want to focus on academic study or innovate new prescriptions. They would turn angry at those who try to persuade them to work harder" (Liu Wansu: *Su Wen Bing Ji Qi Yi Bao Ming Ji ·Author's Preface.* Proofread by Wu Mian: *Liu He Jian Yi Xue Liu Shu* (Six Books on Medicine School of Liu Hejian), block print). He believes that studying medicine must start from the fundamentals and learn the source of the theory, instead of being a "prescription doctor" who is familiar with hundreds of prescriptions. "What people learn today are all modern prescriptions. But they don't seek and understand the origins of those prescriptions" (Liu Wansu: *Su Wen Bing Ji Qi Yi Bao Ming Ji ·Author's Preface*. Proofread by Wu Mian: *Liu He Jian Yi Xue Liu Shu*, block print). Therefore, he began to study the *Nei Jing* at the age of 25 day and night for more than 30 years, before he became fully enlightened. He really advocated the Nei Jing and the *Shang Han Lun* (Treatise on Warm Diseases) and called Zhang Zhongjing "A Sage." Meanwhile, he said, "If you stick to the old books, insist that ancient sages' books cannot be changed and believe in that, you will have a limited outlook" (Liu Wansu: *Su Wen Bing Ji Qi Yi Bao Ming Ji ·Author's Preface*). Bearing in mind this thought, he refused to absorb the achievements of his predecessors, but endeavored to supplement the classic works like the *Su Wen*, especially some of the classic articles, "Nineteen Pathologies," which constitutes the theoretical core of his "Fire Theory."

Liu Wansu's medical works mainly include the *Su Wen Xuan Ji Yuan Bing Shi (素问玄机原病式)* 1 volume, the *Yi Fang Jing Yao Xuan Ming Lun* (The Essentials of Medical Prescriptions Theory) 15 volumes, the *Su Wen Bing Ji Qi Yi Bao Ming Ji (素问病机气宜保命集)* 3 volumes, the *Shang Han Biao Ben Xin Fa Lei Cui* (Collection of Treatise on Febrile Specimen) 2 volumes, of "Cui," the *Shang Han Zhi Ge* (伤寒直格) 3 volumes by his disciple, the *San Xiao Lun* (Three Dissipation Theory) 1 volume, and the *Bao Tong Mi Yao* (Secrets of Keeping Young) and the *Yun Qi Yao Zhi* (The Essentials of Applying Qi) mentioned in the book. In terms of the academic proposition, Liu Wansu centers around the "Fire." But not all the diseases are of cold and cool nature. He advocates the Yun Qi but the essence is not actually about Qi. For such concepts, one needs to read carefully in order to fully understand. The main points are selected below.

Since Liu Wansu emphasized that "When teaching medicine, one needs to center on the Five Yun Six Qi, ...Few people seek medical advice but do not know Yun and Qi, otherwise they would suffer losses" (Liu Wansu: *Su Wen Bing Ji Qi Yi Bao Ming Ji ·Author's Preface*). He wrote the *Nei Jing Yun Qi Yao Zhi Lun* (The Essentials of Yun and Qi in the Nei Jing) to illustrate that "the changes of life depends on the rise and fall of Qi" (Ma Zongsu: *Shang Han Yi Jian (Differentiation of Cold Damage)*. Proofread by Wu Mian: *Liu He Jian Yi Xue Liu Shu*, block print). Therefore, some people think that the theory of Yun Qi is a main part of Liu Wansu's academic views. Even they believe that "Although Liu's theory is derived from the Nei Jing and the *Shang Han Lun*, it is actually influenced by the theory of Yun Qi. Its spirit and essence are drastically different from the previous theories.... There is more subjective reasoning than practical experiences. Especially the nature and function of Six Qi is derived and extended infinitely according to the relationship of generation-inhibition in five elements. This not only brings more mystery to the theory of traditional Chinese medicine, but this style of random inference also brings negative influences on later generations." However, most scholars still noticed that the theory of Yun Qi advocated by Liu Wansu is not the real theory of Yun Qi.

According to the theory of Yun Qi, the heavenly stems and earthly branches are matched with the five elements (Five Yun) and Six Qi. It is believed that the morbidity of each year is governed by this law. There has always been a big controversy over this theory with mixed comments of praise and criticism. Among them, only the relationship between solar terms and dominant Qi objectively reflects the universal law of people suffering from the effects of climate change throughout the year, which has certain scientific principles. However, the morbidity rule based on heavenly stems and earthly branches must meet the requirements of repeated cycles of 12 years, 6 years, and 5 years. Now no credible evidences have been found to prove there is a phenomenon in nature and the human body that conforms to this numerical law. The masterpiece of the theory of Yun Qi is generally recognized as the seven monographs written and supplemented by Wang Bing of Tang Dynasty when he annotated the *Su Wen* with his family's "old collection scroll." This theory seems to be left unnoticed for hundreds of years after Wang Bing's book was completed. It was not until the Song Dynasty that scholars such as Shen Kuo mentioned this theory. In the 2nd year of Yuan Fu in the Northern Song Dynasty (1099), Liu Wenshu wrote the Su Wen Lun Ao (素问论奥), which specifically discussed the Five Yun and Six Qi with illustrations. It was submitted to the imperial court and then gradually noticed by the public. After Wang Anshi's reform, the theory of Yun Qi became one of the subjects for the examination of physicians by the Imperial Medical Bureau. Naturally, all the doctors must study this theory. It should be noted that the theory of Yun Qi, like the doctor's discussion about the *Yi* (Book of Changes), is a manifestation of social thoughts in the medical field during the prosperous development of Confucianism in the Song and Ming Dynasty. Since Liu Wansu was born between the Song and Jin Dynasties, he was naturally exposed to this social thought, so he developed great interests in the theory of Yun Qi. But in the long-term medical practice, Liu clearly realized that there are many far-fetched and unrealistic contents in the dogmatic theory of Yun Qi, so he made many new explanations and interpretations. In a word, it is no longer based on the heavenly stems and earthly branches to estimate the Qi of that year, but depends on the nature of the clinical disease manifestation.

The core of He Jian Theory is "Fire Theory," namely, the Six Qi all come from Fire and the excess five cause fever. In his *Yuan Bing Shi* (Original Disease), the 30 diseases summarized in the "Nineteen Pathologies" in the *Su Wen · Zhi Zhen Yao Da Lun* are expanded to over 90 types, among which the majority is Fire-related, or fever-related diseases (Table 4.11).

Table 4.11 Liu Wansu's "Pathogenesis"

中文	英文
病机	Pathogenesis
肝	Liver
心	Heart
脾	Spleen
肺	Lung
肾	Kidney
风	Wind
湿	Wet
燥	Dry
寒	Cool

(continued)

Table 4.11 (continued)

中文	英文
上	Up
下	Down
火	Fire
热	Heat
《素问·至真要大论》	The *Su Wen·Zhi Zhen Yao Da Lun*
诸风掉眩，皆属于肝。	Light-headedness belongs to liver pathogenesis
诸痛痒疮，皆属于心。	Pain, itchiness, and wound belong to heart pathogenesis
脾诸湿肿满，皆属于脾。	Dampness, swelling, and excess belong to spleen pathogenesis
诸气膹郁，皆属于肺。	Qi stagnation belongs to lung pathogenesis
诸寒收引，皆属于肾	Contracture and tautness caused by cold belong to kidney pathogenesis
诸暴强直，皆属于风。	Sudden and serious onset of disease belongs to wind pathogenesis
诸痉项强，皆属于湿。	Neck stiffness belongs to damp pathogenesis
诸病水液澄澈清冷，皆属于寒。	Running stool and diarrhea belong to cold pathogenesis
诸痿喘呕，皆属于上。	Impotence, wheezing, and vomiting belong to upper energizer
诸厥固泄，皆属于下。	Frequent micturition and excretion belong to lower energizer
诸热瞀瘛，诸逆冲上，诸禁鼓慄，如丧神守，诸燥狂越，诸病胕肿，疼酸惊骇，皆属于火。	Sudden and serious fever, dizziness, and convulsive limbs, logagnosia, chill, and mild mental disorder belong to fire pathogenesis
诸胀腹大；诸病有声，鼓之如鼓；诸转反戾，水液浑浊；诸呕吐酸，暴注下迫，皆属于热。	Sign of ascites, drum belly, yellow urine, and acid regurgitation belong to heat pathogenesis
《素问玄机原病式》	The *Su Wen Xuan Ji Yuan Bing Shi*
诸风掉眩，皆属肝木。	Light-headedness belongs to liver-wood
诸痛痒疮疡，皆属心火。	Pain, itchiness, and wound belong to heart-fire
诸湿肿满，皆属脾土。	Dampness, swelling, and excess belong to spleen-earth
肺诸气膹郁，病痿，皆属肺金。	Qi stagnation belongs to lung-gold
诸寒收引，皆属肾水。	Contracture and tautness caused by cold belong to kidney-water
诸暴强直，支痛緛戾，里急筋缩，皆属于风。	Sudden and serious onset of disease belongs to wind
诸痉强直，积饮痞隔中满，霍乱吐下，体重，胕肿肉如泥，按之不起，皆属于湿。	Neck stiffness, vomiting, and body edema belong to damp
诸涩枯涸，干劲皴揭，皆属于燥。	Frequent micturition and excretion belong to dry
诸病上下所出水液，澄澈清冷，症瘕疝，坚痞，腹满急痛，下利清白，食之不饥，吐利腥秽，屈伸不便，厥逆禁固，皆属于寒。	Running stool, diarrhea, hernia, and abdominal pain belong to cold

(continued)

Table 4.11 (continued)

中文	英文
诸热瞀瘛, 暴喑冒昧, 躁扰狂越, 骂詈惊骇, 胕肿疼酸, 耷呕涌溢, 食不下, 目昧不明, 暴注瘛, 暴病暴死, 皆属于火。	Visual distortion and tugging, manic-depressive psychosis, acute gouty arthritis, convulsive limbs, sudden onset and deterioration of disease, and sudden death all belong to fire
诸病喘呕吐酸, 暴注下迫, 转筋, 小便浑浊, 腹胀大, 鼓之如鼓, 痈疽疡疹, 瘤气结核, 吐下霍乱, 瞀郁肿胀, 鼻塞鼽衄, 血溢血泄, 淋, 身热, 恶寒战栗, 惊惑悲笑, 谵妄, 衄蔑血汗, 皆属于热。	Vomiting, acid regurgitation, turbid urine, swelling abdomen, drum belly, dizziness, stagnation, nasal obstruction, blood spill, blood leakage, fever, chill, emotional instability, and delirium all belong to heat

病机	《素问・至真要大论》	《素问玄机原病式》
肝	诸风掉眩，皆属于肝。	诸风掉眩，皆属肝木。
心	诸痛痒疮，皆属于心。	诸痛痒疮疡，皆属心火。
脾	诸湿肿满，皆属于脾。	诸湿肿满，皆属脾土。
肺	诸气膹郁，皆属于肺。	诸气膹郁，病痿，皆属肺金。
肾	诸寒收引，皆属于肾。	诸寒收引，皆属肾水。
风	诸暴强直，皆属于风。	诸暴强直，支痛緛戾，里急筋缩，皆属于风。
湿	诸痉项强，皆属于湿。	诸痉强直，积饮痞隔中满，霍乱吐下，体重，胕肿肉如泥，按之不起，皆属于湿。
燥		诸涩枯涸，干劲皴揭，皆属于燥。
寒	诸病水液澄澈清冷，皆属于寒。	诸病上下所出水液，澄澈清冷，症瘕癞疝，坚痞，腹满急痛，下利清白，食之不饥，吐利腥秽，屈伸不便，厥逆禁固，皆属于寒。
上	诸痿喘呕，皆属于上。	
下	诸厥固泄，皆属于下。	
火	诸热瞀瘛，诸逆冲上，诸禁鼓慄，如丧神守，诸躁狂越，诸病胕肿，疼酸惊骇，皆属于火。	诸热瞀瘛，暴喑冒昧，躁扰狂越，骂詈惊骇，胕肿疼酸，气逆冲上，禁慄如丧神守，嚏呕，疮疡，喉痹，耳鸣及聋，呕涌溢，食不下，目昧不明，暴注瞤瘛，暴病暴死，皆属于火。
热	诸胀腹大；诸病有声，鼓之如鼓；诸转反戾，水液浑浊；诸呕吐酸，暴注下迫，皆属于热。	诸病喘呕吐酸，暴注下迫，转筋，小便浑浊，腹胀大，鼓之如鼓，痈疽疡疹，瘤气结核，吐下霍乱，瞀郁肿胀，鼻塞鼽衄，血溢血泄，淋闷，身热，恶寒战栗，惊惑悲笑，谵妄，衄衊血汗，皆属于热。

On one hand, Liu Wansu's academic thoughts were spread through his works. For example, Zhang Congzheng called him Liu Shouzhen in the *Jin Shi* (History of Jin Dynasty); on the other hand, his ideas were disseminated through the disciples like Mu Dahuang, Ma Zongsu, Jingshan Futu, and others. Jingshan Futu passed on to Luo Zhiti and then to Zhu Zhenheng, so the Hejian doctrine was spread from the north to the south of the Yangtze River.

4.2.2 Zhang Congzheng and His "Theory of Attacking Pathogenic Factors"

Zhang Congzheng, courtesy name Zi He and art name Dai Ren, hometown Kao Cheng City, Sui Zhou State, was born during the twelfth century to the thirteenth century in the Jin Dynasty. His family was engaged in pharmaceuticals. He once said: "Since my ancestors, I've been giving prescriptions until now for over 50 years" (Zhang Congzheng: "Nue Fei Pi Han Ji Gui Shen Bian Si" (疟非脾寒及鬼神辩四), *Ru Men Shi Qin* (Confucians Duties to Their Parents) (Volume 1), The Imperial Library Wen Yuan Ge Si Ku Version of *Si Ku Quan Shu* (Complete Works of Chinese Classics)). In early years, he followed and adopted the predecessors' common methods of treating diseases, "I knew these prescriptions were not effective, so I started to make changes." After the in-depth study of the *Nei Jing*, he was greatly inspired (Zhang Congzheng: "Wu Ji Liu Ju Zhi Tong Yu Duan Er Shi Er" (五积六聚治同郁断二十二), *Ru Men Shi Qin* (Volume 3)) to innovate new treatments and achieved great success. He was also inspired by Liu Wensu, the representative of Hejian doctrine, who was elder than him and also lived in the Jin Dynasty. In his works, he mentioned directly for several times that he had learnt from Liu Wensu's therapeutic experiences, and many of this thoughts were originated from Hejian doctrine's "Theory of Fire and Heat" (火热论). Zhang Congzheng's medical masterpiece is the 15 volumes of the *Ru Men Shi Qin* (Confucians Duties to Their Parents).

The most prominent feature of Zhang Congzheng's medical theory is "attacking pathogenic factors." He believed that: "Disease is something that is not inherent in the human body. Either it enters from the outside, or arises from the inside. It's called pathogenic factor" (Zhang Congzheng: "Han Xia Tu San Fa Gai Jin Zhi Bing Quan Shi San" (汗下吐三法该尽治病诠十三), *Ru Men Shi Qin* (Volume 2)). Since the disease is caused by pathogenic factors that are not inherent in human body, the therapeutic methods should prioritize the elimination of those pathogenic factors. To that end, he said: "If there were pathogenic factors in the body, they should be attacked quickly so that they could begone soon. Why to preserve those pathogenic factors? Even the fools understand that those pathogenic factors cannot be retained. But they are unhappy when hearing the 'attack' and feel happy when hearing the 'replenishing energy'. Doctors said: one shall replenish the energy and vitality so that the pathogenic factors would be gone automatically. How come there are so many foolish doctors in the world! If not treated properly but merely supplemented with more energy and vitality, when the pathogenic factors can no longer be suppressed, patients would have deteriorated conditions or even died" (Zhang

Congzheng: "Han Xia Tu San Fa Gai Jin Zhi Bing Quan Shi San" (汗下吐三法该尽治病诠十三), *Ru Men Shi Qin* (Volume 2)).

In terms of the therapeutic methods, Zhang Congzheng emphasized on sweating, vomiting, and purgation to cure diseases. In his opinion, "Only when pathogenic factors are eliminated can health be restored. And those methodologies have been practiced and applied for a long time with proven track records. Therefore, I would like to share these with later generations." His three methods of sweating, vomiting, and purgation actually cover a wide range of contents and rules of treatment. He said: "The three methods could be applied to treat many diseases. For example, the vomiting method means drooling, sneezing and shedding tears; The sweating method means moxibustion, steaming, fumigation, precipitation, washing, ironing, burning, acupuncture, stone needling, physical and breathing exercises, massaging and the like; The purgation method means hastening parturition, promoting lactation, grinding and etc. These can be regarded as common therapeutic methods." Given the broad contents of the three methods, the above is only a brief introduction.

The sweating method: it is used to treat exogenous diseases. The wind-cold symptoms should be treated with pungent-natured and warm-natured drugs, while wind-heat should be treated with pungent-natured and cool-natured medicines. But the scope of Zhang Congzheng's sweating method is broader than this. He incorporated acupuncture, moxibustion, stone-needle bloodletting, massage, physical, and breathing exercises into this method. Therefore, the concept of sweating method is extremely abstract and cannot be explained literally.

The purgation method: as one of the 8 rules of treatment in Chinese medicine, if used properly, it can bring surprising effects and bring back life. For example, the purgation method could be used to relieve the symptoms of cerebrovascular diseases. Patients with acute abdomen do not have to go through surgical operation: the manic can immediately calm down; even the dysentery with continuous diarrhea must be treated with the purgation method before it turns to chronic dysentery. Because ordinary people are afraid of the attacking methods and doctors have not grasped the tricks, the purgation method has not been widely used. That's why Zhang Congzheng repeatedly emphasized the adoption of purgation method.

The vomiting method: in the *Su Wen·Yin Yang Ying Xiang Da Lun* (素问·阴阳应象大论), although there is a concept of "If the disease is on the upper part of body (like throat), then it's better to use the vomiting method for treatment." But clinically this method was seldom used for the situations like taking rotten or poisonous food by mistake or committing suicide by taking poison. There are some records of 瓜蒂散吐伤寒 (《伤寒论》), 稀涎散吐膈实中满 (《本事方》), 郁金散吐头痛眩晕 (《万全方》) in ancient medical books, which are rarely used nowadays. Zhang Congzheng is the only one who proposed this vomiting method and compared it to the most common clinical methods like sweating and purgation. According to his medical cases, the vomiting method has been widely used in them.

Zhang Congzheng summarized his knowledge in an article "Sweating, Vomiting and Purgation Methods are Enough to Treat all the Diseases," which becomes his theoretical masterpiece and is selected to be the textbook for Chinese medicine schools. Although his ideas were very hard to understand, the actual efficacy of those therapies made him so famous in the country that he was appointed as imperial physician during 1217–1222.

Although it is possible to "unify" different theories, clinical treatment should be "customized." In essence, Zhang Congzheng's therapeutic methods are extremely flexible and diverse. He created the "exquisite stove steaming method"; he made a chicken feather brush to wash the sores, sharpened the blade with a long clam shell, wrapped the tip with paper, and poured medicine into his nose. When the fetal position is not correct, the stillbirth is taken by hook, etc.

In addition, he is particularly good at psychotherapy. For example, a woman lost appetite and often yelled at others. She even wanted to kill people. There was no way to cure her sickness for a long time. Zhang Congzheng ordered two servants to wear exaggerated makeups like drama actresses to make that woman laugh. The next day, he ordered a game of wrestling, trying to make her laugh again. Meanwhile, he arranged a person to eat and praise the delicious food beside her when she enjoyed the game and show, so the woman was intrigued to eat it as well and her illness was gradually gone.

There was another woman who was frightened after being robbed. Every time she heard some sound, she was shocked and fell in a faint. Therefore, her family all walked gently and cautiously and dared not make any noise. She has taken ginseng, pearl, and other drugs but they were not effective at all. When Zhang came to treat her, he adopted an extraordinary method, asking her to sit on the chair with two people holding her hands and placing one table in front of her. He said: "Please look at this table." Suddenly he hit the table with a stick and startled the woman. He said: "I hit the table with a stick. Why you are so scared?" Later, he hit it again and she was not that frightened. After he hit it several times, she was not shocked any more. Then he asked someone to hit the door and window with a stick and make big noises. After some days, she was not frightened even when hearing the thunder.

Such cases are not few in the *Ru Men Shi Qin*. But unfortunately, this part of knowledge was not emphasized by Zhang Zihe himself. If it can be summarized as a treatment method, he could be said to create a unique "psychotherapy." Another case is worth studying:

> The son of Chou Tianxiang from Wuyang had chills and fever. Several doctors treated it as Gu Zhen Lao Re (A concept in TCM, bone-heat syndrome and fever in chronic consumptive diseases), but his symptoms worsened after half year's treatment. Zhang Congzheng took his pulse and felt a big Guan pulse. He thought this was the symptom of carbuncle. He then asked the foster-nurse: "Is there any pain in his body?" She answered: "None." Zhang ordered servants to take his clothes off and raise his arms, found that his right rib was higher and pressed it with hand. The son leaned away to avoid his press, but didn't move when Zhang pressed the left rib. Zhang said: "His lung has carbuncle, but it's not lung carbuncle. If it's lung carbuncle, then it would have pyemesis." (Zhang Congzheng: "Fei Yong Si Shi Liu" (肺痈四十六), *Ru Men Shi Qin* (Volume 6))

From this case, it can be seen that when Zhang diagnosed, he already noticed the patient's compulsive position and comparison between the two sides of ribs, and came up with a correct judgment: the carbuncle is extrapulmonary, the hydrops caused by pleurisy in modern medicine. This diagnostics method is nearly the same as that in modern medicine, but quite different from the methods of pathogenesis, eight principal symptoms, and Six Qi in traditional Chinese medicine. Why such a unique genre leader like Zhang Congzheng has not been able to make these treatment experiences that are different from traditional medicine, such as psychotherapy and diagnostic methods, into a new theory? Instead, he tried to look for a uniform theory that covers everything – the vomiting, sweating, and purgation methods – to be included in the theory of "Attaching Pathogenic Factors." This remains to be analyzed in the discussion of "The Similarities and Differences of the Four People."

4.2.3 Li Gao's "Theory of Spleen and Stomach"

Li Gao, courtesy name Ming Zhi, was born in Zhen Ding (Now Zheng Ding in Hebei province). Since the previous name of Zhen Ding was Dong Yuan before Emperor Gaozu of Han, Li Gao's art name is Dong Yuan Oldman.

Yuan Haowen (1190–1257), a literati in the Jin Dynasty, is Dong Yuan's good friend, who once wrote the "Biography of Dong Yuan Oldman" and mentioned that Li Gao died in the year of Xinhai at the age of 72. According to that, Li Gao was born in the 20th Year of Emperor Shi of Jin (1180). When the Jin Dynasty perished, he was 55 years old. He died in the 17th Year of the Yuan Dynasty (1251).

Li Gao was born in an affluent family, "which enjoyed great reputation and wealth in Dong Yuan for generations" (Li Gao: *Yi Xue Fa Ming* (Medical Inventions) (Preface), People's Medical Publishing House, 1959 photocopy edition). His ancestors have "donated wealth to earn the title of officials for supervising and managing taxation-related affairs" (Danbo Yuanyin: *Zhong Guo Yi Ji Kao* (Chinese Medical Records), People's Medical Publishing House, reprinted "Huanghan Medical Series" edition (without year), p. 660). At first, he studied the Confucianism. But later when his mother fell ill, he had been attending to his mother "day and night without sleep." However, since he never studied medicine, he could not help treating his mother's disease. Though a large fortune was spent to invite famous doctors for diagnosis, they held different opinions on the disease and argued with each other whether it was cold or fever. Until the death of his mother, the cause of illness was still unknown. Li Gao really regretted and made up his mind to study medicine. He has written many books including the *Nei Wai Shang Bian Huo Lun* (Differentiation on Endogenous and Exogenous Injuries), the *Pi Wei Lun* (Theory of Spleen and Stomach), the *Yi Xue Fa Ming* (Medical Invention), the *Lan Shi Mi Cang* (Secret Record of the Orchid Chamber), and the *Yong Yao Fa Xiang* (Key Principles in Medication).

As known to all, people will have symptoms like cold when going to high-altitude areas with hypoxia. In severe cases, they will turn into pneumonia like a cold (actually pulmonary edema). At this time, taking the "isatisroot granules" and other medicines for clearing away heat and toxic materials for colds is definitely not the "correct treatment." But first of all, the patient needs to "inhale oxygen."

Similarly, due to long-term hunger and "internal injury to the spleen and stomach," symptoms similar to "exogenous wind-cold" can also appear, and the "cold medicine" cannot be used for treatment. Li Dongyuan's the *Nei Wai Shang Bian Huo Lun* shows the fundamental differences between two different causes of the same disease. The *Pi Wei Lun* explains the importance and specific methods of invigorating the "spleen and stomach" in the case of "internal injury," based on the specific objective situation. However, later generations fail to consider such unique objective situations and turn the *Pi Wei Lun* to the "Study on Spleen and Stomach." Furthermore, according to the "five elements attributes" of the spleen and stomach, it was called the "School of Supplementing Earth Element," which evolved from a "concrete" to an "abstract" theory.

Li Gao, Liu Wansu, and Zhang Congzheng were famous doctors in the Jin Dynasty. They all studied the *Nei Jing*, but their theories were drastically different. It's worth noting the specific historical conditions where Li Gao lived. Among them, Liu Wansu is the eldest one. Though it's said that he was born in troubled times, the Jin Dynasty was developing and becoming more prosperous during the Emperor Shi's reign when Liu was middle aged. At that time, the central government stipulated that tax would be exempted in the case of drought, locust plague, and flood, in order to reassure the livelihood of public. The mining rights of gold and silver mines were open for citizens without tax collection. The immigrant old and sick people will be supported by the officials ("The 6th Basic Annals ·Emperor Shizong Part 1," "The 7th Basic Annals · Emperor Shizong Part 2," the *Jin Shi* (History of the Jin Dynasty) (Volume 6, Volume 7)). His theory of fire and fever disease mainly focuses on exogenous feelings. Zhang Congzheng was younger than him, but he lived in Zhongzhou, a place far away from the northwest frontline of Jin and Yuan war. Therefore, he had a carefree life and travelled around, and never suffered hardships of war. But Li Gao was different. He was born in the peak time of Emperor Shi's reign and died in the gradual decline of Emperor Ai's reign, amidst the flames of war between the Jin and Yuan Dynasties. At the age of 41, the Jin Dynasty was besieged on all sides "Xia attacked the Kangu. Song attacked the Qizhou. Red-coated robberies plundered the Suzhou. Yuan soldiers invaded in the capital Yan An" ("The 16th Basic Annals ·Emperor Xuanzong Part 2," the *Jin Shi* (History of the Jin Dynasty) (Volume 16)). By the 1st Year of Emperor Tianxing of Jin (1232), when the so-called Renchen Rebellion took place, the Bian Jing city was on a lockdown. Rice was so scarce and expensive that people died of hunger. Men and women begged on the street. People ate each other. Even someone killed his wife and daughter and ate them. All the leather-made objects were cooked and eaten as well ("The 53rd Ranked Biography ·Wan Yan Nu Shen," the *Jin Shi* (History of the Jin Dynasty) (Volume 115)). As for the imperial court, "officials' thin horses were slaughtered for food." That year, "the epidemic broke out in Bian Jing. Within 50 days, more than 900 thousand people died. This number didn't include those poor dead people who could not be buried" ("The 17th Annals·Emperor Aizong Part One," the *Jin Shi* (History of the Jin Dynasty) (Volume 17)).

Li Gao experienced the Renchen Rebellion himself and wrote in his book that: "After the rebellion finished, almost none of the capital citizens were not sick. There were 12 clinics in the city, receiving 1000 to 2000 people either sick or dead every day." He believed that hunger caused the injuries of spleen and stomach. "Just within three months, why were there hundreds of people with wind-cold symptoms?" He pointed out that "This situation did not only happen in Da Liang, but also in Zhen You, Dong Ping, Tai Yuan and etc. After the war, there were always plenty of the injured and dead." He witnessed that physicians applied exogenous medicines like croton for purgation. "Soon the patients have chest binding syndromes and jaundice. After given the Xian Xiong Wan (pill for severe phlegm – heat syndrome in chest) and Yin Chen Tang (decoction for liver injury), they all died" (Li Gao: "Differentiating Yin and Yang Syndrome," *Nei Wai Shang Bian Huo Lun* (Distinguishing Internal and External Injury) (Volume 1), Wensheng Bookstore, Guangxu Xinsi Nian edition). After the Rencheng Rebellion, Yuan Haowen and Dongyuan left Bianliang, and travelled to Liao Cheng, Dong Ping, and other places. During those years, he saw how Dongyuan treated diseases for people and understood why he was called "National Physician": "He removed the prejudice that doctors are stubborn and use prescriptions without paying attention to the changes of symptoms" (*Dongyuan Trial Prescription· Wang Bowen's Preface.* Quoted from Danbo Yuanyin's *Zh*ong Guo Yi Ji *Kao* (Chinese Medical Records), p. 665).

Due to this historical background, Li Gao wrote the *Nei Wai Shang Bian Huo Lun* based on his practical experiences to show the differences between endogenous and exogenous injuries. He also wrote the *Pi Wei Lun* to furthermore illustrate the specific rules of treatment and prescriptions for spleen and stomach diseases.

However, on the other side, the significance of distinguishing the endogenous and exogenous injuries for treating spleen and stomach diseases is the internal driver to turn it into a kind of abstract theory.

4.2.4 Zhu Zhenheng's "Theory of Nourishing Yin" (养阴论)

Zhu Zhenheng, courtesy name Yan Xiu, was born in Year XinSi of Emperor Gao of Yuan and died in Year Zheng WuXu of Emperor Hui of Yuan (1281–1358) at the age of 78. He was from Yi Wu, Wu Zhou (Now Yi Wu, Zhejiang people). Because there was a "Dan Xi" river in his hometown, Zhu Zhenheng was also called "Dan Xi Weng" or "Mr. Dan Xi" by scholars.

Zhu's family has "generations of descendants with great reputation and contributions to society." Since the late Southern Song Dynasty, Zhu's ancestors opened local schools and taught the Six Classics. Since the early age, Zhu Zhenheng had "great talents in studying and writing poems." At the age of 36, he heard that Zhu Xi's fourth generation of disciple Xu Qian lived in Ba Hua Mountain in Dongyang, so he went to pay respect to Xu and seek for knowledge. He was taught "the secrets of heaven and people, the morality and statecraft of sages." After years of study, he became highly accomplished in academy. One day, the local officials held a banquet to welcome the candidates for the imperial examination. Zhu Zhenheng wanted to take the exam of the *Shu Jing* (The Book of History), but he came across a fortune-teller who gave him ominous divinations twice. He believed in that and gave up taking imperial examinations. He thought that "If I teach the politics in local places, is it also a kind of official job?" He returned to hometown and build several memorial halls on the ruins of the "Shi Yi Ting" (pavilion) by his ancestors, where "he taught the theory of *Zhu Zi Jia Li* (Zhu Zi's Family Rites)." He also built another replicate "Shi Yi Ting" in the south of the memorial hall for clansmen to study there (Song Lian: "Gu Dan Xi Xian Sheng Zhu Gong Shi Biao Ci" (故丹溪先生朱公石表辞). Zhu Zhenheng: *Dan Xi Xin Fa*, Shanghai Science and Technology Press, 1959, pp. 389~395. All citations without citations are based on this).

There are several reasons why Zhu Zhenheng studied medicine. Firstly, he wanted to benefit the people: "As a poor civilian, I could not influence many people. The only way to contribute to the society and influence more people is through providing medical services." Secondly, due to his mother's sickness that cannot be cured by doctors, he determined to study medicine. "He has studied the *Su Wen* for three years and seemed to master the skills." Thirdly, his teacher Xu Qian taught students not for fame and gain, but "according to their talents, provided customized teaching. He didn't teach the test-taking skills for imperial examination and said: righteousness and material gain are different" ("The 76th Ranked Biography ·Confucianism One·Xu Qian," *History of the Yuan Dynasty* (Volume 189)). Zhu Zhenheng's theories were also influenced by him. At that time, Xu Qian fell ill. He had great expectations for Zhu Zhenheng and said: "I've been sick in bed for a long time. Only the great physician could treat my disease. Your intelligence is one of the kind. Do you want to study medicine?" Therefore, Zhu "gave up other studies but only focused on medicine" (Dai Liang: "The Biography of Danxi Weng." Zhu Zhenheng: *Danxi Xin Fa*, Shanghai Science and Technology Press, 1959, pp. 396~403).

At that time, the *He Ji Ju Fang* (Heji Pharmacy Prescriptions) made by Chen Shiwen and others during the Daguan period of Song Dynasty was popular. Zhu Zhenheng studied it day and night, but thought that "the ancient prescriptions might not be suited to cure the present diseases. In terms of dosage and therapeutics, shall we all follow the ancient books like the *Su Wen*?" (Dai Liang: "The Biography of Danxi Weng") However, the rural doctors rarely knew it, so he pretended to travel and visit famous physicians, "Each time I heard there was a respected doctor, I would go and visit." He traveled across Zhejiang, passed Wuzhong and Wanling, and reached Nanxu, Jianye, and Dingcheng before obtaining Liu Wansu's the *Yuan Bing Shi* and Dongyuan's prescription (Zhu Zhenheng: *Ge Zhi Yu Lun·Zhang Zihe Gong Ji Zhu Lun*, People's Medical Publishing House, 1956 pirated edition). However, he never met an ideal teacher. It was not until the 2nd year of Yuan Taiding (1325) that he learnt there was a famous doctor called Luo Zhiti (also known as Mr. Taiwu) in Wulin (Hangzhou), "who was specialized in medicine during the reign of Emperor Li of Song and mastered the theories of Liu Wansu from the Jin Dynasty, Zhang Congzheng and Li Gao." But he was narrow-minded and arrogant about his medical skills, who was very difficult to approach. Zhu Zhenheng visited him several times and wanted to learn skills from him, but failed to meet him. For over 3 months, Zhu stood in front of Luo Zhiti's gate, regardless of wind and rain. Finally, Luo agreed to meet him and they felt like old friends at the first meeting. Afterwards, Zhu officially became Luo's disciple and learn everything from him.

According to historical records, there are more than 10 people who directly learnt from Zhu Zhenheng. There are no fewer than dozens of books in the catalog of medical records titled Zhu Zhenheng. Some books' titles were formerly crowned with the name Danxi, or they were mistakenly believed to be written by Danxi. According to Song Lian's the *Shi Biao* (石表), there are seven types of Zhu Zhenheng's works: one volume of the *Song Lun* (宋论), several volumes of the *Ge Zhi Yu Lun* (格致余论), several volumes of the J*u Fang Fa Hui* (局方发挥), several volumes of the *Shang Han Lun Bian*, several volumes of the *Wai Ke Jing Yao Fa Hui* (外科精要发挥), several volumes of the *Ben Cao Yan Yi Bu Yi* (本草衍义补遗), and several volumes of the *Feng Shui Wen Da* (Q&A on Geomancy). In addition to the *Song Lun* and the *Feng Shui Wen Da*, there are five types of medical books, which are basically the same as the titles of works recorded in Dai Liang's the *Danxi Weng Zhuan* (Biography of Danxi Weng). Besides, the works titled Danxi probably have the following (The following titles are based on Danbo Yuanyin: *Chinese Medical Records*):

The *Dan Xi Ben Cao* (Danxi Materia Medica) The *Dan Xi Mai Jue* (Danxi Meridian Technique)
The *Dan Xi Mai Fa* (Danxi Meridian Method) The *Dan Xi Yi An* (Danxi Medical Cases)
The *Dan Xi Yi Lun* (Danxi Medical Theory) The *Guai Ke Dan* (Strange Disease List)
The *Huo Fa Ji Yao* (The Secret of Living Law) The *Dan Xi Ji* (The Collection of Danxi Theories)
The *Dan Xi Mai Yin Zheng Zhi* (Danxi Pulse Treatment) The *Dan Xi Shou Jing* (Danxi Hand Mirror) The *Dan Xi Mi Chuan Fang Jue* (Danxi Secret Formula)
The *Dan Xi Zhi Fa Yu Lu* (Danxi Therapy Quotation) The *Dan Xi Xin Fa* (Danxi Mastery of Medicine) The *Dan Xi Xin Fa Lei Zhuan* (The Compilation of Danxi Mastery of Medicine)
The *Dan Xi Yao Yao* (Danxi Medicine) The *Dan Xi Zhuan Yao* (The Compilation of Danxi Medicine)
"Danxi Shixuan," "Danxi Heart Needs," "Hundred Questions about Zhu Zhenheng's Production Treasures," "Danxi Living Young Hearts"
The *Dan Xi Shi Xuan* (Dan Xi Medical Methods)
The *Dan Xi Xin Yao* (Summary of Dan Xi Treatment Methods)
The *Zhu Zhen Heng Chan Bao Bai Wen* (Zhu Zhenheng's Methods of Gynecology and Obstetrics)
The *Dan Xi Huo You Xin Fa* (Danxi Methods of Treating Children's Disease)
The *Dan Xi Sui Shen Lue Yong Jing Yan Liang Fang* (Dan Xi Prescriptions and Medical Experiences)
The *Zhu Shi Chuan Fang* (Zhu Family's Prescriptions)
The *Zhu Zhen Heng Zhi Dou Yao Fa* (Zhu Zhenheng's Methods of Treating Acne)

The reason why those works are titled Zhu's name is that maybe his disciples compiled Zhu's monographs, sayings, medical cases, and their own opinions, or that the publisher changed the name when issuing the second edition, or in someone else's name, just to name a few. But so many works were titled the name of Danxi, which in essence showed his influence.

The core of Zhu Zhenheng's medical theory is "Superabundance of Yang and Deficiency of Yin" (阳常有余, 阴常不足) (Zhu Zhenheng: *Ju Fang Fa Hui*, People's Medical Publishing House, 1956 photocopy edition), which could be understood as heaven-yang and earth-yin, sun-yang and moon-yin, sun-solid and moon-wane, and other natural phenomena. Such analogy may not be very thorough. But in fact, as far as human science is concerned, the word "yang" in Chinese medicine since ancient times mainly refers to the intangible function, and correspondingly "yin" refers to tangible material. Since all external functional performance must be based on objective form and material, "yin" is very important. It is the same as Zhu's therapeutic principle of "Replenishing yin and blood to restore yang." Under the guidance of this thought, the medicine he used is in sharp contract with those in the *He Ji Ju Fang*. Such as backache, "patients with big pulse and deficiency of the kidney should take the powered Du Zhong, Gui Ban, Huang Bai, Zhi Mu, and wolfberry and the swine spinal pill; patients with uneven pulse and blood congestion should take the yin-replenishing pill, peach kernel, and safflower (Zhu Zhenheng: "Backache 73," *Dan Xi Xin Fa* (Volume 4)). On one hand, clear away heat and purge pathogenic fire. On the other hand, replenish and nourish yin. This is not only different from Liu Wansu's direct repulsion with bitter cold, but also different from Zhang Zihe and Li Dongyuan's treatment methods, but is a concrete manifestation of Zhu Zhenheng's school of "nourishing yin."

Zhu Zhenheng's theory of nourishing yin is especially suitable for the elderly and vulnerable people, so it was exclusively used for "senior care." It should be pointed out that when people are in their twilight years, their essence and blood will be exhausted, which are reflected in symptoms like dizziness, muscular itching, frequent micturition, falling teeth, excessive salivation, lack of sleep, tinnitus, forgetfulness, baldness, etc. These are all due to the "lack of blood." The treatment should be nourishing yin.

Among the four great medical schools in the Jin and Yuan dynasties, Zhu Zhenheng was the latest one. But he absorbed the essence of the three schools of Liu, Zhang, and Li from his teacher Luo Zhiti and combined them into one. This cannot be achieved without the historical background and Luo's influences. In Zhu Zheheng's works, we could often see the quotations from the Yuan Bing Shi and other books, as well as the treatment experiences of Yuansu, Zihe, Dongyuan, and Luo. His theory of "Superabundance of Yang and Deficiency of Yin" has something in common with that of Liu Wansu, but they have totally different rules of treatment. Liu tried to reach a balance through lessening yang, while Zhu Danxi opted for replenishing yin for balance. For example, when he applied Liu Wansu's representative prescription "Fangfeng Tongsheng San" (Miraculous Power of Ledebouriella), he "removed Ma Huang, Da Huang and Mang Xiao, and added Dang Gui and Di Huang" (Zhu Zhenheng: *Ju Fang Fa Hui*). Liu Wansu attributed all sorts of pain, itch, and wound to internal heat, whereas Zhu regarded them as deficiency of blood.

Regarding Zhang Zihe's theory of attacking pathogenic factors, Zhu Zhenheng commented it as "overdose" (Meng Lang 猛浪) (Zhu Zhenheng: *Ge Zhi Yu*

Lun·Zhang Zihe Gong Ji Zhu Lun), which was closely related to his Confucian doctrine of the mean. Therefore, he tended to adhere to the principle of "Wang Dao" (王道) in the treatment of diseases. Take Jun Chuan San (浚川散) and Yu Gong Wan (禹功丸) as examples, Zhang Zihe's using these two drugs as purgation to attack the pathogenic factors was a rash and reckless decision (Zhu Zhenheng: *Ge Zhi Yu Lun·Gu Zhang Lun*). Another example is female mastitis. The right treatment should be draining the pus through surgery. But Zhu Zhenheng still regarded it as an "impetuous method that would cause unnecessary pain by needle and knife" (Zhu Zhenheng: *Ge Zhi Yu Lun·Ru Ying Lun*).

He also selectively absorbed Li Dongyuan's theories and rules of treatment. For instance, in the case of "uterine bleeding," he once said: "Li Dongyuan has therapies for for metrorrhagia, but they are not targeted the heat cause but mainly the cold cause. We should be cautious when studying his therapies." As for the specific rules of treatment, Li Dongyuan's method is "If the cause is fatigue, the patient should take restoratives like ginseng and astragalus," while Zhu Zhenheng's method is "If there are purple lumps, the patient should take the decoction of four drugs and coptis" (Zhu Zhenheng: "Uterine Bleeding 89," *Dan Xi Xin Fa* (Volume 5)).

Zhu Zhenheng's academic thoughts occupied an important position in the medical circle in the end of the Yuan Dynasty and the beginning of the Ming Dynasty. So, some people said: "Zhu Zhenheng absorbed the essences of different theories from others, like Zhang's exogeneous injuries, Li Dongyuan's internal injuries and Liu Hejian's heat" (The *Dan Xi Xin Fa Fu Yu·Fang Guang Preface*. Quoted from Danbo Yuanyin: *Chinese Medical Records*). His works were brought to Japan by Tashiro Sanki, a monk who studied in China. At first, because Tashiro lived in the remote place of Kanto, Zhu's theory was not spread widely. After that, Tashiro's disciple Manase Dosan promoted this theory in Kyoto, the political and cultural center at that time. Due to the large number of disciples, it became a major school of thought.

4.2.5 Comprehensive Comments

4.2.5.1 The "New Wine in Old Bottle" Style of Inheritance

The major four schools in the Jin and Yuan Dynasties established their own theories, but also emphasized that their doctrines were based on the important classics of the received version of the *Huang Di Nei Jing*. On one hand, it could be said that they originated from the same source but developed in different directions with different focuses. But in essence, they still belonged to the style of "new wine in old bottle," just like the so-called New Text Confucianism: promoting one's own proposition through the ancient sages' classic works. Therefore, if such statements represent their own thoughts, it could be considered as medical theories and practices had some changes – innovation.

When we figure out the kind of "innovation" method implied in the development of traditional medicine, we could naturally reach a fundamental conclusion: the existing traditional medicine in modern world is not the "clone" of the ancient traditional medicine. Hence on the surface, it seems that only the modern science dares to use the "progress" and "exceeding the ancestors" as their ideals, aspirational targets, and criteria of value judgment. However, the traditional medicine was dominated by blindly "respecting the ancient" and never dared to talk about "innovation," or even think about it. In fact, be it "respecting the ancient" or "reviving the ancient," it's only a mindset not objective fact.

4.2.5.2 The Ultimate Truth of "Etiology"

Since the Song Dynasty, three schools of philosophy have been integrated to form the "New Confucianism." Just as the other two names – "Neo-Confucianism" and "Neo-Confucian Education" – showed, there were more speculative contents compared with the inherent Confucian system; with the idea of "investigating things and verity," it aims to explore the basic laws and ultimate principles of the occurrence, development, and evolution of universal things.

Concerning that, the four major schools proposed an "ultimate reason" for the cause of disease, respectively, which was quite similar to the illustration pattern of "Taiji, Vitality → Yin Yang → The Five Elements → Universe" established by the "Neo-Confucianism." From this commonality, we could see the universal influences of the current thoughts at that time. But its specific establishment process is different from the "external causes." For example, the academic propositions of Liu and Zhu contain the core of "Yin" suppressing "Yang," which is opposite to the establishment process of "Tao." As known to all, Taoism emphasized on Yin. As a "Tong Xuan Chu Shi" who specialized in Taoism theories, Liu Wansu adopted the method of cutting what he called the basics of the Six Qi and origin of all diseases, "Fire and Heat" (Yang). But Zhu Danxi who mastered and inherited the Confucianism theories opted for a drastically different method – replenish the insufficient side (Yin). For the same specific methods of "Balancing Yin and Yang," Liu Wansu's way is like "Fox shares pancake" – cut the larger part. If we regard "human body" as a "Taiji" constituted of Yin and Yang, then Liu Wansu's treatment method characterizes "attacking," whereas Zhu Danxi's method of replenishing the insufficient part is called "Expanding and Enlarging Taji."

In terms of the development trajectory of herbalism, similar performances were not shown clearly until the Ming Dynasty – after the *Ben Cao Gang Mu* (Compendium of Materia Medica), nobody pursued the "large and comprehensive" as target again but turned to research on the theories of pharmaceuticals. If someone proposed the question "why modern science didn't originate from China" – Why the transition to herbalism didn't happen in the Jin and Yuan Dynasties? Then the answer is very simple: only what has happened has a cause. What has not happened has no cause. This is where the different perspectives of history and philosophy lie.

On the other hand, from the above specific introductions we could see: although they proposed a theory of "ultimate truth" for etiology, respectively, it's not possible to use only one method or one way to handle all the diseases in real treatment.

4.2.5.3 The Quality and Creativeness of Doctors

Although the motivation and process of Liu, Zhang, Li, and Zhu for studying and practicing medicine are different, these are closely related with "benevolence" and "filial piety." Liu and Li both studied medicine because of their mother's illness; Zhang Congzheng used "Ru Men Shi Qin" as the title of the book; Zhu Danxi gave up his previous career because of his mother's illness, teacher's illness, and his value of life "benefiting people with medical service." However, most importantly, since the Song Dynasty, the knowledge and management of life has been regarded by Confucians as "one part of investigating things and verity," which caused an unprecedented change in the composition of medical researchers who were previously regarded as "inferior" and led to great improvements in the quality of those practitioners. The scholarly attainments of Liu, Zhang, Li, and Zhu were closely related with the unorthodox "Four Major Schools" at that time. This fact can also be applied to "analyze the conditions for the establishment of schools from a historical perspective." The conclusion is simple: the key to establish a new school or theory is undoubtedly talent, rather than other factors. Liu, Zhang, Li, and Zhu all had disciples, but none of them stood out to match up with or exceed the achievements of the four masters – school cannot create talent. Only talent can create school.

4.3 The Medical Schools That Were Founded Early but Flourished Late: "Cold Damage" and "Warm Disease"

"When a book was written" and "when it became a classic" are not the same thing. In addition to the classic works such as the *Huang Di Nei Jing*, the *Nan Jing* (Difficult Issues), and others, it's actually applicable to the analysis of school problems, and the most typical example is the so-called Cold Damage School, based on the common academic propositions of the *Shang Han Za Bing Lun* (Treatise on Warm Diseases) by Zhang Zhongjing, a physician in the Eastern Han Dynasty, and the Warm Disease School that has no classics at all.

4.3.1 Cold Damage and Cold Damage School

The "Cold Damage" referred in Chinese medicine is different from the "typhoid" (typhoid) caused by typhoid bacilli in modern medical names. In terms of its original meaning, it should be based on the discomfort of the human body after catching cold. But with the development of medicine, the disease has broad and narrow meanings. According to the *Nan Jing·Wu Shi Nan* (Difficult Issues ·The Fifty Difficulties): "There are five types of cold damage, including stroke, cold damage, damp-warm symptom, heat disease and warm disease." The above first "cold damage" is the broad sense and encompasses everything, including all the exogenous heat diseases that originated from the *Su Wen· Re Lun*, "People who suffer from heat diseases fall into the category of cold damage," and "the cold damage is also a kind of heat disease." However, in the *Su Wen*, "cold damage" is the cause, which means "a disease caused by cold," but in the name of "heat disease." The above second "cold damage" is the narrow sense, listed in the category of broad "cold damage with stroke, damp-warm symptom, heat disease and warm disease." However, its connotation and extension vary based on different time and event. Although there are five types of cold damage in the *Nan Jing* "with different symptoms," the book only listed the differences of the five pulses without mentioning what are their "symptoms." The rules of treatment are still limited to the broad sense of cold damage, namely, the sweating and purgation methods in the *Su Wen· Re Lun*. It was not until the late Eastern Han Dynasty when Zhang Zhongjing's *Shang Han Za Bing Lun* defined stroke, cold damage, and warm disease:

> Taiyang disease affected by wind with the symptoms of fever, sweat and moderate pulse is called stroke.
>
> Taiyang disease with fever or no fever, affected by cold, with the symptoms of body pain, vomiting, and tight yin and yang pulses is called cold damage.
>
> Taiyang disease with fever and thirst, not affected by cold, is called warm disease.

However, in the *Shang Han Za Bing Lun*, these three definitions are actually only applicable to the first stage of exogenous heat disease, which is the "Sun disease" in the so-called "six meridian syndrome differentiation" system based on the "three yin and three yang" theory. Afterwards, there are five stages closely related to this: "Yangming disease," "Shaoyang disease," "Shaoyin disease," "Taiyin disease," and "Jueyin disease" ("阳明病", "少阳病", "少阴病", "太阴病", "厥阴病"). The clinical manifestations and nature of the disease are completely different from the "Sun Disease," but these six syndrome complexes belong to the *Shang Han Za Bing Lun*, so it is not possible to define the cold damage based on these three definitions, or explain the nature of the *Shang Han Za Bing Lun*.

Another explanation during the same period is based on etiology. Since the law of nature is cold winter, warm spring, and hot summer, the disease caused by cold in winter is called cold damage. If the disease is caused by cold in winter without immediate onset, and the pathogenic cold is hidden under skin, it would become warm disease in spring and heat disease in summer. Since it's caused by cold in winter, the exogenous heat disease caused by sudden climate changes like cool spring and summer is called "Shi Qi" (prevalent epidemic pathogen, 时气) (This can be seen in the "Experiments on Cold Damage" in the Shang Han Lun and in the *Wai Tai Mi Yao* by Tang Wangtao. Both books mentioned: "The Theory of Yin and Yang," but his book was lost). This theory insists that cold is the normal Qi in winter, so it claims that since the disease is due to "cold," it must be caused by the cold in winter. And then the pathogenic cold hides in the skin and becomes warm disease in spring. In summer, it becomes heat disease that is related to the theory of "attacking pathogenic factors," constituting a self-justified theoretical system. But it's not clinically desirable. Third, since the medicine used in the *Shang Han Za Bing Lun* is mostly pungent and warm, after the emergence of the pungent-cool exterior-releasing medicine, it started to distinguish cold damage and warm disease through the proportion of cold and heat in clinical symptoms. The former refers to cold more than heat, applicable to exogenous disease of warm-pungent diaphoresis and "restore yang in warm." The latter refers to heat more than cold, applicable to diseases of cold-pungent diaphoresis and "subduing endogenous Wind." Therefore, in order to understand "cold damage" and its derivatives, we should pay attention to three points:

1. The broad sense of "cold damage" has not disappeared due to the appearance of the narrow sense. Although the study of warm disease developed rapidly in the Ming and Qing dynasties with the emergence of a considerable number of famous doctors and treatises, there was still only the "cold damage" in the medical branch but no "warm disease."
2. The narrow meaning of "cold damage" was a relative concept to warm disease in the Han Dynasty, but it never referred specifically to exogenous diseases of cold deficiency. For example, the "Yangming disease" in the *Shang Han Za Bing Lun* is mainly characterized by fever, thirst, sweating, bulging pulse, and induration of stool, which are no different from warm disease. Its rule of treatment – heat-clearing and purgation representative prescriptions – Baihu Decoction and Chengqi Bei are also adopted by later generations of warm disease school. Hence, some people think that the treatment method for cold damage can be used for warm disease, and there are no essential differences between the two. As for why the later generations established an opposite new theory of warm disease, and why the *Shang Han Za Bing Lun* is regarded as a specialized works on the treatment of cold damage, there are many factors involved. Please see the following for details.
3. The *Shang Han Lun* was published and printed by the Medical Publishing House in the Northern Song Dynasty, and then spread to the world. Since then, there was an increasing number of scholars working on the treatment of "cold damage." They gradually formed a so-called school. Until the late Qing Dynasty and the early Republic of China, the "Classical Prescription School" was officially established. The "Cold Damage School" has both broad and narrow meanings: the broad sense refers to Zhang Zhongjing's *Shang Han Lun* and System of Six-Channel Differentiation, as the general principle of treating exogenous diseases and various miscellaneous diseases. The narrow sense refers to the use of different methods to study and annotate various schools of the *Shang Han Lun*, such as the "Three Programs Compilation School," the "Maintaining Old Theory School," the "Prescription Certification School," and "Treatment Method Certification School." But the "Classical Prescription School" refers to Zhang Zhongjing's "classical prescription" and "medical prescription," observing strictly the original prescription, even the original amount and the original method of clinical application.

4.3.2 Warm Disease and Warm Disease School

The name of "warm disease" is firstly seen in the *Su Wen*. There are two theories concerning the pathogenesis: First, "if caused by cold in winter, it must become warm disease in spring" in the chapter of "Sheng Qi Tong Tian Lun" (生气通天论

篇); second, "if essence is well stored, people will not get warm disease in spring" in the chapter of "Jin Kui Zhen Yan Lun" (金匮真言论篇). The former refers to the theory of "attacking pathogenic factors." The latter demonstrates an important pathology, namely, the warm disease might not necessarily be caused by exogenous feelings, while the root cause is "insufficient yin and internal heat" (not store essence). The symptoms of warm disease in the *Su Wen* are rapid pulse, fever, perspiration, inability to eat, etc. (The *Su Wen· Ping Ren Qi Xiang Lun*: "Each time people exhale, three pulses would move. Each time people inhale, if the pulses are agitated, then its warm disease." The *Su Wen· Ping Ren Qi Xiang Lun*: "Patients with warm disease sweat and have fever again, with agitated pulse. They say they cannot eat any food."), mainly heat-related syndromes. But compared with the *Shang Han Lun*, these symptoms have no essential differences. As a result, though called "warm disease," they still follow the rules of "cold damage." Approximately since the Song Dynasty, the treatment methods for exogenous diseases were divided into "cold" and "heat." According to Zhu Gong, the doctors at that time:

Those who prefer cold medicines don't use drugs like aconite and sulfur. Though in winter, they ask patients to drink cold medicine and take San Huang Yuan (Coptis, Scutellaria, Rhubarb). Those who prefer hot medicines dare not to use rheum officinale and mirabilite, but persuade patients to roast and calcine the drugs and take the elixir of liquefied gold in hot summer.

In this situation:

> Sometimes the patient's relatives would pay attention to the medical formulary and tell the disease from yin and yang; If it's hot disease, they would invite doctor specialized in treating yang symptoms; If it's cold disease, they would call doctor specialized in treating yin symptoms.

It seems like a separation between cold school and hot school, but in essence it shows that distinguishing yin and yang, understanding cold and hot, and adjusting deficiency and excess – the biggest advantage and foundation of Chinese medicine – are not generally applied to physicians who should have identified the disease and provided targeted treatments. Instead, the patients took the initiative to distinguish diseases and then chose appropriate physicians! The phenomenon not only occurred in ancient times but also in modern world.

The famous doctor Liu Wansu during the Jin and Yuan Dynasties is regarded by later generations as an important figure in the development of the warm disease. But since he admired Zhang Zhongjing as his model, he dared not to create something new and original. In terms of theory and rule of treatment, he just emphasized that cold medicine should be applied for hot diseases. Until then, he has not established his own theory of warm disease. However, by the Ming Dynasty, the situation became totally different. The author of the *Wen Yi Lun* (The Plague Theory) (1642) Wu Youxing held that despite the similarities between plague and cold damage, the pathogenesis, onset and process of disease, and therapy are completely different. He believed that plague is caused by the "Evil Qi" of heaven and earth, with strong infectiousness, "No matter old or young, strong or weak, once people contact the plague, they will fall sick. The evilness comes from mouth and nose." Therefore, it is beyond the *Shang Han Lun*, and even the theoretical framework of conventional exogenous hot disease in Chinese medicine. Since then, the warm disease school gradually became an independent one, while at the same time the "Current Prescription School" emerged in correspondence to the "Classical Prescription School."

Given that the warm disease school lacked a founder like Zhang Zhongjing, also known as "Medical Sage" of the cold damage school, as well as a classical works such as the *Shang Han Lun*, its development is free. For example, one of the "Four Major Warm Disease Doctors," Ye Tianshi in the Qing Dynasty, founded a "defense, qi, nutrient and blood differentiation method" out of the "six meridian syndrome differentiation" system. Another "master" Wu Tang followed the so-called Top, Middle, and Bottom "Triple Energizer Syndrome Differentiation" as the rule of treatment. Concerning whether "warm disease" and "plague" are the same meaning, there are two different opinions. For example, Xue Shengbai believed there is definitely no boundary between "warm disease" and "plague"(Xue Shengbai: the *Ri Jiang Za Ji*. Compiled by Tang Lishan: the *Wu Yi Hui Jiang*, Shanghai Science and Technology Press, 1983), the same idea as Wu Youxing. But the author of the *Shi Bing Lun* (Treatise on Seasonal Disease) Lei Feng advocated that "The warm means warm and hot disease; The plague means infectious disease." These two "share the same pronunciation but refer to different diseases." The former indicates "normal qi," while the latter is "evil qi." Also, some said "the treatment methods for warm disease cannot be used for plague and vice versa" (Lei Feng: "The Different Views on Warm Plague," the *Shi Bing Lun* (Volume 8), People's Medical Publishing House, 1956).

What is exactly warm disease? This might be the most difficult question encountered by warm disease experts. Though in the Ming and Qing Dynasties, there emerged a large number of monographs on warm-hot disease or plague, there was still not a unified understanding of the "warm disease." For example, those who believed "warm disease" is "plague" thought the difference between cold damage and warm disease lied in that "cold damage" is an exogenous feeling and caused by the external pathogenic factors entering the skin, while warm disease (plague) is caused by the pathogenic factors entering the mouth and nose. But some people believed that "the so-called evil qi is nothing but stagnated heat" (Zhou Sizhe: "The Epidemic." Compiled by Tang Lishan: the *Wu Yi Hui Jiang*), which denied the viewpoint of evil qi entering the mouth and nose. In addition, Wu Tang suggested "plague" is just a manifestation of "warm disease" and said there are nine types of "warm disease," namely, wind-warm, warm-hot, plague, warm-toxin, summer-warm, damp-warm, autumn-dry, winter-warm, and damp-malaria (Wu Tang: "The Top Energizer Chapter," the *Wen Bing Tiao Bian* (Volume One), People's Medical Publishing House, 1963). But its difference from "cold damage" is that "cold damage starts from taiyang-food merdian, from superficies to interior"; "warm disease starts from taiyin-hand merdian, from up to down." In summary, the common features of the warm disease school – they all believed that there are two different natures of hot disease: cold damage and warm disease, with different pathogenesis, onset, and process of disease and rule of treatment. Through such debates, they tended to challenge the definition of "hot disease belongs to the category of cold damage" in the *Su Wen*.

Actually, the essence of "warm disease" is that damp-hot is disease. Its onset is different from "cold damage" due to the internal cause of damp-hot syndromes. This might be the core of the so-called attacking pathogenic factors. Because of the existence of damp-hot syndromes, no matter it's caused by exogenous cold or "evil qi" in heaven and earth, the onset is doomed to take place sooner or later. Therefore, its clinical manifestation might be exogenous or not, which requires the same therapy. This is because the fundamentals of Chinese medicine treatment are aimed to address the internal cause not the external cause.

4.3.3 Zhang Zhongjing and the Shang Han Za Bing Lun

The founder of cold damage school Zhang Zhongjing, courtesy name Ji (The literature before the Tang Dynasty mentioned "Zhang Zhongjing," such as *Tai Ping Yu Lan* (Volume 722) quoted "He Yong Biography": "Zhang Zhongjing in the same county"; "In 17th Year of Zhongxuan, he came across Zhongjing." Western Jin Dynasty Huangfu Mi's Zhen *Jiu Jia Yi Jing: Preface*: "Han Dynasty has Hua Tuo and Zhang Zhongjing." Zhang Zhongjing is also mentioned in the *Sui Shu·Ji Ji Zhi.* The preface to the book of Song Linyi and others quoted Tang Dynasty Ganbozong's *Biography of Famous Doctors*: "There is no biography of Zhang Zhongjing in the History of Han. According to the *Record of Famous Doctors*: Zhang Zhongjing is from Nanyang. Zhongjing is his courtesy name. He is a filial and incorrupt official in Changsha."), was from Nanyang county (now Nanyang city, Henan province) (There are several different versions of Zhang Zhongjing's hometown. According to the *Xiang Yang Fu Zhi*, he was from Zaoyang, Nanyang. However, Nanyang was the name in the Sui Dynasty. There was no such name in Zhang Zhongjing's era. According to the *He Nan Tong Zhi*, he was from Nieyang, Nanyang, or Nieyang, Nanjun. Nanjun is now located in Hubei province, while Nieyang county is in Henan province. So Nanyang and Nanjun are the same place. Nanyang was a county in the Han Dynasty, now covering the Ye county, Neixiang county in the south of Xionger mountain in Henan, and yingshan county, yun county in the north of Dahong mountain in Hubei province. Although Nieyang belonged to Nanyang country, there are still no reliable literature to prove Zhongjing's hometown). Due to lack of biography in official history, according to other historical materials, he lived from the middle second century to early third century (Zhang Zhongjing's date of birth and death is unknown. Referring to the date of birth and death of He Yong, Wang Can and others who are related to him and his preface to the *Shang Han Lun*, we could estimate that he lived from the middle second century to early third century. According to the history of famous doctors, he was born approximately in 142–210 year, 150–211 year, and 150–219 year. There are no historical records). Gan Bozong in the Tang Dynasty commented on him in the *Ming Yi Zhuan* (Famous Doctors Biography) "He is a filial and incorrupt prefecture chief in Changsha" (The signature in the preface to the *Shang Han Lun* is: "written by Zhang Ji, an Changsha official from Nanyang in the Han Dynasty." This was again seen in Tang Dynasty Gan Bozong's *Ming Yi Zhuan* proofread by Lin Yi and others. Later generations of historians have never seen any Changsha official named Zhangji, so they held different opinions on this: First, they believe this is Zhang Xian, a Changsha official from Nanyang, because the word "xian" and "zhongjing" share similar meanings. So, "Zhang Xian is actually Zhongjing." (Guo Xiangsheng: A Study on the Name of Zhang Zhongjing). Second, they think Zhang Xian and his father lived and died of disease in Changsha. Xian's son used the name Zhongjing. (Zhang Taiyan: "A Study on Zhang Zhongjing"). Third, there was no such name at all. For more information, please read the *Song Yi Qian Yi Ji Kao*, People's Medical Publishing House, 1958, pp. 517–527).

Zhang Zhongjing's medical achievements are reflected in his works the *Shang Han Za Bing Lun*. According to the preface of this book, Zhang's family previously had many members, but "since the Jian'an year of Han till the next 10 years, two thirds of the family members died and 7 out of 10 had cold damage." Despite the sadness, he "started to search and study the ancient prescriptions," referring to many classical medicine books like the *Su Wen*, the *Jiu Juan* (Nine Volumes), the *Ba Shi Yi Nan* (Eighty-one Difficulties), the *Yin Yang Da Lun* (Treatises on Yin and Yang), the *Tai Lu Yao Lu* (Tai Lu Medicine Record), etc.

The content related to "cold damage" in the *Shang Han Za Bing Lun* was compiled by Wang Shuhe, the imperial physician of the Western Jin Dynasty, and spread to later generations. By the Song Dynasty, The Bureau for Revising Medical Books proofread "ten volumes of Zhang Zhongjing's *Shang Han Za Bing Lun*, altogether 22 chapters, 397 treatment methods and 112 prescriptions" (Gao Baohe, Sun Qi, Lin Yi: "Preface of Shang Han Lun. Gangxi Weiren: the *Song Yi Qian Yi Ji Kao*, p. 554). Afterwards, it proofread the *Jin Kui Yu Han Jing*, the same book as the *Shang Han Za Bing Lun* but with a different title, "eight volumes, 29 chapters, 115 prescriptions" (Gao Baohe, Sun Qi, Lin Yi: "Jiao Zheng Jin kui Yu Han Jing Shu". Gangxi Weiren: the *Song Yi Qian Yi Ji Kao*, p. 577). These are the two editions of Zhang Zhongjing's cold damage works in the Song Dynasty. However, only cold damage is mentioned in them, not the miscellaneous diseases. After that, Wang Zhu, an imperial scholar of Hanlin Academy, discovered three volumes of the *Jin Kui Yu Han Yao Lue Fang* among the ancient books. Volume One is about identification of cold damage; Volume Two is about miscellaneous diseases; Volume Three is about prescriptions for female diseases. The Bureau for Revising Medical Books deleted the cold damage part and attached the prescriptions after the miscellaneous diseases to be the *Jin Kui Yao Lue* we see now. How Zhang's original works look like is unknown to all.

The "Six Meridians" differentiation refers to the categorization of hot disease into three yangs (taiyang, yangming, shaoyang) and three yins (taiyin, shaoyin, jueyin), these six stages, according to the process of hot disease. Since the six names of three yangs and three yins are the same as those of TCM meridian theory, they are called the "Six Meridians." The hot disease was divided into three yangs and three yins, which originated from the *Su Wen·Re Lun*. Therefore, most of the later generations thought the *Shang Han Lun* is a theory of syndrome differentiation based on the Six Meridians from the *Su Wen·Re Lun*. But if comparing the symptoms of the six meridians in the two books, we could find that the three yangs part is basically the same, but the three yins part is totally different.

The *Shang Han Lun* including the system of "Six Meridian Symptoms Differentiation" was not given much attention before the Northern Song Dynasty. Sun Simiao in the Tang Dynasty regarded it as one kind of prescription book (Sun Simiao said: "There are only three kinds of prescriptions for cold damage: 1. Guizhi, 2. ephedra, 3. Qinglong." These three prescriptions can cure any cold damage. Other prescriptions such as radix bupleuri can only trigger vomiting, purgation, and sweating but cannot cure the cold damage. (See "Shang Han Part One," "Qian Jin Yi Fang" (Volume 9), People's Medical Publishing House, 1955 photocopy)); Zhang Zhongjing's theories were not even included in the article "eight schools of cold damage" in Volume One of the *Wai Tai Mi Yao* (Medical Secrets of an Official) compiled by Wang Tao. The name of "Zhang Zhongjing's Cold Damage Theory" was just mentioned in the excerpt of different schools' prescriptions. Since the Song Dynasty, the *Shang Han Lun* started to be considered as a work that taught therapies and was studied from various angles. Through the compilation and annotation of articles, and interpretation of the basic concepts like yin and yang, exterior and interior, cold and hot, deficiency and excess, and three yins and three yangs, the core of "Six Meridian Symptoms Differentiation System" was fully elaborated. Therefore, the status of the *Shang Han Lun* has been transitioned from "prescription book" to "classical book." And Zhang Zhongjing started to be called the "Second Sage," and finally was addressed respectfully the "Medical Sage." From the Southern and Northern Song Dynasties to the Jin and Yuan Dynasties, there were at least over 80 schools based on the study of the *Shang Han Lun*. So far, there are 400 to 500 schools exploring the therapies and prescriptions recorded the *Shang Han Lun*, which left nearly thousands of monographs and treatises to later generations and constituted a unique cold damage school in the Chinese medical history. Although their research methods and angles vary from each other, the Six Meridian Symptoms Differentiation system remains to be the outline of the *Shang Han Lun*.

4.3.4 Masters and Masterpieces of Warm Disease

4.3.4.1 Wu Youxing and the *Wen Yi Lun*

Wu Youxing, courtesy name You Ke, was from Wu county, Jiangsu province. In 1642 Emperor Chongzhen's reign of Ming, he wrote the *Wen Yi Lun*. Each time the warm disease school is mentioned, people would rely on this book as basis. There are two reasons: Frist, Wu Youxing was the earliest among the warm disease school in the Ming and Qing Dynasties. Second, "there was no specialized plague books until Youxing's works was written" (The 289th Ranked Biography·Wu Youxing," the *Qing Shi Gao* (Volume 52)).

Wu Youxing is an extremely rational physician. His critical thing was influenced by the principle of investigating things and verity in the theory of the Song and Ming

Neo-Confucianism. He "earnestly studied the law of things and investigated the Qi of different creatures" (Wu Youxing: the *Wen Yi Lun·Preface*, People's Medical Publishing House, 1977). Finally, he realized that "different injuries will bring about various qi" (Wu Youxing: "Different Qis Caused by Different Injuries," the *Wen Yi Lun* (Volume Two)) from "Cows are sick but sheep are not sick, chickens are sick but ducks are not sick, humans are sick but animals are not sick." Imagine if at that time the natural science was so advanced that he had a microscope, he might further his research on this regard and discover different forms and essences of this pathogenic qi. Moreover, he would imagine if he could know what (drug) could control the qi, then "a disease could be treated with one medicine without bothering the emperor and his ministers to add or subtract the taste" (Wu Youxing: "Different Qis Caused by Different Injuries," the *Wen Yi Lun* (Volume Two)). When antibiotics was discovered, to certain extent this assumption became a reality. If we look back at Wu Youxing, one will be surprised to find that his thinking methods are closer to modern Western medicine.

4.3.4.2 Ye Gui and the *Wen Zheng Lun Zhi* (Diagnosis and Treatment of Warm Disease)

Ye Gui, courtesy name Tian Shi, also known Xiang Yan, was from Wu county Jiangsu province. According to the research and investigation by later generations, the Ye family lived in the Emperor Kangxi and Qianlong period of the Qing Dynasty (1667~1746). His father and grandfather generations were both professional doctors. After the death of his father, he studied medicine from his father's disciple Mr. Zhu. Each time he heard there were people specialized in medicine, he would visit and learnt from them. Therefore, altogether he had more than 17 teachers. There were many anecdotes about how he leant skills from masters and how he brought the dying back to life with his miraculous cure. Ye had devoted his whole life to medical treatment. The book *Wen Zheng Lun Zhi* was written by the disciple Gu Jingwen and polished by Tang Dalie (According to the *Wu Yi hui Jiang* compiled by Tang Dalie: "Ye Tianshi, courtesy name Gui, also known as Xiangyan, lived in Xiatang outside Changmen. When he travelled in Dongting mountain with his disciple Gu Jingwen, he talked about the diagnosis and treatment of warm disease symptoms, which were then recorded by his disciple and compiled into twenty articles of 'Wen Zheng Lun Zhi'. Since there are the transcriptions of what he said without changing a word, the words might sometimes seem not that reader-friendly; But Ye's arguments have been loyally recorded without any modification"), which served as the main basis for studying Ye Gui's academic thoughts.

According to Ye's remarks, theory is not the focus, but the specific identification methods and treatment methods in each stage of disease are the most important things. Moreover, the *Wen Zheng Lun Zhi* was not published until Emperor Qianlong's Ren Zi Year (1792) when it was incorporated in Tang Dalie's *Wu Yi Hui Jiang* (Doctor's Medical Journal). As a result, other medical masters obviously didn't understand Ye's remarks at that time. It was not until the later generations that his four stages of "Defense, Qi, Nutrient, Blood" was regarded as the outline of warm disease differentiation. As a clinical therapist, Ye Gui often gave prescriptions without specific names. However, Wu Tang collected those un-titled prescriptions from his medical case records and compiled several famous and effective prescriptions, such as Sang Ju Yin (Mulberry Leaf and Chrysanthemum Beverage), a varied prescription for wind-warm disease; Qinggong Decoction, a varied prescription for warm-hot disease; Coptis and mume decoction, a varied prescription for summer disease, etc.

4.3.4.3 Wu Tang and the *Wen Bing Tiao Bian* (Detailed Analysis of Warm Diseases)

Wu Tang, courtesy name Ju Tong, was from Huaiyin, Jiangsu province. He was born in 23rd Year of Emperor Qianlong of Qing (1758) and died in 16th Year of Emperor Daoguang (1836). He absorbed and learnt various schools of thoughts, ranging from the Su Wen, Zhang Zhong, Wu Youxing, to Ye Tianshi. He believed: "If cold damage could be truly identified, then no one would suspect whether Gui-Ma Integrated Decoction could be used for treatment; If warm disease could be truly identified, then no one would use the pungent and warm-natured therapy for cold damage to treat the warm disease."

Speaking of warm disease, Wu Tang believes that there are three stages: Top, Middle, and Bottom, namely, the "Triple Energizer Syndrome Differentiation." "The warm disease enters from the nose and mouth, passing from the nose to the lungs, passing from the mouth to the stomach. And the lung disease is reversed to the heart. If the top energizer syndrome is not treated, it would pass to the middle energizer of the stomach and spleen. If the middle energizer syndrome is not treated, it would pass to the bottom energizer of the liver and kidney. The syndrome starts from the top and ends at the bottom" (Wu Tang: the *Wen Bing Tiao Bian* (Volume Two)). Based on this, he elaborated on the process and changes of 11 types of warm diseases including wind-warm, warm-hot, plague, warm-toxin, summer-warm, damp-warm, autumn-dry, winter-warm, warm-malaria, summer-heat, and cold-damp from the perspective of the top, middle, and bottom three stages. Therefore, when later generations mentioned the differentiation of warm diseases, Ye's "defense, qi, nutrient, blood" and Wu Tang's three energizer syndrome differentiation are most often used as examples, which serve as the theoretical system for the differentiation, diagnosis, and treatment of warm diseases.

4.3.5 The Times Produce Heroes

When human's life expectancy hasn't reached the current level, there would be so many chronic diseases and senile diseases; and people's understanding of diseases is rather limited. Therefore, the "exogenous diseases" including cold and other epidemic diseases must be the main focuses of mankind. The cold damage school represented by Zhang Zhongjing and the *Shang Han Lun* was formed in early years and thus received widespread attention, which obviously was closely related to this. After thousands of years, due to the changes of objective environment like the urban development, increasing population, and more convenient transportation, the nature of disease also witnessed all sorts of changes. The so-called warm disease school represented by Wu Youxing, Ye Tianshi, and other masters was founded during the Ming and Qing Dynasties, in response to the new demands. In addition, researchers noticed two peak periods of plague and epidemics, namely, from the late Eastern Han Dynasty to the Three Kingdoms, and from the late Ming Dynasty to the Qing Dynasty, when the climates were relatively cooler with lower temperatures that might have caused multiple factors for frequent epidemic diseases. Undoubtedly, the establishment of "cold damage school" and "warm disease school" was driven by internal factors like cognitive development and external factors like social environment.

Knowledge of Medicine and the Development of Materia Medica

5

Yuqun Liao

Contents

Abstract

Both "medicine" and "Materia Medica" are concepts difficult to define. This chapter tries to offer comprehensive explanations to these two concepts. By introducing some famous Chinese classical books on medicine, medical organizations and drug processing methods are illustrated.

Keywords

Medicine · Materia Medica · Wu Pu Ben Cao · Ba Jia Zhi Shuo · Drug processing

Those which could treat diseases are not necessarily medicine – acupuncture needles and scalpers are some examples. Things that are not made for use in medical services are not necessarily called Yao (literally meaning medicine in Chinese), as is the case with Huo Yao (the gunpowder). When wormwood is used in moxibustion, it is regarded as a tool rather than a topical medicine; however, more animals, plants, and

Y. Liao (✉)
The Institute for the History Natural Sciences, Chinese Academy of Sciences, Beijing, China
e-mail: liaogg@ihns.ac.cn

X. Jiang (ed.), *The High Tide of Science and Technology Development in China*, History of Science and Technology in China,
https://doi.org/10.1007/978-981-15-7847-2_5

minerals that are used on the body surface are called "externally applied medicine." If wearing a certain gem is for health care, do we call it an ornament or a drug? If all those which are applied to prevent diseases or promoting health are called drugs, could umbrellas, clothes, and housing be counted as cures? As for food and drug, the boundaries between their definitions have become vaguer. So "medicine" is actually a very abstract and difficult to define concept. The so-called medicine and its related knowledge are gradually developed in the course of trying to use various materials for the purpose of treating and preventing diseases, forming a relatively independent abstract summary and knowledge system.

As things like clothing, holding an umbrella or a stick, and taking a bath are closely related to health, there were monographs on them in the ancient medical classics in India. It was also because of the uncertain relations between food and medicine that there have been terms of food therapy or medicine diet. In the ancients' view, wearing various substances has the function of driving away evil substances or pathogen and healing diseases. Therefore, when "pathogenic factor" or Xie Qi as Chinese called it invades the body, it is necessary to take something orally (or Kou Fu in Chinese) to expulse the pathogen. For this reason, clothing in Chinese (Fu) is a homonym for wearing or taking something orally. Taking medicine orally is also called Fu Yao (Yao means medicine in Chinese). Many similar records can be seen in the early literature *Shan Hai Jing* 山海经, which was completed in the late Spring and Autumn Period to the Han Dynasty. For example, "Fu" a certain bird (wearing its feathers) could make people be not afraid of thunder. There are more records as "Fu Chinese apple cherry that is light and thin, you can protect against water and not drown easily." It is difficult to define whether "Fu" is about "wearing" or "taking orally" to cure a disease. All the abovementioned could demonstrate that had been the complex relationship between "witchcraft" and "science" in the early development of knowledge. Only by understanding these analytical methods in anthropology can we understand theoretically why the origin of medical knowledge such as "Shennong tasted hundreds of herbs" was only a legend.

5.1 The Establishment of Medicine as a Concept

It is generally believed that the Chinese character "medicine" has not been found in Oracle. In 1973, a variety of plant species, mainly "peach nuclei," were found in a site in the Shang Dynasty in Hebei Province (see Picture 5.1). Based on the medicinal records of peach kernels and other medicinal works in later books of herbal medicine, researchers believed that these plant species were all medicinal. (Geng Jianting, Liu Liang: Peach Kernels and Chinese dwarf cherry seeds unearthed from the relics of the Shang Dynasty in Gaocheng, Hebei. In: Cultural Relics. Vol. (8):54. 1974. Xue Yu ed. Chinese Pharmaceutical History. pp.17. People's Medical Publishing House, Beijing (1984)) But this kind of hindsight reasoning based on later pharmacological works obviously is not reliable as it could not be ruled out that these plant species were only the remains of the funeral fruits at the time. Another example is the red mineral cinnabar and grinding tools that appeared in the early tombs (see Picture 5.2). Despite that they were widely used drugs and pharmaceutical tools in later generations, it cannot be proved that

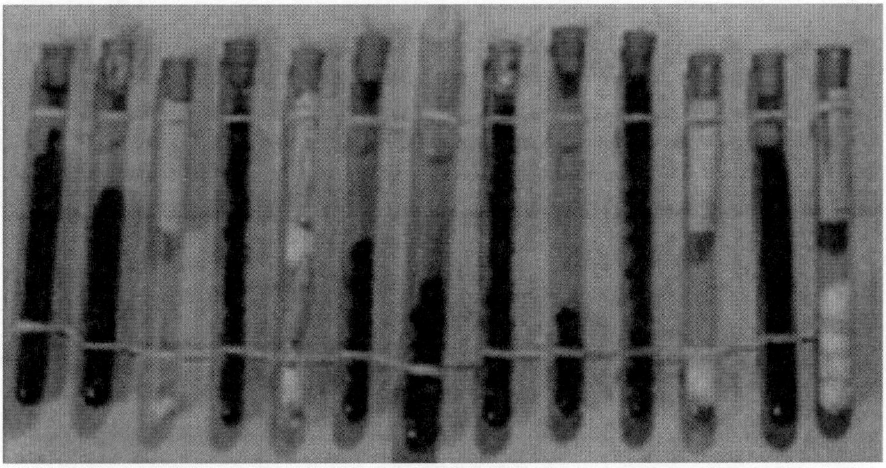

Picture 5.1 Plant seeds unearthed from the tombs of the Shang Dynasty

Picture 5.2 Jade mortar and pestle (玉杵臼) of the Shang Dynasty

they were used for medical purposes. Some researchers believed that the combination of "疒" (the Chinese character component meaning illness) and "water" in oracle bone script made up the Chinese character "medicine"(Wen Shaofeng et al. Studies of the Oracle Inscriptions of the Shang Dynasty on Tortoise Shells or Bones: Science and Technology. Sichuan Academy of Social Studies); when the ancient people tried to predict about the disease, they often asked whether fish should be used. It was based on such a tradition that some researchers believed fish was also used as a kind of medicine, the conclusion of which was not universally recognized though. All the outstanding questions were derived from the vague definition of medicine. The first collection of poems in the history of our country, the Book of Songs, contains a large number of folk poems from the beginning of Western Zhou Dynasty to the middle of Spring and Autumn Period. According to the records in the Books of Songs, Xanthium was used to cure common cold due to chili coldness, the snakeweed to relieve swollen wound, the Chinese herbaceous peony to promote blood circulation and to relieve pain, and Chinese wolfberry to nourish yin and reinforce yang.

Unearthed from No. 14 Tomb of the Shang Dynasty in the Taixi Village of Gaocheng, Hebei. (Fu Weikang, Li Jingwei, and Lin Zhaogeng ed. *Zhong Guo Yi Xue Tong Shi Wen Wu Tu Pu Juan* (《中国医学通史·文物图谱卷》) *General History of Chinese Medicine. Atlas of Cultural Relics*, the People's Medical Publishing House, 2000, P. 19)

Unearthed from the tomb of Fu Hao at Yin Ruins in Anyang with traces of cinnabar around the inside wall. It is common to discover the traces of cinnabar in the jade ware and carved oracle bones of the Shang Dynasty. However, it is very rare to discover the traces of cinnabar on mortar and pestle for grinding purposes. (Edited by National Cultural Relics Bureau: *Zhong Guo Wen Wu Jing Hua Da Ci Dian Jin Yin Yu Qi Juan* (《中国文物精华大辞典·金银玉器卷》) *The Great Dictionary of Chinese Cultural Relics*, Volume of Gold, Silver and Jade, jointly published by Shanghai Lexicographical Publishing House and the Commercial Press, 1996, p. 21)

All these plants are regarded as Chinese herbs and used as medicine. However, back then, they were recorded and described as food. Therefore, researchers attribute it to the theory of food and medicine sharing the same origin.

Since then, the Chinese character "medicine" had definitely been included in the inscription on ancient bronze objects (Rong Geng. Jin Wen Bian《金文编》, pp.30. Science Press: Beijing (1959)). In the *Book of Songs Shi Jing* (《诗》), there were discussions of incurable diseases by medicine (Annotated by Zheng, Commented by Kong Yingda, Da Ya Ban (大雅·板). Mao Shi Zhu Shu《毛诗注疏》Vol.24). In *Shang Shu Shuo Ming* (《尚书·说命》) *Book of History Yue Ming Shang Shu Yue Ming* (尚书·说命), there was a statement widely known as "it is difficult to cure the disease if there is no vertigo and detoxification reactions during the treatment." (Shang Shu Zhu Shu Vol.(9). *Shuo Ming*《说命》is the speech of Wu Ding when he appointed Fu Shuo as the prime minister in the Shang Dynasty. In the two classics in ancient and present times, there is only the ancient version without the present version. Therefore, when it was written cannot be determined. However, the quote was from Meng Zi Teng Wen Gong (part I), which went on like this, according to *Shu*《书》, there is little doubt regarding the statement that it was written in the pre-Qin period. The meaning: if the medicine could not cause the patient to feel dizzy, it would not be able to heal syncope). In *Shuo Wen Jie Zi* (《说文解字》) *Word and Expression*, the author Xu Shen explained medicine (see Picture 5.3) as having the upper component meaning the herb to cure disease with the pronunciation of the lower component pronouncing Yue (乐 (yuè)). In modern simplified Chinese, Yue (乐) has been replaced as Yue (约), but it still means happiness. In a word, medicine is an abstract concept (The modern people's definition of "medicine": "Substances used to diagnose, treat, prevent diseases, and restore, correct, and change organ functions." Concise British Encyclopedia, Compiled Chinese version, Vol. 8, pp. 854. China Encyclopedia Press). It symbolized people's common understanding of the substances that had been used to cure different diseases after they witnessed how

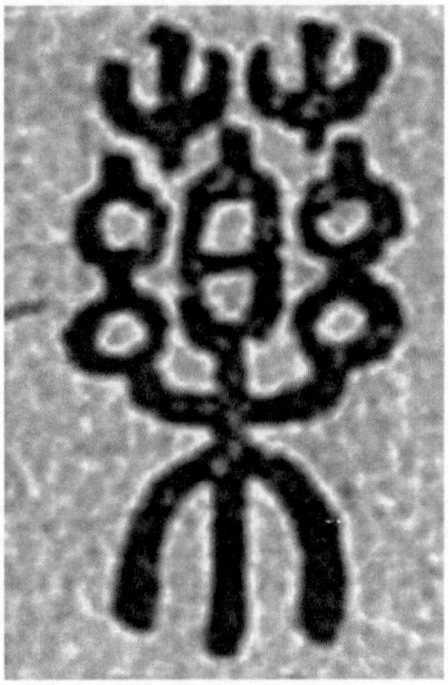

Picture 5.3 Chinese character "medicine" in the calligraphy style of small seal

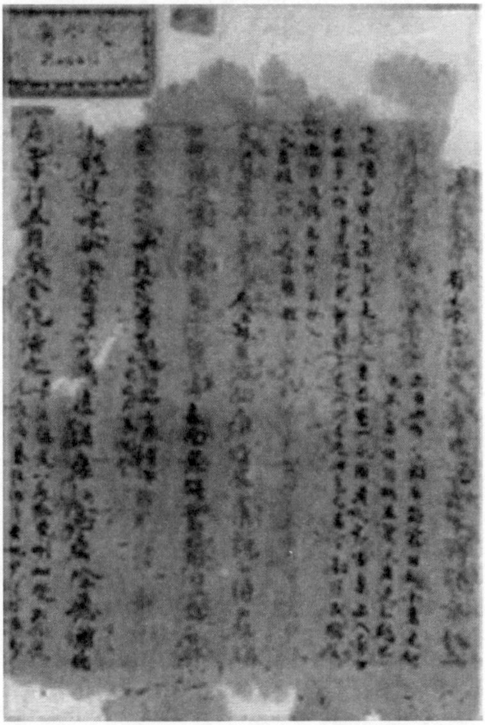

Picture 5.4 Photograph of Ben Cao Ji Zhu unearthed in Dun Huang, courtesy of Zhen Chengliu

the substances worked against the diseases. Therefore, the Chinese character medicine evolved from the practices of using certain substances to treat certain diseases accordingly with an increasingly comprehensive understanding of the abstract concept and the same pronunciation as that of joy or happiness yue (乐). It follows that the two components of the Chinese character had their functions of implying the meaning and suggesting the pronunciation, though the occurrences of both would not have happened simultaneously.

In addition, from the official medical activities recorded in the Rites of Zhou. Minister of State, Physician *Zhou Li Tian Guan Zhong Zai Yi Shi* (《周礼·天官冢宰·医师》):

> Master Physician was in charge of medical affairs, gather all the medicine, including those poisonous, for the disposal of the medical practitioners.
>
> General physician was in charge of treating the diseases of all peoples by using the five flavors, five kinds of grains and five medicines. Promoting health by taking nourishing food and rest is more important than treatments.
>
> Royal surgeon was in charge of applying medicine to patients with ulcers, ulcers, golden ulcers, and folded ulcers according to a certain dose and proportion, as well as scraping off blood stasis and destroying rotten meat. For the treatment of ulcers, use five poisons to attack it, use five Qis to nourish the health and apply five kinds of strong medicines to treat illness, use grains to nourish, use five medicines to treat, and use five flavors to adjust the efficacy.
>
> For those who take medicine, nourish their bones with sourness, nourish their tendons with pungent taste. For those who suffer from ulcers, nourish their pulses with salty taste and nourish their muscles with sweetness. Nourish the seven apertures in the human head with smooth circulation (Annotated by Zheng, Commented by Jia Gongyan: Tian Guan Zhong Zai Xia, Zhou Li Zhu Shu (《周礼注疏》. Vol. 5).

Here nourish is used many times. In the entry of royal surgeon, besides the description of the concept of using the five Qi to nourish health, nourish mostly means treating. It is important to be able to be aware of this. It is prone to trace the origin of science of health maintenance of traditional Chinese medicine to *Zhuang Zi Yang Sheng Zhu* (《庄子·养生主》) *Zhuang Zi Health Maintenance*, which was cited mainly out of convenience of politics. The parable that Lord Wenhui knew how Pao Ding dissected the ox was mainly to tell how the rulers should follow the laws of treating the civilians by promoting their livelihoods. In the meantime, the abovementioned records also foretold the theories of five phases or elements in the discussions of health promotion.

5.2 The Establishment of Materia Medica as a Concept

The medicines used in the traditional Chinese medicine include animals, plants, and minerals, among which plants account for the most proportion. In ancient Chinese pharmacological works, the name of "Pharmacopoeia" was often used. Therefore, "Pharmacopoeia" can be regarded as the traditional Chinese "pharmacology." A common explanation of its meaning is that Chinese medicines are mostly herbaceous plants, so they are called "materia medica." What is questionable in this statement is that the difference between herb Cao Ben (草本) and materia medica Ben Cao (本草) does not only lie in the reversal of the order. The "ben" of the herb means essence

Zhi (质), and the term wooden essence Mu Ben corresponds to it. When the ancient Chinese created the self-explanatory character essence Ben (本), they added a horizontal stroke to the character wood (木), to show the part of root. Therefore, it is proper to explain materia medica as the roots of plants and other parts.

In many historical literature, it was recorded that Zi Yi (子仪), an apprentice of Bian Que's (扁鹊), the famous doctor in the pre-Qin period once wrote one volume of Herbal Medicine. Chun Yuyi, a medical practitioner, obtained the Theories of Medicine *Yao Lun* (《药论》) from one of his masters. However, there were not any works on pharmacology with the title of materia medica in Qi Lue (《七略》) or *Han Shu Yi Wen Zhi* (《汉书·艺文志》*History of the Han Dynasty Art and Literature*. In the notes to the ancient Chinese medical classics, there were descriptions like adjusting and combing a variety of medical agents to make it appropriate for treatment. Regarding the classical prescriptions or formulas, there was *Shen Nong Huang Di Shi Jin* (《神农黄帝食禁》), the notes to which there was statement that bitter agents shall be applied to the five internal organs like the liver, heart, spleen, lungs, and kidney while the pungent agents shall be applied to the six organs including the stomach, large intestine, small intestine, triple focus, bladder, and gallbladder. These are obviously related to the knowledge of pharmacology. Such a phenomenon shows that the rich pharmacological knowledge which had been existing failed to get formed into an independent system for a long historical period. According to the records in *Han Shu the History of the Han Dynasty*, since the late Western Han Dynasty, there had been a post in charge of compiling the Pharmacopoeia. After Wang Mang's administration, he also called for scholars who knew the Five Classics, Calendars, and Materia Medica to serve the royal palace by conducting research and writing works. There were various signs that Independent herbal medicine-pharmacology gradually came into being for the first time. Correspondingly, the medical administrators of the central government in the Eastern Han Dynasty still adhered to the medical order left behind by the predecessors of the Pre-Qin period and Western Han dynasty featured with the imperial physician and aide to the imperial physician. The traditional system was different from that of Western Han Dynasty; the aide to imperial physician further fell into two categories, i.e., aide to the imperial physician in charge of materia medica and aide to the imperial physician in charge of prescriptions. As recorded *Hou Han Shu Bai Guan Zhi* (《后汉书·百官志》) *The History of Later Han Dynasty Officials*:

The monthly salary for one imperial physician is 600 Dan of grains. He was in charge of medical affairs. There were two aides to the imperial physician. One was in charge of the materia medica, and the other was in charge of the prescriptions (*Bai Guan Zhi No. 26. Bai Guan No.3.* (百官志第二十六·百官三) *Officials Part 23. No. 3. History of the Late Han Dynasty*. Vol. 36).

The reason for such a small change to the official position that requires special attention is that the medical staff specialized in medicine preparation was added, which was closely related to the fact that Materia Medica had become an independent discipline at this time. In short, the science of herbal medicine, i.e., ancient Chinese pharmacology, contained the description of the names, characteristics, functions, indications, origins, and the knowledge of retrieving and extracting, processing, and preservation of various drugs. In later books regarding Materia Medica, there are also many examples of coordinated use of various herbs, called the classical prescriptions or formula to Materia Medica.

As medicine plays an extremely important role in the medical process, once they become specialized, various types of "Pharmacopoeia" would emerge one after another. Generally speaking, the important works bearing epoch marking significance presented below are some of the examples.

5.3 Wu Pu Ben Cao (《吴普本草》) Wu Pu's Pharmacopoeia and Ba Jia Zhi Shuo

Wu Pu was a disciple of Hua Tuo. According to the records in Hou Han Shu (《后汉书》) History of the Late Han Dynasty and *San Guo Zhi Wei Shu* (《三国志·魏书》) History of Three Kingdoms Book of Wei, Wu Pu once studied medicine from Hua Tuo who taught Wu Qin Zhi Xi (五禽之戏), a way of body building by imitating the movements of five wild beasts. After 10 years of exercising Wu Qin Zhi Xi, even when Wu Pu reached 90 years old, he could still have sharp sight and hearing with a full set of strong teeth. In the Jin Dynasty, in his works *Bao Pu Zi Zhi Li Pian* (《抱朴子·至理篇》) Ge Hong said, Wu Pu enjoyed his 100th birthday. Therefore, it is generally estimated that Wu Pu was born in the middle of the second century and died in the middle of the third century.

In *Sui Shu Jing Ji Zhi* (《隋书·经籍志》) the *History of the Sui Dynasty Chinese Classics & Culture*, Hua Tuo had a disciple named Wu Pu who wrote six volumes of Wu Pu Ben Cao (吴普本草). Unfortunately, the book was lost long time ago. Research could only be carried out based on citations kept in books such as *Tai Ping Yu Lan* (《太平御览》) *Tai-Ping reign-period Imperial Encyclopaedia [lit. the Emperor's Daily Readings]*. The most typical feature of Wu Pu Ben Cao was that it described the medical theories of the eight famous medical practitioners, i.e., Shennong (神农), Huangdi (黄帝), Qibo (岐伯), Bianque (扁鹊), Yihe (医和), Tongjun (桐君), Lei Gong (雷公), and Li Dangzhi (李譡之). Wu Pu Ben Cao emerged later than *Shen Nong Ben Cao Jing (《神农本草经》) Shennong's Pharmacopoeia*, which was regarded as the first classics on materia medica. However, its contents could well match, if not exceeded, those of Shen Nong Bai Cao Jing. Based on the existing literature citing the theories of the eight famous medical practitioners, it was impossible to cover the overall scope of all materia medica including second names, place of origin, properties and flavor, function, and processing methods or tell their amounts from what was recorded. It was mainly because the original book of Wu Pu Ben Cao was lost, and the quotations from the book as listed in other literature could only tell us what was included in the medical theories of the eight famous medical practitioners while it was hard to tell what could have been omitted or missed. As a result, we could only get some hints regarding the medical principles from Wu Pu Ben Cao and the relevant literature.

(1) Toxicity of medicine. In the early discussions on medicine, the more unified views were on the toxicity and non-toxicity of medicine (Table 5.1). This may be seen as the first step that ancient people experienced when they learned about materia medica. Regardless of whether it was a kind of food or a kind of medicine, no matter how it tasted, what functions or effects it had, which were secondary, whether it was toxic or edible was the first concern.

Table 5.1 Comparison of eight famous medical practitioners' Theories of Toxicity as cited in *Wu Pu Ben Cao* (《吴普本草》)*Wu Pu's Pharmacopoeia*

中文	英文	中文	英文
神农	Shennong	雷公	Leigong
黄帝	Huangdi	李氏	Li
岐伯	Qibo	说明	Notes
桐君	Tongjun	相同	Similarities
医和	Yihe	不同	Differences
扁鹊	Bianque		

	神农	黄帝	岐伯	桐君	医和	扁鹊	雷公
黄帝	▲▲▲▲▲▲▲▲ ▲▲▲▲▲▲▲▲ ▲▲▲▲▲▲▲▲						
岐伯	▲▲▲▲▲▲▲▲ ▲▲▲▲▲▲▲▲ ▲▲▲▲ ××	▲▲▲▲▲▲▲▲ ▲▲▲▲▲▲▲▲ ▲▲ ×					
桐君	▲▲▲▲▲▲▲▲ ▲▲▲▲▲▲▲▲	▲▲▲▲▲▲▲▲ ▲▲▲▲▲▲▲	▲▲▲▲▲▲▲▲ ▲▲▲▲▲▲▲				
医和	▲ ×	×	▲▲				
扁鹊	▲▲▲▲▲▲▲▲ ▲▲▲▲▲▲ ×××	▲▲▲▲▲▲▲▲ ▲▲▲▲▲▲▲▲ ××××	▲▲▲▲▲▲▲▲ ▲▲▲▲ ××	▲▲▲▲▲▲▲▲ ▲▲▲▲▲▲ ××	▲▲		
雷公	▲▲▲▲▲▲▲▲ ▲▲▲▲▲▲▲▲ ▲▲▲▲▲▲▲▲ ▲▲▲▲▲▲▲▲ ▲▲▲ ×	▲▲▲▲▲▲▲▲ ▲▲▲▲▲▲▲▲ ▲▲▲▲▲▲ ××	▲▲▲▲▲▲▲▲ ▲▲▲▲▲▲▲▲ ▲▲▲▲▲ ×	▲▲▲▲▲▲▲▲ ▲▲▲▲▲▲▲▲ ▲▲▲▲ ×	▲ ×	▲▲▲▲▲▲▲▲ ▲▲▲▲▲▲▲▲ ××	
李氏	▲	▲	▲▲				▲▲
〔说明〕相同：▲；不同：×							

(2) Medicine flavors: the flavors of medicine (pungent, sweet, sour, bitter, salty, and so on) could be regarded as the observations of physicians and medical persons over different medicines. They were supposed to be quite unified. However, that was not the case in reality. The unified opinions regarding the flavors of medicines were quite rare. It suggested that when recording data of observation, the observations were listed as a part. However, the personal conceptions of some medical people regarding the relationship between the flavors and effects of materia medica were embedded. For example, the pungent flavors and sweetness were to expel, therefore regarded as Yang, and the sour and bitter flavors were to smooth circulation, therefore regarded as Yin. Even the theories of five elements such as five Qi, five flavors, and the needs and wants of the five internal organs were reflected. In addition, the inconsistent records of odors clearly showed the independence of eight medical practitioners' opinions. By comparing them, we can see that not even two of them were regarded as consistent (Table 5.2).

Table 5.2 Comparison of eight famous medical practitioners' Theories of Herbal Medicines' odors as cited in *Wu Pu Ben Cao* (《吴普本草》) *Wu Pu's Pharmacopoeia*

中文	英文	中文	英文
神农	Shennong	雷公	Leigong
黄帝	Huangdi	李氏	Li
岐伯	Qibo	说明	Notes
桐君	Tongjun	不同	Differences
医和	Yihe		
扁鹊	Bianque		

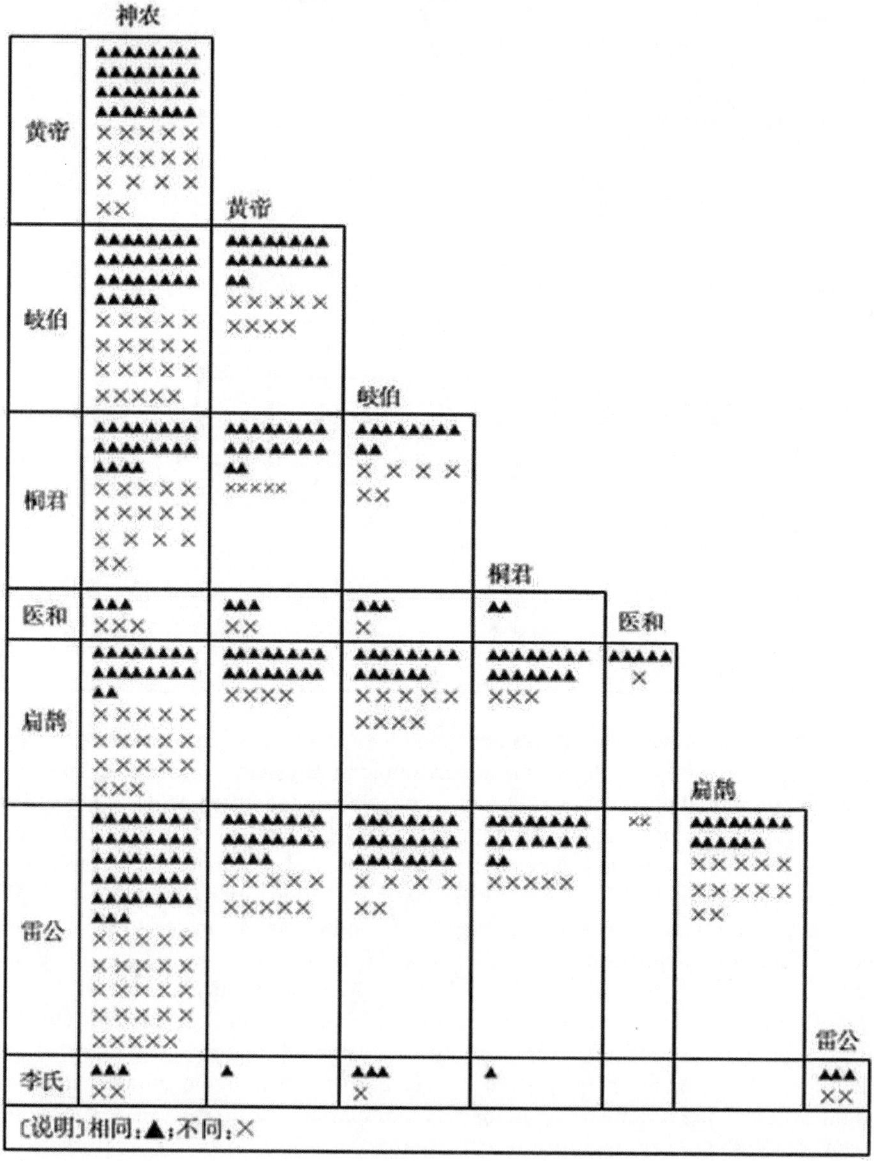

(3) Properties of medicines: the definition of medical properties (cold, hot, warm, cool, and mild) could be regarded as the further perception of medical knowledge in addition to their toxicity and odor. This kind of understanding had actually completely evolved from the primary stage of descriptive science with the establishment of abstract concepts. It was based on the fact that the disease had been identified with the abstract properties of cold or heat. That is to say, the medicinal records were mainly based on the basic principle of cold medicine for fever and hot medicine for cold diseases, the functions of which were told by the demonstration of symptoms of coldness or heat. Without the abstraction of the diseases, symptoms, and properties, there were no discrimination between the cold or hot medicine. In the theories of the 8 famous medical practitioners, the historical data on medicinal properties were far less abundant than those on toxicity and odor, for which the author could not be blamed. It could only be considered as the result of underdevelopment of in-depth theoretical understanding. For example, in the existing historical data of "Wu Pu Ben Cao," there were 108 references to the medicinal taste and the medicinal properties as told by Shennong. There were 22 references to the medical properties. In the medical works that emerged later, each kind of herbal medicine was referred to with description of properties and flavors.

In *Wu Pu Ben Cao* (《吴普本草》) *Wu Pu's Pharmacopoeia*, a lot of theories of the eight medical practitioners were quoted. It suggests that the works demonstrated the continuity of medical science. The role of the works also helped to develop the science of materia medica, which was revealed that in the Liang Dynasty, when Tao Longjing (陶弘景) modified Ben Cao (本草), besides the 365 kinds of medicines illustrated in *Shen Nong Ben Cao Jing* (《神农本草经》) *Shen Nong's Pharmacopoeia*, another 365 kinds of medicines and the materials complementary to those mentioned in *Shen Nong Ben Cao Jing* were listed, many of which were cited from *Wu Pu Ben Cao*.

5.4 The Milestone Work of Materia Medica

Several important books on Materia Medica that are mentioned below are veritable landmarks and milestones. However, most of them had been lost except for *Zheng Lei Ben Cao* (《证类本草》) *Classified Materia Medica* and Ben Cao Gang Mu (《本草纲目》) *The Great Pharmacopoeia* in the Song and Ming Dynasties. Most of what we see today has been compiled by modern or contemporary scholars using texts preserved in handed down documents. From the standpoint of evolutionary history, and from the perspective of practicality, the reason why the early works of pharmacology had not existed was precisely because its essence had been absorbed by subsequent works. If a certain category of knowledge was epoch-making and enduring, it might just have proved that there failed to be development and progress in the regard. Viewed from the standpoint of historical research, new light can be shed on the old issue. The reproduction of various transcripts, after all, has provided a great convenience for a glimpse of how these medical works might have looked like in the long ancient history.

1. *Shen Nong Ben Cao Jing (《神农本草经》) Shennong's Pharmacopoeia* in the Han Dynasty

There is a legend in China that "Shennong tastes Baicao and encounters 72 poisons in one day." Shen Nong (神农氏) Mythical Agriculture was to identify whether a plant was edible or not. For this reason, Shen Nong was known as the mythical agriculture. However, since *Huai Nan Zi* (《淮南子》) in the Han Dynasty described it as the origin of medical knowledge, Shennong became the originator of medical knowledge. Moreover, as long as the sages (or Shennong) were replaced by "the vast number of working people," the empirical knowledge acquisition mode of "testing in person-getting poisoned-recognizing medicine" contained in was consistent with historical materialism, which helped to provide a solid foundation for its being true and disseminated and carried down from generation to generation and a wide citation by modern scholars at present. Most ancient scholars believed that the Book of Materia Medica in the name of Shennong survived the book burning movement at the beginning of the Qin Dynasty. However, according to the common place names in Eastern Han Dynasty and the time line regarding the formation of independent and specialized discipline of pharmacology, it can be inferred that the book must be compiled in the early Eastern Han Dynasty. Although the book is often referred to as "the earliest existing book on Materia Medica," it actually resulted from a compilation initiative in modern times. The reason why people in the modern times have made modification to it and rendered it to the readers at present lay in the fact that in compiling most of the important books of ancient times, the compilers regarded it as a classic passed down from ancient times and made a lot of citations. When the scholars of future generations wanted to edit it, they had to respect what was mentioned as requested by the needs of historical research. In addition, it was also due to the scholars' respect of the medical classics.

By comparing the discussions of Shennong cited in *Wu Pu Ben Cao* (《吴普本草》) *Wu Pu's Pharmacopoeia* and the corresponding descriptions in *Shennong Ben Cao Jing* (神农本草经》) *Shennong's Pharmacopoeia*, it is obvious to see the interconnection and sharp differences between them. For example, in terms of the records regarding the odors of medicine, 83 kinds of materia medica were the same while 17 were different. As for the records regarding the properties of medicine, 15 kinds were the same and 4 were different. However, the records of the toxicity were quite different. Therefore, it can be considered that *Shen Nong Ben Cao Jing* was made by the school that regarded "Shennong" as the originator in the pharmacology system. It was a masterpiece which symbolized that the thoughts of the school had reached a certain stage of maturity. However, it could not be regarded as a comprehensive summary of pharmacology before the Western Han Dynasty, especially given the fact that in the works, the sacred scriptures and alchemy flavors of "light body and longevity" and "immortal immortality" may not be able to represent the medical knowledge system of clinical doctors who deemed disease treatment as a task rather than effects of alchemy. This is the reason why in this article, before introducing this earliest important book on Materia Medica, the author has had a separate section dedicated to the introduction of *Wu Pu Ben Cao* and its citations of the thoughts of the eight famous medical practitioners.

The school advocating Shennong did not leave behind medical classics or works on prescriptions or formulas. However, they had made great achievements in the studies of medicine. As recorded in *Sui Shu Jing Ji Zhi* (《隋书·经籍志》) *the History of the Sui Dynasty, Chinese Classics & Culture*, there had been at least 3 versions of *Shen Nong Ben Cao* circulated at the time, including the version of 8 volumes, that of 3 volumes, and the 4-volume version with the notes made by Lei Gong. In the 8-volume version, there were notes suggesting the existence of 5-volume version in the Liang Dynasty. It is hard to know whether these versions were different or not or how they were similar or different. It is easy to tell that the masterpiece has been modified by scholars or medical men of the future generations for more than time. It had also been regarded as a classical medical scripture in the works on materia medica in the future, thus occupying an unshakable position. Based on the works on materia medica emerging afterwards, it is known that 365 kinds of medicine were recorded, listed in the upper, middle, and lower grades. For example, the medicines of upper grade could be taken to promote health and prolong life. Those of the middle and lower grades had toxicity corresponding to their medical effects which could not be taken in too much dosage. In each entry of medicine, the descriptions of name, properties, odors and flavors, functions, place of origins and alias, and so on were contained.

2. *Ben Cao Ji Zhu (《本草集注》) Annotations on The Pharmacopoeia* of the Southern and Northern Dynasties

After the formation of science of Materia Medica, its development was extremely rapid. When the catalogue of classics appeared for the second time in the history, not only a variety of books on herbal medicine appeared, but also in a considerable number. *Sui Shu Jing Ji Zhi* (《隋书·经籍志》) *the History of the Sui Dynasty, Chinese Classics & Culture* listed 28 medical works ranging from *Shen Nong Ben Cao* (《神农本草》) *Shennong's Pharmacopoeia* to *Zhong Shen Zhi (《种神芝》) Planting Sacred Lucid Ganoderma*, which reflected the emergence of various specialized works *on Materia* Medica at the time, such as the interpretation of sound and meaning, research on drug aliases, the time for herb gathering, artificial cultivation, and the sketches of materia medica to describe the specific forms of each medicine. The notes on these works also revealed information about a large number of previous herbal works. For example, the notes included in the 8-volume Shen Nong Bai Cao include 18 types of works on Materia Medica that had been included in this system, totally accounting for 86 volumes. In the 3-volume Tong Jun Yao Lu, a list of works subordinates to this system such as a variety of *Yao Lu* (《药录》) *Pharmaceutics*, *Yao Fa* (《药法》) *Pharmaceutical Law, Yao Lv* (《药律》) *Rules of Medicine*, *Yao Dui* (《药对》) *Medicine Pair, Yao Mu* (《药目》) *Medicine Catalogue*, *Yao Ji* (《药忌》) *Taboos and Avoidance in Applying Medication*, and other works.

Annotations on Pharmacopoeia compiled by Tao Hongjing based on the four volumes of *Shen Nong Ben Cao* (《神农本草经》) Shennong's Pharmacopoeia (Figure 4 also known as "Pharmacopoeia"). The reason why the works recording 730 kinds of herbal medicine is regarded as epoch-making medicinal works after the

Shen Nong Ben Cao Jing (*Shennong's Pharmacopoeia*) firstly lies in that the contents of *Shen Nong Ben Cao Jing* were written in red, making it possible to preserve this classic. Secondly, besides the records of some new medicines, the works took the notes and medical experiences of the previous medical practitioners and offered many explanations and notes to the medicines as recorded in *Shen Nong Ben Cao Jing*. Thirdly, the three-grade classification typical of the ancient alchemy was replaced with the classification based on the natural properties and origins of materia medica that had widely been adopted by other medical practitioners.

The important books of later generations basically follow this system for further development in the abovementioned three aspects. The use of different fonts or labeling methods has made it possible for people to understand the contents of the important early medical classical works such as Shen Nong Ben Cao Jing Shennong's Pharmacopoeia and Ben Cao Jing Ji Zhu and Pharmacopoeia, though the original works had been lost.

Tao Hongjing, with a style name of Tongming, was born in Moling, Danyang (today Jurong Jiangsu). He was born in 456 A.D., the third year during the Xiaojiang period of the Song Dynasty, and died in 536 A.D. at the age of 81. (See Nan Shi Vol.76 Lie Zhuan No. 66 Yin Yi Xia Tao Hong Jing (《南史》卷七十六 "列传第六十六·隐逸下·陶弘景") *the History of the Southern Histories.* Vol.76: *Collected Biographies. No. 66. Life in Seclusion Part II.* Tao Hong Jing. According to Liang Shu Vol.51 Lie Zhuan No. 45. *Chu Shi Tao Hongjing* (《梁书》卷五十一"列传第四十五·处士·陶弘景"). "He died in the second year during the period of Datong at the age of 85.") It is said that Tao Hongjing was clever and had been eager to learn since his childhood. He was said to have read 10,000 books, good at playing the musical instruments and craftsmanship. He mastered the calligraphy, in particular good at the cursive and scripture styles. He once served the kings and the lords as a reading peer in the late years of Song period of the Southern Dynasties. When he was 10 years old, he got Shen Xian Zhuan (《神仙传》) "Biography of the Immortal" and read it admirably. Since then, he had been keen on maintaining health. In the tenth year (492) of Emperor Qiwu's dynasty, he hung his court dress at the Shenwu Gate and wrote a letter to the king to resign from his position. He then lived in seclusion in Juqu Mountain, Jurong. He put up a pavilion in the mountain and renamed himself Huayang specializing in writing and alchemy. Tao Hongjing loved landscapes and cared about the scenery. When he was in the government, he stayed in the royal court but did not interact with people from the outside. He only reviewed the articles of the king and officials as his first duty. After he resigned from the government, he travelled afar to seek for herbal medicines and elixir. He had conducted deep research regarding the theories of Yin and Yang and five elements, the astronomical observations and predictions, geography, land survey, medicine, and materia medica. He wrote 100 volumes of *Xue Yuan* (《学苑》) Academic Garden, *Xiao Jing Ji Zhu* (《孝经集注》) *Annotations on the Book of Filial Scriptures*, *Lun Yu Ji Zhu* (《论语集注》) *Annotations on the Analects of Confucius*, *Di Dai Nian Li* (《帝代年历》) *The Calendar of the Emperor Dynasty*, *Ben Cao Ji Zhu* (《本草集注》) *Annotations on Pharmacopoeia*, *Jiao Yan Fang* (《效验方》) *Efficacy Prescriptions*, *Zhou Hou Bai Yi Fang* (《肘后百一方》) *One Hundred Prescriptions for Emergency*, *Gu Jin Zhou Jun*

Ji (《古今州郡记》) *Books of Ancient and Modern Prefectures*, and *Tu Xiang Ji Yao* (《图像集要》) *Image Collection*, and other books (As for the stories of Tao Hongjing's life, all can be seen in Nan Shi Vol.76 Lie Zhuan No. 66 Yin Yi Xia Tao Hong Jing (《南史》卷七十六"列传第六十六·隐逸下·陶弘景") the History of the Southern Histories. Vol.76: Collected Biographies. No. 66. Life in Seclusion Part II. Tao Hong Jing).

As Tao Hongjing enjoyed a high popularity in Chinese history and the Ben Cao Jing Ji Zhu, "Annotations on Pharmacopoeia" compiled by Tao occupies an important position in the history of pharmaceutical development. The research on this book has always been a popular topic in medical history research. In my opinion, textual research is only fundamental. It is more valuable to explain the development of medicine on top of it.

(1) Book Title: In *Sui Shu Jing Ji Zhi* (《隋书·经籍志》) *the History of the Sui Dynasty Chinese Classics and Culture*, in the Liang Dynasty, Tao Hongjing compiled 7 volumes of Ben Cao Jing Ji. Afterwards, though there were inconsistency in the records of *Jiu Tang Shu Jing Ji Zhi* (旧唐书·经籍志) *the History of the Tang Dynasty Chinese Classics and Culture* and *Tang Shu Yi Wen Zhi* (《唐书·艺文志》) *the History of the Tang Dynasty, Art and Literature*, what was recorded in the book is convincing. However, Tang's own description of the book was different from the abovementioned description. In the first paragraph in remains of *Ben Cao Ji Zhu Xu Lu* (《本草集注序录》) *Preface to the Annotation of Materia Medica unearthed in Dun Huang* (This volume is collected at Ryukoku University, Kyoto, Japan. There are photocopies of *Ji Shi An Cong Shu* (《吉石盦丛书》) in China. In 1955, *Ben Cao Jing Ji Zhu* (《本草经集注》) *Tao's Pharmacopoeia* was published by Shanghai Qunlian Press. Later, it can be seen in Ma Jixing ed. *Dun Huang Gu Yi Ji Kao Shi* (《敦煌古医籍考释》) *Examination and Explanation of Ancient Medical Books in Dunhuang*. Jiang Xi Science and Technology). (hereinafter referred to as Preface), i.e., the Author's Preface made by Tao Hongjing, it was pointed that the preface was combined totally in 3 volumes. It was also mentioned in the text that "now the author compiles the three volumes of works." We can infer that there were totally 3 rather than 7 volumes of the book. As a circumstantial evidence, in the *Preface to Bu Que Zhou Hou Bai Yi Fang* (《补阙肘后百一方》) *Supplementary to One Hundred Prescriptions for Emergency*, all the abovementioned laws have been included in the beginning scroll of Materia Medica. "If the original book had been divided into 7 volumes, it would be more appropriate to say it was included in Volume 1 of 7." As for why 7 volumes were first brought up, it will be discussed as follows:

The beginning part of the Preface was lost. It is unknown whether the original book had the title or not. In the Epigraph at the end of the volume, there were words Ben Cao Ji Zhu Volume 1 Preface, under which there were six Chinese characters compiled by Hua Yang Tao the Hermit. It was consistent with the records in *Nan Shi Tao Hong Jing Zhuan* (《南史·陶弘景传》) *History of the Southern Dynasty Biography of Tao Hongjing*. However, in *Su Shu Jing Ji Zhu* (《隋书·经籍志》) *the History of the Sui Dynasty Chinese Classics and Culture*, the title was *Ben Cao Jing Ji Zhu*. The main difference lies in whether there is a word Jing, the meanings are different. Ben Cao Jing Ji Zhu, Annotation on Pharmacopoeia, means it was annotations on the book *Shen Nong Ben Cao Jing* (《神农本草经》) *Shennong's Pharmacopoeia* with 50 percent of the contents mentioned in both. In contrast, the book entitled *Ben Cao Ji Zhu* did not involve annotations. Such differences might result from the mistakes in transcription or the opinions that the works *Shen Nong Bai Cao Jing* had been regarded as the most fundamental classics. The books such as *Wu Pu Ben Cao* by Wu Pu, or *Yao Lu* by Li Dangzhi or the thoughts of Qi Bo and Bian Que among other famous medical figures were omitted directly even the statements of Tao Hongjing made in the preface to the effect that the book was compiled with an aim to record all the relevant literature regarding the wide variety of medicines. The contents of *Ben Cao Jing Ji Zhu* were omitted to such a degree that even the newly added 365 medicines listed under the item of Supplementary Medicines used by the famous physicians were also regarded as part of the annotations. If viewed from the perspective of studying the medical thoughts of China, it was until then that *Shen Nong Ben Cao Jing* gained its supremacy in the establishment of pharmacological knowledge. However, *Ben Cao Ji Zhu*, the title was not the original title used by Tao in his Pharmacopoeia, which contained a preface by the author and the table of contents as follows:

Ben Cao Jing (《本草经》) Annotations on Pharmacopoeia "Beginning Scroll"

This book is to trace the origins of medical properties of materia medica to provide reference for the diagnoses and treatment of diseases. For details regarding all the medicines listed here, please review it before you apply them.

Ben Cao Jing Ji Zhu (《本草经》) Annotations on Pharmacopoeia "Middle Scroll"

Three grades of jade and stone, herbs and woods, in total 365 kinds.

Ben Cao Jing (《本草经》) Annotations on Pharmacopoeia "Last Scroll"

Three grades of insects and beasts, fruits, and vegetables as well as grains and other food, totally 195 kinds. With three grades classified for those in name only, i.e., those with names but no reference to specific medicines, totally 179 kinds. The two categories added up to 374 kinds.

Among the three scrolls, the middle and the last scrolls recorded over 730 kinds of medicines with their own medicine directory. The original text of Ben Cao Jing was written in red ink while the original text of Bie Lu was written in black ink. Both had the annotations made by other medical professionals. The body part of the series contained 7 volumes.

In *Zheng Lei Ben Cao* (《证类本草》) *Classified Materia Medica*, they were recorded with white characters against black background, meaning they were original text of *Shen Nong Ben Cao Jing* (《神农本草经》) *Shennong's Pharmacopoeia.* Ma Jixing made the notes which read "the above mentioned three items were the original medicine directory in *Shen Nong Ben Cao Jing* (《神农本草经》) *Shennong's Pharmacopoeia*" (Ma Jixing ed. *Dun Huang Gu Yi Ji Kao Shi* (《敦煌古医籍考释》)

Examination and Explanation of Ancient Medical Books in Dunhuang. Jiang Xi Science and Technology. pp. 338. 1988). However, such a statement could not hold for the following reasons. First, in the Preface, *Shen Nong Ben Cao Jing* (《神农本草经》) *Shennong's Pharmacopoeia*, Tao said the currently preserved four volumes were the original text of *Shennong's Pharmacopoeia.* From this we know that despite there were different versions of *Shennong's Pharmacopoeia, Tao's Cao Ben Jing Ji Zhu* (*Tao's Pharmacopoeia*) was based on the 4-volume version. In addition, from the small-size notes in the text or the big-size text afterwards, the medicine directory contained 730 kinds of medicine, which could only be regarded as that of *Tao's Cao Ben Jing Ji Zhu* rather than the 365 kinds of medicine as listed in *Shennong's Pharmacopoeia*. Therefore, Tao mentioned for many times the volume number of *Shennong's Pharmacopoeia* was the same as that of his own Ben Cao Jing Ji Zhu. As for why the medicine directory turned out to be in red ink or written with white characters against the black background as shown in *Shennong's Pharmacopoeia*, it might be attributed to the mistakes in transcriptions for 400 years as no two versions of *Shennong's Pharmacopoeia* were identical. The fundamental reason behind such a phenomenon lay in people's conception of *Shennong's Pharmacopoeia* as the medical classics with predominant influence to such a degree that whenever Pharmacopoeia was brought up, they would assume it to be *Shennong's Pharmacopoeia*. When it came to the Six Dynasties, the situation got changed. For example, in Sui Shu Jing Ji Zhi the History of the Sui Dynasty Chinese Classics and Culture, there were Wang Jipu's *Ben Cao Jing* and Li Dangzhi's *Ben Cao Jing* as well as Cai Ying's *Ben Cao Jing*. Tao's works might not be named as Ji Zhu (annotations) but Ben Cao (as shown in *Sui Shu Jing Ji Zhi* (《隋书·经籍志》) *the History of the Sui Dynasty Chinese Classics and Culture* and *Preface to Bu Que Zhou Hou Bai Yi Fang* (《补阙肘后百一方》), there were ten volumes of Tao's works or Ben Cao Jing (as shown in the table of contents.) Tao's aim and thoughts could be well shown in his *Preface to Ben Cao Jing Ji Zhu* (《补阙肘后百一方》). He did not feel that his works came second when compared to Shennong's Pharmacopoeia. Instead, his work was to provide a review of all the relevant literature regarding herbal science and medicine so people could discard the dross and select the essential. Tao's work well revealed his own scientific thoughts, which could be carried down together with the knowledge of various herbal formula and prescriptions of materia media even after Tao's death. Upon reviewing the works titled *Mai Jing* (脉经) *the Science of Pulse* written jointly by Wang Shuhe and Huangfu Mi in the Jin Dynasty, Jia Yi Jing (《甲乙经》) and the other works as recorded in *Sui Shu Jing Ji Zhi* (《隋书·经籍志》) *the History of the Sui Dynasty Chinese Classics and Culture*, the medical science was developed not on the basis of respecting or advocating the existing or previous classical works. The works were not merely notes to the previous works; instead they were infilled with the reflections and studies of the previous literature. The aim and effect of the works was to build up a rather complete and independent system of discipline with subbranches covering the knowledge from the ancient.

(2) Volume number. In the table of contents in Tao's *Ben Cao Jing Ji Zhu*, the notes in smaller size say "the body part was divided into 7 volumes." From the text,

the description of 730 kinds of medicines was listed in the beginning and middle scrolls. There might be two possibilities for such a way of compilation. First, in terms of the studies of materia medica, Tao made the notes to the medicine and was planning to include the notes in 7 volumes according to the classification method based on natural properties such as jade, grass, and wood. Then the seven volumes were recompiled into the middle and last scrolls of the works. Second, the scholars afterwards compiled the parts into 7 volumes. No matter which held water, the works titled *Ben Cao Ji Zhu* in 7 volumes had the contents exactly the same as those of the middle and last scrolls of Tao's works, which was consistent with what was mentioned in *Sui Shu Jing Ji Zhi*《隋书·经籍志》*the History of the Sui Dynasty Chinese Classics and Culture*, in the Liang Dynasty; there were Tao's works in ten volumes if counted together with one volume of the Preface and two volumes of Tai Qing Cao Mu Ji Yao. Therefore, the volume numbers were not the only factor to be verified. Another work written by Tao titled *Tai Qing Cao Mu Ji Yao* (《太清草木集要》) should be reviewed as well. It was classified as the classical medical works in the History of the Sui Dynasty, Chinese Classics and Culture.

In addition, in the abovementioned table of contents, the sentences that the original text of *Ben Cao Jing* was written in red ink while the original text of *Bie Lu* (别录) was written in black ink, both of which had the annotations made by other medical professionals mainly referred to the contents in the middle and the last scrolls. It did not suggest that in the Preface, the beginning scroll, there were words in two inks and their annotations.

(3) Preface. This was the beginning scroll of Tao's works, which offered a general description of Tao's opinions of science of materia medica. There were no annotations on the text in the beginning scroll. In the middle and last scrolls of Tao's works, besides the notes made by the author, there were opinions of Huangfu Mi, Xu Yi, Guo Jingchun, Li Dangzhi, Tongjun, and the thoughts of *Xian Jing* (仙经) Immortal classics, Taoism, Taoist scholars, astrologists, medical professionals, and civilians as the annotations. However, the Preface made by Tao was about the thoughts of the author with no annotations, which was different from the middle and last scrolls with the function to solve the issues of clarification and interpretation. In the passage named *Qi Qing Biao* (七情表), besides the names of medicine, all the texts were in small size. If the small characters were deleted, all the contents were lost. Therefore, the small characters were not annotations. They could only represent a way of writing.

It is more important to study whether Article 10: Ben Shuo, the opinions of the author were the original texts of *Shen Nong Ben Cao Jing* (《神农百草经》) *Shennong's Pharmacopoeia*? According to *Shuo Wen Jie Zi* (《说文解字》) *Word and Expression*, opinions refer to the thoughts expressed by people. According to the History of the Han Dynasty, when speaking of notes, "a sentence of five Chinese characters can be explained with 20 or even 30 thousand words. The medical opinions might not require such long expressions. However, the parts of opinions could outnumber those of the original text in *Shennong's Pharmacopoeia*. The opinions could be well understood and different. For example, according to *Ling Shu Xiao Zhen Jie* (《灵枢·小针解》), the four Chinese characters Xu Ze Shi Zhi

(虚则实之), first cited in *Ling Shu Jiu Zhen Shi Er Yuan* (《灵枢·九针十二原》 Spiritual Pivot), means all the syndromes of physical deficiency, including weight loss, loss of appetite, malnutrition, general weakness, numbness in the limbs, sweating, blood loss and pulse weakness, were not suitable for circulation promotion but for complementary filling. In *Su Wen Zhen Jie Pian* (《素问·针解篇》), it said acupuncture was used for the treatment of deficiency syndrome, and there should be a sense of heat under the needle as there was energy; the needle would become hot. When acupuncture was used for the treatment of pathogens, there should be a sense of coldness under the needle as the pathogens were expelled; the needle would become cold. The explanations were quite different in these two kinds of medical classics.

In Tao's *Xu Lu* (《序录》) *Preface to Tao's work*, the first paragraph of Ben Shuo is as follows: the 120 kinds of medicines of upper grade were first choices. They were to promote the health and not poisonous. They would not harm health with an aim to promote health by providing more energy, relieving discomfort and make people live longer. What came next was the interpretations of the medicines of middle and lower grades. We can notice that the words were opinions, instead of original text in *Shennong's Pharmacopoeia.* It is prone to think of works like Tai Ping Yu Lan (《太平御览》) *Tai-Ping reign-period Imperial Encyclopaedia [lit. the Emperor's Daily Readings]* and Yi Wen Lei Ju (《艺文类聚》) and their citations of *Ben Cao Jing* (《本草经》) or *Shennon's Pharmacopoeia* (《神农百草经》): all the medicines of upper grade could help prolong people's lives; all the medicines of middle grade could help promote health and all the medicines of lower grade could help recuperate health, which was more like original texts of the medical classics. In the times when people could read both opinions and original text in *Shennong's Pharmacopoeia*, why did Tao cite the first rather than the second type? The factors that the opinions were richer and easier to understand with more specific reference could help justify such a choice. Besides, the opinions of medical professionals and the original text in Shennong's Pharmacopoeia were both regarded as classical. In *Huang Di Nei Jing* (《黄帝内经) *Pharmacopoeia*, all the sentences initiated with the so-called words could be regarded as the explanations of or opinions regarding the original scripture. However, no one ever doubts their significance.

Two more examples are illustrated. Paragraph 7 in Ben Shuo is cited here: when using a toxic drug, its dosage should be based on the condition, and it should be started from a small dose, and gradually increase the amount only when small amount could not work. In the explanations provided by Tao, the original texts were cited, "as recorded in *Shennong's Pharmacopoeia*, the drug compatibility and dose control were important to avoid toxic effects and ensure drug efficacy during clinical application." The toxicity of medicine should be considered first to decide the dosage. Another example also showed the relationship between original texts in the medical classics and the opinions.

> ...The original text was not necessarily to be followed as timing is very important as what was mentioned in the text should be applied with consideration of time. According to the Chinese era, or Heavenly Stems and Earthly Branches, when gathering the herbs and

> selecting the parts to make medicine, we have to follow the season and calendar to know the property of the herb parts so as to determine whether the select the part exposed to the sun or hidden in the shade.

As mentioned here the part hidden in the shade was cited from the original text. However, the explanations regarding the herbal parts that were supposed to be gathered to make medicine were based on the astrological theories which were not necessarily true. Therefore, Tao found it not convincing. Therefore, the original text was not to be followed.

In Paragraph 10 of Ben Shuo, the contents were to identify about 40 kinds of diseases. Generally speaking, the diseases were named with two Chinese characters, which were not in line with the features of early naming system of diseases. For example, Zhang Zhongjing put forward the diseases related to liquid, rather than edema with swelling syndromes. It was not until the Sui Dynasty that people raised the specific diseases in Zhu Bing Yuan Hou Lun (《诸病源候论》) Origins and Syndromes of Diseases (Volume 21 the Swelling Syndromes or Shui Zhong Hou (水肿候). Besides, there were other diseases like allotriophagia (癖食), prolonged fluid retention (留饮), and dysuria and constipation which were not named in the ancient times listed in *Zhu Bing Yuan Hou Lun* (《诸病源候论》) Origins and Syndromes of Diseases. From this, we can infer that the formation of Ben Shuo (本说) could not be in ancient times.

In a word, besides the descriptions of the 365 kinds of medicines in Shennong's Pharmacopoeia, there were legends regarding Shennong, which might be the original part of the medical classics. Due to the lack of medical value, Tao deleted it in his own works. Scholars of later generations recorded the legends of Shennong in the literature or the lost articles of *Ben Cao Jing* (本草经》) *Shennong's Pharmacopoeia.* In addition, there were other parts of the original text from *Shennong's Pharmacopoeia* including the rich knowledge of medicine usage which had reached to a certain level of theory generalization. Tao took the part he believed to be true and included it into the Preface, i.e., the beginning scroll of his works.

(4) *Ming Yi Bie Lu* (《名医别录》) *Informal Records of Famous Physicians*

With regard to the relationship between Tao Hongjing's works on materia medica and *Ming Yi Bie Lu* (《名医别录》) *Informal Records of Famous Physicians*, scholars had been carrying out verification for a long time. In the contemporary time, Shang Zhijun (尚志钧) stood out as an earnest compiler and editor to Ming Yi Bie Lu (《名医别录》) Informal Records of Famous Physicians. In addition, Shang also elaborated on the origin, formation, contents, features, and implications of Ming Yi Bie Lu (《名医别录》) *Informal Records of Famous Physicians* in detail. Shang argued that Ming Yi Bie Lu (《名医别录》) *Informal Records of Famous Physicians* was written before Tao's Ben Cao Jing Ji Zhu. It contained the materials in Shennong's Pharmacopoeia added by famous medical professionals. Upon the completion of Tao's Ben Cao Jing Ji Zhu, Tao compiled all the materials in *Shennong's Pharmacopoeia* added by famous medical professionals into the works titled Ming Yi Bie Lu (《名医别录》) *Informal Records of Famous Physicians* (Shang Zhijun, Ming Yi Bie Lu (名医别录) Informal Records of Famous Physicians, People's Medical Publishing

House, 1986, P318). However, such a statement could not stand because of the following reasons.

First, *Ming Yi Bie Lu* (《名医别录》) *Informal Records of Famous Physicians* was written at the time earlier than that of Tao Hongjing. For this reason, Tao mentioned in the notes to the two medicines lamprobolite and *Clematis apiifolia* that "the magnetic rock in Ben Jing was also called lamprobolite. It was also listed in Ming Yi Bie Lu (《名医别录》) *Informal Records of Famous Physicians.*" "In Ben Jing, there were records of Clematis apiifolia (女萎). But there were no records of Polyghace Seche (萎蕤)." "*In Ming Yi Bie Lu* (《名医别录》) *Informal Records of Famous Physicians*, there were records of Polyghace Seche (萎蕤), but there were no records of Clematis apiifolia (女萎)."

Second, Ming Yi Bie Lu (《名医别录》) *Informal Records of Famous Physicians* was not medical classical works, but a book of prescriptions or formulas. Therefore, in the History of the Sui Dynasty, Chinese Classics and Culture, Ming Yi Bie Lu was combined into the category of formulas and prescriptions first in the title of *Ming Yi Ji Yan Fang* (《名医集验方》) and then in the title of Shan Fan Fang (删繁方). Its rank was way behind those of the medical classics on materia medica. All those who studied Ming Yi Bie Lu (《名医别录》) Informal Records of Famous Physicians took the characters in black ink in Zheng Lei Ben Cao (《证类本草》) Classified Materia Medica. Therefore, Shang Zhijun took the example of Ai (艾) Chinese mugwort and offered explanation from those listed in Ming Yi Bie Lu (《名医别录》) Informal Records of Famous Physicians including the name, properties and odor, toxicity, functions to cure symptoms, other name, place of origin and time of gathering, and so on (Shang Zhijun ed. Ming Yi Bie Lu (名医别录) Informal Records of Famous Physicians, People's Medical Publishing House, 1986, P319). The records in Ming Yi Bie Lu (《名医别录》) Informal Records of Famous Physicians regarding Ai (艾) Chinese mugwort were as follows.

Ai (艾) Chinese mugwort is cold in its raw form and becomes hot when heated. It can help promote the blood circulation and stop bleeding from five sense organs or subcutaneous tissue. It can be boiled or made into pills for oral take (see the description in Ming Yi Bie Lu (《名医别录》) Informal Records of Famous Physicians as cited in *Tang Ben Zhu* (《唐本注》) *Annotations on the Tang Dynasty's Version*, the notes made in the Tang Dynasty in Zheng Lei Ben Cao (《证类本草》) Classified Materia Medica). Take a look at the contents of Ming Yi Bie Lu in various literature. They were mainly about the treatment and medical uses of drugs with the corresponding prescriptions and medical formula. Therefore, Ming Yi Bie Lu was categorized as the book of formula or prescription in *Sui Shu Jing Ji Zhi* (《隋书·经籍志》) the History of the Sui Dynasty Chinese Classics and Culture.

Third, since Ming Yi Bie Lu (《名医别录》) Informal Records of Famous Physicians was not related to Tao's works titled Ben Cao Jing Ji Zhu, it is hard to identify the author named Tao listed beside the title of Ming Yi Bie Lu. In Sui Shu Jing Ji Zhi (《隋书·经籍志》) the History of the Sui Dynasty Chinese Classics and Culture, all the works by Tao Hongjing were listed with the author's name. From this, we can also infer that Ming Yi Bie Lu and Tao's Ben Cao Jing Ji Zhu were not related.

Fourth, the confusion regarding Ming Yi Fu Pin (名医副品) and Ming Yi Bie Lu (《名医别录》) Informal Records of Famous Physicians. Tao Hongjing used the miscellaneous medicine to generally refer to the medicine other than those listed in Shennong's Pharmacopoeia. As mentioned in the Preface, "there were miscellaneous medicines, totally 365 kinds." Such information was also shown in the notes to the item Abalone Shell: "this was a kind of abalone shell that was quite common. Shells were categorized as other kind of medicine. They shall be classified as miscellaneous medicine. In the item of clam, Tao made notes that said, clams were part of seashell. They should be categorized as miscellaneous medicine, which were limited in the list." Before that, Tao did not include the miscellaneous medicine as those listed in Ming Yi Bie Lu. The issues or topics of articles written by Wu Pu (吴普) and Li Dangzhi (李讜之) were referred to the issues of Ming Yi Bie Lu.

When it came to the Tang Dynasty, the emperor asked about the changes of materia medica. Yu Zhining answered, "in the old times, Tao Hongjing compiled *Shen Nong Ben Cao Jing* (《神农经》) *Shennong's Pharmacopoeia* and other articles written by the famous medical professionals. Then the emperor asked again, "why there were Ben Cao and He Lu two kinds of literature." Yu Zhining answered, "for those listed in Bie Lu, since the Wei and Jin Dynasties, there were passages written by Wu Pu and Li Dangzhi. The descriptions regarding the plants, flowers, and so on were also taken together. So Hong Jing compiled both in his works" ("列传第二十九·于志宁," 《新唐书》 (卷一四0)。Lie Zhuan No. 29 Yu Zhining (列传第二十九·于志宁) Collected Biographies. No. 29. Yu Zhining. In: The New History of the Tang Dynasty. Vo. 140). Yu Zhining did not refer to it as the miscellaneous medicine, but as Bie Lu, or the Supplementary Records. The miscellaneous medicines could cover the medicine which were not listed in Shennong's Pharmacopoeia, but not cover the additional materials listed under the title of Shennong's Pharmacopoeia. Besides, the words like supplementary records or the collection of medical professionals' passages were not the titles of books. Yu Zhining participated in the compilation of the *Newly Edited Pharmacopoeia*. He also read the copy of Ming Yi Bie Lu. Based on such a reason, the two Chinese characters Bie Lu or Supplementary Records were misused, helping us to figure out the two wrong courses of deduction.

Bie Lu mentioned by Yu Zhining was the book titled *Bie Lu* (《别录》), which was exactly the book titled *Ming Yi Bie Lu* (《名医别录》), i.e., the contents in black ink written by Tao in his works titled Ben Cao Jing Ji Zhu.

The newly edited version of Ben Cao, Pharmacopoeia in the Tang Dynasty, was completed with citation of Bie Lu (《别录》) Supplementary Records, which shall then in turn be used to verify the contents in black ink in Tao's Ben Cao Jing Ji Zhu (Shang Zhijun, Ming Yi Bie Lu (《名医别录》) Informal Records of Famous Physicians. pp.381. People's Medical Publishing House, 1986).

5) Classification of Medicine. Tao's reformation regarding how to classify the medicine in his works was mostly praised by those who have been studying Tao's Works Ben Cao Jing Ji Zhu. In a word, the classification method of three grades, namely, the upper grade, the middle grade, and the lower grade of medicine, were further classified according to the natural origin or natural properties of the medicines. They included jade, grass, wood, insects and beasts, fruits, vegetables, grains, and those medicines in name only, i.e., those with names but no reference to specific

medicines. Such a statement was not correct. As pointed out by Shang Zhijun, "we can infer that in Wu Pu Ben Cao, the classification framework had come into being. Or maybe it was Wu Pu Ben Cao that had started the classification of medicines according to their natural properties" (Shang Zhijun, Wu Pu Ben Cao (吴普本草) pp.91 People's Medical Publishing House, 1986).

As a matter of fact, since the ancient times, people always followed the natural properties of medicine to classify the medicine.

First, classification of medicine according to natural properties is the most basic and natural way of thinking and classification method adopted by ancient people which could be well seen by reviewing the shapes and roles of the radicals of Chinese characters. Thus it is unnecessary to further elaborate the point.

Second, as a systematic manifestation of this kind of classifying system, "Er Ya (尔雅)" can be taken as an example, in which "from Thirteenth to Nineteenth" are "interpretation of grass, wood, insects, fish, birds, beasts, and livestock." Even Zhang Hua's Bo Wu Zhi (博物志) **Naturalis Historia** is described according to the method of "categorized as another beast, bird, insect, fish, and grass and wood." This shows that such a method of classification is the most basic and universal classification method.

Finally, within the medical field, "Zhou Li (周礼) The Rites of Zhou" began to record "five medicines to cure diseases." Zheng Xuan noted: "five medicines, namely grass, wood, insects, stones, and grains." It was an unquestionable classification based on natural properties. Later in "Su Wen (素问)" and "Ling Shu (灵枢) Spiritual Pivot," there were five grains, five fruits, five animals, and five vegetables. Although the numbers mentioned were all five due to the influence of the thoughts of five elements, the classification methods were based on the natural properties.

In a word, *Shen Nong Ben Cao Jing* (《神农本草经》) *Shen Nong's Pharmacopoeia* " is of course very important works on Materia Medica. If we regard it as a comprehensive summary, we would ignore the problems of the school of thoughts in the development of medicine in the Qin and Han dynasties. It would be a shame because we might have ignored the whole while touching the parts. It should be noted that "Shen Nong's Pharmacopoeia" had been a masterpiece of theory in the history of medicine development in the Qin and Han dynasties. The many immortals in the record of medicinal effects and the basic classification of three grades were not enough to reflect the overview of the previous development of pharmacology in the Western Han Dynasty. There is no doubt that medicine with clinical treatment as its top priority cannot follow the three-grade classification. Tao Hongjing's classification method was a combination of the following two classification methods: the first is the categories including jade, grass, wood, insects and beasts, fruits, vegetables, grains, and in name only. The second is the upper grade, middle grade, and lower grade. This classification method can also be said to reflect Tao Hongjing's dual personality as a medical scientist and Taoist from one side – he did not give up the inheritance and pursuit of immortality through being a Taoist alchemist. He also had the consciousness of adhering to the scientific classification.

The above analysis of Tao Hongjing's Pharmacopoeia in such a detailed way is to explain that in the history of ancient medical development, "Shennong's

Pharmacopoeia" had not been the supreme. Tao Hongjing would produce a more comprehensive *Pharmacopoeia*, even if it was very likely that Tao's works were also based on Shennong's Pharmacopoeia. Compared with ancient and modern times, people today respect the ancients even more. At the same time, the "milestone" for comprehensively summarizing the predecessor's medical knowledge was not compiled or written in the Han Dynasty. *Shennong's Pharmacopoeia* mistakenly won the honor of being the most fundamental and classical. However, it had not been the representative of the comprehensive summary of pharmacology. Therefore, before the North and South Dynasties, pharmacology could actually be said to have developed along multiple paths; it was only through Tao Hongjing's Materia Medica that the first systematic and comprehensive summary was achieved.

3. *Xin Xiu Ben Cao (新修本草) The Newly Edited Pharmacopoeia* of the Tang Dynasty

In Tao Hongjing's *Ben Cao Jing Ji Zhu*《本草经集注》*Tao's Pharmacopoeia*, he often mentioned the problem of poor understanding of medicines and medications in the north due to the geographical segregation between the north and the south. He also mentioned that the prosperity of the Tang Dynasty, when the country was unified and the culture was developed, provided a comprehensive summary of drug knowledge condition. In the second year of Xianqing Period (657) in the Tang Dynasty, Su Jing and other officials suggested that the imperial court revise the herbal medicine to change the situation of drug confusion and incomplete description at that time. With the approval of the imperial court, a team of more than 20 people including Su Jing was formed, creating a precedent for collective editing of medical classical works.

The official Pharmacopoeia consisted of 3 parts: 20 volumes of the text, 1 volume of the catalogue; 25 volumes of medicine pictures, 1 volume of the catalogue; and 7 volumes of illustrated album of medicine. The existing *Xin Xiu Ben Cao*《新修本草》Newly Edited Materia Medica had only the main body among the three parts mentioned above. Its content was expanded on the basis of *Ben Cao Jing Ji Zhu*《本草经集注》*Tao Hongjing's Pharmacopoeia*, with 114 new medicines added, and the total number of medicines contained in them was 850. In the process of editing, the imperial court "universally awarded the world and sought drugs. "(Tang Shenwei. *Xu Li Part One Tang Ben Xu (序例上·唐本序). In Zheng Lei Ben Cao* (《证类本草》) Classified Materia Medica of the Song Dynasty), "the imperial court collected drugs from the world and counties, and illustrated them extensively." (Wang Pu, Yi Shu (医术), Art of Healing, Tang Hui Yao (《唐会要》), Volume 82). According to the statistics available, there were medicines transported along thirteen paths from 133 cities or towns which were combined into the works. The large-scale medical collection and census could be regarded as a heroic undertaking in the history of science and technology in China. But unfortunately, under the historical conditions at that time, it was impossible to spread the illustration of medicines with the detailed descriptions.

In the fourth year during the Xianqing period in the Tang Dynasty, i.e., 659 A.D., the newly edited Compendium of Materia Media was completed. When the emperor asked about the differences between the newly edited version and the previous versions, the chancellor Yu Zhining answered this way:

The old version of Pharmacopoeia was compiled by Tao Hongjing with combination of *Shennong's Pharmacopoeia* and the notes of *Ming Yi Bie Lu* (名医别录) *Informal Records of Famous Physicians.* Hong Jing was living in the south. He could not identify all the materia medica. He also made mistakes in referring to medicines. All the medicines he recorded were about 400 with verifications made by contemporary scholars. Besides those recorded in Tao's version, there are another 100 kinds of medicine that work well added to the newly edited version.

The emperor was very pleased with the report and the work. He ordered the newly edited version was hidden in a place unknown to the world (Wang Pu, Yi Shu (医术), Art of Healing, Tang Hui Yao (唐会要), Volume 82).

As a new-generation milestone in the history of medicine, the later generations often use the name Tang Ben Cao (《唐本草》) Tang Dynasty's Pharmacopoeia to show the characteristics of the times. Therefore, the book was an official pharmaceutical work, so people today would call it the first pharmacopoeia in the history of China.

Another feature of the Tang Dynasty's Materia Medica was that against the background of frequent overseas traffic at that time, many foreign medicines appeared. In the middle of the eighth century, Zheng Qian wrote 7 volumes of *Hu Ben Cao* (《胡本草》) *Compendium of Foreign Materia Medica*, although it had disappeared. But as the name may have suggested, Hu Ben Cao had been known as a monograph reflecting the knowledge of medication of ethnic minorities outside the territory. Table 5.3 lists 10 kinds of medicines that had long existed in the Indian medical classics *Du Luo Jia Ji* (《阇罗迦集》) *Charaka Samhita* and *Miao Wen Ji* (《妙闻集》) *Sushruta Samhita* and were found in the *Newly Edited Pharmacopoeia*. You can feel the prosperity of Sino-Indian cultural exchanges in the Tang Dynasty and the sources of new medicines in the *Newly Edited Pharmacopoeia*.

Table 5.3 Examples of foreign medicines included in the *Newly Edited Pharmacopoeia*

中文	英文	中文	英文
梵文药名	Medicine name in Sanskrit	石榴	Pomegranate
学名	Scientific name	苏合香	Styrax
中文植物名/药名	Chinese herbal name/ medicine name	槟榔	Betel nut
蓖麻	Castor	合欢	Albizia
阿魏	Chinese Asafoetida	诃梨勒	Haritaki
沉香	Chinese eaglewood	始见本草文献	Firstly seen in the literature of literature of Materia Medica
胡椒	Pepper	《新修本草》	The Newly Edited Pharmacopoeia
姜黄	Turmeric		

梵文药名	学名	中文植物名/药名	始见本草文献
eranda	Ricinus communis	蓖麻	《新修本草》
hingu	Ferula asafoetida	阿魏	《新修本草》
aguru	Aquilaria agallocha	沉香	《新修本草》
marica	Piper nigrum	胡椒	《新修本草》
haridra	Curcuma longa	姜黄	《新修本草》
dadima	Punica granatum	石榴	《新修本草》
turuska	Liquidambar orientalis	苏合香	《新修本草》
kramuka	Areca catechu	槟榔	《新修本草》
sirisa	Albizzia lebbeck	合欢	《新修本草》
haritaki	Terminalia chebula	诃梨勒	《新修本草》

4. *Zheng Lei Ben Cao (《证类本草》) Classified Materia Medica* of the Song Dynasty

There were as many as 80 works on the materia medica in the Northern and Southern Song dynasties, which recorded 1883 kinds of medicines, 1000 more compared to those listed in the *Tang Dynasty's Newly Edited Pharmacopoeia*. The pictures of medicines were mostly sketches from objects [2].

The outstanding achievements of pharmacology in the Song Dynasty were determined by many factors. Among them, the imperial court's attaching great importance to medicine and the development of the printing industry constituted the two very important aspects. Soon after the establishment of the Northern Song Dynasty, Song Taizu ordered gathering "the senior medical professionals with excellent medical skills," and stipulated that those who offered more than 200 medical books would be rewarded (*Song Da Zhao Ling Ji* (《宋大诏令集》) Collections of Great Imperial Orders in the Song Dynasty. In the fourth year during the Dabao period, it was ordered to "gather the senior medical professionals with excellent medical skills." In the period of Tai Ping Xiong Guo (Peace Brought Forth Prosperity of the State), it was stipulated that those who offered more than 200 medical books would be rewarded). Many new herbal books were compiled by government organizations. Many large masterpieces of medicine in the Song Dynasty came from collective wisdom and were published by the central government or local government, forming the mainstream of the development of herbal works during the Northern Song Dynasty. At the same time, the advancement of shipbuilding technology and the use of compasses in navigation had promoted foreign trade and the exchange of science, technology, and culture. For example, countries like India, Vietnam, North Korea, Japan, and Arab countries had traded with China and imported many medicinal materials, which further enriched the drug varieties and knowledge of medical uses on the original basis. The charitable act, the act of benevolence, which took medical assistance as an important form of expression, and the advancement of clinical treatment technology mutually promoted treatment activities, making treatment activities more common, and the demand for drugs was naturally increasing. As a result, medicinal plants had been cultivated as an important economic crop at that time, and large-scale pharmaceutical factories had also emerged. All these factors had led to the inevitable progress of Materia Medica at that time, in order to meet the objective needs. Some foreign scholars had been evaluated in certain books of the Song Dynasty as "more brilliant than European botanical works of the fifteenth and early sixteenth centuries." (Needham Joseph. Science and Civilization in China. Vol.1 Part.1, Chinese Version. pp.289. jointly published by Science Press. Shanghai Ancient Literature Press, 1975).

In terms of the abovementioned official revision of the materia medica by collectives, the version *Kai Bao Xin Xiang Ding Ben Cao* (《开宝新详定本草》) the *Detailed Kaibao Pharmacopoeia* (referred to as Kaibao Materia Medica in short) was compiled jointly by the Confucian officials and the medical officials with the imperial order in the first 6 years after the establishment of the regime Kaibao (973). Nine officials including Liu Han, the Taoist Ma Zhi, and Zeng Ji Ze took the reference of many versions on the basis of Tang Dynasty's *Xin Xiu Ben Cao* 《新修本草》 *Newly Edited Pharmacopoeia*. They selected many effective and precious medicines, such as Clove, Black medicine, Toad, *Gastrodia*, and rhizoma corydalis, which were included in the official herbal medicine for the first time. *Kaibao Pharmacopoeia* contained 983 kinds of medicines, 133 kinds more compared with Tang Dynasty's *Newly Edited Pharmacopoeia*. After the book was completed, it was prefaced by Song Taizu himself and published by The Imperial College, which became the first printed classical works regarding materia medica in China. During the following year, it was revised again and called *Kaibao Revised Pharmacopoeia*. With the change from copying to printing, the editors used the engraved version of Yin (white on black background) and Yang (black on white background) to replace the past two color red and black characters. All the newly added materials were indicated with the sources so that readers could clearly understand the source of each text. At the same time, it also reflected the historic turning from copying to printing of the classical works of Materia Medica.

When it came to the second year during the reign of King Jiayou in the Song Dynasty, i.e., 1057 A.D., the Academy of the Scholarly Worthies established the Office of Revising Medical Books. The first task was to edit the Pharmacopoeia as requested by the imperial order. This is because after only a few decades, because of the continued development of pharmacology, the once-completed herbal medicines that were once imperially compiled and considered perfect were outdated. So the government ordered the medical officials and scholars like Zhang Yuxi, Yu Xi, Su Song, and Lin Yi among others to revise. It took 4 years to compile 21 volumes of Jia

You Bu *Zhu Shen Nong Ben Cao* (《嘉祐补注神农本草》) *Jiayou-Reign Period Replenished Shennong's Pharmacopoeia* with 99 new medicines added, totaling 1082 kinds.

During the period, the Office of Revising Medical Books applied to the imperial court for the collection of new medicines and the illustration of medicines by taking the successful experiences of the Tang Dynasty from all over the country.

> All the states, cities, prefectures and towns shall trace the origins of all medicines and sent medical professionals to identify the size, shape and colors of the stems, leaves, roots of the plants as well as other materials including insects, fish, birds, beasts and jades and minerals for medical purposes before illustrating those materia medica and the descriptions of their properties as well as the gathering methods, time and purposes. For the medicines collected from foreign territories, all the merchants, shippers and transporters concerned shall be inquired to gather more specific information and the samples. All the foreign medicine samples shall be sent to the capital with illustrations and descriptions in detail accordingly. (Su Song. Ben Cao Hou Xu (本草后序) *Epilogue to the Pharmacopoeia*. In. Collected Works of Su Wei Gong. Vo. 65. *Si Ku Quan Shu* (《四库全书》) *Complete Library in the Four Branches of Literature)*

After the materials were sent to the capital, the officials including Su Song compiled them into 21 volumes of *Tu Jing Ben Cao* (《图经本草》) *Illustrated Pharmacopoeia* and published it in 1061 together with *Jia You Bu Zhu Shen Nong Ben Cao* (《嘉祐补注神农本草》) *Jiayou Replenished Shennong's Pharmacopoeia.* If the official revision of the Materia Medica was sufficient to reflect the collective creative power, then *Tu Jing Ben Cao* (《图经本草》) *Illustrated Pharmacopoeia* could be regarded as the result of collective wisdom in the largest scale. At that time, 933 medicine illustrations were presented in more than 150 states and counties, and 103 kinds of herbal medicines for civilian use were added. Thus *Tu Jing Ben Cao* (《图经本草》) *Illustrated Pharmacopoeia* became the first known medicine atlas in China [1].

Interestingly enough, the two Compendiums of Materia Medica produced in the period of Jiayou in the Song Dynasty, namely, *Jia You Bu Zhu Shen Nong Ben Cao* (《嘉祐补注神农本草》) *Jiayou-Reign Period Replenished Shennong's Pharmacopoeia* and *Tu Jing Ben Cao* (《图经本草》) *Illustrated Pharmacopoeia*, with the collective wisdom and the support from all over the country could represent the mainstream and advances of medical development. However, over 20 years after their birth, in Sichuan the place of talents and outstanding natural resources, there were two great classical medical works of materia medica written by individuals, the common features of which were the combination of the two official versions with the addition of more notes. They were 23 volumes of *Chong Guang Bu Zhu Shen Nong Ben Cao Bing Tu Jing* (《重广补注神农本草并图经》) *The Widely Supplemented Shennong's Pharmacopoeia* and *Jing Shi Zheng Lei Bei JI Ben Cao* (《经史证类备急本草》) *the Newly Edited Pharmacopoeia for Political, Economic and Historical Verification*, also known shortly as Classified Materia Medica. The former was written by Chen Cheng during the period of Yuanyou when reigned by King Zhezong of the Song Dynasty (1086 ~ 1093), and was lost. The latter was quite influential as seen in Picture 5.5.

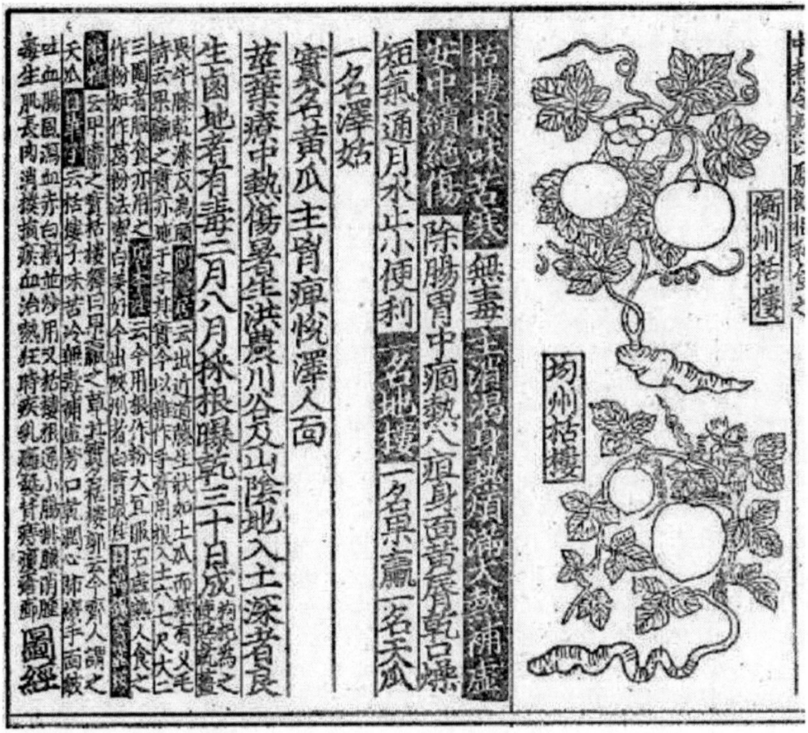

Picture 5.5 The photographed *Newly Edited Pharmacopoeia for Political, Economic and Historical Verification*
(Published by the People's Medical Publishing House, photographed in 1957 of Zhang Cunhui's original curved Mingxuan Version)

The author of *Zheng Lei Ben Cao (《证类本草》) Classified Materia Medica* was a medical practitioner who practiced in Chengdu, Sichuan, for a long time. He was known as the famous doctor Tang Shenwei in Shuzhou (today's Sichuan Province). He was kind, cautious, and skillful. According to Yuwen Xu, Tang's friend, Tang Shenwei was ugly and inward, and treated the disease without hesitation. He would treat patients regardless of their social statuses or personal financial conditions. He didn't charge for treating scholars. Instead he would ask them to provide famous or secretive formulas or prescriptions so he was very popular among scholars. Whenever those scholars saw information about medicine in the classics, they would transcribe it to Tang Shenwei, who would then collect the segmented information and compile them into a medical masterpiece (*The Newly Edited Pharmacopoeia for Political, Economic and Historical Verification)*.

Zheng Lei Ben Cao (《证类本草》) Classified Materia Medica had totally 32 volumes in over 600, 000 Chinese characters, including 1 volume of catalogue. It described 1748 kinds of medicine with reference of over 500 kinds of literature covering the fields of scriptures, classics, history, legends, Buddhism and Taoism, and materia medica of various versions. Among them, 554 kinds of medicines were added with the elaboration of herbal theories and their applications before the Song Dynasty. Li Shizhen of the Ming Dynasty once said, "Classified Materia Medica helped preserve various versions of medical classics and prescriptions and formulas by taking them as the reference when compiled."(Li Shizhen. The Great Pharmacopoeia. Vol. 1).

Zheng Lei Ben Cao (《证类本草》) Classified Materia Medica, written by one person, though, collected the experience and development of its predecessors to the greatest extent. It had also been repeatedly edited and engraved by the government and thus became a private work on materia medica. First, in the second year of Daguan Period during the reign of King Huizong in the Song Dynasty (1108), a number of notes were added by Ai Sheng a Hangzhou local official, which became *Da Guan Jing Shi Lei Zheng Lei Bei Yong Ben Cao* 《大观经史证类备用本草》 *Daguan Pharmacopeia for Political, Economic and Historical Verification*. On top of it, in the sixth year during the period of Zhenghe (1116), the medical official Cao Xiaozhong edited it into *Chong Xiu Zheng He Jing Shi Zheng Lei Bei Yong Ben Cao* 《重修政和经史证类备用本草》 *Revised Pharmacopeia for Political, Economic and Historical Verification* as requested by the imperial order. In the 27th year of Shaoxing period (1157), medical officer Wang Jixian of the Southern Song dynasty re-edited the block copy, which was called *Shao Xing Jiao Ding Shi Zheng Lei Bei JI Ben Cao* (《绍兴校订经史证类备急本草》) *Shaoxing Pharmacopeia for Political, Economic and Historical Verification*. Although the names were different and the words were slightly different, they were perfunctory works of Tang Shenwei's *Jing Shi Zheng Lei Bei Ji Ben Cao*《经史证类备急本草》 *Classified Materia Medica*.

5. *The Great Pharmacopoeia* in the Ming Dynasty

Among all the medical classics in the ancient times, *The Great Pharmacopoeia* might be the most well-known. It was accomplished by Li Shizhen, the medical scientist in the Ming Dynasty.

Li Shizhen (李时珍) with the style name Dongbi (东璧) was called Bin Hu Shan Ren (濒湖山人) in his late years. In the 13th year during the period of Zhengde in the Ming Dynasty, Li was born to a family of doctor in Qichun, Hubei. His father, Li Yanwen, was considered a Confucian doctor, who had served as the official of the Imperial Hospital; in addition to clinical treatment, he was also said to have written *Si Zhen Fa Ming* (《四诊发明》) *Inventions of Four Diagnoses*, *Ai Ye Zhuan*《艾叶传》 *Biography of Folium Artemisiae Argyi*, and *Ren Shen Zhuan* 《人参传》 *Ginseng Biography*. He originally hoped that his son would be able to stand out in the imperial examination and thus bring honor to his ancestors and the whole family. Therefore, Li Shizhen rarely obeyed his father's order to devote himself to studying the four books and five classics. Who would know that after passing the imperial exam at the county level, he failed three times to pass to imperial exams at the provincial level. As a result, Li Shizhen completely gave up the idea of advancing the imperial examinations, and determined to inherit the family business and use the medicine to help the world. Since then, he had been recommended to serve at the

Lord Chu's Mansion and the Imperial Hospital. By taking advantage of these opportunities, Li was able to read a large number of books on medicine and other aspects not readily available to the public, further enriching his experience and cultural literacy.

It had been the consistent aim of the ancient Chinese to edit pharmacological works to draw on the strengths of the families and strive to be comprehensive; Li Shizhen was no exception. He also believed that the predecessors' work had the issue of being not able to tell the differences between plants and trees or those between insects and fish. Therefore, he determined to exhaust the literature, correct errors and misconducts, and add new content. Since the 31st year during the period of Jiajing (1552), he had begun to reedit *The Great Pharmacopoeia*. It is commendable that he paid great attention to the knowledge of books and practice, so he asked humbly for the opinions of farmers, hunters, woodcutters, fishermen, craftsmen, medical farmers, and bell doctors (visiting doctors) in the process of compiling the books to get more suggestions and learn about their experiences. This determined that the editing process could not be carried out entirely in the study. He had visited and inspected mountains and villages many times, and his footprints had covered the regions of Hubei, Jiangxi, Anhui, Jiangsu, Hebei, and Henan. In addition to field investigations of medicinal plants, dissection of medicinal animals, and mining of medicinal minerals, pharmacology was also analyzed through a large number of actual observations, and even the effectiveness of a certain drug was verified by taking it in person. For example, he had dissected the carp (Pangolin) and saw that there were "liters of ants " in its stomach, only to affirm the predecessors' opinions on this animal's eating habits. So after decades of unremitting efforts, with the enthusiasm and help of many people, they accumulated a large amount of information, and after repeated revisions of their own manuscripts, they finally completed the final version of *The Great Pharmacopoeia* in the year of the 16th year of the Wanli Period in the Ming Dynasty (1578). In his preface to this book, Wang Shizhen recalled and summarized this difficult journey in the following words: "during the 30-year period, more than 800 books had been verified to compile the book, and the manuscripts had been modified for three times to delete the mistakes and unnecessary parts and add the missed parts." When this great work was finally published in Nanjing in 1596 (Jinling Edition), Li Shizhen had already died. His son, Li Jianyuan, devoted the book to the king and wrote the Memorial to the Emperor when submitting the book, in which he emphasized its nature of concerning physical knowledge though being a medical classical work. Li Shizhen had been able to focus on the value of practical experience in the editing process, rather than confined to book knowledge and written records. For this reason, the people in modern times have given him the reputation of "being an ancient scientist." However, on the other hand, it is impossible for anyone to completely break away or to transcend the era in which they live. For example, in the *The Great Pharmacopoeia*, we can still see the description that spear wedge could help fix "difficult labor" (the fire effect could promote the delivery of babies) and the comb could cure obstruction in breastfeeding (the effect of circulation was to be transferred to chest) (Li Shizhen. Entry of Chong Xie and Shu Bi In Fu Qi Bu Fu Qi Er Qi Wu Lei (服器部·服器二·器物类

之“锍楔”与“梳篦”条) of *the Great Pharmacopoeia*. Vol. 38). All of the similar descriptions were based on analogous thinking or utility explanations of witchcraft principles.

The 52-volume *The Great Pharmacopoeia* has 1.9 million Chinese characters with the structure as follows.

Volume 1 and 2 are the preface and introduction, offering a comprehensive overview of the part of pharmacological theory, including an introduction to the herbal medicines of various dynasties, the ancient and modern medical literature and the bibliography of classics, based on the seven prescriptions, ten doses, odor, yin and yang, the rise and fall of medical effect, the theories that certain drugs can guide the strength of other drugs to the diseased site or a certain meridian, acting as a guide, and other related concepts. It also introduced the relationship and contraindications between drugs, with the addition of Li Dongyuan's Examples of Drug Use with Syndromes, Prescriptions, and the three methods of treatment in the beginning, middle, and end stages of diseases.

Volume 3 and 4 introduce the cures for all kinds of diseases. 113 diseases were listed with indications of syndromes and the detailed usage of medicines.

From Volume 5 on, the specific descriptions of various drugs were made, and the classification method adopted was "the matter gathers in the same class, and different categories follow the overall arrangement of the schema." That is to say, water, fire, soil, gold stone, grass, grain, vegetables, fruit, tree, clothing and instrument, insects, scales, medium, poultry, beasts, humans, and other totally 16 classes constitute the whole schema, which was further divided into 62 orders. Such a schema applied to the whole text of *The Great Pharmacopoeia*. There were 1892 kinds of medicines collected, of which 374 kinds were newly added by Li Shizhen, 1109 illustrated with pictures and more than 10,000 prescriptions attached, and more than 800 kinds of references cited.

Looking at the achievements and contributions of *The Great Pharmacopoeia*, it can be said that the track of the development of Materia Medica is completely in the same vein:

1. Organize and expand medicines. The book includes the contents of 41 books of herbal medicine ranging from Shennong's Pharmacopoeia to contemporaneous Chen Jiamo's *Ben Cao Meng Jian* (《本草蒙荃》) *Materia Medica*. It reorganizes the drugs contained in it, eliminates the complexity, merges various types, deletes 228 types with 1518 kinds of medicines listed, and their original provenances indicated respectively; 374 kinds of medicines that had not been collected by the predecessors were added, accounting for about one fifth of the total medicines in the book, totaling 1892 kinds. Therefore, it has been the medical works with the largest variety of medicines in history.
2. The importance of illustration was valued. The method of illustration adopted in the Tang Dynasty's *Newly Edited Pharmacopoeia* was taken in compiling *The Great Pharmacopoeia*. There were 1109 illustrations which could help people a lot in identifying, gathering, and verifying medicines.
3. Important prescriptions in the appendix to medicine. Li Shizhen also attached great importance to the collection and arrangement of representative medical

prescriptions. From the *Inner Canon of the Yellow Emperor* to the Song and Ming dynasties, 361 medical books had been enlisted. In the appendix to medicine, there were more than 10,000 medical prescriptions, including almost the experiences and prescriptions mentioned in the existing medical books. Therefore, from a certain perspective, *The Great Pharmacopoeia* is not only a collection of medicine, but also a comprehensive medical book. This method of taking medicine with prescriptions attached can not only closely link medicines with clinical applications, but also deepen readers' understanding of the uses of medicines, offering great implications to improve the practical value of pharmaceutical works.

4. Strive to further improve the classification system. Based on the predecessor's methods or the natural properties of medicines or the three grades of "upper, middle, and lower," a classification method combining "classes" and "orders" was created to be able to sort out the complicated natural products into a system containing different division standards and grades. Regardless of whether the classification standards were consistent with modern natural sciences or whether the specific classification was completely reasonable, such a spirit of pursuit, exploration, and practice itself has been consistent with the so-called scientific spirit and scientific method.
5. Discriminate the right and wrong of previous descriptions. This is also the common pursuit and common practice of the editors of *The Great Pharmacopoeia* over the ages. As for the correctness of discrimination and revision, it depends on the proficiency and the amount of knowledge owned by the editors. The reason why Li Shizhen paid attention to field trips and could openly ask all those who had practical experience was not just to obtain varieties of new medicine or knowledge of their applications, but also out of the good intentions of revising the false or false theories.

5.5 The Changing Academic Atmosphere

Despite the evolution of the times and the influence of cultural trends in different times, the development of pharmacology has shown several trends of different values. Zheng Jinsheng, a well-known scholar in the field of Materia Medica, has repeatedly stated the opinions in his treatises, as follows.

First, based on the so-called mainstream system that continued to supplement the abovementioned major milestone works in order to complete a broad and extensive compilation work, Li Shizhen's *The Great Pharmacopoeia* in the Ming Dynasty reached its peak as it was almost impossible to see anyone working on such a huge project with such a great success.

Secondly, under the influence of Neo-Confucianism and Taoism research initiatives, books focusing on pharmacology had emerged. For example, Kou Zongxi's *Ben Cao Yan Yi* (《本草衍义》) in the Northern Song Dynasty incorporated the theory of qi and taste in Huang Di Nei Jing (黄帝) *Inner Canon of Yellow Emperor* into the herbal medicine as a basis for explaining its curative effect. The doctors in the Jin and Yuan Dynasties followed with more detailed elaborations on odor, yin and yang theory, thus deriving the description of the rise and fall of medical effects. Then there came another theory of channel tropism by leading the medical effect. Then there came the theory that certain drugs can guide the strength of other drugs to the diseased site or a certain meridian, acting as a "guide." The author even related the medicine with the hexagram to explain the principles of medical effects. To a certain degree, such a shift suggested the experiences regarding medical effects could be generalized into theories. However, some of the statements in the works of Materia Medica were far-fetched. The authors of some medical classical works could not avoid the stereotypical thoughts. For instance, Sheng Ji Jing Yao Li (《圣济经·药理》) Classics of Holy Benevolence Pharmacology initiated by King Huizong of the Northern Song Dynasty and Methods of Medicine Use (《用药心法》) by Li Dongyuan, who enjoyed the fame of one of the four great pharmacologists in the Jin and Yuan Dynasties, were based on some stereotypical thoughts: the medicinal parts of herbal medicine should be in alignment with the disease site. For example, the top of herbal medicine should be used to cure the head disease, the body part of the herbal medicine should be used to cure the diseases of the body, and the bottom part of the herbal medicine should be used to target the diseases in the lower part of body, the hips, or the ends of the limbs.

Third, as long as there are different doctrines or schools in any era, there will be compromises. After entering the Ming Dynasty, the above two styles gradually merged. Wang Lun's *Ben Cao Ji Yao* (《本草集要》) *Wang Lun's Pharmacopoeia* combined the Preface to Classified Materia Medica with Jin Yuan's knowledge of pharmacology as "General Theory." Each kind of medicine is not divided into "three grades." The prescriptions and formulas are arranged according to the order of diseases. Chen Jiamo's *Ben Cao Meng Jian* (《本草蒙筌》) *Materia Medica* furthered the systematic way of compilation. In the General Theory, the pharmacology and production among other practical issues were discussed one by one. All of these demonstrated the combination of two styles. They were very popular among readers back then.

Fourth, when the wind of Confucian classics was introduced into the field of medicine, there was also a tendency to return to *Shennong's Pharmacopoeia* in the Materia Medica system. The first was Miao Xiyong of the Ming Dynasty, whose "Pharmacopoeia" (ca. 1622) combined the actual interpretation of the scriptures with clinical medicine; then, there were *Shen Nong Ben Cao Jing Bai Zhong Lu* 《神农本草经百种录》 *A Hundred Kinds of Records of Shennong's Pharmacopoeia* by Xu Dachun of the Qing Dynasty; Ben Cao Jing Du (《本草经读》) *Studies of Pharmacopoeia* by Chen Xiuyuan, and so on. These works either respected the original texts without changing the word, or sought fair and pragmatic uses. On this basis, the *Shennong's Pharmacopoeia* appeared from the compilation of scriptures in the past.

Fifth, it was a popular book for reading in the style of songs that went hand in hand with those of respecting the classics style. In the Yuan Dynasty, Hu Shike's *Ben Cao Ge Kuo* (《本草歌括》 *Songs of Materia Medica*) was illustrated with pictures and poems. Li Dongyuan's *Yao Xing Fu* (《药性赋》) *Verses of Medical Properties* was divided into four parts, warm cold, hot, and cool, which was easy to recite, and had become a medicinal enlightenment reading book for people for hundreds of years. On the one hand, the increase in popular books was directly related to the increase in population during the Ming and Qing dynasties, and the corresponding increase in the number of medical practitioners. On the other hand, when the drug knowledge is accumulated to the vastness of the ministry, it will inevitably end with the return of the simplicity and the choice. For example, in Yao Xing Fu, the medical effects were summarized in several words. The rhino horn could relieve heart heat, the antelope could pacify the lungs and liver, and rhizoma alismatis was diuretic and well complement the insufficiency of Yin. The statement is catchy and clarified in expressing the uses of each kind of medicine.

Since the Ming and Qing Dynasties, the influence of foreign cultures had been increasing. In the *Shi Wu Ben Cao* 《食物本草》 *Materia Medica*: the Diet of the mid-Ming Dynasty, there were American plants such as groundnut introduced into China after Columbus discovered the New World; the *Supplement to The Great Pharmacopoeia* (《本草纲目拾遗》) by Zhao Xuemin in the Qing Dynasty began to cite western pharmaceutical literature and recorded cinchona and other drugs. Since then, Chinese herbal medicine works had gradually contained more knowledge of modern natural sciences. All the medicines began to gain their scientific names in Latin and descriptions of effective ingredients and their chemical structures. It suggested that China had entered a new era with people's more comprehensive scientific understanding of crude drugs on top of inheriting traditional knowledge.

5.6 Medical Organizations and Medicine Market

For quite a long period of time, medical practitioners may need to identify medicines and collect medicines in person. Before the urbanization reached a certain level, it was not easy to travel far with a medical kit filled with the common medicine and treat people in the remote rural area that was far away from the central commercial area, not to speak of travelling freely near or afar. However, such a situation would gradually change with the separation of medicine and pharmacy caused by the increasing specialization of work and the development of cities and commerce. Here, we only take Pharmacy of the Court of the Imperial Treasury and Public Pharmacy that emerged during the Northern Song Dynasty as examples to illustrate.

Pharmacy of the Court of the Imperial Treasury emerged as a result from the revolution and reform in the Northern Song Dynasty. The materia medica were included or listed as the medical items for sales organized by the country. Pharmacy of the Court of the Imperial Treasury could be taken as the earliest state-owned drug store and drug manufacturer. It was affiliated to the Imperial Medical Services and established in June 1076. During the first year after its establishment the imperial court gained a profit of 25,000 strings of copper coins by selling the finished drugs (Li Tao. Shen Zong Yuan Feng Yuan Nian. In: Xu Zi Zhi Tong Jian Chang Bian《续资治通鉴长编》History as A Mirror Continued Chronology Vol.289). The business of Pharmacy was very good and more and more branches were set up. In the period of Chongning (1102–1106) seven pharmacies were set up in the name of drug compounding and benefiting the people. In addition, the places for purchasing, trading, and logistics were set up. In the sixth year of the Shaoxing period in 1136, the Drug Administration Bureau was set up to replace the four branches in the east, south, west, and north of the Imperial Medical Service. Inside the imperial court, there was a bureau set up in the name of Bureau for Compounding (和剂局) (Fu Dayong, Public Pharmacy (惠民司). Gu Jin Shi Wen Lei Ju Xin Ji (《古今事文类聚新集》) New Collections of all the Things in Ancient Times and Now. Vol. 36). In the 18th year during the reign of Zhaou Gou, King Gaozong of the Southern Song Dynasty, the drug institute was changed into Tai Ping Hui Min Ju (太平惠民局) Pharmacy of Song (Li Xinzhuan. Annals Since the Period of Jianyan. Jian Yan Yi Lai Xi Nian Yao Lu (《建炎以来系年要录》) Vol. 158).

Zheng Jinsheng, a historian studying medical and pharmacological history, said the following when introducing and evaluating the official medical services:

> Pharmacy of the Court of the Imperial Treasury was established at the moment in order to reveal the government's concern over people's health care and medical conditions. In fact, it was also one of the measures for financial management. In the second year during the period of Chongning (1103), the Pharmacy of the Court of the Imperial Treasury which had been affiliated to Imperial Medical Services, was expanded to 7 pharmacies affiliated to the Court for the Palace of Revenues. From the title, we know that the Pharmacy of the Court of the Imperial Treasury was built mainly for trading and commerce. It also implied the beginning of separating the organizations for medical services and pharmaceutical services. Among the seven pharmacies of the court in Bianliang (today Kaifeng) the Northern Song Dynasty, 2 were for pharmaceutical manufacturing and 5 were for drug sales. In the period of the Zhenghe, in order to show that the government did not treat the pharmacies as the organizations for merely selling drugs, the Bureau of Compounding was renamed as Medical and Pharmaceutical Institute and the pharmacy that sold medicine was renamed as Public Pharmacy (Liao Yuqun, Fu Fang, Zheng Jinsheng: Science and Technology in China. Part VI. Medicine. pp.303. 1998).

Besides the capital Bianliang (today Kaifeng) in the Northern Song Dynasty and the capital Lin An (today Hangzhou) in the Southern Song Dynasty, there were officially medical services bureaus locally. For example, in the historical literature, there were records that all cities and prefectures shall have their own Public Pharmacies built up with the medical works provided officially as ordered by King Gaozong Zhao Gou in the 21st year during the period of Shaoxing (in 1151) (Biographical Sketches of Emperors No. 30. King Gaozong No.7. In the History of the Song Dynasty. Vol. (30)).

Edited by Editorial Committee of the Atlas of Ancient Chinese Civilization: Atlas of Ancient Chinese Civilization Volume VIII Yi Shou, jointly published by the People's Daily Press, Lotte Culture Company, Yixin Culture Enterprise Co., Ltd. 1992, P45)

The medical works officially provided as ordered by King Gaozong refer to Tai Ping Hui Min He Ji Ju Fang (太平惠民合剂局方) Prescriptions of the Pharmacy Bureau as edited by the Imperial Medical Services in the Song Dynasty (see Picture 5.6). It was often referred to by the medical practitioners shortly as the Prescriptions. Normally, it was believed that the works were based on the *Tai Yi Ju Fang* (《太医局方》) *Prescriptions of the Imperial Medical Services*, which was compiled with the

Picture 5.6 Prescriptions of the pharmacy bureau

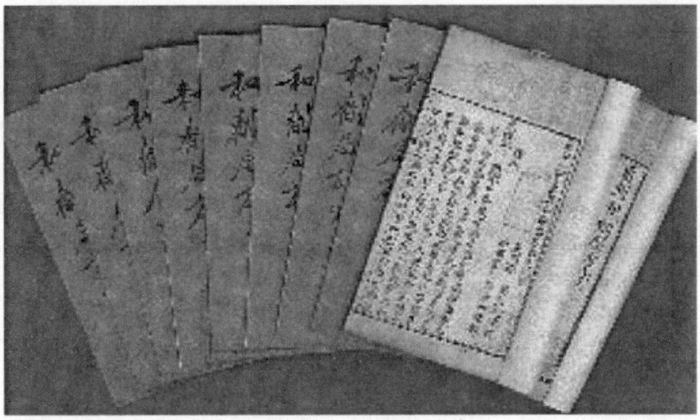

imperial order of King Shenzong in the period of Yuanfeng (from 1078 to 1085) through contributions from all the outstanding doctors with their secretive formulas and prescriptions as verified by the Imperial Medical Services (Ma Duanlin. Jing Ji Kao No. 50 Zi. Yi Jia. Tai Yi Ju Fang Vol.10. (经籍考五十·子·医家·太医局方十卷) Historical Documents No.5. *Medical Professionals. Prescriptions of the Imperial Medical Services*, Wen Xian Tong Kao (《文献通考》) Comprehensive Textual Research of Historical Documents Vol. 223). Later, in the period of Daguan from 1107 to 1110, the works were modified by three medical officials Chen Cheng, Pei Zongyuan, and Chen Shiwen before it was published and printed as the model of prescriptions and classical formulas for all the local medical services bureau. Many famous or popular prescriptions that were used for the people afterwards were cited from the book, such as Zhibao mini-pills (至宝丹), purple snow elixir (紫雪丹), oriental sweetgum (苏合香丸), bezoar sedative pill (牛黄清心丸), shenling baizhu powder (参苓白术散), Xiaoyao powder (逍遥散), sijunzi decoction (四君子汤), and huoxiang zhengqi powder (藿香正气散). Prescriptions of the Pharmacy Bureau were so famous that Xu Hong, the famous medical practitioner in the Southern Song Dynasty, once highly praised it when modifying the works as follows: "I have taken the family business as a third-generation successor. I have reviewed many prescriptions from the ancient times to those of today. Those with the greatest medical effect are from Prescriptions of the Pharmacy Bureau." (Xu's Annotations on the Prescriptions of the Public Pharmacy Preface. Dan Bo Yuan Ying, Investigations of Pharmacology in China. pp.591.the People's Medical Publishing House. Beijing,1983. Ed.2). On the one hand, the most important effect of the works lay in the fact that the finished drug could be produced with the formulas and quality assurance as recognized by the whole country. On the other hand, all the benefits made it natural for medical practitioners to follow the prescriptions and formulas listed in Prescriptions of the Pharmacy Bureau. However, many patients would take the book as their reference when they got ill instead of asking for help from medical practitioners. Against the background of the flourishing spice trade, the frequent use of fragrant medicine in this book affected many problems such as medicine and home medicine. Different as they might seem, all of these could be attributed to the powerful influence of Prescriptions of the Pharmacy Bureau. The book had become popular classical works known to patients and doctors for a time. Scholars called it medicine based on Prescriptions of the Pharmacy Bureau. As to how Zhu Danxi, who was known as one of the four famous medical practitioners in the Jin and Yuan Dynasties, criticized it for its frequent use of fragrant medicine in his works entitled the Applications of Prescriptions of the Pharmacy Bureau, it was an academic issue.

At the same time, the private pharmaceutical industry was also very prosperous. There were many Chinese medicine stores in the city with abundant financial resources. Medicinal materials were traded far overseas. One example was the various drugs contained in the shipwreck of the Song Dynasty unearthed in Quanzhou. At the same time, various cultivated products of authentic medicinal materials also appeared in this period, such as aconite from Zhangming, Sichuan, yam from Bianjing, *Achyranthes bidentata* from Xijing, Peony from Hangyue, and so on. In the late Southern Song Dynasty, *Poria cocos* had been successfully cultivated in the mountainous area of Zhejiang.

In the Ming and Qing Dynasties, the pharmaceutical industry had not only become more prosperous but also had formed a certain industrial scale. The "medicine market" at the end of the Ming Dynasty was best known as Anguo in Hebei and Zhangzhang in Jiangxi. The medicinal materials all over the country concentrated in the drug market for trading. Some well-known pharmacies that upheld the ethics of Confucian businessmen to gain righteousness and benefit and be honest to clients and themselves had become famous and profitable. Tongrentang, a century-old pharmacy in Beijing, took the motto that no steps of complicated processing should be skipped to save cost and no efforts should be spared to gather the most precious raw medicinal materials that were recorded in 1706 in the Preface to the Catalogue of Tongrentang carried down through the five generations of medical practitioners from the ancestor Yue Wugang. Hu Qingyutang in Hangzhou, adopting the motto of being sincere and honest without ever cheating customers, is still in business today popular among their clients. The Ginseng Store specialized in dealing with precious tonics such as ginseng and antler are in business, too.

5.7 Drug Processing with Supplementary Materials

Ru Fa Pao Zhi (如法炮制) is a frequently used term in modern Chinese. It first appeared in the literature of the Song Dynasty. The drug processing, as the literal meaning of the term, had been revealed in the medical classical book unearthed in Mawangdui. At that time, the internal medicine was basically taken by swallowing the powder or medicine itself, which was literally eating the medicine. In the late Eastern Han Dynasty, the method of drinking the decoction of the medicinal ingredients came to be widely used in practice.

Pao or baked means burying the medicine in the waging ashes till it gets dark. Zhi means heating it over the fire. Pao Zhi came to be known as the processing of drugs. In Chinese Zhi (炙) (literally means torched) was also replaced by Zhi (制) (literally means making). The most frequently used eight methods of processing include baking, cooking, frying, washing, soaking, bleaching, steaming, and boiling. Just as shown literally, the first four methods required the use of fire, and the latter four methods required the use of water, or the agent with use of water. In addition, there were processing methods involving the use of liquid, honey, vinegar, and table salt among other things. It is noteworthy that in a time where analogy was often resorted to, the processing ways including grinding, pressing, powdering, cutting, or steamed with liquid or fried within earthy utensils on fire also referred to the medicinal effects. For instance, in Yellow Emperor's Canon of Medicine, there was a formula of herb broth that could cure the sleep loss caused by the blocked pulse channels and the abnormal circulation of Qi of Yin and Yang. It required the use of flowing water and hollow reed to extract the decoction. The flowing nature of the water and the hollowness of the reds are supposed to help dredge the pulse channels and circulation system so as to achieve better medicinal effect. Viewed at present, we could summarize the purposes of drug processing as follows:

1. It would be easier to store or preserve such as drying plants.
2. It was requested by use, for example, the antlers are hard and require bound to be shredded so as to obtain the effective ingredients during steaming and extracting. The alum can be powdered only after being calcined.
3. It was for detoxification. For example, the plants of the Araceae family such as aconite have certain toxicity, and it is much safer to use after being cooked and steamed.
4. It was to enhance the effect of medicine, for example, treating *Pinellia ternata* with alum can enhance the effect of phlegm reducing.
5. It was to change the medicine properties, for example, after steaming and drying the rice wine for nine rounds, the rice wine would be matured, and its nature of cooling blood could be changed to nourishing blood.
6. It was to remove the coarse and keep the fine, for example, talc can be taken as the finest powder after being ground while adding water.

Methods also signified the importance of following rules. In the early medical classics that were mostly about drug processing, *Lei Gong Pao Zhi Lun* (《雷公炮炙论》) Leigong Treatise on the Preparation written by Lei Xiao in the Southern and Northern Dynasties with modifications by scholars afterwards was one example. *Ben Cao Jing Ji Zhu* (《本草经集注》) *Tao Hongjing's Pharmacopoeia* had a lot of discussions regarding drug processing in the preface. As a matter of fact, nearly all medical classics regarding materia medica would contain the description of medicine name, properties and odor, purpose and diseases which could be cured, and the place of origin which would be followed with the instructions of drug preparing and processing, most of which were concerned with the methods or rules.

In order to process drugs as requested by obeying the rules or laws, we have to realize that the methods or rules could change as time passed. From the perspective of continuity, we should respect the experiences of the ancestors. The new anti-malarial drug "Artemisinin" developed in the twentieth century uses the method of soaking as described in *Zhou Hou Fang* (《肘后方》) *the Handbook of Prescriptions for Emergency* by Ge Hong in the Jin Dynasty without steaming or boiling so that the effective ingredients in it can be preserved. In terms of development, after the founding of the People's Republic of China, on the basis of inheritance, scholars from all provinces and cities have compiled *Zhong Yao Pao Zhi Gui Fan* (《中药炮制规范》)the *Chinese Medicine Processing Standards*, which has led to the development of regularized standardized and scientific Chinese medicine processing procedures. These contents have also become part of the new Pharmacopoeia.

Throughout the study of Materia Medica, we can see some of the most basic characteristics: First, the development of medical knowledge is indeed closely related to the accumulation of life and medical practice experience. Therefore, when summarizing through the ages, the varieties of materia medica used were increasing with each passing dynasty. Secondly, in the ancient Chinese medical knowledge system, analog always played an important role in the explanation. For example, the healing effect of many natural products was not interpreted or derived from experience, but by analogy. After continuous "trial and error," those "pseudo-knowledge" that had proven to be ineffective was eliminated gradually. However, it did not prevent medical practitioners from continuing to "discover" new medicines by analogy.

References

1. Zheng, J. (1982a). Song Dai Ben Cao Shi (宋代本草史) *The history of materia medica in the song dynasty. Chinese Medical Journal, 4*, 204–208.
2. Zheng, J. (1982b). Song Dai Ben Cao Shi (宋代本草史) The history of materia medica in the song dynasties. *In Chinese Journal of Medical History, 4*, 204–208.

Selected Works of Natural History in Ancient China

6

Huajie Liu

Contents

Abstract

Natural history is closely linked to the livelihood of the people. Development level of natural history in a society can be inferred from the colorfulness in its daily language. Highly developed natural history could also be one of the reasons why the Chinese civilization, among others, lasted thousands of years uninterrupted. But in all existing works of history, natural history, as a domain of inquiry, was not singled out as it should be. Why, with all the books on scientific or general history, has not natural history in ancient China been proudly mentioned? The reason, more or less, is that it is not significant enough according to some standards. It is safe to say that ancient China did not lack its share of excellent works of natural history. We can easily provide many examples. Those books involve disciplines including geography, astronomy, Chinese medicine, and botany. In this book, a selected few are discussed.

H. Liu (✉)
Department of Philosophy and Religious Studies, Peking University, Beijing, China
e-mail: huajie@pku.edu.cn

X. Jiang (ed.), *The High Tide of Science and Technology Development in China*,
History of Science and Technology in China,
https://doi.org/10.1007/978-981-15-7847-2_6

Keywords

Natural history · Herbal medicine · Regional botany · Essential techniques · Famine relief herbal · Plants · Western sciences

Natural history is closely linked to livelihood of the people. Development level of natural history in a society can be inferred from the colorfulness its daily language. For example, here are some Chinese words about plantation, geng (plough), pa (rake), chao (puddle), yun (weed), and tang (loosen the soil), and some about cooking: jian (pan fry), chao (sauté), dun (stew), zha (deep fry), wei (simmer), and bian (stir fry). Natural history, as a culture, was highly developed in ancient China. It is an unquestionable fact that knowledge and techniques related to natural history were ubiquitous in the Chinese society. From the court to the country, the Chinese lifestyle was deeply linked to natural history. For the people, this rather "shallow" knowledge had brought much convenience as well as flavor to their lives. From *Shi Jing (Classic of Poetry)*, *Bo Wu Zhi (History of Natural History)*, and *Nan Fang Cao Mu Zhuang (Plants of the South)* to *Meng Xi Bi Tan (The Dream Pool Essays)*, *Jiu Huang Ben Cao (Famine Relief Herbal)*, and *Ben Cao Gang Mu (Compendium of Materia Medica),* even in literary classics such as *Hong Lou Meng (Dream of the Red Chamber)*, *Jing Hua Yuan (Flowers in the Mirror), and Xian Qing Ou Ji* (*Pleasant Diversions*), the presence of natural history can be easily detected. For example, here is a line from a poem titled *July* in *Shi Jing*: "In July the fiery star sets into the west and in September we prepare clothes for the winter." The "fiery star" refers to Antares, which appears at due south every year in June on the summer calendar. It is a perfect example of the natural connection between astronomy and agriculture. Or, "Thick grow the rush leaves; Their white dew turns to frost. He whom I love; Must be somewhere along this stream." The original text has only 16 characters, but they painted a vivid picture of seeing off a lover by the river in autumn. Without careful observation of the rush or personal experience of sending off loved ones in a somber season, it is hard for the author to compose such beautiful sentences, which created a lively scene as that in *Song of the Boat*, Tayu Lo's famous ballad. It's even more ingenious than the similar scene from *The Izu Dancer.*

Highly developed natural history could also be one of the reasons why the Chinese civilization, among others, lasted thousands of years uninterrupted. But, in all existing works of history, natural history, as a domain of inquiry, was not singled out as it should be. Why, with all the books on scientific or general history, hasn't natural history in ancient China been proudly mentioned? The reason, more or less, is that it's not significant enough according to some standards (modern, western, natural science). Or, only a fraction of it is considered important while the majority are merely garbage. We should seek the cause in historiography itself, which decides what kind of history will be written for the development of science.

And this is only stage-one natural history. When it comes to stage two, natural history in ancient China became much more extensive and interesting, enabling relevant researches to break the barrier between science and general history. Just as Mr. Ji Xianlin wrote in *Zhe Tang Shi (A History of Cane Sugar)*, his most ambitious masterpiece: "I know not much about science and technology. My passion lies in the history of cultural exchange. Sugar, being something that people eat daily but deem insignificant, provides a specific and vivid example for cultural exchange. That's why I find it fascinating." and "What few people have noticed or considered or even guessed is that behind lots of seemingly common things in life such as animals, plants and minerals, lies a complicated yet brilliant history of cultural exchange. Sugar is one of those things." (Same book, page 5).

The *Zhe Tang Shi (A History of Cane Sugar)*, Mr. Ji Xianlin's magnum opus consisting over 800,000 characters, is a typical work of natural history. It was published in 2009, clearly not ancient. But the two quotes from the book as cited above have, on some level, implied the vast scope of the neo-style study on the history of science in ancient China and its transcending yet practical tendency. "Vast scope" means it can encompass trivial things and matters that normally would be neglected by science historians. "Transcending" means people today need not to think from the perspective of narrow patriotism, blindly trying to discover more fields in which China exceled in ancient times. And "practical" means we should focus on the history of lifestyle and culture so that we can reestablish our knowledge of the ancient society and its everyday life through a more pragmatic approach.

Suffice to say that ancient China didn't lack its share of excellent works of natural history. We can easily provide many examples: *Shan Hai Jing (Book of Mountains and Seas), Bo Wu Zhi (History of Natural History), Kai Yuan Zhan Jing (The Treatise on Astrology of the Kaiyuan Era), Shen Nong Ben Cao Jing (Classic of Herbal Medicine), Nan Fang Cao Mu Zhuang (Plants of the South), Zheng Lei Ben Cao (Classified Materia Medica), Xu Xiake You Ji (Xu Xiake's Travels), Meng Xi Bi Tan (Dream Pool Essays), Ju Pu (Book of Chrysanthemum), Shen Pu (Book of Bamboo Shoots), Zhu Pu Xiang Lu (Detailed Records of Bamboo), Jiu Huang Ben Cao (Famine Relief Herbal), Ben Cao Gang Mu (Compendium of Materia Medica), Zhi Wu Ming Shi Tu Kao (The Illustrated Book of Plants), Mu Dan Pu (Book of Peonies), and Zhi Wu Xue(Botany).* Those books involve disciplines including geography, astronomy, Chinese medicine, and botany. It is impossible to discuss them all in this chapter. Limited by length of text and to avoid repetition, the definition of natural history in this book is to be narrowed down to involve only contents related to botany, and only a selected few will be discussed.

However, before we start to formally examine how ancient Chinese learned and utilized plants, let's first argue how the non-intensive natural history is different from modern mathematical\physical science and reductionism.

6.1 The Non-intensive Natural History and Its History

Quoting the *Zhou Yi (Book of Changes)*: "Sheng is the ultimate virtue of the universe." The word "Sheng" means creation and evolution, and it's not limited to things related to human beings. It showed that our ancestors had a profound comprehension of how the world was shaped and transformed. Creation and destruction don't necessarily work against each other. They can be mutually reinforcing and complementary. Without destruction there is no rebirth. It is but one of the cold hard facts in natural philosophy that the entire history of man will inevitably be a flash in the pan, however unwilling we might be.

If the earth or the solar system has a "lifespan," then what could possibly affect it or shorten it? Surely, a great many of species today are in danger of annihilation, or, in other words, being destructed. But by whom? Certainly, nature and man are both to be blamed.

Modern technology is unlikely to extend lifespan of the earth, our home planet, yet it has provided multiple means for its early perish. And technology will, undoubtedly, invent more efficient means of destruction, where criminals and politicians have tried and failed. Obviously, this is not really a compliment on technology and scientists. Nor should it be viewed as insult or defamation. Humans, of all living beings on earth, solely have at their disposal modern technology, which evolves on a daily basis. Carrying forward the development of technology has become an institutional requirement of the modern society, to an extent that it's almost considered unquestionable. Any skepticism against the progress brought by technological advance is regarded as ridiculous, anti-intellectual, anti-civilization, even a sociopathic crime against humanity. But is it unquestionable? On what ground can a scholar raise his questions, and how will the energy of human civilization be released after the questioning?

The following are excerpts from an article titled *Son of Voldemort on the Pace of Development of Pure Science:*

> Alice: First, people all over the world all agree (as supported by poll numbers) that the development of pure science should be accelerated and investment on it should be increased; Second, there is not one government or community of scientists would want to stop.
>
> Seth Riddle: That's only because of the long-term propaganda of how great science is. People made the mistake of thinking of science as innocent. They even believe however much science advances, it will only benefit us.
>
> Alice: It surely seems you're against science.
>
> Seth Riddle: Wrong, or inaccurate. I'm only against excessive science. Science is good and necessary with the appropriate amount, no less, no more. Otherwise it would cause trouble. As a species, human beings are still too young to really learn to acclimatize through evolution. The so-called environment problems happen to be caused by ill acclimatization. Human beings have a long way to go and, unless we wish to plunge into extinction, learning to adjust and accommodate is a wise choice. Many species before us have become extinct after failing to adapt. At first, science was developed as a way to help us adapt better. But then its development was twisted and caused maladaptation.
>
> Alice: So, humans are not rational enough? Excessive development of science is irrational?
>
> Seth Riddle: Absolutely. All human communities, should they be rational enough, would sit down to negotiate on developing science at a controlled pace. Only by controlling, from the source, the pace of development of pure science, can we truly control technology as well as man's power to manipulate and conquer.
>
> Alice: In that case, not only are you not anti-science or anti-reason, but you even have a stronger scientific spirit than most, if the word science is still in use.
>
> Seth Riddle: Whatever you say. ("阿里巴巴,伏地魔之子论纯科学推进的速度" (*Alibaba: Son of Voldemort on the Pace of Development of Pure Science*),《我们的科学文化:科学的算计》(*Our Scientific Culture: Calculation*), edited by Jiang Xiaoyuan and Liu Bing, East China Normal University Press, 2009. page 256–260)

As a matter of fact, long since the two world wars, many scholars have realized the problem of the unstoppable momentum of modern technology. Phenomenology the Frankfurt School's critical theory is the living proof. However, the introspection as such did not provide the proper prescription. Nor did it suggest to restrain the development of science or find an outlet for the mankind's intense desire of knowledge. Perhaps those scholars did not open their minds enough to realize the growth of human knowledge had far surpassed that of our ability to harness it. And, they didn't appropriately distinguish different cognitions while ignoring the difference between natural history and other knowledges (mathematics, physics, reductionism).

Natural history tries to understand the physical world on a macro level, including the concept of nature, methods of description and classification, practical knowledge and skills, as well as understanding evolution of natural objects and their internal connection.

Historically, natural history had had its share of vicious deeds such as being involved with imperialism, invasion, plunder, and destruction. But, generally speaking, natural history is still a relatively "perfect science," with flaws that can be fixed. In order to discern the difference between the non-intensive natural history and the intensive modern sciences like mathematics and physics, the following arguments shall be discussed:

Argument 1: At most times in our history, human beings' very livelihood depended on knowledge within the scope of natural history instead of advanced technology. In this aspect, Clive Ponting's *A Green History of the World* remains an all-time classic. People might question the depth of knowledge of natural history, but the idea that "as shallow as it is, knowledge of natural history is still worth bragging" can easily be justified. Nevertheless, it is necessary to know that "acquiring knowledge is not the goal. Nor is seeking understanding. Survival is what matters. How should an individual live his or her life, in good times or bad? And how should he or she handle the relationship between material life and the spiritual one?" Natural history offers a survival philosophy as well as a practice. The unity of knowledge and actions along with perceiving people, nature, and knowledge under the frame of "lifeworld" in Phenomenology might contribute to sustainability and co-existence of the man-earth system.

Argument 2: Comparing to knowledges of mathematics, physics, and reductionism, knowledge of natural history is not as powerful therefore is less destructive. Borrowing Mo Yan's words from his acceptance speech at the Nobel banquet, we can argue that: "Compared with science, natural history is indeed useless. But, being useless may well be its greatest value." Isn't natural history science? A part of it, maybe. On the whole, it's unnecessary to put the science label on it. Just because something is not science, doesn't mean it's unworthy of respect and inheritance, for example, literature, art, etc. It's the same with natural history. Calling it science could mean taking the risk of humiliation. On the other hand, it could also damage the reputation of natural history itself.

Argument 3: Damage of the ecosystem is directly related to the expansion of human civilization, with science and technology being the ultimate instrument, especially mathematics, physics, and reductionism. Too often is the latest development in science and technology to be used in warfare or further exploitation of mother nature and, of course, environmental protection. So far, people have only spoken highly of the progress of civilization brought by technological advance, instead of carrying out in-depth analysis of the other side. It's like appreciating the engine for pushing the car to speed up and expecting it to be faster without knowing how to brake or turn it off. Sitting in a car like that can be very risky. But in reality, when sitting together on the train of modern technology, we merely feel a bit of nervousness. For the most part, we're either filled with optimism or complaining the fact that it's not going fast enough. That is the truth.

Argument 4: Many of us prefer a "fast life" and never want it to stop. But the real sustainable development is about raising individual and collective awareness so as to restrain our desires as well as control the pace of development to live a "slow life" where man and nature exist and evolve in harmony. Currently, the majority of mankind is unlikely to agree to this. Even if they might approve the idea, they would be unwilling to take any real actions. Today, productivity is at a rather high level so the general public don't have to work for five days a week to live a comfortable life. Is it okay to work only three days? But there are always some in the society, the so called trouble-makers, who resent idleness and want to get others busy, too. It's the kind of behavior that benefits themselves instead of others while harming society and nature. And they always have a fancy excuse for their trouble-making: that people would starve if it weren't for them. In fact, they could've simply shared extra wealth with those in poverty. The real motive, and actual effect, for that matter, for their trouble-making are to enlarge the gap of development so they can stay on top. The consequences, in a larger sense, are moral corruption and overburdening of nature.

Argument 5: Studying natural history will not only satisfy our thirst for knowledge, but it will also enable us to better fathom the complexity of the ecosystem so we'll realize the significance of co-existence therefore arouse in us awe and humbleness and gratitude for the nature and fellow humans.

Argument 6: Restraining the development of modern technology doesn't mean shutting it down totally. The key is pace-control. Opposing a specific technology is not the same as opposing all technologies.

Argument 7: Restraining the development of technology will not be an easy task. "Not being able to do it now" shouldn't be the decisive reason for "maybe we shouldn't be doing it." A professor from university Q claimed that ethics won't stop popularity and application of new technologies (human cloning, transgenic crops, etc.) so we shouldn't control it at all. Throughout history, laws and ethics have been against murder and theft yet never completely eradicated those crimes. So, we should just let them happen?

Argument 8: Through calm deliberation and negotiation, mankind could find wise solutions to discrepancies in knowledge growth. Nowadays, competition among nations has become the competition of technologies, which none would relinquish because it'd mean surrender. This is a difficult problem but, smart as humans are, we're capable of solving it. It takes only an average mind to see that the current cut-throat competition benefits no one. Countries can't help but keep the investment in arms escalating. The national perspective and standpoint, though important, are not the only option. Evil minds and arms dealers are the ones who wish to keep things the way it is or make it even worse so they'll benefit from the chaos. Segregation and opposition among nations can be broken. "To ordain conscience for heaven and earth; To secure life and fortune for the people; To continue lost teachings for past sages; To establish peace for all future generations." Can't Chinese thinkers, who are believers of such great creed, think beyond nationalism? Intellectuals who contribute to escalating the competition of military technologies are either foolish or plain evil. Others who possess both "virtues" have even become role models for generations to come.

Argument 9: Natural history has made tremendous contribution to the development of modern science. "As early as in the 16th and 17th century, natural philosophers and natural historians began to establish learned societies where rational exploration and experience study combined to produce new scientific methods. They, as a science community, also put forth their own ideas, beliefs and principles, giving birth to the scientific culture, of which scientists were leaders and practitioners."However, with division and in-depth development of disciplines, natural history has been gradually excluded from contemporary scientific inquiry and education system. It seems necessary for it to be restored. The biosystem and the natural world are irreversible in both structure and nature, which determines the indispensability of natural history when it comes to understanding how the world forms and evolves. In this regard, Michael Polanyi's *Personal Knowledge* and his article titled *Life's Irreducible Structure* on the *Science* magazine are still worth reading (*Science*, 160(3834):1308—1312). John Pickstone, the science historian, has written a new-style history of science and technology, in which study of natural history is deemed necessary at all stages in the development of human society. It represents a newly formed recognition in the intellectual circle.

Argument 10: A life dedicated to the study of natural history is a life worth pursuing. Not only does this endeavor ensure happiness for the individual, but it also barely damages the environment. It is not the only option but it is certainly a great one.

It is not my intention to expound and prove all these arguments in this chapter. But a smart read would come to the realization that "it's simply not worth it to give up current development pattern and the enjoyable lifestyle for a groundless conjecture on the future." And he's right. Standards for good or bad as well as purpose in life can vary with each individual's knowledge, faith, and will. Should the world end tomorrow or the humans become extinct in a few generations, it'd be pointless to contemplate that much. Everything will be as simple as it is. Indeed, people can, like the intellectual in a news story I read who believed that 2012 was the end of world, impawn their houses and make donations or indulge their appetite. Although, even in that case, you can still discern the difference between good and evil.

For those who sympathize or support, and want to carry out on their own, study of natural history, the next concern is approach, which involves stage-one natural history.

Step one would be to answer the call of Confucius himself: "Know more about the names of birds, animals and plants," or "know more."

Today, in the twenty-first century, the first suggestion for studying natural history is just to know more about things of nature, living and non-living? Absolutely correct! Do it and you'll see what an appropriate suggestion that is.

Names are the index, the key, the means. In the age of the Internet, everyone is familiar with the importance of search engines, but few care about verification of names. "Know more" does not mean finding some names on books.

"Know more" is just a means through which one can achieve the state of personal knowing by learning names. The purpose is to comprehend nature by analogy. "Know more," so we live a natural life and follow the way of the nature after understanding everything in it. "Look up to the heavens and down at the earth to see the pattern on birds and animals as well as their compatibility with nature. Observe all things around you as well as those are far away…so one shall understand the divine virtues and grasp the secret of all creation" *(Zhou Yi(Book of Changes))*.

Specifically, as a path to natural history, "know more" encompasses the following: match names to what they represent and integrate different learnings; observe and investigate, absorbing common knowledge till it becomes your own; persist and you may make discoveries, turn personal learning into common knowledge through writing. During the process, the idea that the human society is part of nature is to be consolidated, mentally and emotionally.

Natural history puts a premium on the importance of knowledge being the individual experience. For example, the award-winning *A History of Plants in China*, which consists of over 100 books, was published, but theoretically it had nothing to do with the general public. For the average person, it is not knowledge. Only through the "know more" process of matching names to the things they represent can a small fraction of the knowledge in *A History of Plants in China* become useful to the individual. In the words of Aldo Leopold, studying natural history can help prevent two types of risk: "One is to think your breakfast comes from a grocery store, the other is to think that heat comes from the stove." To avoid the first risk, people should grow vegetables themselves or at least go down to a farm. And to avoid the second, they should try chopping some firewood or take a walk in the forest.

When carried out in such manners, what would the study of natural history become or what would it produce? Plainly speaking, it allows us to "conform to the nature or preserve life," so that humans can live better individually and collectively. It was so in a primitive society and it is so in a modern one. The same goes with living in the wild and in the cities. With a certain amount of knowledge of natural history, people tend be more sensitive to nature, so they can be more poised when encountering the threat of natural disasters.

For the average person, studying natural history will bring the following advantages: First, it satisfies one's curiosity while enhancing his or her faculty of understanding; Second, it improves one's ability to appreciate the beauty of nature therefore helps to cultivate the bond between man and nature; Third, it eases one's mind to avoid unnecessary conflicts so that life becomes more interesting and happier. For the mankind as a whole, natural history reminds us of what it is like being one of the species on planet earth. Man is a creature of nature and the earth is our only home. Flying into the depth of the universe is merely a myth sugarcoated in new technology. Only by using knowledge of natural history can the human civilization last on earth.

It's noteworthy that studying natural history shouldn't be regarded as the opposite of studying other knowledges. Instead, from a natural history point of view, all knowledges must be respected. Meanwhile, all knowledges should be handled and utilized in the spirit of natural history. That's why natural history does not reject findings of molecular biology.

Because of its accessibility, natural history is close to the public so it's practicable in building a better society. On the other hand, it is difficult for mathematics, physics, and reductionism to be integrated in the process. Promotion of natural history should be top priority. Revitalizing natural history among the general public has a solid connection to the construction of "eco civilization." I find it hard to believe that the public will wholeheartedly help preserve biodiversity had they not truly understood the world around them and realized what sustainable development really means. "If a man can't love the land where he lives then all his talks about loving the country are most likely void and false" (John Muir). An article on a magazine mentioned how the author, when seeing water hyacinth in Heilongjiang, was worried for it might induce ecological problems. The truth is it won't. The author probably doesn't really understand how water hyacinth grows, let alone the fact that it can't survive the winter in north China, where the real hazard comes from alien species such as ragweed, giant ragweed, the coastal plain yellowtops, fevervine, and the torch tree, instead of crofton weed, bitter vine, and water hyacinth. On the other hand, the bigger problem is that even a biology major in college can't tell common alien plants from local ones. And identifying plants is only the tip of the iceberg. To quote Mr. Liu Yiqing: "Natural history isn't just about knowing your flowers and trees. The path to natural history lies in the attention to the fertility of the land and the passion towards mother nature."

Revitalization of natural history in a civil society requires some essential conditions (especially economic conditions), which now are provided. The only things missing are guidance and instruction.

The value of natural history can't be simply understood from the perspective of popularization of science, though it is acceptable. Today, when we reintroduce natural history to the people of a well-off society, the unity of knowledge and action must be emphasized. And the focus is not on the volume of knowledge but in nurturing a "new sensibility," in recreating the dialogue between each individual and nature, in improving our status and the quality of living, as well as in prolonging the human civilization.

The Natural History of Chili is a work of natural history in conception which has obvious cultural significance. The authorities of Sichuan and Hunan or even the central government will do no wrong setting up a special project for it. It would definitely be more interesting and valuable than some projects established out of shortsightedness. Another work worth expecting is the *Natural History of Soybeans.* Tofu is absolutely a major invention of which our people should be proud. Its importance is no less than that of the Gougu Theorem, the calculation of Pi, the invention of saddle and stirrup, or the Maxim machine-gun. However, account of tofu is barely found in any work of history of science and technology.

The natural history culture and natural history study have only become popular over the last 20 years. Cambridge University published the *Cultures of Natural History*, a collection of essays. The *History of Science* magazine published a special edition of natural history. Fa-Ti Fan, the science historian, wrote in the foreword of the Chinese version of *British Naturalists in Qing China* that "In recent years, the history of natural history, as part of the history of science, has become a prominent discipline. In comparison, it is quite neglected in China."

6.2 Er Ya: An Encyclopedia

Er Ya was China's first comprehensive encyclopedia arranged by definition. *Er* means near, close, approach and *Ya* means proper, refined, correct. Through research, scholars found that the book was finished between the last year of the Warring States Period and the first year of the Western Han Dynasty An original draft was written and then it was revised several times.

The original *Er Ya* consisted of 20 articles in three volumes, of which 19 were remained to this day. The articles can be categorized into two parts. The first part, which offers explanation for common words, includes three articles, respectively, titled *Shi Gu*, *Shi Yan*, and *Shi Xun* that provided literal meaning of common words. The second part, consisting of 16 articles, respectively, titled *Shi Qin*, *Qin Guan*, *Shi Qi*, *Shi Yue*, *Shi Tian*, *Shi Di*, *Shi Qiu*, *Shi Shan*, *Shi Shui*, *Shi Cao*, *Shi Mu*, *Shi Chong*, *Shi Yu*, *Shi Niao*, *Shi Shou*, and *Shou Chu*, provided literal meaning of words in social and natural inquiries. Er Ya offered an explanation from everything from the universe to everyday life. In ancient times, it was a necessary tool for intellectuals to understand documents of old. In late Tang dynasty, it was officially established as a classic, even before Mencius' works. And it was listed before *Mencius* in the *Thirteen Classic*, a solid proof of its standing.

Er Ya is unquestionably an important document of natural history. It provided a clear summarized description of all aspects of the "lifeworld." The enormous amount of knowledge it contains makes it a crucial material in learning geography, social structure, natural knowledge, and practical arts of the ancient world. Today we may re-learn knowledges of sociology, astronomy, geography, botany, and zoology in different parts of the *Er Ya* and even pass judgment. It is okay to do so but it's merely a shallow entry. In *Er Ya*'s time, there was no such thing as classification of disciplines. It might feel awkward when trying to understand their classification from a modern viewpoint. For example, *Shi Tian* includes knowledges of sacrifice ceremonies, military strategies, and banners; *Shi Qi*, clothing and food; *Shi Guan*, roads and bridges. Classification of *Er Ya* is a reflection of the social structure and knowledge system of its time. That's the obvious fact. Therefore, researchers of the history of science, like their counterparts in archeology and anthropology, should, as much as possible, put themselves in the context of the ancient society and lifestyle, instead of trying to understand *Er Ya* from a one-way historiographical perspective that narrows everything down to the present.

Er Ya holds great value because (1) it initiated the critical interpretation of ancient texts; (2) it was the key to all ancient documents; and (3) it painted, with an abundant vocabulary, a picture of the "lifeworld" in ancient China. Those three aspects are the concerns of natural history.

Let's take a glimpse into *Er Ya*'s explanation method with the following examples.

Example 1: "Chu, zai, shou, ji, zhao, zu, yuan, tai, jiao, luo, quanyu, all share the same meaning as the word shi." In this entry, eleven characters are defined by one: shi (beginning); zai, beginning of plants; ji, beginning of construction; jiao, beginning of a motion; luo, beginning of defoliation; quanyu, beginning of growth.

Example 2: "Hunting in spring is called sou, in summer miao, in autumn xian, and in winter, shou." No further explanation is needed. This entry is about different names of lie (to hunt, hunting) in all seasons of a year.

Example 3: "Hua is on the south of the river. Yue is on the west. Dai is on the east and Heng (Shaanxi) north. Heng (Hunan) is on the south of the other river." This entry explains the naming of the five great mountains according to their position to the two rivers, Yellow and Yangtze. Mount Hua is on the south of Yellow river. Mount Yue (Wu Yue) is on the west. Mount Tai is on the east and Mount Heng (Shaanxi) north. Mount Heng (Hunan) is on the south of the Yangtze River.

Example 4: "A mountain that's large and tall is called song. A mountain that's small but tall is called cen. One that's sharp and tall is called qiao."

Example 5: "The fruit of guoluo is called gualou." This entry explains the name of a plant in the gourd family mentioned in the *Book of Songs*.

Example 6: "Cheng, willow trees growing by a river. Mao, willow trees growing in a marsh. Yang, willow trees growing in a thicket."

Example 7: "Trees with slim branches growing upwards are called qiao. Trees with no branches are called xi and trees with no trunk are called guan." In modern botany, there are arbor and shrub. And Er Ya listed another type called "xi," which refers to trees that have no obvious branches, such as those in the palm family, the bellflower family, and the lobelia family.

Example 8: "A four-chi-tall dog is called ao." (chi is a traditional Chinese unit of length. Its present value is standardized around one-third of a meter.)

6.3 Shen Nong's Classic of Herbal Medicine

The *Shen Nong Ben Cao Jing*, also known as *Shen Nong Ben Cao, Ben Cao Jing* or *Ben Jing*, was an early work of medicine and natural history that was constantly cited by books in later generations. Now it is determined as being written by an anonymous herbalist in the Han dynasty. "Shen Nong" is another name for the legendary Emperor Yan, who, as legend goes, knew the art of farming at the age of three. He was a pioneer of ancient agriculture and the founding father of Chinese medicine. A wildly popular folktale tells how he once tasted a hundred herbs to find out which ones can be used as drugs. The word "bencao" means herbal drugs, or, if further extended, "plants with medicinal uses," or even "anything that can be used as drugs."

The *Shen Nong Ben Cao Jing* was possibly finished in Qin or Han dynasty or the Warring States Period. In fact, no one of our time has seen the full original content. The earliest account of this long-lost classic was found in *Book of Sui.* The version available to us today was collected from works of herbalism throughout history. That is, restoring the original text based on later materials. Now there are several versions of the *Shen Nong Ben Cao Jing*, mostly of two types: "before-Tao" and "after-Tao." The "after-Tao" versions are based on text from the original book as sorted by Tao Hongjing, and these were included or cited by most works of herbalism since then. Tao Hongjing (456–536 AD) was a Taoist and herbalist in the time of the Southern Dynasties. Unfortunately, the original text of his work was lost too. The "before-Tao" versions are based on variations of the text of the *Shen Nong Ben Cao Jing* before Tao Hongjing's time. The most commonly used ones today are the "after-Tao" versions. In this chapter we use the version with collation and annotation by Mr. Shang Zhijun.

The preface of *Shen Nong Ben Cao Jing* listed the types of drugs in the book along with the system of classification, a characteristic approach of Chinese natural history. Some description might see less convincing, if not absurd, by modern standards, but the public attention is on the history of science itself, as well as the amount of solid knowledge the classic can provide for contemporary textbooks. And the public pay closer attention to how people thought and survived in ancient China, that is, how they dealt with the relationship between man and nature and how they handled all sorts of problems in daily life. *Shen Nong Ben Cao Jing* divided drugs into three classes: shang (upper), zhong (middle), and xia (lower). Each plays a different role, respectively, known as jun (sovereign), chen (minister), zuo (assitant), and shi (messenger), serving to nurture vitality and temperament as well as to cure diseases.

The 120 shang drugs are sovereigns, whose function is to nurture vitality so as to adapt to the way of the heaven. Long-term and large-quantity uses of these drugs are safe. Those who desire for a stronger body with sufficient chi (inner energy) and a long healthy life should stick to the shang drugs. The 120 zhong drugs are ministers, whose function is to nurture temperament so as to adapt to the way of people. Some of them are poisonous and others are not. Be cautious while using them. Those who desire to contain illness and strengthen their weakness may benefit from the zhong drugs. The 125 xia drugs are assistants and messengers, whose function is to cure diseases so as to adapt to the way of the earth. Most of them are poisonous so long-term use should be avoided. Those who desire to get rid of cold or fever or vicious energies and those who desire to break a long illness can use the xia drugs. The 365 drugs of three classes represent the 365 degrees, or days, of a year.

From which we can see that, for a long time, the best result pursued by Chinese medicine wasn't curing the disease afterwards but to keep the body healthy before falling ill. Curing the disease is a last resort. And the best drugs aren't the ones with the most lethal poison or the most powerful effect. The passage above shows that the Shen Nong Ben Cao Jing was under the influence of Confucianism and medicinal methods of Taoism. The classification of the three types of drug is a highly artificial system unlike those of *Er Ya* or *Zhou Li (Rites of Zhou)*. The system had dominated

Chinese botany for over a thousand years. Yet no blame should be placed upon it whatsoever. The system was merely the reflection of the ancient Chinese people's thoughts and feelings as well as their needs in life. Passing judgment on the classification system with no regard for the reality is a typical outdated thinking and attitude in historiography.

Why are there 365 drugs in total? It's more or less a farfetched attempt to match the 365 days of a year. In fact, some entries include more than one drug. For example, "qingshi(blue stone), chishi(red stone), huangshi(yellow stone), baishi (white stone), heishizhi(black grease stone), etc." If we are to count all in, then the total number will surpass 356. According to Tao Hongjing, there were three versions of the book, respectively listed 595, 441, and 319 drugs. And when citing the *Shen Nong Ben Cao Jing*, books such as the *Baopuzi (Master Who Embraces Simplicity)*, *Bo Wu Zhi*, and *Tai Ping Yu Lan (Taiping Imperial Encyclopaedia)* didn't even mention the number 356 and its significance. It was not until Song dynasty that the number became a constant expression.

The preface then goes on to talk about some "common knowledge" of Chinese medicine. In fact, the reason why the *Shen Nong Ben Cao Jing* was later considered common knowledge was because the book actually had a huge influence in history.

> Under the "jun-chen-zuo-shi" rule, drugs work to enhance or restrain each other. For the same purpose, drugs can be divided by their nature, like yin or yang. Other properties of drugs include taste(sour, salty, sweet, bitter, spicy), chi (cold, hot, warm, cool) and toxicity (poisonous, non-poisonous) can also be used to classify them. To treat a disease, first observe the source then intervene at the right stage. Drugs work the best when the patient's internal organs are not failing, his blood circulation not disordered and spirit not beaten. Drugs are half as effective when the disease settles in and they are of no use once the fever is over.

Drugs of the three classes include plants (polygala root, ginseng, *Eucommia*, Chinese magnolia vine, monkshood, *Pinellia ternata*, elegant jessamine), animals (bear's grease, bee, rhinoceros horn, canine penis, centipede), and minerals (mercury, hollow azurite, orpiment, gypsum, creta). Most of the drugs are plants, but each class consists of drugs from the three sources. Statistically, the shang drugs include 107 plants, zhong 72, and xia 76, totally 255. So, the majority of drugs are plants. And these 200 some plants are still the most common herbs in Chinese medicine. Two conclusions can be drawn from that fact: one, the development of Chinese medicine is steady and continuous; two, the foundation of herbalism was laid as early as more than 2000 years ago, with details being added and revisions being made over time.

Let's take a glimpse into the specific content of the *Shen Nong Ben Cao Jing* with a few examples.

Scutellaria baicalensis among the xia drugs (it's listed as a zhong drug in some versions) is a plant of the Lamiaceae family that is still widely used in medical practices today. "Scutellaria baicalensis has a bitter taste and mild nature. It can be used to treat fever, jaundice, diarrhea, edema, thrombus, ulcer and acute scleritis. It's also known as fuchang (rotten intestines). The plant originates in Chuangu, Zigui." This entry listed the drug's properties, function, and place of origin.

Another example: "Dipsacus asper has a bitter taste and mildly warm nature. It can be used to treat typhoid, malnutrition, metal cuts, bone fractures, women lacking breast milk. Long-term use of the drug can enhance your general vitality. It's also known as longdou(dragon bean) or zhuzhe(connecting the broken)." This entry is about Dipsacus asper, which is mostly used for metal cuts and bone fractures. That's how the plant got its name. "Zhu," as in "zhuzhe," means "to connect." It is a shang drug.

The book's description of *Hyoscyamus niger*, a plant in the Solanaceae family, about it's causing hallucination is quite accurate, not so much so when it comes to its other properties. "Hyoscyamus niger has a bitter taste and cold nature. It can be used to treat saprodontia, arthralgia. It can make people walk fast and see ghosts. Large dosage of it will cause hyperactivity. Long-term use of the drug can improve agility, making the user run as fast as a horse. It enhances mental and physical strength, generating psychic abilities. The drug is also known as hengtang." This is a xia drug, which, with a large dosage, can cause people to see ghosts, go maniac, and run fast. But it also enhances memory and strength? All these magical effects are included in volume 17 of the *Ben Cao Gang Mu(Compendium of Materia Medica)*: "Long-term use of the drug can improve agility, enabling the user to run as fast as a horse. It enhances mental and physical strength, generating psychic abilities." The *Ben Cao Gang Mu* also provided a rather elaborate "affiliated prescription," for conditions such as "episodes of mania," "lasting diarrhea," "intestinal bleeding," "protruding haemorrhoids," "teeth decaying," "hardened acute mastitis," "aggressive rabies," "stuck arrowheads," etc.

Limited by its time, some descriptions in the *Shen Nong Ben Cao Jing* are inaccurate. Constant and continuous citing of these descriptions had generated bad influence. However, intellectually and culturally speaking, the book was one of the four symbols of the establishment of Chinese medicinal theory. Its success in this aspect should be recognized. As a classic in both the history of natural history and the history of sciences, the *Shen Nong Ben Cao Jing* fully deserves its reputation. Mr. Zhang Dengben had tried to expound its value in 12 respects, such as all information about the drugs are solid and reliable; it initiated classification of drugs; it recorded effects, place of origin, and processing of the drugs; it regulates dosage form of drugs; it emphasizes the dialectical use of drugs; it divided drugs by their taste and chi; it introduced the idea of the seven basic emotions.

6.4 Nan Fang Cao Mu Zhuang (Plants of the South), a Work of Regional Botany

The *Nan Fang Cao Mu Zhuang* is an authentic work of botany, allegedly written by Ji Han in the Jin Dynasty. Ji Han, also known by his courtesy name Jundao and self-claimed title Boqiuzi, was a native of Qiao county. He could be from Anhui or Henan according to different sources, but generally he was believed to be from Luzhuang Village of Luzhuang Town in Gongyi City, Henan. Ji Han had once held the position of prefecture chief. According to a speaker introducing the 2009 edition of the book, published by Guangdong Provincial Publishing Group based on the 1592 carving copy from the South China Agricultural University library, the author of the Nan Fang Cao Mu Zhuang was Ji Han, one of the prestigious Seven Sages of the Bamboo Grove in Jin Dynasty. But in fact, Ji Han was not among the Seven Sages, namely, Ji Kang, Ruan Ji, Shan Tao, Xiang Xiu, Liu Ling, Wang Rong, and Ruan Xian. Ji Kang, who was ranked the highest among the seven, was Ji Han's (262–306 AD) granduncle. Ji Han's grandfather Ji Xi was Ji Kang's older brother. Ji Fan, Han's father, died at an early age and he was raised by Ji Shao, son of Ji Kang.

There are several theories concerning the time when the book was written. One version of the book specifically printed: "written by Ji Han, Prefecture Chief of Xiang Yang and General Prestige, on November 13th, the first year of Yongxing period (304 AD)." However, according to *Book of Jin*, the official historical text, there was no November in the first year of Yongxing. Some blamed the error on people who made copies of the book. November could be the miswriting of December. Ji Han was killed at 44 and it was uncertain whether he had ever been to Guangzhou. One investigation led to the conclusion that the book as we know it today was actually written by an anonymous author in the Song Dynasty. In 1983, an international seminar on the *Nan Fang Cao Mu Zhuang* was held in Guangzhou. A book of collected essays from the seminar was published by China Agriculture Press in 1990. Scholars exchanged ideas over the book pertaining to authorship and time of completion, yet no agreement was reached. Decades later, the intellectual circle, after much deliberation, is still inclined to believe that the *Nan Fang Cao Mu Zhuang* was written by Ji Han in an early period. And it is quite a thin argument to disclaim Ji Han's authorship simply based on the possibility that he might have never been to Guangzhou.

The *Nan Fang Cao Mu Zhuang*, consisting of three volumes in total, provides an introduction to plants in the Guangdong region, which are divided into four types: herbs, trees, fruits, and bamboo. A review in the past praised it for being "complete, concise and elegant."

In volume one, a total of 29 kinds of herbs are listed; volume two, 28 trees; and volume three 17 fruits and 6 kinds of bamboo, adding up to an even 80. In fact, there are more than 80 plants mentioned in the book. While introducing these plants, the author cited other plants people were more familiar with at that time. For example, when introducing banana, the author described its shape and taste as being similar to those of lotus, taro tuber, grape, and lotus root. And with "qiansuizi," the author referred to nutmeg (which was mentioned in the *Ben Cao Gang Mu* by its "foreign name." The Ben Cao Yan Yi, a book written by Kou Zongshi in the Song Dynasty, also mentioned this plant and Li Shizhen cited the example.) Analogies were also made between yege and basil, sweetgum and poplar, jaba and hollyhock, as well as henna and elm. Describing the unknown with the known and comparing differences are both common approaches in botanic history writing.

The second plant mentioned in the book was name "yeximing," an alien species judging by its name. "The yeximing and jasmine were both flowers transplanted to South China from the west by foreigners. Later, the locals who loved their fragrance started growing the flowers, too. According to Lu Jia's *Nan Yue Xing Ji*, a book recording his travels, in the Guangdong region, cereals have no flavor and flowers are scentless. These two new flowers, with enchanting fragrance, were transplanted here from foreign lands and their properties were unaltered by relocation, unlike oranges, which developed a different taste when grown in the south instead of the north. Local women pull colored ribbons through the flowers and wear it like a jewelry." This paragraph is a distinct account of international exchange of plants. The "yeximing" is in fact *Jasminum grandiflorum*, a plant in the Oleaceae family. There were also some who believed it was *J. officinale*. According to this piece of information, botany historians were able to draw a preliminary conclusion (not absolutely) that it was an alien species. This paragraph also mentioned another function of the flower: being stringed together and worn as jewelry, a custom we can still see today in tropical areas, like the widely popular lei in Hawaii. Here are more examples:

Menyanthes trifoliata, a plant in the Menyanthaceae family: "Zhuocai grows among swamps in summer. The leaves are like those of arrowhead and the roots are similar to lotus' roots. Locals experienced sleepiness after eating it and they call it mingcai (sleepy vegetable)." Contemporary botanic works describe *Menyanthes trifoliata* in more or less the same way, if not less vivid.

Here is the description of water spinach, or *Ipomoea aquatica*, a plant in the Convolvulaceae family: "the water spinach bears leaves similar to those of red vine spinach but smaller. It has a cold nature and sweet taste. Locals would build a raft with reeds and then drill a hole in it before setting it afloat on water. And then they'd plant seeds, which float like duckweed. Once it's grown, the stem and the leaves would reach out from the hole in the raft and drift with the flow. It is quite a marvelous vegetable of the south. Yege is highly toxic yet it withers immediately when its seedling is touched by fluids of the water spinach. Legend has it that Cao Cao, emperor of Wei, once ate a great deal of yege after eating some water spinach." This passage described a process that is no different than the water planting method in "floating garden cultivation" or soilless culture. The raft is a tool for fixation, much like any of the floating-board devices being used today. "Yege" is in fact *Gelsemium elegans*, a plant in the Loganiaceae family, which can be detoxicated by the water spinach. The book also mentioned that yege's poisonousness is rather relative for "goats that eat its seedling grow bigger and fatter." Rats can eat crotons without having to worry about diarrhea because "chemistry vary with species." For humans chocolate is a delicious treat whereas for dogs it's lethal.

The description of *Hibiscus rosa-sinensis*, a plant in the Malvaceae family, was quite accurate; even the name is still used today. "The zhujin flower has stem and leaves similar to those of mulberry. The leaves are smooth and thick. The tree grows to four to five chi and is luxuriant. The flower blossoms since February and will not perish until mid-winter. The flow has a dark-red color and five pedals, with size as large as hollyhock. It grows one single stamen longer than the leaves with golden powders on it, which give off a fiery sheen in direct sunlight." A cluster can have several hundreds of flowers that blossom in the morning and wither in the evening. It grows even from a cut twig. The plant originated in Gaoliang County and is also known as "chijin" or "riji." The part about its stamen being longer than the leaves and having golden powders on it is a precise observation of the flower's monadelphous stamen structure. Making an analogy between zhujin and hollyhock, both from the same family, is a logical approach, too.

The book also gave a detailed description of how plants in the *Ficus* genus can "grow from one tree to a grove" with aerial roots: "A soft ivy-like branch, reaches downward till it touches the ground. Then, after the ivy has just penetrated the soil, new roots or a big stem start growing. There would be four to five roots that reach out, attaching to neighboring trees and form a connection."

Most plants mentioned in the book can be matched to their contemporary forms. Still, there are questions. For example, what exactly is the plant named "qiansuizi"? "Qiansuizi has ivies that grow out of the soil, with seeds underneath the root. It has green root hair that's intertwined. One bud normally produces some 200 seeds wrapped in a blue-yellow shell, with pulp shaped and tasted like chestnuts. Once dried, the shell separates itself from the pulp and makes a crisp sound when shaken, much like nutmeg. The plant originated in Jiaozhi (north Vietnam)." Based on the description in this passage, it is almost undoubtedly *Arachis hypogaea* from the Leguminosae family, more commonly known as peanuts.

San Fu Huang Tu, an ancient book of geography, mentioned that many plants were introduced from the south in Han Dynasty. Those plants were included in the *Nan Fang Cao Mu Zhuang*, and qiansuizi is among them: "The Fuli Palace is located in the Shanglin Garden. In the sixth year of Yuanding, under the rule of Emperor Wu of Han, the palace was built for growing exotic plants after the conquering of Vietnam. There were 100 plants of calamus, 10 of Alpinia japonica, 12 of banana, 10 of Rangoon creeper, 100 of laurel, tambac and henna, longan, lychee, areca-nut, olive, qiansuizi and citrus. Arbors grow differently in south and north so most of the trees perished the next year. A hundred lychee trees were transplanted from Vietnam and none survived. Yet, more were transplanted year after year." This shows there is connection between the two books, time of completion for each awaiting further study. The author and time of *San Fu Huang Tu* has yet to be determined. Allegedly it was written during the Six Dynasties, and the earliest account of the book can be found in *Book of Sui*. Professor Chen Zhi of Northwest University believes the copy as we know today was supposedly written after mid-Tang Dynasty.

Hua Jing, a book in Qing dynasty, adapted the description of qiansuizi from the *Nan Fang Cao Mu Zhuang*, only with slight alteration, for example, adding a line that says "it is extremely effective for relieving alcohol intoxication and heatstroke."

However, peanuts are universally believed to be from the Americas. How was China even related to the Americas back then? Possibilities are as follows: (1) peanuts used to be a domestic plant of China; (2) the American peanuts were introduced to south Asia in an earlier time; (3) qiansuizi is not peanut; (4) the qiansuizi in the *Nan Fang Cao Mu Zhuang* is not the peanut we know today; (5) the book was actually written in a later time (the earliest account dates back to Song dynasty. Some even believed it's a counterfeit). The first two possibilities aren't necessarily inconsistent. The key is the length of time. In the 1960s, archeologists discovered peanut fossils in a Neolithic stratum in Jiangxi. All the evidence is worth considering. It'd be quite an interesting task trying to prove or disprove the possibilities above.

As for what "hecao (crane grass)" refers to, there is no final conclusion to this day. It is certainly unrelated to Silene fortune, a plant in the Caryophyllaceae family. It could either be *Mucuna sempervirens*, *Strongylodon macrobotrys*, or Crotalaria agatiflora, all from the Leguminosae family. It could also be a plant in the Orchidaceae family. The description is basically the same as that of the "hezicao" in *Ling Biao Lu Yi*, a book of random geographic notes.

> Hecao is a rampant plant with dusty-colored flowers and a light-purple base. It has short leaves similar in shape to those of the willow. It blossoms in summer with flowers that look like flying cranes, down to every detail from beak to claw. It grows in coastal areas of the south and locals call it the enchanting grass. Insects feeding on it grow and metamorphose into butterflies of a reddish-yellow color. Local women collect it and call it the enchanting butterfly, the magic of which is said to make their husbands tenderer.

> Hezicao is a rampant plant with dusty-colored flowers and light-purple base. It has short leaves shaped like those of the willow. It blossoms in summer with green flowers and leaves. Locals call it the enchanting grass and collect it to dry, wearing it as a facial accessory. The flower is shaped like a flying crane down to every detail from beak to claw. In springtime, insects would be drawn to its ivy and eat its leaves only. Local women collet the insects and raise them like they do silkworms. They feed the insects with leaves of Hezicao and when the insects grow old thy would stop eating and become butterflies of a reddish-yellow color. The women wear it and call it the enchanting butterfly. (*Ling Biao Lu Yi*)

In comparison, the *Ling Biao Lu Yi* was more specific and provided more information. The book was allegedly written by Liu Xun in Tang Dynasty.

Besides introducing form, origin, and function of plants, the *Nan Fang Cao Mu Zhuang* is a remarkable work overall. It broke through the limit of herbalism in terms of botanic classification. Also, the writing was concise, natural, and transcending just practical use. The classification of herbs, trees, fruits, and bamboo was completely different than the old "three classes" method. It is closer to the "natural classification" system used in modern botany. Meanwhile, it focused on endemicity and practicality as well as geographical differences.

Chen Demao's *Zhong Guo Zhi Wu Fen Lei Xue Shi (A History of Plant Taxonomy in China)* calls this book "the earliest Chinese thesis in botany." It was somewhat a rebel in botanic inquiries through Chinese history, therefore lacked due influence. Some had said that "it was submerged in the ocean of herbalism."

Others tried to explain the significance of the Nan Fang Cao Mu Zhuang from the perspective of national botany. When describing the tangerine, the book vividly expounded a classic case of pest control:

> People of Jiaozhi keep ants in a nest made of straw and sell it. The nest is as thin as cotton with twigs and leaves in it. The ants kept inside are sold along with the nest. The ants, reddish-yellow in color, are slightly larger than regular ants. Without these ants, tangerine fruits in the south would be harmed by all kinds of pests, leaving none intact.

Using ants to prevent pests from harming fruits was quite remarkable, not to mention selling it as a merchandise. Even to modern standards, it was a money-saving and environment-friendly model of eco-agriculture. It was very likely the earliest account of biological pest-control in the world. The Nan Fang Cao Mu Zhuang could be written in a later time, but it was unquestionable that this kind of biological technology occurred in a very early period of ancient China, given the fact that there were tons of other documents containing the similar content all the way from Tang to Qing Dynasty.

Duan Chengshi, a scholar in the Tang dynasty, wrote in his famous *You Yang Za Zu (The Miscellaneous Morsels from Youyang)* that: "There are ants in Guangdong that are larger in size even than the ants in Shaanxi. They would build their nest on a tangerine tree and often crawl on the fruits. That's why the surface of tangerine peel is thin and smooth. Sometimes fruits grow inside the ant-nest and those fruits, when being picked in winter, taste much sweeter than usual."

Liu Xun also wrote in the *Ling Biao Yi Lu* that: "There is a variety of ants in Guangdong. Locals keep ants in a nest made of straw and sell it. The nest is as thin as cotton with twigs and leaves in it. The ants kept inside are sold along with the nest. The ants, yellow in color, are slightly larger than regular ants and have longer limbs. Without the ants, tangerine trees in Yunan are often harmed by pests. So, people there would buy the ants for safekeeping their tangerine grove." Again, this account resembles that in the *Nan Fang Cao Mu Zhuang.*

The *Tai Ping Huan Yu Ji (Universal Geography of the Taiping Era)*, a book written in late tenth century, went further by mentioning two types of ants: "According to a local tale of Cangwu County, tangerine fruits were often eaten by black ants so people bought some yellow ants and released them on the trees. The two colonies fought each other and the black ants died, leaving the tangerine fruits well-grown." Apparently, in Wuzhou, Guangxi, people used to control black ants with yellow ants.

According to *Ji Lei Pian*, a book written in the twelfth century, people used to collect tangerine ants with animal fat. "Farmable land is scarce in Guangzhou so locals grew tangerine trees instead to earn some extra profit. The trees were harmed by pest and the loss was great, till they discovered that ants could help exterminate the pest. So, grove owners would go buy ants from vendors who collected and sold them. They put fat in the bladder of a sheep or a pig then placed it in front of the ants' nest, leaving it open. Then they waited till the ants crawled into the bladder and they picked it up to go. This is called tangerine-ants keeping."

Relevant accounts can be found in the *Zhong Shu Shu* (fourteenth century), the *Ling Nan Za Ji* (late Ming dynasty), the *Wu Li Xiao Shi* (written by Fang Yizhi in the seventeenth century), the *Hua Jing* (early Qing dynasty), and the *Guangdong Xin Yu*, which was written slightly later, as well as the *Nan Yue Bi Ji* (late eighteenth century). In China, using ants as pest-control has a history longer than a thousand years (For further details, see 李约瑟:《中国科学技术史·植物学》*(Science and Civilisation in China: Botany)*, Science Press, Shanghai Classics Publishing House, 2006, page 444-155). H.C. McCook, an American naturalist, was the first in the west to notice that ants were being used as pest-control in Guangzhou, yet his report didn't interest entomologists and horticulturists back then. It was not until the twentieth century did the tangerine ants draw the attention of western scientists, who at first didn't deem the method effective.

Certainly, ancient Chinese weren't the only ones who discovered useful tricks like pest-control. People in other parts of the world had accumulated similar wisdom of natural history in their own right, as well. All these practical methods were specific, fully verified, and local, even more important than standardized knowledge of science in our textbooks today, which science historian should spend more time rediscovering and sorting out. In *Seeing Like a State*, American anthropologist James C. Scott (1936~), passionately described a case in a Malaysian village where locals used black ants to control red ants in order to protect mango trees. He wrote: "It is hard to imagine this knowledge being created and maintained except in the context of lifelong observation and a relatively stable, multigenerational community that routinely exchanges and preserves knowledge of this kind. One of the purposes of the story is to remind us of the necessary social conditions based on which similar practices are created. These conditions require, at least, an interested community, accumulated information and continuous experiments." Such natural history wisdom of survival was well-matched with agricultural civilization. It is completely different from the lab-based science and technology of reductionism in mode of thinking, standard, and objective, and it serves a different demographic group.

6.5 Qi Min Yao Shu (Essential Techniques for the Welfare of the People)

The *Qi Min Yao Shu*, written by Jia Sixie in Northern Wei Dynasty, was the first fully intact comprehensive agriculture book in the world. "qi min" means the people, and "yao shu" means essential methods for a good living. Combined, the title translates into essential techniques for the welfare of the people. Jia Sixie, the author, was the prefecture chief of Gaoyang in Northern Wei Dynasty. Gaoyang is generally believed to be a county in Shandong instead of Hebei, and he was a native of South Shouguang, Shandong. Scholars believe this book was written around the sixth century, and it wasn't until the Tiansheng Period in Northern Song Dynasty that the first carving prints were released. Today, the only remaining copy is in Japan and severely mutilated. In Ming dynasty, the book was published in massive numbers and poor quality. A better proofread version of the book came to be widely spread by the Qian-Jia Period in Qing dynasty.

The *Qi Min Yao Shu*, composed of about 115,000 characters, has 92 chapters in 10 volumes. The book mainly discusses techniques, production, and management of dry farming in North China. According to the author's self-written foreword, materials he used came from four sources, which, when translated into plain Chinese, were the following. One, historical documents related to agriculture, which were passed on through his quotation, including the famous *Fan Sheng Zhi Shu* and *Si Min Yue Ling*. Two, common sayings related to agriculture that he collected extensively, much like the deeds of a modern-day folklorist, sociologist, and anthropologist. Those sayings were crystallized traditional wisdom that encompasses all-inclusive long-tested knowledge. Three, personal experiences of farmers and experts. And four, the author's practice in agricultural work. These sources are quite convincing even by modern standards.

Jia Sixie was a strong believer of the fundamental importance of food supply in a society. His idea of agriculture, however, is rather extensively general. It includes farming, forestry, husbandry, fishery, and sideline products, from cultivating, planting, breeding, to products-processing and "huozhi," or merchandising in plain language, but profiteering wasn't part of it. According to Jie Sixie: "Abandoning essentials for trifles is not the doing of the wise. It is a time of poverty and suffering so I shall not write about endeavors in commerce." Jia's proposition of stressing agriculture while restraining commerce was reflected by his differentiating "merchandising" and "commerce." The real focus of *Qi Min Yao Shu* is on multi-operations based on farmland. The book was an embodiment of the typical ideology of agricultural civilization. Product processing, for instance, includes brewing liquor, vinegar and soy sauce, fermenting soy beans, making malt sugar, pastry, and all sorts of cuisine as well as cultural appliances.

The using of green fertilizers mentioned in the *Qi Min Yao Shu* was called "field-beautifying." Here's the method: "Densely plant mung beans, red beans and sesame in the fields during May and June then plow the soil in July and August. By next spring, when it is time to sow, the previously-planted grains will become fertilizers, enriching the soil for a good harvest. And that "the beauty of this method is as that of utilizing the feces of silkworms."

The *Qi Min Yao Shu* provides an extremely elaborate description of horses, down to the smallest detail. Here's a passage on the art of equine physiognomy: "A horse' head is the king, so it should be square and noble. Its eyes are the prime minister, so they should be bright. Its back and waist are generals, so they should be powerful. Its chest and stomach are walls of the city, so they should be wide and firm. Its limbs are like local officials, so they should be far-reaching."

In equine physiognomy, inferior horses must be eliminated before judging others. Here's the method for judging a horse' internal organs: "The ears must be small for small ears means a small liver and a small liver means it's easier to tame. The nose must be large for a large nose means large lungs and large lungs means fast-runner. The eyes must be big for big eyes mean a big heart and a big heart means bravery, poise and powerful stride." Horse-judging begins with head: "The head must be sharply cut as it is carved. And it must be heavy with few muscles like a rabbit's." "The eyes must be high and neatly-shaped. It's best they are triangular with pupils like hanging bells giving off purple flash light." "The ears must be close to each other, forwardly-erected, small and thick." "The nostrils must be large." "Color of the inside of its mouth must be reddish-white like flame, a symbol of talent, vitality and longevity." The treatment for when a horse gets fever is: Feed the horse with a mixture of boiled soybeans and cooked rice. It will be healed after three feeds. And in case of mange: Treat it with realgar and human hair. Slowly cook both with lard of winter till the hair melts. Then, scrape the pus off with a brick so the lesions appear red in color. Apply the potion while it's still hot and the horse will be fine. And the prescription for edema is: take two handful of salt and stuff it into the horse' nostrils then pinch them till tears come out of the horse' eyes. Let go and it's cured. As for constipation: apply oil to your hands and reach them into the horse' rectum. Find the clotted feces and pull it out. Put salt in the horse' urethral and it will pee in a short while. It is an excruciating disease that could kill the horse within a day, so it must be treated as soon as possible.

The *Qi Min Yao Shu* even gives an account of what a horse' teeth are like from age one to thirty-two, which is vital information for horse-buyers. The book also provides a large volume of information about raising horses and treating ill horses. On the issue of hybridizing horse and donkey, the book says: Normally, a male donkey mating with a female horse will produce a mule. And a male horse mating with a female donkey will produce a hinny. It is bigger and stronger than a regular horse. A seven-to-eight-year-old female donkey with wide pelvis is the ideal choice for mating with the male horse. Wide pelvis means it's easier to be impregnated and a tall male horse will produce a strong offspring. Female mules are mostly infertile, even when they do give birth, they'd die soon after the labor.

The *Qi Min Yao Shu* also gives a detailed description of making fermented beans, listing every step of the procedure. For example, about location choosing: Prepare an enclosed warm hut and dig in the ground a hole that's two to three chi deep. The hut must be built by thatches instead of tiles. Seal the windows with mud to keep out wind, insects, and rats. Leave a small opening wide enough for one person to enter and exit. Then cover the door with a curtain woven with thick straws. And about time of making: The best time is around April and May. From July 20th to August would be the next best period. Other times are manageable except for winter because of the cold (Same book, page 582-589).

Here's the procedure for making Yu Chun Geng (fish and water shield soup), a traditional dish, excerpted from the *Qi Min Yao Shu*: In April, water shield will grow a new stem from its root but leaves are yet to sprout. At this time, it is called the "zhiweichun" (pheasant tail) and it's the best ingredient for the soup. When the leaves unfold to full length it is called "sichun"(silk), which is best to be picked in May and June. It won't taste as good till July and August because of the parasites, which are tiny and hard to discern while clinging to the leaves. People will get sick eating this parasite. However, by October the water will become so cold that it will freeze the parasites to death and the water shield is again edible. Put both the fish and the water shield into the pot while the water is still cold and add only a pinch of salt.

Ancient Chinese works of natural history often casually refer to knowledge of regional phenology. When telling of growing a plant named lanxiang, the *Qi Min Yao Shu* first discusses the origin of its name before getting to the time of sowing: Lanxiang is another name for luole (basil), which was renamed because it partly rhymes with the name Shi Le, ruler of Zhao during the Sixteen Kingdoms period. Anyway, the name lanxiang got popular later. In mid-March, when new buds begin to sprout on jujube trees, it is time to plant lanxiang seeds. The seedling won't sprout if planted sooner and seeds would be wasted (Same book, page 213-215).

Sowing of one plant in a certain place depends on the condition of another plant. Does that even make sense? The answer is yes. The jujube tree grows naturally in the wild, and its condition reflects the change in climate. The time when new buds sprout on jujube trees is different in every year so to plant seeds in a fixed date is to work against natural conditions of the region.

The *Qi Min Yao Shu* also mentioned some useful tricks. For example, let pigs help plowing fields: plant some turnip seeds one step away from roots of trees in a mulberry field. Then, after having harvested the turnips, set pigs free in the field so they'd feed on remaining stumps. While they're doing so, the field is plowed, even better than plowing it on purpose (Same book, page 323).

It can be seen from these examples that the *Qi Min Yao Shu* is a rather practical agriculture book. With a stress on the combination of time, location, and manpower, it summarizes, in detail, comprehensive experiences in intensive farming of North China in the sixth century. Its idea of working in line with the laws of nature was ahead of its time and it's inspiring even for people in the modern world. Some 1,000 years after the birth of *Qi Min Yao S*hu, techniques used in agricultural production in North China were still within the scope and direction of this book.

The most important ancient Chinese agriculture books after the Qi Min Yao Shu were *Wang Zhen Nong Shu* and *Nong Zheng Quan Shu.* The former discussed agriculture in both south and north China. Plus, it was especially accomplished in introducing hydraulic engineering and illustrating farm tools. The latter, written by Xu Guangqi in late Ming Dynasty, was the most inclusive agriculture book in Chinese history. It covers twelve aspects such as agricultural documents (classics and miscellaneous essays), land ownership systems, farm work, hydraulic engineering, farm tools, arboriculture (grains, vegetables, and fruits), silkworm breeding, activities relevant to silkworm breeding, growing crops, husbandry, production (food processing and house construction), and policies during times of famine. The *Nong Zheng Quan Shu* has far transcended the realm of agricultural production by involving contents of economic policies and political stability.

6.6 Jiu Huang Ben Cao (Famine Relief Herbal)

Hunger breeds discontent. What can be done when the people don't have sufficient food supply for various reasons? The beautifully illustrated *Jiu Huang Ben Ca*o provided a possible solution. In ancient China, the book fell under the category of "famine control." With it, people dying of starvation could collect and process different parts of wild plants, so the society would remain stable. Though having the word "bencao" (herbal) in the title, the book is different from previous works of herbalism in that it is not meant for people to cure themselves of diseases but to feed them. The focus is on food instead of medicine.

The Jiu Huang Ben Cao was written by Zhu Su (1361~1425) in the 4th year of Yongle period in Ming Dynasty, or 1406 AD. Li Shizhen, Xu Guangqi, and others had mistaken Zhu Su for his second son Zhu Youdun. Zhu Su was the fifth son of Zhu Yuanzhang, the founding emperor of Ming dynasty. He was conferred the title

Prince of Wu, which was later changed to Prince of Zhou because the emperor found the former title inappropriate. After Zhu Su's death, he received the posthumous title Ding and that's why he was more commonly known as Prince of Zhou Ding. Zhu Li, the third emperor of Ming dynasty, was Zhu Su's half-brother.

In spite of being a prince, Zhu Su didn't live comfortable life. He was twice demoted to a commoner and forced to migrate to Yunnan, though he spent most of his life in Henan. Unlike today, government officials at that time weren't allowed to travel freely. Zhu Su committed the cardinal sin when he left Kaifeng for Fengyang to visit Feng Sheng, his father-in-law who fell ill. The emperor then banished him to Yunnan. The second time he was banished, again to Yunnan, was because his son reported to the court that he was plotting treason. Years later, he was transferred to the imperial capital where he was imprisoned. And it was till Zhu Li ascended the throne that he was exonerated and allowed to return to Kaifeng.

The *Jiu Huang Ben Cao* includes a total of 414 edible plants, among which 138 can be found in previous works and the other 276 were newly added. The earliest version that we can find today is the 1525 re-print in Taiyuan, Shanxi.

The *Jiu Huang Ben Cao* is packed with precise illustration, which is the dominant feature of the book, with auxiliary written descriptions. However, most versions today are arranged in quite the opposite fashion, which is against the original purpose of the book. The large-sized and exquisitely drawn illustration, besides the writing, ensures that most plants in the book can be identified to genus, some even to species.

The author of Jiu Huang Ben Cao was not allowed to travel freely so he could just observe plants directly. Zhu Su had to hire someone to grow wild plants in a garden, where he carried out his observation. According to the foreword of the 1406 version of *Jiu Huang Ben Cao*, he "purchased farmland from locals and had some 400 plants grown in a garden, where he went to observe daily. By the time the plants were ripe he hired professional artists to portray each of them. And then he found out which parts of the plants were edible and compiled all the information in this book titled *Jiu Huang Ben Cao*." The 400 plants he mentioned were actually seedlings with newly sprouted buds. He initiated a new method by establishing a mini botanical garden where wild plants are introduced and studied. This method was followed by many in generations to come.

The *Jiu Huang Ben Cao* can be divided, generally, into five categories: herbs, which contains 245 plants; trees, 80; crops, 20; fruits, 23; and vegetables, 46. Furthermore, based on edibility, the plants are subcategorized into the following: edible leaves, edible fruits, edible leaves and fruits, edible root, edible root and leaves, edible flowers, edible leaves, bark and fruits, edible stem, and edible shoot. For each plant, the introduction includes (1) name, (2) portrait, (3) and description, which is essential and normally composed of 30 to 60 words, in some cases even 300; (4) "hunger relief," which offers cooking tips, taste, and effect, written in a concise fashion (e.g., under "hunger relief" of the entry "yeshanyao" (wild yam), there is only a short line that goes: "collect its root and cook it"); (5) "medicinal use," which does not come with every entry. The author would mark those that were recorded in previous documents but not the newly added. Let's take a glimpse into his specific approach with the example of "cijicai" (thistle vegetable):

> **[cijicai]** also called xiaoji, more commonly known as qingciji. The northerners named it qianzhencao (grass with a thousand needles). It originated in the swamps of Jizhou and is now ubiquitous. The seedling could grow to a chi tall, with leave similar to those of endive. There are thorns on both the stem and the leaves, which are not wrinkled. From the center of leaves grows the flower, which is blue and purple in color. It has a cold nature and is non-poisonous. Some say it has a sweet taste and warm nature. [hunger relief] collect leaves from its young seedling and cook them. Soak them in water and wash. Add some flavor with lard and salt and it tastes delicious. It can dispel the common cold. [medicinal use] see "big and small thistles" under "herbs" category.

In general, the *Jiu Huang Ben Cao* is succinctly written, with abundant information in description that is easy to understand. Under "hunger relief," the original text contains a word "za," which means "cook in boiling water or oil." "Za" is a frequently used word throughout the entire book. Given the description and the illustration, cijicai is in fact Cirsium segetum. In today's north China, people still eat young leaves of this plant.

Other plants mentioned in the book are also common food today, such as xuancaohua (*Hemerocallis*), huicai (*Chenopodium album*), gegen (root of kudzu vine), baihe (lily), waitoucai (*Vicia unijuga*), jianpeng (common seep weed), feicai (*Phedimus aizoon*), fenghuacai (*Rorippa globosa*), maoliancai (*Picris hieracioides*), chunshuya (buds from tree of heaven), jiaoshu (pepper), guaizao (honey raisin), gouqi (medlar), cangshu (rhizoma atractylodis), jitoushi (gorgon fruit), and yeshanyao (wild yam).

The book, of course, is not error-free. For example, the madouling (birthwort) was considered non-poisonous, whereas current findings show that it can damage the kidney; therefore the plant is not edible. So is the baiqucai (celandine). Surely, most of the poison would be removed from the leaves through the cooking process recommended in the book: "collect the leaves and cook them in boiling water. Soak the leaves to get rid of the bitter taste before washing. And add some lard and salt for flavor." In the process, the cooking and washing are extremely important. The same goes for some fungus.

The value of *Jiu Huang Ben Cao* can be seen in the following aspects: (1) first-hand observation, concise and accurate writing, and strong verifiability. The book was highly praised by western scholars including Emil Vasilievitch Bretschneider, Bernard Emms Read, Howard Sprague Reed, and Joseph Terence Montgomery Needham. (2) Practicality, fine illustration, a breakthrough in popularization of science. (3) It is a significant work of Economic Botany, having an extensive influence in China and abroad. Different parts of the book had been cited in *Ben Cao Gang Mu*, *Ye Cai Bo Lu*, *Nong Zheng Quan Shu*, and *Zhi Wu Ming Shi Kao*. About 300 years after the completion of the *Jiu Huang Ben Cao*, it was republished in Japan and helped accelerate Japanese herbalism's transformation towards natural history. *Ben Cao Tu Pu* and *Zhi Xue Qi Yuan*, both eminent works of botany in Japan, were inspired by the book.

6.7 Zhi Wu Ming Shi Tu Kao (The Illustrated Book of Plants)

Zhi Wu Ming Shi Tu Kao, written by Wu Qijun (1789~1847) in Qing Dynasty, was the most specialized botanic work in ancient China. Its prominent feature, comparing to most books of herbalism prior to its time, is that it focuses on the plants' description instead of their function to humans. To quote a review on the original foreword: "The book, a unique masterpiece of its time, includes all there is to know about plants, offering a vivid description of its objects."

Wu Qijun, also known by his courtesy name Jishen and self-claimed titles Gulan and Yulounong (the character pronounced "jun" in his name is not to be mistaken for another with the same pronunciation, which was misused by Bai Shouyi in his *Zhong Guo Tong Shi* (*Complete History of China*)). In fact, Wu had a cousin named Wu Qijun (courtesy name Qizhan), with "jun" being the other character mentioned above), was a native of Chengguan, Gushi County, Henan, though his family line originated in Nanchang, Jiangxi. Wu Qijun was born to a family of intellectuals, and he himself became one of the "ten outstanding scholars of Wu." The "yu," as in his title Yulounong, is the name of a place near the town of Liji in Gushi County. Literally, "yulounong" means "a farmer of Gushi" and it was "an implication of modesty."

Wu Qirui was considered "a genius rarely seen in his time." In 1817, he passed the imperial exam with the highest rank among all candidates, thus began his life-long career in politics, during which time he had "traveled to half of the places under heaven." At 30 years old, he was assigned to an official post in Guangdong. Soon after that, he returned to Henan for his father's funeral and stayed there mourning for the next seven years. It was during that time he built a garden named "Dong Shu" (the Eastern Villa), where he grew plants and studied them for his book. Later in his life he had held official positions in Hubei, Jiangxi, Hunan, Yunnan, Fujian, and Shanxi, where he collected various plant samples. From 1838 to 1846, he was fully engaged in the writing of *Zhi Wu Ming Shi Tu Kao*, which was the abridged edition of his other work *Zhi Wu Ming Shi Tu Kao Chang Bian.* One advantage Wu Qijun has on Zhu Su is the first-hand observation came with his travels.

Wu Qijun once discussed the issue of plant-naming with no other than emperor Daoguang, who asked him about the meaning behind the name "wanggua" (literally, the king's melon). And Wu told the emperor everything he knew from rumors to relevant documents. The emperor then asked about the origin of the name and Wu narrated the whole story, from its earliest name in Han dynasty to the variations later. Exchange of botanical knowledge between a monarch and his minister could be a tale worth remembering.

Zhi Wu Ming Shi Tu Kao is composed of 38 volumes including 1714 plants in 12 categories: namely, grains, vegetables, mountain herbs (the only entry that doesn't have "category" in the original title), swamp herbs, stone herbs, water herbs, creeping herbs, fragrant herbs, poisonous herbs, flowers, fruits, and trees. The book consists of over a thousand fine illustrations that are extremely helpful for identifying plants.

Unfortunately, Wu Qijun himself didn't get to see this masterpiece in his lifetime. *Zhi Wu Ming Shi Tu Kao* was proofread and published by Lu Yingu, the governor of

Shanxi, in the second year after Wu's death. Then, in 1880, the Junwen Book Company of Shanxi republished the book based on the original print. In 1915, the Provincial Library of Yunnan published the book in lithographic printing based on a Japanese version from the Meiji period, with foreword written by Long Yun. In 1919, The Commercial Press published the book in letterpress printing and then republished it in 1957.

Er Ya nicknamed the Xuanfuhua (Inula japonica) as Daogeng (time stealer), about which Wu QIiun explained: Yellow flowers of Daogeng tend to blossom earlier, to steal the golden color from autumn. Then he laid out a long argument citing Lieh Tzu. "Man steals from nature the four seasons, why can't a fragile grass? Isn't the creator often revealing on purpose some of his hidden treasure for man to steal from? The creator could be so generous as to let those things to be stolen or he could be so mean as to reject any stealing. Either way, it is not that the creator has shifting attitudes but the stealer fails to understand his true intention. He who is good at stealing is a smart observer who knows the unknown and sees the unseen. A merchant good at stealing would take what others throw away. A farmer good at stealing would build fences and reserve water. An artisan good at stealing would go into the mountains to measure trees. A scholar good at stealing would seek success in pursuing the truth. A real stealer knows not what stealing is. Those who do know what stealing is aren't even qualified to speak of it." What an excellent argument of "stealing"! According to him, "stealing" means something more or different than its literal definition. In fact, when introducing the Xuanfuhua, Wu Qijun gave no description except pointing out that it's mentioned in S*hen Nong Ben Cao Jing* and *Er Ya.* However, the illustration of Xuanfuhua in the book was absolutely first-class, especially the leaves and flowers, which are so authentic that it is easy for anyone to identify the plant in the wild. There are countless plants in the Compositae family, and many of those are hardly distinguishable. Yet, any reader can identify the Xuanfuhua introduced in *Zhi Wu Ming Shi Tu Kao*, all because of the illustration.

Wu Qirun's preference of illustration over words isn't because he's not good at describing. In fact, let's just take a look at how elaborate and precise his observation/description of the *Agrimonia pilosa* is: "The seedling grows to over one chi, with rough downs growing on the stem. The leaves are similar to those of kerria japonica in shape but wider, with neatly-arranged tips. Normally, five or seven leaves grow next to one another on one stalk. There is small leafy outgrowth, appearing in pairs, at the base of the stalk. From among tips of leaves grow the fringe, bearing small round yellow flowers with five pedals. The fruit is a follicle covered with green hair. The seeds, as tiny as millets, have a sweet taste and are often collected as food (after being smashed and grinded)." The description resembles the content of a modern flora, such as the pilose stem, the pinnately compound leaf, the stipule, the raceme, the yellow flower with five round pedals, and the shape of the follicle. The depiction of the stipule, in particular, is extremely life-like and accurate: "there is small leafy outgrowth, appearing in pairs, at the base of the stalk."

Another example would be Wu Qijun's description of Choumudan (*Clerodendrum bungee*), a plant in the Verbenaceae family: "The choumudan, also known as choufenggeng or dahongpao, grows in fields and deserted gardens of Jiangxi and Hunan. They can grow to as tall as three to four chi, with round pointy leaves shaped like those of a judas tree but thinner. They are also similar to the leaves of a tung tree but smaller. The leaves growing near the tip of the branch are slightly redder in color. The flowers, light purple in color with five pedals, grow in clusters among leaves on the branch's tip. They are similar in shape to those of the hydrangea, with spikes slim as needles. Locals of Nan'an would pick its root and boil it to treat swollen feet. The plant has an unpleasant smell and that's why people of the capital call it choubabao (smelly eight treasures). Some claim it to be a foreign species of hydrangea and sell it. Amateur doctors in Hunan use it to treat dizziness by cooking it with a silkie and they say it can also cure carbuncles as well as relieve swelling and pain."(Same book page 296). This entry emphasizes on shape of the leaves, number of pedals, length of the stamen (spike), shape of the inflorescence, and the smell of the plant. All the description is rather accurate, and the illustration is finely painted with precise portraying of opposite heart-shaped leaves, inflorescence terminal in the shape of a hemisphere, and the needle-like stamen.

Not all illustrations in the book are perfect. For example, the fritillary (page 128), the Chinese violet (page 280), the *Entada phaseoloides* (page 356), the *Ampelopsis japonica* (page 400), and the *Rheum officinale* (page 433) are rather poorly illustrated.

Zhi Wu Ming Shi Tu Kao bears the following characteristics: (1) Identifiable with accurate description. The author travelled all over China and saw the plants with his own eyes, which is crucial, for knowledge of natural history is hard to obtain through logical deduction. It relies more on first-hand experience. (2) Through citing local historical documents, the book is rendered more specific and verifiable. Zhi Wu Ming Shi Tu Kao may have put a premium on historical documents, but the author attached greater importance to contemporary experiences and materials, an approach much like that which is taken in modern science. (3) The emphasis on narration and argument in the first person. Often the expression "Yulounong says" is used. Sometimes even personal memories are cited. Modern science books seldom narrate in the first person, which is said to make the text more objective. Indeed, it sure seems so. Science is the work of man. It is done by certain groups or individuals. And the author must be responsible for what he wrote. Narrating in the first person is merely a habit. On the other hand, deliberately avoiding it could be the sign of concealed intentions. Some of Wu Qijun's arguments are worth noticing but others are purely showing off. Those were written in a fashion that is abstruse and allusive, of which traditional Chinese scholars were often accused. Also, some arguments went too far on their subject. Having said that, his writing was a perfect reflection of the author's thoughts, through which readers can see Wu Qijun for whom he truly was. In that sense, the book's shortcomings are also its advantages. Perhaps the seemingly irrelevant content is what makes this masterpiece interesting and charming. Wu Qijun was a botanist, but more importantly, he was a traditional scholar whose qualities should not be questioned. Aren't we trying to combine science with humanism?

6.8 Zhi Wu Xue (Botany): An Approach to Modern Western Sciences

Udagawa Yoan (川田宇榕菴), a Japanese scholar, used the phonetic translation of the word "botanic" at first, then changed it to "zhiwuxue," thanks to the influence of Li Shanlan(1811–1882) and others, who translated the book *Botany.* Li and his colleagues were the first to use the term "zhiwuxue(botany)" when translating the book based on English botanist John Lindley's (1799-1865) *Elements of Botany; Structural, Physiological, Systematical, and Medical: Being a Fourth Edition of the Outline of the First Principles of Botany*, published in 1841.

Zhi Wu Xue held a critical position in the history of science and its spreading for reasons listed below: (1) Modern Chinese botany started with this book, which included knowledges of different branches of botany such as morphology, taxonomy, anatomy, and ecology. (2) The book determined the use of botanical terms such as family, cell, calyx, petal, root, pistil, carpel, ovary, placenta, embryo, and endosperm. Before, words like petal or root were simply used to refer to organs of a plant. And the Chinese word for cell was borrowed from the Japanese translation. It was till then did the Chinese-speaking world establish a direct linguistic connection to the glossary of western sciences.

As far as knowledge is concerned, the *Zhi Wu Xue* is close to the textbooks used for middle schools and colleges today. Surely, it can't possibly be more accurate and sufficient than what the current textbooks have to offer. Listing and commenting its every entry are unnecessary. What needs to be addressed is a more important issue. That is the information "besides science" reflected in the *Zhi Wu Xue*, which was finished in the mid-1800s. The expression "besides science" is not really justified for the information in question was an inseparable part of science at that time.

There have been some comments on the *Zhi Wu Xue* in terms of versions, significance, its connection to foreign botanical works of the same time, and its publication in Japan. However, little attention was paid to the cultural background. It is no surprise because for too long have we been stressing on the universality and pureness of knowledge in works of natural history that we tried as best as we could to exclude relevant cultural background and local knowledge, especially those of religion and theology. The consequences of eliminating context and values are as follows: (1) Certain scientific inquiries were simply removed from its historical and cultural context. This may have rendered the inquiries more universal but it also made them unfathomable to outsiders and future generations. (2) The more practical and utilitarian aspects of motives and goals of scientific inquiries were overemphasized while the aspects concerning sensibility, values, and transcendence were neglected. We tend to evaluate basic scientific research from technical and utilitarian angles so that interest becomes the sole driving force of innovation. (3) The pursuit of absolute objectiveness leads to total separation of instrumental rationality and value rationality from the entire process of scientific exploration. As a result, science as a request is without guidance or a purpose and scientists are no longer concerned with the inherent ethics of their subjects. One prominent feature of modern science is its close ties with natural theology. Today, development in science doesn't have to be bound by natural theology, but it still can't avoid the interference of certain value rationality.

In 1858, Li Shanlan wrote in the foreword of *Zhi Wu Xue*:

> The *Zhi WU Xue* is composed of eight volumes in total and the former seven were translated by myself and Mr. Alexander Williamson (1829-1890). Unfortunately, Mr. Williamson had to return to his home country before we finished due to his illness. So, Mr. Joseph Edkins (1823-1905) and I translated the rest. Totally, there are fourteen categories with some 200 illustrations, covering every detail of various aspects of the plant world. My compatriot scholars can read the book and observe accordingly, so as to make new discoveries for completion of the study.
>
> Mr. Williamson and Mr. Edkins are both Jesuits from the west who are diligent in cleric duties. The reason why they took time in their busy schedule to help me translate the book is because everything in nature was God's creation. One would appreciate a craftsman's work when using fine tools and appreciate a farmer's industry when seeing well-cultivated fields; Certainly, one would admire God's intelligence when observing the beauty and intricacy of plants. In that sense, the two gentlemen were supposed to translate the book so that when scholars read it they'd believe the existence of God and revere his might. The book serves to nurture people's moral virtues as well as good manners for they are in awe of divine providence. That's why the translation of this book was such a beneficial act to all.
>
> On February 5th, the eighth year of Xianfeng, the book was published and that's when I wrote this article. Li Shanlan, of Haining.

A strong presence of natural theology can be seen in the foreword, and the main text as well.

As an outstanding example of cultural exchange between China and the west, the publication of *Zhi Wu Xue* should be considered a cultural phenomenon before it was considered a case in the history of science or the history of the spreading of science. Science, in itself, is a kind of culture, too. And as a culture, science has a rather vague boundary, of which the line is drawn on view of science, idea of historiography, and current interest. Should we see the *Zhi Wu Xue* outside the box of intellectual history, the landscape of botany may appear brand new.

There is considerable amount of natural theology in the *Zhi Wu Xue*. Does it make the book a scientific work or a theological one? Or, was there ever real separation between science and theology during that time? Could natural theology and science be reunited once again? The primary motive of natural theology is to reveal God's presence in the natural world by resorting to rational and experiential proofs, so as to praise the intelligent and all-mighty lord. The tradition of natural history in modern science had always kept a close relationship with natural theology. The Cavendish Lab at Cambridge University was established in 1874. Nine days after its establishment, *Nature* magazine was founded and the first issue published an article that cited the inscription above the lab's entrance, which was a quote from the *Holy Bible*: "Great are the works of the LORD; they are pondered by all who delight in them" (《圣经·诗篇》*(The Holy Bible: Psalms),* Chinese Union Version, page 552) (*Psalms* 111:2). It was a typical expression of natural theology. By inscribing the sentence on its door, the Cavendish Lab made a clear statement of the relationship between modern science and religion.

"Certainly, one would admire God's intelligence when observing the beauty and intricacy of plants." What a classic quote of natural theology! Provided we replace the word God with "nature" or "evolution," would any believer of botany object the sentence? The Zhi Wu Xue is composed of eight volumes in total, with a presence of the natural-theology ideology throughout its main text. A large proportion of the book, six volumes to be exact, involves natural theology.

Volume one of the book is titled "Pandect," and the idea of "everything is God's creation" is cited in many chapters:

> Plants, animals and minerals were all made of basic substances. Minerals are inorganic and they were made chemically. Plants and animals are organic and they were made biologically. The reason why chemistry and biology can turn basic substances into other things is an act of God and not to be fathomed by man.

The following are another four citations of natural theology in the book:

> To understand why plants were made and how and what God intended with them, one must go and observe all plants on earth.

> Rock formation is seen everywhere after lava from volcanic eruption cool down. Over time, lichen grows on them and then trees and other plants. It shows that all God's creation, no matter how small, has a great purpose.

> In regions at a latitude of 72° south and beyond, summertime is short. There fore, leaves, flowers and fruits of plants in those regions all grow fast. For at a slower rate, they wouldn't have grown due to the cold. This is a living proof of God's design, in which all plants grow in an environment that fits them the best.

People living in warm areas crave meat, so grass thrives there to feed enough livestock, whose very meat people eat. People living in cold areas walk upon ice and ride in wheel-less carts pulled by deer. Deer eat lichen; therefore, those areas are abundant in lichen for deer to feed on so that they'd have the strength for pulling carts. God created all plants and put each of them in an appropriate environment for the sake of man.

These passages were written with the following implications: man can understand the act of God through studying plants; God's design is thorough and exquisite; God is omniscient; teleologically speaking, the food chain and God's creation were specially created for human beings.

Volume two is titled "On the Inside," which introduces the microstructure of plants with no reference to natural theology at all.

Volume three is titled "On the Outside," which discusses feature of roots while expounding the "scientific explanation" (often written as SE, a basic concept in philosophy of science) of theological teleonomy. Complexity and adaptability, as results of evolution, are defined as God's design.

> The root has two functions. One, to fixate the trunk. And two, to feed the plant by suckling juices in the soil through a tube on the root's end. The tube doesn't grow long all the way from the beginning. Instead, it grows section by section from the end of the root. That's how it pierce through interspace among gravels to suck juice from farther soil. The end of the tube is its slimmest section, which is called micro sponge in the west. The reason why the tube's mouth grows on the end instead of any of the thicker sections is because so that it would be able to reach farther soil through gaps among gravels. Such is the brilliance of God's design.

It is quite a common tone natural history narrative. Similar examples are often cited in argumentation of natural design, such as clocks or eyes. And this one just happens to go with tube. Also in this volume, a microscopic image of the tube is demonstrated graphically.

Who does the tube's mouth grow on the end? The book's explanation is that it wouldn't be able to reach father soil through gaps of gravels if it had grown on a thicker part of the root and the credit goes to the creator. In fact, biologists who are not religious often use the expression "for the purpose of something" when explaining structure of function of certain creatures. That's a clear sign of teleonomy or purposive explanation. Except those scholars, if questioned, would claim that it's only for the convenience of narration. If need be, the phrase "for the purpose of" can be deleted from the text and "something" can be reduced and simplified to various insignificant interaction. Which means, technically, the "purposive explanation" in biology can be reduced to some kind of "deductive-nomological model" or a simple case of cause-effect. However, the word "technically" can be tricky, for complete reduction is hard to achieve.

Next is a passage on the "mercifulness of God":

> Roots of trees reach out far to all directions underneath the soil. Yet they always keep in perfect alignment with branches and leaves above the ground. Drops of rain or dew trickle down from branches would be absorbed by the root through the tube. Every single leaf has a tube on the end of its central vein. If the tube were even just a bit longer or shorter, it wouldn't be able to absorb the water on the leaf. Such is the mercifulness of God.

Now we know this description is not exactly true. There is no alignment between a plant's roots and its branches and leaves. It is common knowledge that anyone with some life experience can prove. It doesn't require expertise or complicated experiments. The part about how "every single leaf has a tube on the end of its central vein" is even more ridiculous. Perhaps, in this case, empirical study was simply replaced by some teleonomic imagination. It can be inferred from which that not all knowledge in the *Zhi Wu Xue,* a translated and edited book, is reliable. And it doesn't represent the actual level of botanical studies at that time. Certainly, this doesn't mean the book was not important to the Chinese.

Similar description can be found in this volume, too:

> The function of the truck is to hold up branches and leaves so that they can spread out absorbing heat and carbon, which are vital to metabolism. Trunk of a giant tree is usually thick on the upper and the lower sections and slim in the middle. That's how the middle section stays firm and solid. Should all sections be of the same size, the truck would break more easily. Such is the brilliance of the creator.

Then the author goes on to write that the sea area off England's south coast is dense with riffs, which pose a hazard to ships. So, lighthouses were built to provide guidance. However, they were frequently destroyed and rebuilt due to violent storms. Locals tried everything to reinforce the lighthouses and nothing worked. Until one day, someone "saw by chance that a giant oak tree has a truck that's thick in the upper and the lower sections and slim in the middle. He suddenly realized the reason behind it and built a lighthouse in that very shape. And that lighthouse endured, even to this day." In modern science, this is a case of bionics, which normally does not discuss the ultimate cause.

At the end of volume three, the author even cited *the Book of Genesis* when talking about the relationship between branches and thorns:

> Thorns of trifoliate plants aren't like those of the rose. For trifoliate plants thorns are their branches while the rose's thorns are technically prickles. When God first created the earth, none of the plants had thorns. According *the Book of Genesis*, God told Adam: "Because you have done this, thorns and thistles it shall bring forth for you; and you shall eat the plants of the field." And "Thorns shall be no more if your descendants shall change their behaviors to the good." "If all man shall act good and serve the LORD, thorns shall turn to branches or soft hair and no more shall be produced."

Analyzing this passage for the sake of epistemology would be pointless. The mysterious connection between evolution of plants and human activities is hardly verifiable. Plus, existence of thorns has its evolutionary ground or at least a common explanation in the form of "for the purpose of something." However, the ideas of teleonomy and natural ethics expressed in this passage, as a cultural phenomenon, is worth noticing.

As for the Fibonacci sequence existing in modes of attachment for seeds of pine cones, which involves left-hand and right-hand spirals and the angle of intersection, the *Zhi Wu Xue* explains, undoubtedly, with a natural-theology rhetorical question "could it be anything else but the act of God?":

> An average person exhales 110 to 180 kilos of carbon per year. In China's case, if every individual exhales 80 kilos of carbon each year, a population of 360 million, varied in age, would exhale about 28.8 billion kilos of carbon annually. The amount of carbon exhaled from all animals and insects is about twice of the humans' amount, so the number goes up to 57.6 billion kilos. And the amount of carbon released by fire is about the same amount that which is exhaled from humans, so it is another 28.8 billion kilos. The three numbers add up to 115.2 billion and all that carbon is absorbed by plants. Herbivores feed on plants for nutrition and then exhale the carbon from their lungs; Carnivores feed on other animals for nutrition and the carbon in animal fat is then exhaled from their lungs as well. So, carbon circulates around plants and animals over and over. Such is the incredible brilliance of the creator. Plants thrive near the equator so they give off more oxygen whereas plants are scarce near the two poles so they give off less. The creator then made air circulate around the earth so that all animals and plants can adapt to the environment. That couldn't possibly have happened a naturally. In north of the equator wind blows northeast and in south southeast, a phenomenon commonly known as trade winds. Above high, the trade winds blow in the opposite direction as they do near the surface of the earth.

This passage is quite interesting. It estimates carbon circulation on earth while speculates the reason for atmospheric circulation. Also, it gives away a rather important piece of information – in 1857 China had a population of around 360 million. The author attributed the scale and complexity of substance circulation in nature to the creator's power and design, with praises. It is an approach that not only fits the profile of an advocator of natural theology but, in a common non-scientific sense, it's also easy to understand and acknowledge for the general public. Even today, similar expressions are seen in literary works, except some people don't actually believe in the god as defined in Christianity. Mechanism or motive or whatever it is called, people tend to assume, out of perceptual intuition, that there is something behind the natural world.

Volume five talks about flowers, the magnificent colors of which are introduced through a rhetorical question that acknowledges the creator's intelligent design.

> Colors of flowers display themselves on pedals with kaleidoscopic magnificence. There are countless cells in the pedal, and juice, which is the source of colors. Also, there are numerous strands of threads hidden inside. The cells are densely aggregated yet unattached to one another. They are yellow or white or black and all are glittering. All those micro cells, neatly arranged, constitute the splendor of colors. Such a secret is not to be known by even the most proficient artist. How is it possible that it happens naturally instead of by God's design?

The compiler of *Zhi Wu Xue* clearly objected the idea that flowers are a product of nature. Instead, he stressed that flowers were created and designed by God. Then he went on to introduce the three prime colors of light and then, in an abrupt manner, threw in the following statement: The creator followed the same pattern when endowing everything on earth with colors so that they would bring out the best in each other for the pleasure of the beholder:

> Other examples would be pine trees and fir trees, which have genders and needle-like leaves. Pollen from male flowers would fall through the needles to pollinate the pistils of female flowers. Provided the leaves are shaped like regular trees, the pollen would be blocked by those leaves and never reach the female flowers. There are also plants that have genders and regular-shaped leaves. Those plants have flowers growing above the leaves so the male and female flowers can still touch each other, like the filbert. Evidently, the creator had contemplated on every detail to make everything perfectly fit. Those who believe it all happens naturally are but delusional.

According to Chinese culture and traditions, or traditional knowledge, or even western scholars and the general public Darwin's theory of evolution was introduced, autogenesis did make sense and has now come to be known as common knowledge. However, to believers of natural theology, it is ridiculous for one to think that the process of pollination happens naturally.

When discussing fruits, volume six quoted St. Paul, the apostle, and cited resurrection as an analogy. It is obviously intentional preaching:

> Like a dead man, a seed is lifeless after leaving the tree. And when buried in the soil, the seed will rot and then resurrect, growing to be branches and flowers and leaves. The human body

is but a seed. When we are dead and buried we will rot, and then resurrect. Resurrection of seeds can happen slowly or quickly. Resurrection of humans is the slowest. It won't happen until the end of days. St. Paul once said: "What you sow does not come to life unless it dies. When you sow, you do not plant the body that will be, but just a seed, perhaps of wheat or of something else. But God gives it a body as he has determined, and to each kind of seed he gives its own body. So will it be with the resurrection of the dead. The body that is sown is perishable, it is raised imperishable; it is sown in weakness, it is raised in power; it is sown a natural body, it is raised a spiritual body." Truth be told.

St. Paul's original quote can be found in *the New Testaments, First Corinthians 15:36-15:45:*

"How foolish! What you sow does not come to life unless it dies. When you sow, you do not plant the body that will be, but just a seed, perhaps of wheat or of something else. But God gives it a body as he has determined, and to each kind of seed he gives its own body... So will it be with the resurrection of the dead. The body that is sown is perishable, it is raised imperishable; it is sown in dishonor, it is raised in glory; it is sown in weakness, it is raised in power; it is sown a natural body, it is raised a spiritual body. If there is a natural body, there is also a spiritual body." (《新旧约全书》*(The New and the Old Testaments),* Chinese Union Version, Bible Society of Singapore, Malaysia and Brunei, 1989, page 181).

At the end of volume seven, the author concluded with objection to the idea of autogenesis put forth by Confucianists. He then tried to prove God's existence with a series of rhetorical questions of natural theology:

Confucianists argued that everything originated from taiji(the ultimate source of the universe), or simply the natural world. Nature by itself is nonautonomous and taiji insentient, how could either of the two turn nutrition of a seed into sugar to nurture its growth? Or make leaves grow in a spiral pattern so exquisitely? Or give roots tubes to suck juice from soil and dew from leaves? Or make the juice ascend to leaves so that oxygen is released while carbon is absorbed to form the wood? Or make flower pedals absorb colors from sunlight to please our eyes? Or give some plants male and female organs so they would produce fruits? Or give other plants prickles or thorns so they can protect their seeds? Ponder these issues one by one and you'd be certain that there be a creator for everything.

Volume eight, which mainly introduces classification of plants, contains no content of natural theology.

To sum up, all volumes, except for volume two and eight, have sufficient natural-theology material. And a scrupulous reader would find that ideas of natural theology are ingeniously embedded into common knowledge of botany. What effort must the compiler have put into the process!

So much natural-theology content in a book composed of some 35,000 words. It's sure worth analyzing.

Shortly after translating *Zhi Wu Xue*, Alexander Williamson and others co-founded *Liu He Cong Tan,* a magazine first issued on January 26, 1857. The magazine had published 15 issued totally over a course of 2 years, with Williamson being a major contributor. He published, in installments, an article titled *Zhen Dao Shi Zheng* (Factual Demonstration of the Truth), where he delivered eleven essays combining contents of both science and natural theology. Those essays were later gathered to constitute volume two of a six-volume book titled *Ge Wu Tan Yuan*. The book, published in 1910, had its English title printed on the flyleaf: *Natural Theology in Six Books*. The Peking University library has a copy of the book stamped with "School of Theology."

Volume one of the *Ge Wu Tan Yuan* is titled *Fan Li* (General Notices) and it says: "Ever since the *Book of Changes* opened with the sentence 'Great was qianyuan that gave birth to everything' and then defined 'qian' as heaven, Confucianists began to mistake heaven for God without realizing they were wrong. God resides in heaven, with the stars and planets being his palaces and the earth his villa. And he filled those places with his creation. By writing this book, I was merely making the master's glory known to all, like a child introducing every piece of furniture in his home to a visiting guest." In these works of natural-theology characteristics, Williamson talked about chemistry, geology, biology, psychology, philosophy, and theology but not especially botany (except for mentioning botany at several occasions). He discussed, in particular, the STS (science, technology, society), all popular ideas today. According to Williamson, as he wrote in *Ge Wu Qiong Li Lun (On Exploring Origins of Everything)*, an article published on the sixth issue of the *Liu He Cong Tan* magazine, a nation's strength lies in its people. A people's strength lies in its heart. And the heart's strength lies in exploration of truth, vis-à-vis, fundamental sciences.

Wang Yangzong, Tian Yong, and Yatsumimi Toshifumi have studied the science and natural theology in *Liu He Cong Tan.* The main source of knowledge in the magazine was *Chambers' Information for the People.* According to Yatsumimi Toshifumi, at that time, "science was being gradually separated from religion" and this "encyclopedia" kept releasing new editions, which had less and less natural-theology content; Given technology's influence on the power and wealth of a nation, "even individuals like Williamson would support, on some level, the religion-free development of science." But, Williamson and John Fryer had different attitudes. The latter seemed more extreme, for he may have been affected by ideas of cultural relativism earlier in his life, thus put more effort in spreading science instead of religion.

In all, it was like what Yatsumimi Toshifumi said: "Natural theology and natural sciences narrowly co-existed in *Liu He Cong Tan,* which was a reflection of the actual situation of sciences in Britain at that time."(Same book, page 133)

Both *Zhi Wu Xue* and *Zhen Dao Shi Zheng* encompass huge amount of scientific knowledge as well as natural-theology content. To compare the two books in terms of preference, the former focuses more on sciences and the latter natural theology, to the extent of preaching. *Zhen Dao Shi Zheng* stresses, in particular, the unicity of God while repeatedly criticizing the idea of nature according to Confucianism.

In terms of proportion, expertise and systematicness, neither *Zhen Dao Shi Zheng* nor *Ge Wu Tan Yuan* can hold a candle to *Zhi Wu Xue*.

Zhi Wu Xue clearly involves sufficient natural-theology content, accounting for a large percentage. Where, then, did those materials come from? Both Williamson and Edkins were missionaries so they were speculated to have played big roles in it. Given the fact that volume eight involves no content of natural theology whatsoever, Edkins is out of the question. My preliminary speculation would be that the natural theology in *Zhi Wu Xue* might be the work of John Lindley, Alexander Williamson, and Li Shanlan. Based on Li Shanlan's foreword, it's a reasonable deduction that Li himself could intentionally add the natural-theology comments. Further analysis showed that Lindley and Li Shanlan merely contributed little. So, the key was Williamson.

Zhi Wu Xue was compiled based on Lindley's botanical work. Many have already studied on that. For example, Shen Guowei did all-round researches on the English text. Lindley's book was obviously written for his students. It was more like a collection of key points for the inductor theory of botany. About that, Lindley himself had made it very clearly in the foreword of the fourth edition of the book. Normally, each key point is a sentence labeled with a number. For example, part one, titled "*Structural Botany and Physiological Botany*," stretches on for 72 pages, consisting of 615 entries. There was zero mentioning of God or natural theology-related content in the foreword of either of the three editions of the book published respectively in 1841, 1857, and 1859. As far as the main text is concerned, there are no natural theology-related sentences, either.

It can be inferred from this considerable contrast that there is a huge difference between the Chinese version of *Zhi Wu Xue* and the English book it's based on. All of the illustrations were from the original book (with only slight alteration), whereas the text was largely deleted and supplemented. Are the natural-theology ideas in Zhi Wu Xue in accordance with Lindley's? Due to shortage of systematic materials, the question is not yet to be answered. However, one definite fact is that modern western botany and natural history inquiries in general were usually carried out in a natural-theology tradition. When it comes to spreading scientific culture, the Chinese version of *Zhi Wu Xue*, on the other hand, was closer to the original British scientific culture during the first half of the nineteenth century.

Britain has a typical natural-theology tradition, on which *Zhi Wu Xue* definitely drew something, although the book didn't directly borrow natural-theology expressions from Lindley.

According to the traditional classification of modern sciences, natural theology can be roughly subcategorized into two styles: (1) Mathematical natural theology (related to natural philosophy). By using geometry and Newtonian mechanics, they stressed that the world ran on the fundamental rules of math, with God being the ultimate source of which. God was the designer of the "clockwork." William Paley and Richard Dawkins often cited the "blind watchmaker" metaphor. The two also mentioned the forming of complicated organs like the eye, which wasn't part of the mathematic tradition. (2) The natural history theology (related to natural history) gave the credit to God for the exquisite structure and harmony discovered through observation by natural history means. And that God was the designer of "complexity." The latter style represents the real natural history. Strictly speaking, the terms "natural philosophy" and "natural history" mentioned above shouldn't be translated literally.

The enormous amount of experiential materials studied in natural history, at best, can be used to "confirm" the existence and some qualities of God, but they can't prove that God absolutely exists in a certain form. Today, there is an interesting tendency towards reductionism that is quite different than the one in Immanuel Kant's time. Instead of trying to turn faith into knowledge, it's turning knowledge into faith. Nothing escapes faith, including the fundamental premises and rules of natural sciences. All these had combined to provide a window for the rejuvenation of natural theology.

Natural theology had been criticized by scholars, especially Hume and Kant, who differentiated knowledge and faith. But his "making" wasn't as strictly executed as his "breaking." To the SSK (science, knowledge, sociology) scholars today, knowledge (including scientific knowledge) is also a faith.

The twentieth-century-scientific philosophy, which relied mainly on empiricism, had repeated claimed, based on profound studies, that limited evidence couldn't possibly prove (or "explain") laws of universal natural sciences. Experiential evidence can't "prove" common hypothesis. At best, it can continuously "confirm" them. In that case, why do people believe in common rules of nature? There is no longer a fine line between knowledge and faith.

Natural theology once played a significant role in the history of science. Today and in the future, it could still combine with science. On one hand, it could be a source of value for science. On the other hand, it could be an opportunity for the development of theology. According to Brooke, the new natural theology will, based on the belief in God, serve to awaken emotions instead of explaining. It makes no attempt at building new evidence for God's existence by using achievements in natural sciences. Instead, it simply expresses a feeling towards God's presence through abundant materials and new discoveries in natural sciences, so as to cause similar reaction in others.

So far, we are still trying to understand the *Zhi Wu Xue* in a traditional (perhaps with a little modification) scientific perspective. However, given today's diverse views on science as well as scientific historiography, the big picture might as well be different.

For a long time, why haven't we considered the *Zhi Wu Xue* a literary work? Or a piece of art? Why didn't we just call it a book of (natural) theology? And why did we separate science from natural theology in the book and treat each one differently? When writing the history of science and the history of the spreading of science, the focus of attention should be on culture, instead of analyzing an ancient book based on the quantity and solidness of the scientific knowledge it contains.

If, by definition, the *Zhi Wu Xue* is a science book or a book written for promoting science, then should it be all about science? Obviously, this sort of comprehension implies a classification ex post facto. It is an approach to judge a certain phenomenon in history through our eyes and by the outcome of development in natural sciences. For example, text books or popular science books on botany today no longer focus on theology, from which we infer, and judge, that some of the content in China's first modern botanical work of the mid-1800s are science and others are not. And then evaluate its "value" or "fineness" as a science book. With such thinking pattern, we can affirm that Kepler's works were but miscellaneous for they involved both science and superstitions like astrology. One of the tasks when writing the history of science is to eliminate the false and the rough while retaining the true and the fine.

The problem is, did writers and translators back then think in the same way? Dr. Qian Xuesen, one of the most prestigious Chinese scientists, was known to be an enthusiastic advocator of somatic science, which, according to some in the intellectual circle today, is not qualified as science. But to Dr. Qian himself, somatic science was of course a science, even the more important one. He may never have thought of drawing a fine line between traditional sciences and somatic science, pointing out which ones are qualified and which ones aren't. How would the future generations think of that part of history? Is there, or rather, is it necessary for us to assume there is a certain absolute knowledge beyond reality? Is there such a thing as ideal development of absolute science at every stage throughout history? So that we should find the disparity and pass judgment by measuring certain works or activities against that development?

It seems we have reasons to provide a non-Whiggish explanation of the *Zhi Wu Xue* along with the content of natural theology in it according to the following points:

1) Williamson came to China in 1855 and return to his home country in1857 to treat his illness. From 1855 to 1857, he worked in the Mohai Library, during which time he helped Li Shanlan translate the *Zhi Wu Xue*. He was less than 30 years old (being born on September 5, 1829) when he finished the translation of the first seven volumes of the book. Before Darwin's *On the Origin of Species* was published in 1859, the combination of natural sciences (of natural history) and natural theology was a norm in western nations. Separation of the two, on the other hand, was viewed as abnormal.
2) In the mid-1800s, a book titled *Zhi Wu Xue* was published in China, where there were no other botanical works in a modern absolute sense whatsoever. Or, with a little exaggeration, we can say that *Zhi Wu Xue* was all about science (certainly, some of the content are false or even ridiculous by current standards). Even what we considered to be natural theology was part of science back then. After all, the so-called science at that time had not been completely separated from natural theology.
3) Williamson, Edkins, and Li Shanlan not so much "translated" an important science book named *Zhi Wu Xue* as the three of them "composed" a brand-new book in Chinese based on Lindley's book (especially the illustration) in the historical and intellectual context of the mid-1800s. In terms of writing style and overall arrangement, "compose" is no exaggeration. The *Zhi Wu Xue* is a written text and a text doesn't talk. It didn't, by itself, claim to be a work of absolute science or absolute religion. It can serve all sorts of purposes. It is an organic combination of science and natural theology, instead of a sketchy collage of the two. Even today, when reading the *Zhi Wu Xue*, one would be impressed by how ingeniously the author had incorporated natural theology into the general narration without seeming abrupt or annoying science-lovers.
4) So often has the explanation of history been affected by ideas of later generations. Surely, Chinese people today can talk about science-preaching back then with an open mind. It goes without saying that missionaries had their own purposes. We can't choose to see only the science in their activities and works just because we

live in a time when science is popular. Nor should we simply eliminate the religious part just because some of us don't like religion.

5) Throughout history, science books were never simple accumulation of knowledge that was later deemed useful and effective. Instead, they were complicated aggregation bearing the hints of contemporary culture. The *Zhi Wu Xue* was no exception. Was the content of natural theology in the book part of science back then? Or was science part of natural theology? Both statements are possible. There can even be a third: it was a work of "scientific natural theology." The very first task Williamson carried out after coming to China was "composing" the *Zhi Wu Xue*. As a missionary, perhaps he was neither trying to spread pure science nor pure religion. For Williamson, spreading pure knowledge itself was going against his understanding of science and religion as well as the happiness of all humanity. Earlier in his life, he had stated explicitly that natural science couldn't develop independently from the guidance of values. Otherwise the consequence would be tragic. Williamson had said, in quite a solemn tone, that: "To separate science from God will lead to disaster in China." And that "disbelieving God, the Devil, the saints or the ancestors" will cause China to "collapse." As long as we don't confine the word "God" to its literal meaning while observing the Chinese society, the global situation and the orientation of civilization led by modernity, we'd be impressed by how far ahead of its time Williamson's vision was. Without the guidance of value rationality, where would instrumental rationality lead us? The history of the twentieth century had given us a definite answer. Looking into the future, shouldn't we hope the development of science and technology, which is out of control, can be adjusted and restricted by value rationality?

6) It's not necessarily a bad thing when a science book bears some kind of value. According to philosopher of science Karl Popper's idea of "metaphysical research programme" or Thomas Kuhn's "paradigm" theory, science is a colossal evolutionary system combining faith and empirical knowledge. At the same time, it can have a certain mainstream system (nature includes values and ontological hypothesis) and a number of branch systems. On some level, it is the differences among values that constitute the variety of sciences and provide the seed for future innovation. Based on which, we may affirm that natural theology shouldn't be combined with botany in a certain way, but we can't simply draw the conclusion that the combination has never worked out or would never succeed. Let alone saying that knowledge of pure science, once tied to some value or religious view that we don't currently approve, couldn't possibly been or would never be successful, and that we must restrain it or get rid of all the dross before studying it.

Even by today's standards, natural history sciences such as plant taxonomy and the study of pollination aren't directly conflicting natural theology. Now, people are not worried about certain botanist believing in natural theology or revering plants. Instead, they fear that botanist has no shame or discretion, or that he sees in plants only their utilitarian value to people, especially a few people. When we appreciate the beauty of a plant and study its exquisite structure, we'd exclaim with admiration, much like in natural theology, although we may not believe in "the lord" or "God" per se. Naturalists with an authentic passion for their cause would also stand in awe of plants and mother nature. They wouldn't allow abusing of plants, biodiversity, and ecological balance for simple utilitarian purposes. It is the same with natural theology. Most botanical lectures in schools today focus only on knowledge instead of sensibility and value. As a result, students were taught to be indifferent towards actual plants and the natural world. They don't appreciate plants or hold such life forms in awe. Professor Wang Jinwu, a plant taxonomist in his 80s at Peking University, told me in deep regret that biology majors today seldom have real passion in plants. According to the new curriculum standard issued and promoted by the central government, sensibility and value are just as important as knowledge in education. Ecology shouldn't be reduced to ecological engineering. Nor should botany be reduced to specific knowledge of plants. Without knowing the great ideas of Rousseau, Thoreau, Leopold, and Carson, one can't call himself a qualified ecologist regardless of the volume of his knowledge or the proficiency of his engineering skills. Nor can one, even who can tell the names of thousands of plants and has published dozens of papers in botany, be seen as a qualified botanist should he fail to appreciate the beauty of plants or cherish biodiversity, to the extent that he would serve as a guide for those who loot and destruct the plant world. Text books that are suitable for students today, even college students, should follow the same pattern as the *Zhi Wu Xue*.

From the first day it entered China, modern botany was closely connected to natural theology. Today, it might not be a bad idea for the public and for the botanists to think beyond utilitarianism when experiencing and studying botany. With only a little concentration, people can discover beauty in the plant world as well as feel the presence of the great chain of being, while experiencing the incredible harmony and leading an interesting life. "Natural theology today can no longer serve as the bridge between science and religion. And the distance between the two worlds will only grow greater." Such is the awful truth. Maybe we can be optimistic, hoping that some kind of variation of natural theology can still play a key role in communicating science and values.

7) "Preaching the Gospels" and "promoting science" are actually similar in so many aspects. Williamson combined the two and called it "scientific preaching." When praising the positive value of promoting, spreading, and helping the public understand science, we should not forget the ideology and value-related assumption behind traditional science-promotion movement, which used to be, if not still is, viewed as some sort of "Gospel preaching." One of the differences, should there be any, is that we haven't tried hard enough. Not as much as Williamson and Fryer and others. We are not as pious and devoting as they were. And our work didn't produce as much as theirs did.

Today, when discussing the history of science as well as the history of the spreading of science, we should try to recreate the social context back in the mid-1800s. And we should try to understand the *Zhi Wu Xue* from the perspectives of Li Shanlan, Williamson, Edkins, and the general public at that time. Our comments on relevant individuals and matters should stay rational and fair. The *Zhi Wu Xue* was indeed an important book and a great new beginning in the mid-1800s.

To conclude, this chapter merely gives a brief introduction of books related to plants in a natural history perspective, such as *Er Ya*, *Shen Nong Ben Cao Jing*, *Nan Fang Cao Mu Zhuang*, *Qi Min Yao Shu*, *Jiu Huang Ben Cao*, *Zhi Wu Ming Shi Tu Kao*, and *Zhi Wu Xue*. In fact, we've only commented a selected few of books, leaving out so many others that are equally important. For examples, *Zheng Lei Ben Cao* written by Tang Shenwei in Song Dynasty, *Yu Zhi Ben Cao Pin Hui Jing Yao* compiled by Liu Wentai and others in Ming Dynasty, *Ben Cao Gang Mu* written by Li Shizhen in Ming Dynasty, *Jing Zhu Ben Cao* written by Danzengpengcuo in 1736, *Ben Cao Hui* written by Guo Peilan in Qing Dynasty, and *Zhi Wen Ben Cao* written by Wu Jizhi in Qing Dynasty. The books we did discuss were written in various styles and most of them focused on practicality. There are a few that bear the characteristics of pure botany, such as *Nan Fang Cao Mu Zhuang* and *Zhi Wu Ming Shi Kao*. Some of the description or knowledge in these books are unreliable, but basically they all came from actual life experiences, which means they were based on solid facts with the least amount of fabrication. Being brilliant works of natural history, the books were passed on through time and improved with accumulation. They simply aimed to serve the daily needs of the common people in China instead of pursuing some abstract science or innovating for the purpose of snatching excessive profit and taking control of the world. In the mid-1800s, western botany was introduced to China. In the midst of a cultural clash between east and west, the Ben Cao Xue, compiled and translated by Li Shanlan and others, put a premium on the combination of instrumental rationality and value rationality. That approach alone was worth noticing. Botanical studies in ancient China, which focused on herbalism, had obvious disadvantages, for example, the excessive attention on existing knowledge. Sometimes too much effort was wasted in lengthy quotations without actually increasing the amount of practical knowledge. Also, too much consideration of practicality had hampered observing and understanding the natural world. Generally speaking, there had been persistent studying on animals and plants in ancient China, a tradition of natural history that constituted our nation's unique cultural legacy. Today, it bears cultural significance globally therefore will draw more and more attention. These studies can serve as crucial references for future development of Chinese society as well because no civilization can be built solely on copying others while eliminating its own tradition.

References

1. Lindley, J. (1841). *Elements of botany; structural, physiological, systematical, and medical; being a fourth edition of the outline of the first principles of botany.* London: Taylor and Walton.
2. Polany, M. (1968). Life's Irreducible Structure. *Science, 160*(3834), 1308–1312.
3. Alibaba. (2009). "伏地魔之子论纯科学推进的速度" (Son of Voldemort on the Pace of Development of Pure Science), 《我们的科学文化:科学的算计》 (Our Scientific Culture: Calculation). In J. Xiaoyuan & L. Bing (Eds.), (pp. 256–260), East China Normal University Press.
4. Toshifumi Y. (2006). 八耳俊文, *"在自然神学与自然科学之间:《六合丛谈》的科学传道" (Between natural theology and natural sciences: Scientific preaching of the Liu He Cong Tan)* Shen Guowei: 《六合丛谈》 (Liu He Cong Tan) (solution· index), (pp. 117–137). Shanghai Lexicographical Publishing House

5. Demao, C. (1993).《中国植物分类学史》*(A history of plant taxonomy in China),* (pp.180–181).East China Normal University Press.
6. Rongxia, C. (2006).*"对当代自然神学合理性依据的反思" (Retrospection on the logical ground of contemporary natural theology),《上海交通大学学报》* (*Journal of Shanghai Jiaotong University), Philosophy and Social Sciences Edition, 14*(6), pp.56–63
7. Chongming, C., & Yinghui, C. (2011). "*《南方草木状》一书中的民族植物学" (National botany in Nan Fang Cao Mu Zhuang),《中国野生植物资源》(Chinese Wild Plant Resources),* issue 06.
8. Fadi, F. (2011).*《清代在华的英国博物学家》(English Naturalists in Qing-dynasty China),* China Renmin University Press.
9. Qide, H. (2012).*"科学文化的核心是科学精神" (The scientific spirit as the core of science culture),《民主与科学》(Democracy and Science)*, (05), p 2.
10. Pin, H. (2012).*《杂花生树:寻访古代草木圣贤》(A thousand blossoms on a tree: Finding pioneers of herbalism in ancient China),* (pp. 28–32). Henan Wenyi Publishing House.
11. Hu, Q., & Huanhai, F. (2012).*《尔雅译注》(Er Ya in Common Chinese, with Notes),* Shanghai Classics Publishing House.
12. Xianlin, J. (2009).*《蔗糖史》(A history of cane sugar),* China Customs Press.
13. Shiying, J., Jin, P., & Shuting, L. (2011). "《南方草木状》作者、版本与学术贡献的研究" (Studies on authorship, Versions and academic contribution of Nan Fang Cao Mu Zhuang), *Journal of Guangzhou University of Traditional Chinese Medicine, 28*(3).
14. Shizhen, L. (2012).*《本草纲目》(Compendium of Materia Medica)*, edition 2, (pp.1140–1144). People's Medical Publishing House.
15. Needham, J. (2006).*《中国科学技术史·植物学》(Science and civilisation in China: Botany),* (pp.444–155). Science Press, Shanghai Classics Publishing House.
16. Qiyu, M., & Guilong, M. (2009).*《齐民要术译注》(Qi Min Yao Shu in common Chinese, with notes)*, (pp. 2–4). Qilu Press.
17. Guowei, S. (2006).*《六合丛谈》(Liu He Cong Tan) (solution· index)*, Shanghai Lexicographical Publishing House.
18. Guowei, S. *《植学启原と植物学の語彙:近代日中植物学用語の形成と交流》(Origin and glossary of botany: Forming and exchange of botanical terms in Japan and China of modern times).*
19. Scott, J.C. (2011). *《国家的视角》(Seeing As A Nation)*, Social Sciences Academic Press, p. 430.
20. Yong, T. (2006). *"韦廉臣在华的西学传播与传教" (Alexander Williamson's promotion of western sciences and preaching in China),* Capital Normal University Master's Dissertation, pp. 1–53.
21. Xiaoqin, W., & Edkins, J. (2001). "致力于中西科技交流的传教士和学者" (Missionaries and scholars devoted in promoting science exchanges between China and the West),《自然辩证法通讯》. *Journal of Dialectics of Nature, 23*(5), pp. 74–83
22. Zhenru, W. (1988). "关于植物学一词的来源问题"(Botany: Origin of the term)《中国科技史料》. *The Chinese Journal for the History of Science and Technology, 8*(1), p. 88.
23. Zichun, W. (1981). "*李善兰和他的植物学" (Li Shanlan and His Botanical Inquiry)*《植物杂志》*(Plant Magazine), 2*, pp. 28–29.
24. Zichun, W. (1984). "我国传播近代植物学知识的第一部译著《植物学》" (Zhi Wu Xue: The First Modern Botanical Work Introduced to China)《自然科学史研究》. *Studies In The History of Natural Science, 3*(1), pp. 90–96
25. Zichun, W. (1981). "中国早期传播植物学知识的著作《植物学》" (Zhi Wu Xue: An Early Botanical Work in China)。《中国科技史料》. *The Chinese Journal for the History of Science and Technology,* (1), pp. 86–87
26. Jiakui, W. (2007).*《救荒本草校释与研究》(Emendation, Explanation and Study of Jiu Huang Ben Cao),* (pp. 21–22). Chinese Medicine Classics Publishing House.
27. Yangzong, W. (1990). "《六合丛谈》中的近代科学知识及其在清末的影响"(Knowledge of Modern Sciences in Liu He Cong Tan and Its Influence on China in Late Qing Dynasty), 中国科技史料》. *The Chinese Journal for the History of Science and Technology, 20*(3), pp. 211–226

28. Zongxun, W. (1988). *"中国近代植物学回顾" (Review on Botanical Inquiries in Modern China)*《生命科学》*(Bioscience), 4*, pp. 2–4.
29.《植物学》(Botany), edited and translated by 韦廉臣 and 艾约瑟, recorded by 李善兰, Mohai Library, 1858
30. Williamson, A. (1910).《格物探原》*(Ge Wu Tan Yuan)*。Shanghai: Christian Literature Society. Printed at the American Presbyterian Mission Press.
31. Zhengyi, W. (1953). *"中国植物学历史发展的过程和现况" (History and Current Situation of Botany in China)*《植物学报》*(Bulletin of Botany), 2*(2), pp. 335–348.
32. Weiying, X. (1950). "李善兰介绍" (An Introduction to Li Shanlan)《中国植物学杂志》. *The Chinese Journal of Botany, 2*(2), p. 72.
33. Zhipei, Y. (1998). "李善兰和我国第一部《植物学》译著" (Li Shanlan and the First Chinese Version of "Botany")《生物学通报》. *Bulletin of Biology, 33*(9), pp. 43–44.
34. 宇田川榕菴 (うだがわようあん: *"植學啓原" (Origin of Botany)*《文睮源流叢書》*(Origins of Civilization Book Series) (Volume 2),* 圖書刊行會第二工場, 1914, pp. 284–323.
35. Dengben, Z. (2009).《全注全译神农本草经》*(Shen Nong Ben Cao Jing in Common Chinese, with Notes)*, New World Press, pp. 17–30
36. Ruixian, Z. etc. (2008).《植物名实图考校释》*(Emendation and Explanation of Zhi Wu Ming Shi Tu Kao)*, Chinese Medicine Classics Publishing House, p. 1

Song Ci, the *Xi Yuan Ji Lu*, and the Judicial Examination System

7

Yuqun Liao

Contents

Abstract

Through relating the "military accomplishments" of Song Ci, a Northern Song high-ranking judge in criminal case, this chapter points out that Song was recorded in history as a law-abiding government official. Then it discusses the vital function of his Xi Yuan Ji Lu in perfecting and developing the ancient Chinese judicial examination system. Lastly, it analyzes the basic features of the system and the similarities and differences between it and modern forensic medicine.

Y. Liao (✉)
The Institute for the History Natural Sciences, Chinese Academy of Sciences, Beijing, China
e-mail: liaogg@ihns.ac.cn

X. Jiang (ed.), *The High Tide of Science and Technology Development in China*,
History of Science and Technology in China,
https://doi.org/10.1007/978-981-15-7847-2_7

Keywords

Song Ci · The Xi Yuan Ji Lu · The judicial examination system · Forensic scientist · Inquest technology

The *Xi Yuan Ji Lu* (the Washing Away of Wrongs, 洗冤集录) written in the Song Dynasty has been called "the world's first forensic work" by Chinese historians of science and scholars who are interested in studying the history of forensic medicine in China. And its author, Song Ci, naturally enjoys the reputation of "a great forensic scientist in ancient China." Even in the famous western historian of science Joseph Needham's great work *Science and Civilisation in China – Medicine*, Song Ci and his book appeared in the so-called forensic medicine in ancient China in this capacity and occupied a considerable space. However, have the *Xi Yuan Ji Lu* and other similar works before it reflected the basic concept of "forensic medicine?" Should their authors be called "forensic scientists?" It is indeed worth serious consideration. The wisdom and greatness of the ancients and the value and achievements of ancient books may not necessarily be labeled as contemporary science.

7.1 About Song Ci

Song Ci, the author of the *Xi Yuan Ji Lu*, was born in Jianyang, Fujian. There are two statements about the year of his birth and death. First, according to the *Epitaph* (The Song Jing Lue of Mu Zhi Ming. In the Hou Cun Xian Sheng Da Quan Ji, Vol. 159) written by his friend Liu Kezhuang, Song Ci died in the 6th year of Chunyou during the Southern Song Dynasty (1246) at the age of 64. Then it can be inferred that he was born in the 10th year of Chunxi (1183). However, the Preface of *Xi Yuan Ji Lu* was written during the 7th year of Chunyou (1247). One of them must be wrong. According to the *Chronology of the Southern Song Dynasty*, Song Ci was appointed the Prefect of Guangzhou in 1248. He became the Judicial Commissioner of Guangdong in 1249 and died the same year (Wu Tingxie: the Nan Song Zhi Fu Nian Biao. Quoted from the Er Shi Wu Shi Bu Bian (Vol. 6), Kaiming Bookstore, 1936: 7988 & 7985). So modern writers assumed that Song Ci's birth and death years were either between 1183 and 1246 based on Liu Kezhuang's Epitaph or between 1186 and 1249, with the latter being the majority.

Song Ci was born into an official family; his father Song Gong had been a Military Commissioner in Guangzhou. In his youth he was taught by Wu Zhi, one of the followers of Zhu Xi, the greatest Neo-Confucian philosopher. He got to communicate with Confucian scholars and celebrities such as Yang Fang, Huang Gan, and Li Fangzi; he obtained more knowledge day by day. After he entered the Imperial University, whose Chancellor Zhen Dexiu, also known as Master Xi Shan, praised him for his literary ability. Therefore, Song Ci also became Zhen Dexiu's student. However, Song Ci's career was not entirely based on his "literary ability" but to a great extent on his "military strategy."

During the 10th year of the Jiading era in the Southern Song Dynasty (1217), Song Ci started his career as a government official after graduating as a Presented Scholar. In 1225~1227 he was appointed as Registrar at Xinfeng, Ganzhou (now

Xinfeng in Jiangxi Province) (The Zhi Guan. In the Gan Zhou Fu Zhi (Vol. 7), Shanghai Ancient Books Bookstore, photocopied the Ming Jiajing engraving in Tianyi Pavilion in 1962). Although he was only an official in charge of literature, Song Ci was also involved in the military because of the favor of military commander Zheng Xingzhi. During the following years, he made a great name by putting down rebellions, dealing with their root causes by distributing food from government granaries, but his success offended his superiors, and he was repeatedly impeached (Lu Xinyuan: Song Ci, the Song Shi Yi (Vol. 22), in the Qian Yuan Zong Ji). Later, chaos broke out in Fujian. Zhen Dexiu recommended Song Ci to Chen Wei, a famous minister in the Southern Song Dynasty, and cleared him of the previous accusation. When attacking the Tiger Stronghold, the commanding general, Wang Zuzhong, could not help but exclaim, "You have been loyal and brave enough to make a military commander!" (Lu Xinyuan: Song Ci, the Song Shi Yi (Vol. 22), in the Qian Yuan Zong Ji). Since then, many military affairs had been consulted with Song Ci. He always planned well before he fought, therefore winning countless victories (Lu Xinyuan: Song Ci, the Song Shi Yi (Vol. 22), in the Qian Yuan Zong Ji). Chen Wei admired his talent and recommended Song as magistrate of Changting County during the Southern Song Dynasty (1228~1233) (The Zhi Guan Zhi (Vol. 7), Fu Jian Tong Zhi (Vol. 32), 1938).

During the second year of Duan Ping's reign in the Southern Song Dynasty (1235), Palace Secretary Zeng Conglong was supervising operations in the Jianghuai region, and he specially appointed Song Ci as his subordinate. However, Zeng died before Song arrived. Later Wei Liaoweng took charge, making Song Ci his subordinate. Since then, Song's career as an official seemed quite smooth. He successively held the post of the Controller-General of Shaowujun, the Prefect of Ganzhou, Judicial Intendant of Hunan, the Prefect of Guangzhou, and the Military Commissioner of Guangdong (Fig. 7.1).

Song Ci was over 60 when he took up the post in Guangzhou. Although he felt sick, he still insisted on dealing with official business in person. One day, the local academy held the opening ceremony; Song's peers thought that they could go represent him. However, Song Ci, as the local chief executive, went there personally. Since then, he had been mentally exhausted and finally died in Guangzhou office on the 7th day of March this year.

During his administration in various places, Song Ci scrupulously abided by his duties and gave benefits to the people. For example, when he was the Prefect of Changting, he found that the local salt price was very high and the people could not afford it. It was because the salt had to be transported via the Shuiming River in Fuzhou. It was a long journey and would take a year. Song Ci diverted its transportation from Chaozhou, Guangdong, along the Hanjiang and Tingjiang rivers. The waterway reached Changting. It took only 3 months to go back and forth, saving freight and exploitation by officials along the way. As a result, the price of salt was reduced, for both public and private benefits.

Song Ci was sent to Nanjian, where a famine broke forth. He classified the people into five groups according to the degree of wealth and poverty, relieved the extremities of the poorest by redistributing the excess property of the richest. Because Song Ci was able to "treat others with courtesy," the policy of "helping the poor" was carried out, and finally "all the people were instructed and nobody was hungry."

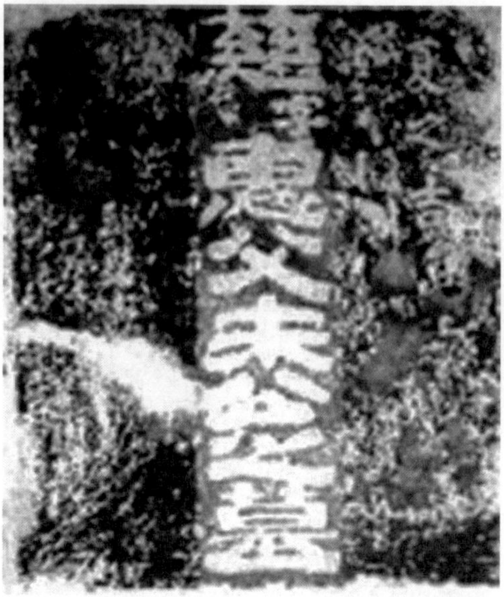

Fig. 7.1 Tombstone of Song Ci. (He Zhongjun, Wu Hongzhou, ed., Atlas of Chinese Medical Cultural Relics, Sichuan People's Publishing House, 2001: 50)

Song Ci was transferred many times in his life, and in his later years he was given priority to judicial punishment. When he was promoted to the Judicial Intendant of Guangdong, he found that most of the local officials did not obey the law, and there were some prisoners who had not been tried for several years in prison, so he concluded a treaty and closed the case within a time limit. He decided over 200 cases of murder, suicide, and accidental death in just 8 months. Everywhere he went, he took "liberating the innocent and prohibiting the violent" as his duty and dealt with the case "by examining it over and over again, not daring to show any signs of being slow to change mind" (Song Ci: Preface to the Xi Yuan Ji Lu, Commercial Press, 1936). If he had any doubt, he would "think it over and over again for fear that the rate will go ahead and the dead will be flooded." Later generations recalled his style and features and called him "hearing the lawsuit is clear and just, and making fair decisions."

When Song Ci was in office to mention Hunan's criminal prisons, he summed up the previous cases of criminal investigation and verdict, combined with his own experience in practice, and wrote the *Xi Yuan Ji Lu*. He hoped that this book would be beneficial to his colleagues and play the role of "participating in and examining each other," which was not different from doctors' efforts to revive the dead according to classical and ancient methods.

What needs to be explained is that most comments and biographies on Song Ci are from the perspective of "great forensic scientists." They either avoided mentioning his "military achievements" or accused him of participating in the suppression of peasant uprisings. This is no doubt that according to today's "good and evil" standard, the ancient people are bound to be influenced. It is not in conformity with the basic principle of co-authorship of the faithful history and stating the actual facts. Naturally, it is impossible to obtain the truth from it – to objectively grasp Song

Ci's outlook on life, values, and life history. In history, Song Ci was recorded as an official who did not make great achievements in his official duties (Lu Xinyuan: Song Ci, the Song Shi Yi (Vol. 22), in the Qian Yuan Zong Ji). What he did was a "political achievement" known to both the government and the people at that time. It is also because of these achievements that Song Ci was able to step into a higher position and occupy an important judicial post because the Judicial Intendant was originally a post held by military officers in the Song Dynasty. Emperor Xiaozong of Southern Song Dynasty (1127~1194) changed to policy of employing civil servants. However, all his actions in the "redress of grievances and prohibition of violence" and the prudent management of prisons undoubtedly adhere to the moral and value concept of the law of serving duties. In the process of realizing his purpose in life, Song Ci was able to realize more deeply than his predecessors the significance of corpse examination, material evidence collection, on-site inspection, etc. in the treatment and judgment of prison cases. He paid attention to observing and summarizing the knowledge in this field and wrote a special book on examination rich in knowledge related to modern forensic science, which is really important.

7.2 History of Inquests

7.2.1 A Long History of Judicial Examination System

The examination work serving the judicial and criminal investigations can be traced back to the pre-Qin era in our history. At that time, the prison governor judged the case based on the results of "observing the injury, investigating the wound, reviewing the damage, and examining," in order to achieve the goal of "deciding the lawsuit in the prison and bringing it to justice"(Annotated by Zheng Shi, Interpreted by Kong Yingda: the *Yue Ling*, the *Li Ji Zhu Shu* (Annotation and Interpretation of the Book of Rites) (Volume 16)). Later, in the Qin Dynasty, which was known for its heavy "criminal law," records of on-site inspections and autopsy were already available; doctors were also ordered to examine the defendant to determine whether he was a leprosy patient who should be sent to quarantine or inspect the fetus aborted due to fighting and identify whether the plaintiff's vaginal bleeding was caused by the abortion, etc. With the passage of time, the inquest system has become more and more stringent, with the definition and classification of injuries, the development of specialized autopsy form, and the clarification of the responsibilities of the autopsy, which has promoted the continuous improvement of inquest quality. At the same time, the selection of judicial officials was given much attention. Since the Tang Dynasty, most of them have passed the imperial examination, so "The Subject of Law" was included in each year's civil examination for government degrees. Emperor Taizong of Tang also specially appointed 1 doctor of law and 50 students to train judicial officers.

Based on the Tang Dynasty's system, the imperial courts of the Southern and Northern Song dynasties had regulations on inspectors, inquest implementation, autopsy documents, etc. and constantly revised and supplemented them, bettering the inquest system of the Song Dynasty. The Song Law clearly stipulates that, except for the deaths with clear causes such as illnesses that can be exempted from autopsy on the premises that the relevant personnel could guarantee there are no other causes and that the case has been thoroughly examined, all the other deaths must undergo the procedures of initial inquest and re-inquest. In addition, the Tang and Song dynasties had extremely strict penalties for inquest errors, which led to the continuous improvement of the level of judicial examination. "The inquest system of the Tang and Song Dynasties was the most advanced and most complete in the medieval era. At that time, Europe was still in the dark age of religious rule. No country could establish a systematic and rigorous inquest system like ours."

The essence of the system is to provide better-quality evidences for law enforcement. Objectively speaking, it improves the quality of inquest, promotes the advancement of inquest technology, and constitutes the basic motivation for the generation of systematic inquest works.

7.2.2 Doctor Participated in Judicial Examination

Since the Qin Dynasty, Shang Yang has begun to lay great emphasis on the rule of law, followed by Li Si. Therefore, "The law is enforced and Qin people are regulated" (*The Records of Historian ·The Fifth Book of History of Qin Dynasty*). But people were not clear how many laws there were and how they were enforced. In 1976, more than 1,150 bamboo slips were unearthed from the tomb of Qin in Shuihudi, Yunmeng County, Hubei Province. Most of the bamboo slips were related to the laws of the State of Qin (Fig. 7.2). The content related to "medical" issues was mainly in the part called "Feng Zhen Shi" (封诊式, a collection of regulations and

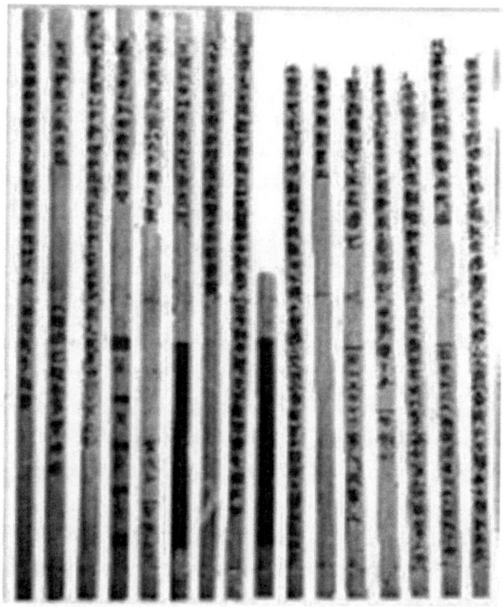

Fig. 7.2 Bamboo slips in Shuihudi, Yunmeng County. (Compiled by the Institute of Archaeology, Chinese Academy of Social Sciences: "The Essence of Archaeology-The 40th Anniversary of the Founding of the Institute of Archaeology, Chinese Academy of Social Sciences", Science Press, 1993, p. 374)

cases concerning the principles of trial and investigation, inquest, interrogation, and sealing of cases). "Feng" (封) has the meaning of sealing and confinement, which can be understood as consciously protecting and restricting the scope of the scene, physical evidence, and related persons; "Zhen" (诊) means investigation and inquest, so there are detailed records of on-site investigation, inquest, and autopsy; "Shi" (式) is the standard for performing government affairs, namely, the meaning of "style" (procedure). Researchers believe that the legal documents in this batch of Qin bamboo slips have one thing in common: for each relevant system, one example is cited to illustrate how people should abide by it, so that officials at all levels can follow the "style" to implement it in a similar way.

In a case where it is necessary to confirm whether the defendant is a leprosy patient who should be sent to the quarantine area, it could be clearly seen that the doctor was instructed to examine the defendant, which was recorded in the case note. Since the defendant was suspected to be a patient with "leprosy" disease, he was then interrogated by the judicial officer; the defendant replied: "I had the disease of 'arthritic psoriasis' at the age of three. And my eyebrows were bald, but I didn't know why it was caused." Then the doctor was ordered to diagnose his illness. After examination, the doctor replied: "The defendant has no eyebrows and the root of eyebrows is completely gone; the nasal cavity is damaged. When his nose is pricked, he doesn't sneeze; there are ulcers in his hands and feet; he is asked to shout but cannot speak normally. It is indeed 'leprosy'." But there is an issue that often arises unconsciously in the narration of the historian: after citing such examples, the role of doctor in judicial examination is overgeneralized. However, the fact is that many autopsies were conducted by those people called "Ling Shi" (令史, director); the examination of the fetus aborted due to fighting and the identification of whether the plaintiff's vaginal bleeding was caused by the abortion were also executed by "Ling Shi" and "subordinate concubines" who had rich maternity experiences (The collection team of Bamboo Slips of Shui Hu Di Qin Tomb: the *Bamboo Slips of Shui Hu Di Qin Tomb*, The Cultural Relics Publishing House, 1978, Page 263 and 275).

Also noteworthy is that the *Xi Yuan Lu* (*The Washing Away of Wrongs*) stipulated: upon receipt of the inquest task, they are not allowed to contact all kinds of people including "Shu Ren" (术人, people who possess extraordinary skills). Although the scope of "Shu Ren" was not clarified, throughout Chinese history the status of doctor was usually very low, and doctors could basically be regarded as a kind of "alchemist" or "warlock." This is because only those with a Confucian background and certain official rank were considered as reliable people. It might have never come to readers' mind that when Empress Dowager Cixi in the 6th year of Emperor Guangxu's reign of the Qing Dynasty (1880) was unwell, she could not recover after several months of treatment by the imperial physicians. Then the Grand Minister of State informed the governors from all over the country to visit people who were proficient in medical skills and invite them to Beijing for treating Cixi. Li Hongzhang, the governor of Zhili Province, recommended Xue Fuchen from Shandong Taiwu Lindao. Also recommended to Beijing were Wang magistrate Shouzheng of Yangqu County, Shanxi Province; educational inspector Xue Baotian of Shangyuan County, Zhejiang Province; educational officer Zhong Xuezheng of Chunan County, Zhejiang Province; deputy magistrate Cheng Zhao Tianxiang of

Jiangxi County; magistrate Lian Zihua of Xinning County, Hunan Province; and candidate officer Cheng Chunzao of Hubei Province. But the real professional physician was only Ma Wenzhi from Wujin, Jiangsu Province, recommended by Jiangsu Governor Wu Yuanbing. Similarly, when Emperor Guangxu was critically ill, the Privy Council called the governors of Zhili, Liangjiang, Huguang, and other places to recommend outstanding physicians to Beijing. As a result, magistrate Lu Yongbin of Yushan County, Jiangxi Province; Chen Bingjun, the principal of the Ministry of Penalty; physician Cao Yuanheng; magistrate Zhou Jingtao of Funing County, Jiangsu Province; candidate magistrate Du Zhongjun of Zhejiang Province; candidate magistrate Shi Huan of Jiangsu Province; candidate officer Zhang Pengnian; and others were sent to Beijing. Here is a typical example of a Confucian official who was trusted by the emperor and replaced the full-time imperial physician to deal with medical problems: Emperor Qianlong's concubine Dun was suspected to be pregnant, but the imperial physicians' memorials to the throne were always ambiguous. Hence, Emperor Qianlong commanded Yu Wenyi, minister of the Ministry of Penalty, to diagnose and treat the concubine. Despite the difference between Yu's prescription and other physicians,' Emperor Qianlong still 100% trusted Yu (For details of the process of this concubine Chun's "suspected pregnancy case", see Chen Keji, editor-in-chief: the *Research on Qing Palace Medical Records* (Volume 1), Ancient Chinese Medicine Publishing House, second edition, 2006, Page 88–94). It is really incomprehensible and unimaginable for modern people: treatment of illness is the job and expertise of physicians. But why a civil officer is appointed to tackle the difficult and miscellaneous diseases?

Through these interesting examples and differences between the past and the present, I would like to make readers understand more about the status of physicians in the history and why they could not play a pivotal role in judicial examination. In fact, apart from the example of "leprosy inquest" recorded in Shuihudi Qin bamboo slips, among the various so-called forensic medicine works in the following eras, such examples of inquests with the involvement of physicians were not only increases but rather barely seen.

7.2.3 The Accumulation of Inquest Experiences

In ancient China, there are a few books that record law enforcement cases. The *Si Ku Quan Shu* (Complete Works of Chinese Classics) compiled them into the "Law Category." In modern time, because of some contents about inquest, people referred to those books as "forensic classics" of ancient China. Apparently, this was not accurate because among those classics "inquest" only accounted for a minimal proportion, whereas most of the content was description of the reasoning process of lawsuit settlement that served as references for other peers. For example, in the Five Dynasties He Ning's *The Yi Yu Ji* (The Suspected Case Collection), the famous case "Zhang Ju and Buring Pig" – putting two pigs alive and dead together in a firework. There is ash in the nose of the pig that died of a firework. And there is no ash in the nose of the pig that is placed in firework after death. This shows people how to identify the case of faking people's death in fire. However, there are no more than 3 cases concerning such inquest content in the 79 cases in the book.

In those works, there are some cases to show people how to adjust the relationship of "emotion" and "reason" flexibly. For instance, in the Ming Dynasty Zhang Jing's *The Bu Yi Yu Ji* (Supplement to the Suspected Case Collections), some works even have no plots of criminal investigation and inquest at all, but merely extol the virtues of officials, such as the example in Volume 4 of Zheng Ke's *Zhe Yu Gui Jian* (Collection of Judging Lawsuits).

Although it is not appropriate to use these as forensic classics, the valuable inquest experience recorded in them cannot be ignored. For example, in Zhang Jing's *Bu Yi Yu Ji* mentioned earlier, senior official Yang Gong in the Song Dynasty inspected the fatal wound under the rib – "The wound length is 1.2 inch with a white road in the middle" – and regarded it as a stick stroke. This is what later generations call "the strike of bamboo" (竹打中空), meaning when the round stick acts on the soft tissues of the body, two parallel subcutaneous bleeding bands will be formed, and the skin in between will be white. It is called "double streaks" in modern western forensic examination.

Another example is the case "Li Gong Inspects Beech" recorded in Gui Wanrong's *Tang Yin Bi Shi* (棠阴比事) in the Song Dynasty. In short words, the case is about two people fighting. A is strong and B is weak. Both of them have wounds on their bodies. After Li Gong pinched their wounds by hand, he concluded that B's wound was real, while A forged the wound with some kind of leaf coloring. The basis is: "The wound caused by fighting feels hardened due to blood coagulation while the fake one is opposite." This is a well-known case of examining real injuries. "Blood gathers and the wound touches hard" is a correct description of subcutaneous hematoma; the faker has no subcutaneous bleeding, so it just looks similar. The *Xi Yuan Ji Lu* absorbed this valuable experience.

According to Song Ci's "Preface," the *Xi Yuan Ji Lu* is "a complete collection of books in the recent generations including the *Nei Shu Lu* (内恕录) and other classics." This demonstrates the importance of ancestors' experiences.

7.3 The *Xi Yuan Ji Lu*

The historical value of the *Xi Yuan Ji Lu* could be judged from two aspects: first, the special attention given to inquest system; and second, the experiences of inquest constitutes the technological advancement.

7.3.1 Respect Life and Strive for Fairness

Civilization is composed of many elements. The establishment of social order, the accumulation of material wealth, the ability to understand and leverage nature, and even the ideologies of wise sages and the customs of citizens are all integral parts of civilization. However, the status of a person – respect life and basic rights – is also an important part of civilization and critical indicator for measuring the level of civilization. Respect for life, on one hand, requires the awareness on the ideological level and at the same time requires institutional guarantee. Specifically concerning the aspect of justice, since resurrection does not exist – once the death penalty is wrongly sentenced

Fig. 7.3 The Preface of the *Song Ti Xing Xi Yuan Ji Lu* (Yuan Dynasty Edition). (Collected in Peking University Library)

and executed – it cannot be reversed and corrected. As a result, Song Ci firstly emphasized in the *Xi Yuan Ji Lu* that those who have the authority to decide life and death should be cautious about the "death penalty"; in order to achieve that, the most important thing is to have a standardized inquest system as guarantee.

Although the Preface (Fig. 7.3) of the *Xi Yuan Ji Lu* is not long, it clarifies the motive, purpose, and process of compiling this book.

In terms of the main text, first of all, the "regulations" of the strict inquest system, such as "Should examine but not examine," "Being assigned the tasks for two hours but not yet start working," "Not make personal visit to the site," "Cannot determine the death cause," or "Improperly determine the death cause," are all regarded as "violation" with the caning punishment of "100 sticks." After finishing the initial inquest and re-inquest of the corpse in accordance with the regulations, one should fill in the "autopsy form" in triplicate made and printed by judicial officers, one copy for the subordinate state and county, one copy for the victim's family, and one copy submitted via "express delivery."(Express delivery: a postal express that travels four hundred miles a day)

Then, one should examine the corpse thoroughly from head to toe, check the wounds, and determine the details of sentence and punishment, which constitutes the book of the *Xi Yuan Ji Lu*.

7.3.2 The Features of Inquest Technology

Just as scientific research requires equipment, and the solution of problem requires the design of necessary experiments, it is also essential to use knowledge, technology, and methods to redress mishandled cases or find out the real culprits. Otherwise, judicial justice will be just empty talk. The *Xi Yuan Ji Lu* basically includes most of the contents related to the examination of corpse appearance in modern forensic science. Due to the limitation of social and historical conditions and the constrained development of natural science, during the era when Song Ci lived, it was impossible to have the contents of modern forensic medicine, such as autopsy, pathological analysis, and determination of chemical properties of toxicants. Therefore, in terms of inquest technique, it could be considered that the *Xi Yuan Ji Lu* is a relatively comprehensive and systematic works summarizing the experiences of corpse appearance inspection. As a scholar who specializes in the research of forensic medicine history said: "The Xi Yuan Ji Lu is, in essence, the forensic medicine that guides the inquest of corpse appearance." Its main achievements are shown in the following aspects:

7.3.2.1 Postmortem Phenomena

The postmortem phenomena refer to the fact that after the death of a person, since the objective matter is still in constant motion and change, there occur various phenomena different from the time of death. Based on this, inferences can be made about the cause of death and the time of death. For example, the mechanism and distribution characteristics of "livor mortis" were basically recognized at that time:

> Every dead person has a light red color on the back neck, back, ribs, lower waist, inner legs, upper arms, back legs, two bends, upper and lower calves. If the person lies on his back when dead, the light red color is caused by the fall of blood vessel. Then this person's death is not caused by others.

Normally 1–3 h after death, blood circulation stops and blood gradually sinks. Capillaries expand and blood accumulates at the lower part of corpse to form the stains. The appearance of livor mortis is one of the indications of irreversible death in modern clinical medicine; and its color can be diverse with the color of hemoglobin. For example, it is generally dark purple-red; the color of people poisoned by carbon monoxide is dark cheery, and the color of those poisoned by aniline is gray-blue. Although there are some descriptions about the livor mortis phenomena and correct analysis of the causes in the *Xi Yuan Ji Lu*, its significance is merely about explaining that death spot is a natural phenomenon and not caused by beating.

Concerning the putrefaction of corpse, the book points out: First, there will be some "slight greenish flesh color" on the mouth, nose, abdomen, two ribs, and chest, which is called "greenish discoloration" in modern terms (the hydrogen sulfide produced by bacterial decomposition combines with hemoglobin to form hemoglobin sulfide that causes the green spots on the surface of corpse). Then there will be the outflow of stench liquid from the ear and nose with maggots, swelling body, everted lips, peeling skin, herpes, hair losses, and other phenomena appearing one after another. There are some arguments on the occurrence time of these putrefaction phenomena, as well as the influence of factors such as season, physique, and region (see Table 7.1). It shows that it was already possible to estimate the time of death based on the degree of corruption of the body.

Table 7.1 The relationship between putrefaction phenomenon and time of death

中文	英文
季节	Season
春	Spring
夏	Summer
秋	Autumn
冬	Winter
中文	英文
时间	Time
现象	Phenomenon
2~3日	2–3 days
口鼻、肚皮、两胁、胸前肉色微青	Slight greenish flesh color on the mouth, nose, abdomen, two ribs, chest
10日	10 days
鼻、耳内有恶汁流出，膨胀	Outflow of stench liquid from the ear and nose and swelling body
1~2日	1–2 days
先从面上。肚皮、两胁、胸前肉色变动	The fresh color starts to change from the appearance of abdomen, two ribs, and chest
3日	3 days
口鼻内汁流、蛆出、遍身胖胀、口唇翻、皮肤脱烂	Outflow of stench liquid from the ear and nose with maggots, swelling body, everted lips, peeling skin
4~5日	4–5 days
发落	Hair losses
2~3日	2–3 days
先从面上。肚皮、两胁、胸前肉色变动	The fresh color starts to change from the appearance of abdomen, two ribs, and chest
4~5日	4–5 days
口鼻内汁流、蛆出、遍身胖胀、口唇翻	With maggots, swelling body
6~7日	6–7 days
发落	Hair losses
4~5日	4–5 days
身体肉色黄紧，微变	The flesh color of the body started to change slightly to yellow
半月以后	After half month
先从面上。肚皮、两胁、胸前肉色变动	The fresh color starts to change from the appearance of abdomen, two ribs, and chest
影响因素	Influence factors
肥胖人变化快	The changes are faster for fat people
因存放地不同而受影响	Influenced by different storage locations
南北方气候不同有影响	Influenced by different weather in South and North
山区不同于平原	Mountain area is different from plain

季节	时间	现象	影响因素
春	2～3 日	口鼻、肚皮、两胁、胸前肉色微青	1. 肥胖人变化快 2. 因存放地不同而受影响 3. 南北方气候不同有影响 4. 山区不同于平原
	10 日	鼻、耳内有恶汁流出，膨胀	
夏	1～2 日	先从面上。肚皮、两胁、胸前肉色变动	
	3 日	口鼻内汁流、蛆出、遍身胖胀、口唇翻、皮肤脱烂	
	4～5 日	发落	
秋	2～3 日	先从面上。肚皮、两胁、胸前肉色变动	
	4～5 日	口鼻内汁流、蛆出、遍身胖胀、口唇翻	
	6～7 日	发落	
冬	4～5 日	身体肉色黄紧，微变	
	半月以后	先从面上。肚皮、两胁、胸前肉色变动	

In addition, in the relevant discussion about the inquest of women corpses, Song Ci also mentioned the phenomenon of "childbirth in coffin":

> A pregnant woman was killed or died due to childbirth. Her corpse was buried in the cellar, but a dead fetus was found at the time of inquest. After investigation, it was found that since the water, fire, fire and wind blowing into the corpse buried in the cellar, its head was swelled and the joint was cracked open. And the fetus inside the abdomen came out. The umbilical cords were found under the feet of the corpse; Blood water flowed out from the vaginal orifice.

This paragraph is considered as the earliest record of childbirth delivery in coffin in forensic medicine around the globe.

7.3.2.2 Mechanical Asphyxia

In modern forensic medicine, mechanical asphyxia is divided into "external asphyxia" and "internal asphyxia." The former one means oxygen in the air cannot enter the pulmonary alveoli and carbon dioxide in the body cannot be exhaled; the latter one means lack of oxygen in blood or tissues cannot be used, mainly seen in disease, intoxication, etc. Mechanical asphyxia accounts for the largest proportion in asphyxia. Its causes include non-external compression (neck, chest, or abdomen) and blockage (pneumostome, respiratory tract). The *Xi Yuan Ji Lu* basically discusses the various causes of mechanical asphyxia, including self-hanging, strangling, drowning, blockage of mouth and nose by external compressions, and throttling.

Since self-hanging is the top one in suicide, it is extremely common. As a result, the identification of self-hanging and strangulation is always involved in the inquest. The *Xi Yuan Ji Lu* has full details about the various circumstances of self-hanging. The main point is to examine the ropes used by the hanger, the way of buckling, the hanging position, the body position, the rope groove, etc. At the same time, one should investigate the neighbors and figure out the cause and effect so as to reach the conclusion of self-hanging. In case of any suspicious things, the possibility of homicide should be considered. For example, pay attention to the "urine and feces to see whether there are swelling and stamped marks," which is also emphasized in modern forensic medicine, because attacking the vulva first to make people lose their defense is often seen in many homicide cases.

7.3.2.3 Mechanical Injury

In terms of inquest, the *Xi Yuan Ji Lu* basically follows the code of law since the Tang Dynasty that classifies injuries into three types: hand and feet, other objects, and weapon. Regarding the examination of injury, this book focuses on the identification of "before death" and "after death," which is one of the most important achievements of inquest. For example:

> If one were cut by a knife before death, the scar is broad with crossed irregular strips; If the scar is orderly, it is just a fake wound after death.
>
> If a living person were killed by a blade, the skin and flesh of the cut is tightened surrounded by blood stains; If a dead person were beheaded, the skin and flesh is like what they used to be, without blood stains. The skin of the cut is not tightened. There is no blood flow at the end of the blade cut and the color is white.

Even if it's injury of the hands and feet, it should have the blood stain near the bruise; then it could be determined as "beating injury before death."

As for the identification of self-injury and injury by others, this book points out that if there is evidence of fighting, it must be homicide. If it's homicide without fighting traces, then it must have hurt the vital party of the body. On the other hand, the injuries of suicide feature "heavy start and light end."

"Bone examination" is an important part of autopsy, mainly applied for skeletal remains that have decayed for years. In the relevant records about human bones in the *Xi Yuan Ji Lu*, there are some correct understandings obviously based on real observation but also mixed with many fallacies that should not have been generated. For example, the book mentions "There is one bone in the waist of men and women respectively, which is as big as palm with 8 holes and 4 lines," attached with a picture of that. This is obviously a correct description of the form of sacrum. But why those experienced in autopsy for years would say, "men's bone is white and women's bone is black" or "Men have 12 left ribs and 12 right ribs while women have 14 left ribs and 14 right ribs"? Also they said, "Cai Zhou people have one more piece of skull than those from other places." What are the reasons for such mistakes? So far there are no satisfying analyses yet.

During the process of bone examination, the bones must be steamed or boiled first, which triggers many different opinions. Jia Jingtao regarded it as a "cleansing method of filthy bones"; Ji Qingyi criticized this method as "unscientific" in the entry of "the Xi Yuan Ji Lu" he wrote for the *Encyclopedia of China·Volume of Law.* Afterwards, "the method of red umbrella" is used to inspect the bone of corpse: "If a part of bone is beaten, there will be a shade of red stain; at the broken part of the bone, the two connecting ends will both have bruises; if the scarred bone is shined in the sun and shows red color, then it shows that this person has been beaten before death." Jia Jingtao commented: "It proposed the concept of bone shade that still has research value so far"; Cai Jingfeng believes: "Umbrella can absorb some optical waves of the sunlight. So the light waves that pass through the umbrella are selective, which could have greater effects on the examination of the bone injuries." In addition, the book also mentions using oil, ink, cotton, etc. to check whether the bones are damaged and the "bone test" (the well-known method of "blood test" – if there is a blood relationship, the blood of the living person can penetrate into the bone of the deceased; if there is no blood relation, then the blood cannot penetrate). What needs to be clarified is that when the researchers commented on the methods of testing recorded in the *Xi Yuan Ji Lu*, except the "poison test method by silver hairpin" that can be empirically proved and rationally explained, or said certain method is reasonable or unscientific, be it commendable or derogatory, it seems not to be based on experimental evidences. Therefore, it remains to be seen whether there are empirical methods to justify those methods in the future.

After all, concerning the examination of corpse appearance and biopsy, the *Xi Yuan Ji Lu* does fully consider various possibilities of certain phenomenon. Therefore, it is necessary to refrain from making assertions. Instead, the inspector is required to conduct a comprehensive survey of the scene, interview the insiders, and then combine the comprehensive analysis of the inquest results to reach a correct judgment. Only by reading the whole book one could fully understand this point.

7.4 Comments on "Achievements"

Medical jurisprudence refers to the use of theories and technologies of medicine, biology, and other natural sciences to study and solve various problems such as violent death or injury encountered in judicial work, thereby forming a scientific and relatively independent discipline. The basic content and purpose of forensic inquest is to provide scientific basis for investigation and trial through on-site inspection, corpse check, bioassay, and physical evidence.

The inception of independent modern forensic medicine is generally believed to trace back to the sixteenth century when King Karl V of Germany issued a criminal order in 1532, stipulating light punishment for injuries, expressly requiring doctors to participate in inquest as experts, and allowing doctors to perform autopsy. Then the doctors of the university would discuss and evaluate the appraisal of the participating doctors and publish the results publicly, so that the medical examination gradually forms a unique knowledge system based on the integration of various medical disciplines. However, in ancient China, the situation was of course different. The history of judicial "examination" was very long, but it was always carried out without the participation of doctors. Therefore, in the mid-thirteenth century, the *Xi Yuan Ji Lu*, a more systematic work containing rich contents equivalent to forensic examinations, was produced. In addition to Song Ci's personal contributions, it is closely related with the time-honored inquest system, accumulated experiences, and objective needs in ancient China. From the content to the compilation style of the book, all the elements clearly serve the purpose of "inquest." In other words, the *Xi Yuan Ji Lu* is better to be regarded as a compilation of cases rather than an "ancient forensic works." As it were, the *Xi Yuan Ji Lu* marks the ancient Chinese judicial examination system reached a mature level. This judicial examination system has the following basic characteristics.

7.4.1 "Inquest System" Is Not "Medical Jurisprudence"

Although there are records of allowing doctors to identify patients in the *Shui Hu Di Qin Mu Zhu Jian*, this is actually only a relatively special case – because it does require very professional medical knowledge, which needs doctors to participate. For this specific case, the professional medical knowledge is required, and even the behavior in the entire process of judgment is no different from the doctor's daily diagnosis and treatment – if the doctor participated in the autopsy and the identification of blood stain, then the nature would be totally different, and this type of work could satisfy the definition of "forensic science" at that time. Therefore, generally speaking, medical jurisprudence in ancient China (especially autopsy) was basically conducted by officials and full-time inspectors – "post-mortem examiner." Therefore, although from the perspective of modern "forensic science," this knowledge system can only be incorporated into the scope of medical jurisprudence. Considering the real situation in ancient China, be it the subjective consciousness of judicial personnel or objective facts, there was no direct and clear connection between forensic science and medicine in terms of both form and content. Some knowledge related to human physiology and postmortem change is independently developed and gradually accumulated in the "inquest" system. Since researchers crowned the *Xi Yuan Ji Lu* as the world's first forensic works, all the ancient case judgment works and case compilations have been included in the "forensic classics," which is really far-fetched. If we always follow the classification rule of modern Western natural sciences to analyze and study the ancient Chinese historical materials, we could naturally see a development sequence of "forensic medicine"; but if we view the inherent Chinese culture as a whole, it is not difficult to see the basic characteristics of this judicial inquest system. Although the experiences of judicial examination are getting more and more abundant, and some knowledge of human anatomy, physiology, and pathology is actually involved, these are merely from judicial inspectors' common senses and have nothing to do with the medical system that aims for treatment. Therefore, it is not possible for the judicial inspectors to have the desire and motivation to merge the two systems and establish a new knowledge system.

7.4.2 Limit "Surface Inspection" Not "Feudal Ethical Restriction"

The ancient Chinese judicial examination system mainly focused on appearance inspection. Although the examination of bones was attached great importance, it normally happened after the corpse decayed. To some extent, it was still a kind of appearance evidence collection. It's generally believed that it was restricted by feudal code of ethics, but this was not the root cause of the problem. In the relevant discussions about judicial examination, the ideas of respecting corpse or not injuring the corpse arbitrarily were not seen. Since the bones could be boiled for inspection, why are there any feudal ethical restrictions at all? In addition, the correct definition of "feudal ethical codes" is not as superficial as "covering the corpse with clothes"; instead, it should be seen that the true implementation of "feudal ethical codes" includes conducting stringent inspection, readdressing mishandled cases, and punishing the culprits. Plus, even in the *Xi Yuan Ji Lu*, a masterpiece that emphasizes on detail orientation, personal experience, and hands-on experiment, there still remain many errors concerning the description of skeleton morphology! For example, "Men's bones are black and women's bones are black"; "Men have 12 left ribs and 12 right ribs; Women have 14 left ribs and 14 right ribs". Also, Cai Zhou people have one more piece of skull than those from other places and so on. For the inspectors who always opened coffins to examine the bones, how could they make such mistakes? It's really worth wondering. This also implies the potential influence of Chinese traditional culture mindset, namely, paying too much attention to "appearance" and believing "the inside content will be reflected in appearance". Priority was given to search the cause (injury) directly related to the result (death), instead of emphasizing on the knowledge system that aims for pure basic morphology research like anatomy – this is the ultimate reason why the judicial inquest system in the past dynasties mainly focused on the "appearance inspection."

7.4.3 The Conscious Amendment of "Presumption of Guilt"

The development of judicial inquest system implies the amendment tendency of "presumption of guilt" in lawsuit settlement. The so-called presumption of guilt means the judicial officers in their subjective consciousness determined that the defendant was guilty from the beginning. Accordingly, during the trial process, they would try every means to make the defendant confess guilty. Even if "the defendant's confession was extracted under torture," it was also considered as extremely reasonable, and the case was ended successfully. During the early years of Western bourgeois revolution, the "presumption of guilty" was proposed to target the arbitrariness and imperiousness of feudal judges and religious courts, namely, the principle that the defendant in criminal proceedings should be temporarily regarded as innocent before the court makes a final verdict of the crime. Since the Yi Yu Ji, such as the Nei Shu Lu, the *Xi Yuan Lu*, the *Ping Yuan Lu*, the *Wu Yuan Lu*, and other works, those titles are obviously opposite to the "presumption of guilty," indicating the authors' request that judicial officers should judge the case through careful inspection and substantial evidences in order to punish the culprits and readdress the mishandled cases. In other words, the amendment of "confession to false charges under torture" and "presumption of guilty" in ancient Chinese judicial system is different from that of the Western world driven by external revolution. Instead, such amendment in China was gradually materialized through the establishment, improvement, and scientific development of judicial examination system. This is actually the ultimate value of ancient Chinese judicial system for social progress and civilization development.

7.4.4 The Development After "Peak" Awaits "Revolution"

The relevant works after the Song Dynasty were basically based on the *Xi Yuan Ji Lu*. So it is undoubted that Song Ci's contributions to the judicial inquest in the Yuan, Ming, and Qing dynasties should be commended. This phenomenon illustrates two problems: Firstly, the *Xi Yuan Ji Lu* marked the establishment of the independent knowledge system of the ancient Chinese "judicial inquest science". These experiences sorted by Song Ci were no longer scattered in case records, but compiled into judicial inquest books after the Song Dynasty. Secondly, the reason why the *Xi Yuan Ji Lu* became the main content of judicial examination works in the later dynasties is that this book represents the culmination of the development of ancient Chinese judicial examination technology. If it does not break through the barriers of appearance inspection, and if there is no introduction of modern western forensic knowledge that truly has the inherent meaning and definition of "forensic medicine," namely, without the involvement of modern anatomy, physiology, pathology, and medicinal chemistry knowledge, there will be no possibility to "aim higher." Therefore, although there were various collections of evidences, detailed definitions, supplementary notes, and sorting works such as verses and record tables in the Qing Dynasty, some of which were quite innovative, the basic model was still the *Xi Yuan Ji Lu* (Fig. 7.4).

7.5 Different Versions and Their Influence on Later Generations

After the *Xi Yuan Ji Lu* was compiled, it was initially published in Hunan in the in the 7th year of Emperor Chunyou's reign (1247). However, the Song Dynasty's edition failed to be handed down from past generations. The Yuan Dynasty's block-printed edition of the *Song Ti Xing Xi Yuan Ji Lu* collected by Peking University is the existing earliest version. Plus, Sun Xingyan in the Qing Dynasty copied this book according to the Yuan block-printed edition in the 12th year of Emperor Jiaqing's reign (1807) and compiled it into *the Collection of Dai Nan Ge* (岱南阁丛书). This book was re-published by the Commercial Press in 1937 and compiled into the "First Book Series Collection," which became a widely circulated and easily available version.

Fig. 7.4 Several annotated editions of the *Xi Yuan Ji Lu* in the Qing Dynasty. (Collected by the Natural Science History Research Institute of Chinese Academy of Science)

In the Qing Dynasty, Wu Na published the *Song Yuan Jian Yan San Lu* (Inquest Records of the Song and Yuan Dynasties) including the *Xi Yuan Ji Lu* in the 17th Year of Emperor Jiaqing's reign (1812). Its original version also came from Sun Xingyan. In 1981, Shanghai Science and Technology Press published Jia Jingtao's proofread and annotated edition of traditional Chinese characters based on the Yuan block-printed version. All of the above are the readable texts for the study of the *Xi Yuan Ji Lu*. The reason for explaining the version issue is that, although the *Proofread Version of the* Xi Yuan Ji Lu issued in the 33rd year of Emperor Kangxi's reign (1964) of the Qing Dynasty was mainly based on the *Xi Yuan Ji Lu*, the content was actually slightly different and was often cited as Song Ci's works by mistake. The most common mistake is the mentioning of the achievements in toxicology and first aid of the *Xi Yuan Ji Lu*. In fact, these contents were not originally included in Song Ci's works.

About the influences of the *Xi Yuan Ji Lu* on the later generations, there are two aspects: domestic and abroad. In China, during the Song and Yuan dynasties Zhao Yizhai's *Ping Yuan Lu* was published. In the Yuan Dynasty, Wang Yu's *Wu Yuan Lu* was printed and issued (1308). According to the investigation, there are as many as 272 indications of "the same as the *Xi Yuan Lu* and the *Ping Yuan Lu*" among 42 items in the second volume of the *Ping Yuan Lu*. As can be known, the *Ping Yuan Lu* is undoubtedly based on the *Xi Yuan Lu Ji*. Of course, the *Wu Yuan Lu* also included a large number of the *Xi Yuan Ji Lu*'s contents. In the Ming Dynasty, there emerged all kinds of inquest works which were mainly based on the *Xi Yuan Ji Lu* and supplemented by the *Wu Yuan Lu*. The *Lu Li Guan Jiao Zheng Xi Yuan Lu* in the

Qing Dynasty was based on the *Xi Yuan Ji Lu* and added other experts' experiences after Song Ci, which became the official book issued by the court.

Outside China, although there are many translations of the *Xi Yuan Ji Lu* in multiple languages, those versions basically adopt the *Jiao Zheng Xi Yuan Lu* (the proofread version) by Wang Youkui in the Qing Dynasty and compiled into the *Xi Yuan Ji Zheng* (1796). Therefore, the spread of Song Ci's *Xi Yuan Ji Lu* abroad is not that straightforward as many medical works or article mentioned. The ancient scholars have already conducted researches on this. I would like to quote Jia Jingtaoshi's article "Ancient Chinese Forensic Medicine Abroad" as a reference for all the readers.

> The spread of Chinese forensic works abroad firstly started in the Ming Dynasty. The republished Edition (the 17th Year of Emperor Hongwu's Reign in the Ming Dynasty, 1384) of the *Wu Yuan Lu* by Yang Jiao Shan Sou was first introduced to North Korea. Due to the language relationship, the third generation of Sejong of the Li Dynasty ordered the civil minister Cui Zhiyun and others to add phonetic annotations for this book, which was completed in the third year of emperor's reign (1438). This is the Korean version of the *Xin Zhu Wu Wu Lu* (the newly annotated Wu Yuan Lu), which was applied for over 300 years in forensic inquest in North Korean. In the 20th year of Emperor Yingzu's Reign (1744), Ju Zhaikui added and deleted some annotations referring to the *Lu Zheng Guan Jiao Zheng Xi Yuan Lu* and named the new version as the *Zeng Xiu Wu Yuan Lu* (the *Wu Yuan Lu* with additional revisions). In the 16th Year of Emperor Zhengzu's Reign (1792), there was a Korean translation of the *Zeng Xiu Wu Yuan Lu* by Chu Youling. In the 20th Year of Emperor Zhengzu's Reign (1796), Ju Yunming and others re-edited and published the *Zeng Xiu Wu Yuan Lu*. The *Wu Yuan Lu* is not only a specialized book for judicial inquest in North Korea, but also an examination subject for the appointment of judicial officials. From the 5th Year of Emperor Taizong's Reign (1405) when this examination subject was established to the Founding Year of Emperor Xilong's Reign (1908) when the *Wu Yuan Lu* was listed as invalid decree, it's been approximately 500 years.
>
> The development of Japanese forensic medicine was later than that of North Korea. It was indirectly influenced by China through North Korea. The *Xin Zhu Wu Yuan Lu* was introduced into Japan during the Tokugawa era. The earliest forensic medicine book written by Japanese people was published during the 5th Year of Emperor Minghe's Reign (1768), which is the translation version of the *Wu Yuan Lu Shu* by Kawai Jinbei and Gen Naohisa. Afterwards, it has been re-published for multiple times and renamed as the *Bian Si Shang Jian Shi Bi Xie Wu Yuan Lu Shu* (The *Wu Yuan Lu Shu* – A Must for Inspecting deaths and injuries) in the 24th Year of Emperor Meiji (1891). At that time, there were many European forensic medicine works published in Japan, but that book was reprinted for as many as 6 times during the decade till the 34th Year of Emperor Meiji.
>
> Japanese forensic historian Yamazaki (1941) pointed out that the *Wu Yuan Lu Shu* is "an important book in the history of Japanese forensic medicine, which played a significant role in the history of trial and judgement." The *Wu Yuan Lu* and its translated version written by Japanese inspectors were circulated among the public in Japan. The most famous one is the *Jian Shi Bian Yi* (The Prosecutor's Discrimination), which was written approximately in the Yuanwen era (1736~1740). Different from the *Wu Yuan Lu*, the *Lu Li Guan Jiao Xi Yuan Lu* was directly introduced to Japan, and the Japanese translation was published in the 10th Year of Emperor Meiji's reign (1877), called the *Jian Shi Kao* (Autopsy). Among the Japanese translations, there are the *Fu Hui Quan Shu* (1850) translated by Obata Xingxun, the *Fu Hui Quan Shu He Jie* (1874) translated by Keizo Kondo, and the *Ping Yuan Lu* (1931) translated

by Dong Bingran. The latter is based on the translation of the *Ping Yuan Lu* in the *Song Yuan Jian Yan San Lu*, which is actually the second volume of the *Wu Yuan Lu*.

The *Xi Yuan Lu* was the main ancient Chinese forensic works introduced to Europe mostly after the Opium War, when the European translated or introduced it based on Wang Youkui's (1796) *Xi Yuan Ji Lu Zheng* or *Bu Zhu Xi Yuan Ji Lu Zheng*. Among them, there are 6 translations in 4 countries including Holland (De Grijs' translation, 1862), Germany (Breitenstein's transaltion, 1908; Hoffman's translation), France (abridged translation, 1779; Martin's translation, 1882), and the UK (Giles' translation, 1873). Plus the abovementioned 3 North Korean versions and 7 Japanese versions, Chinese forensic medicine works altogether have 16 versions and translations in 6 overseas countries.

One problem should be paid attention to: in the Chinese character culture circle, the purpose of North Korean and Japanese translating Chinese books is drastically different from that of European translating this kind of works after the Opium War. The former is for study and use, while the latter is mainly for introducing the oriental culture.

Physics Knowledge and Experiments in Ancient China

8

Shuyong Liu and Zengjian Guan

Contents

S. Liu
Department of Physics, Capital Normal University, Beijing, China

Z. Guan (✉)
School of History and Culture Science, Shanghai Jiao Tong University, Shanghai, China

X. Jiang (ed.), *The High Tide of Science and Technology Development in China*, History of Science and Technology in China,
https://doi.org/10.1007/978-981-15-7847-2_8

Abstract

In ancient China, there was no physics in the contemporary sense. However, in the long process of exploring nature, the ancients gradually developed some technologies related to physics today and did some distinctive experimental work, which promoted the development of related knowledge and deepened people's understanding of nature. These technologies and experiments played different roles in the corresponding research fields of modern physics. This chapter presents a brief introduction based on the classification of modern physics research fields.

Keywords

Technologies and experiments · Mechanics · Acoustics · Optics · Electromagnetics · Thermal science

In ancient China there was no physics in the contemporary sense in ancient China. However, in the long process of exploring nature, the ancients gradually developed some technologies related to physics today and did some distinctive experimental work, which promoted the development of related knowledge and deepened people's understanding of nature. These technologies and experiments played different roles in the corresponding research fields of modern physics. This chapter will present a brief introduction based on the classification of modern physics research fields.

8.1 Mechanics

8.1.1 Testing of Bow Elasticity

The bow (including the crossbow) was an important weapon in ancient times. Given the importance of the bow and crossbow, the ancients tirelessly explored ways to make them and developed methods of quantitatively testing the elasticity of the bow. These methods contain a rich vault of physics knowledge.

The ancients measured the elastic force of the bow body for two purposes. One was to obtain the rated elastic force of the bow body (i.e., the elastic force produced by the bow body when the bowstring was pulled to a specified length), which was used as a quantitative standard to characterize the bow's performance. The ancients often denoted the elasticity of the bow in units of weight such as jun or stone. Shen Kuo of the Song Dynasty said: "The ancients used stone as the rate of turning the crossbow."(Shen Kuo: *Meng Xi Bi Tan* (《梦溪笔谈》) (Vol.3)). Another purpose was to use the bow force measurement to ensure that the production of the crossbow would meet the requirements. "Kao Gong Ji" (《考工记》) had already dealt with the measurement of bow elasticity. The entry of "Bow Producer Makes Bows" (Shen Kuo: *Meng Xi Bi Tan* (《梦

溪笔谈》) (Vol.3)) said: "Jiao cannot be more elastic than gan, and gan cannot be more elastic than jing, which is called Trinity Equilibrium. If these three are equally elastic, this status is called Nine Harmonies." The so-called jiao (or angle), gan (or trunk), and jing (or tendon) were all materials for making the bow. Upon making, stiff wood or bamboos were used to make the stem, with the horn as the inner lining, which is wrapped with tendons outside, and strings were added to complete the bow. The horn, the stem, and the tendon all have impact on the elasticity. "Jiao cannot be more elastic than gan, and gan cannot be more elastic than jing" means the ideal condition of equilibrium of the three ("Trinity Equilibrium"). When the three contribute equally to the bow's elasticity, the bow reaches the optimal performance. Measurement is required to ensure this could be achieved through bow making. "Measuring the force" suggests people in those times had mastered the method of testing bow elasticity.

How to ensure bow production could achieve the above two purposes by determining bow elasticity? When annotating this entry in Kao Gong Ji (《考工记》), Zheng Xuan of Eastern Han Dynasty provided the clue to answer the first question. According to Zheng Xuan's annotation, when measuring the elasticity of the bow body, ancients first loosened the tight string on the bow to keep the string in a loose state, tied the two ends of the bow with a rope to keep the bow free from force, hung a heavy item on the rope, and adjusted the weight of the item to stretch the bow string by around 93 cm and the weight of the object at this time represented the elasticity of the bow. If the object is three stones (c.19 kg), then the elasticity of the bow is three stones, which was the "bow of three-stone elasticity" called by Zheng Xuan. In this way, the rated elasticity could be ensured.

Jia Gongyan of the Tang Dynasty added a further explanation to Zheng Xuan's annotation: "When the bow is not complete, there is no jiao, and its elasticity is one stone; after jiao is mounted, its elasticity is two stones; afterwards, jing is mounted, and its elasticity is three stones. The three-chi bow is an aggregation of gan, jiao and jing, with elasticity of three stones and a length of three chis. The primitive bow has elasticity of one stone and the length remains three chis; with jiao added, its elasticity becomes two stones and the length remains three chis; with jing added, its elasticity is three stones and the length remains three chis. Zheng said if we make the bow elasticity three stones and make the bow three chis, this is a three-stone powerful bow. To determine whether the bow elasticity is three stones, one can relax its string, tie a rope and add objects. When one stone of weight is added, the string stretches one chi; two stones, two chis; and three stones, three chis."

According to Jia Gongyan's explanation, the determination process is as follows: At the beginning, a suitable bow trunk must be selected so that it will be stretched to 100 cm (three chis) when a heavy object of 6.35KG (one stone) is hung. Then, place jiao on the bow trunk, after which an object of 12.7KG is hung to stretch the bow trunk to 100 cm. Afterwards, jing (string) is attached to the bow trunk, and a 19 kg object is hung, and if the bow trunk is also stretched to 100 cm, it suggests jiao, jing, and gan have the same effect on the rigidity index of the bow, which reaches the Trinity Equilibrium said by the ancients and ensures the bow made meets the quality requirements.

This approach makes sense. We know that when an object undergoes elastic deformation, the amount of external force it receives is equal to the product of its rigidity coefficient and deformation. In the above measurement, the bow's deformation variable remains unchanged, and so the change in the stiffness coefficient of the bow is uniquely proportional to the weight of the suspended object. Therefore, the weight of the suspended object can be used to directly determine whether the combination of the angle, tendon, and stem meets the requirements, so as to ensure the bow made reaches the required performance. The rated elasticity of the bow made through this procedure is certain, which makes it comply with the requirements of standardization and universality. For ancient wars, this is very important.

From the perspective of the history of physics, the ancients' determination of the bow's elasticity is very meaningful. Under normal circumstances, a weighing instrument used to measure the weight of an object actually measures the mass, but what is measured here is the actual force. The weighing of a weighing instrument is based on the principle of leverage, while the measurement of elastic force follows the law of elasticity (also called Hooke's law), which is different.

In addition, the ancients' method of limiting the length of the bow to three chis upon measurement is also very clever. On the one hand, the rated elastic force of the bow can be directly obtained. On the other hand, it also saves the trouble of each measurement when making the bow. The calculation of the deformation variable does not need to consider whether the deformation is within the elastic limit, and it can ensure that the manufactured crossbow meets the requirements, which is very scientific.

It was of particular importance that Zheng Xuan's comment put forward: "With every additional shi in weight, the bow string stretches by an additional chi." Jia Gongyan's annotation further explained: "With one additional shi, there is an additional chi; two shis for two chis; and three shis for three chis." These notes meant they had linked the weight of the bow to the extent to which it was pulled. If they had followed this path, the discovery the law of elasticity would have been possible. However, due to the lack of the tradition of exploring quantitative natural laws in the Chinese traditional culture, the ancients failed to go further in this direction.

It is noteworthy that the ancients also used testing of bow elasticity as a method in bow design. The concrete example is as follows:

> When Wei Pi was in charge of weapon production, the top brass ordered him to make bows with a range of 1000 steps instead of 700 steps of traditional bows, and he sought a ruler from Xin. Xin ordered suspension of the bow at a rack, place heavy objects on both ends. The objects are compared and the weight of one object is increased by a half for the new bow, which should reach a range of 1000 steps. When the bow is made as such, it can truly reach the range of 1000 steps (Jiang Shaoyu: *De Liang Zhi Shi* ("德量智识"), *Shi Shi Lei Yuan* (《事实类苑》) (Vol.14)).

The above is a story that took place during the Northern Song Dynasty period as recorded by Jiang Shaoyu of the Southern Song Dynasty in his Shi Shi Lei Yuan (《事实类苑》). It reflected an attempt of the ancients to improve the bow range by determining bow elasticity. Such an attempt seems logical, as the bow's range is definitely related to its elasticity. To extend the range, you must start with elasticity. We will analyze here whether this attempt by the ancients makes sense.

In this story, the original range of the bed crossbow was 700 steps, and the emperor ordered it to be increased to 1000 steps, or 1.43 times and nearly 1.5 times the original. To this end, the designer's solution was to increase the elasticity of the crossbow by 50% to 1.5 times the original. This means that in the eyes of the designer, the range of the crossbow is linearly related to its elasticity.

What is it like in reality? We know that the range of the crossbow mainly depends on the initial velocity v of the crossbow arrow, which comes from the elastic potential energy of the crossbow being transformed into the kinetic energy of the arrow during the launch process. If the stiffness coefficient of the crossbow is k, the rated deformation is x, and the mass of the arrow is m, then its rated elastic potential energy is W = 1/2kx^2 (assuming that the deformation of the crossbow is completely elastic), and it is assumed this potential energy is completely transformed into the kinetic energy of the arrow E = 1/2mv^2 upon launch. Obviously, when the deformation variable x of the crossbow is constant (because the length of the crossbow arrow is constant), the kinetic energy of the arrow is proportional to the stiffness coefficient k of the crossbow, namely the initial speed v of the arrow is directly proportional to the square root of the stiffness coefficient 1/k. Thus, as long as the relationship between the range of the crossbow and the initial speed of the arrow is known, the relationship between the elasticity and the range can be derived.

Given the testing conditions available to the ancients, the crossbow was supposedly placed on the ground and leaned forward at an elevation angle of α, with its range tested. We know from physics that the theoretical range of the crossbow bolt at this time is:

$$S = \frac{v^2}{g} \sin 2\alpha$$

With the launch angle constant, the range S is uniquely proportional to the square of the initial velocity v of the arrow, that is, S is proportional to the initial kinetic energy of the arrow. According to previous analysis, when the deformation of the crossbow is constant, the kinetic energy E of the arrow has a linear relationship with the stiffness coefficient k of the crossbow. Therefore, the range S and the stiffness coefficient k also have a linear relationship. Obviously, if the elastic force of the crossbow is increased to 1.5 times the original under the condition of constant deformation, it means that k is increased to 1.5 times the original, so the range S will naturally reach 1.5 times, too. In this case, the design described in this historical record becomes feasible.

In the above discussion, we did not consider the air resistance and the inelastic deformation of the crossbow, which is fine in this kind of rough estimation. Through the above analysis, we know that although the ancients did not clearly express and record their understanding of the relationship between the range of the crossbow and its elasticity, they did a quantitative discussion of it after all, found the correct connection between the elasticity and the range of the crossbow when it is launched in the inclined upward direction and verified it in practice, which should be acknowledged. This is also an important theoretical result obtained by the ancient Chinese who measured the elasticity of the bow.

8.1.2 Understanding of the Principle of Leverage

As the mechanics knowledge of ancient Chinese developed, the ancients finally reached an understanding of the very important principle of leverage in physics through a large number of experiments, which is noteworthy. What we want to verify here is when the ancients achieved an understanding of this important principle.

The ancients began to use leverage a long time ago and knew from experience that using leverage could save effort. They called leverage the Shaduf. The pre-Qin work "Zhuang Zi" (《庄子》) recorded such an event:

> Zigong traveled southward to the Chu Kingdom and returned to Jin through Hanyin, where he saw an elderly man making a pond by drilling a tunnel into the well and holding an urn for irrigation. He made a great effort, which did not quite pay off. Zigong said: "there is a type of machinery that can irrigate hundreds of acres and uses little effort for a great effect. Don't you want that?" The pond maker looked up and said: "How?" Zigong replied: "The machinery is made of wood and is heavier in the rear. It pumps water like ladling soup, called a pulley." (*Zhuang Zi – Wai Pian – Sky and Earth No.12* (《庄子·外篇·天地第十二》).

The "pulley" here is the shaduf. According to the "little effort" said by Zigong, people then clearly knew leverage saves strength. But even if they knew leverage saves strength and knew how to use it, it does not mean they had mastered the principle of leverage. The principle of leverage is a quantitative physics principle, and to be considered to have mastered it, one must understand the quantitative relations involved. It was during extensive weighing practice that the ancient Chinese mastered the principle of leverage.

In terms of weighing, the ancients first used the balance, and later the steelyard. The result from these weighing instruments is actually the mass of the material, but people always regard the weighed mass as the weight of the object. In order to correspond to the state of knowledge of the ancients, we do not distinguish between the two in the narrative.

"For balances and weights, balances are about leveling; and weights are about heaviness. A balance is leveled because the objects on both sides are equally heavy." Balances include equal-arm balances, unequal-arm balances, and steelyards. Accordingly, weights are the objects of known weight for balancing a scale in weighing other objects. The three types of balances evolved during a process, which was closely related people's deepened understanding and smart application of the leverage principle.

Weighing instruments appeared very early, and the initial form was an equal arm balance. At present, an earlier and complete weighing instrument from archaeological excavations is the Muheng copper ring weight in a tomb of the Chu Kingdom in the Warring Kingdoms Period in the Zuojiagong Mountain, Changsha, Hunan. It is an equal-arm balance. The rings were made in fine detail and formed a complete set of sizes, obviously used as weights. This suggests that during the Spring and Autumn Period and Warring Kingdoms Period, equal-arm balances were widely used. "Han Shu – Lv Li Zhi" (《汉书·律历志》) said: "When the rings and the item are equal in weight, balance is achieved." If the rings and the item are of an equal weight, the balance will be flat and not inclined. This obviously refers to an equal-arm balance.

Huai Nan Zi – Zhu Shu Xun (《淮南子·主术训》) said: "The balance can be leveled because the weights on the left and right sides are almost equally heavy. . .the difference in weight is negligible." This also refers to an equal-arm balance with high precision. The smallest of the copper balances unearthed at Zuojiagong Mountain weighs less than 1 gram, which suggests the balances at that time did reach a high level of precision.

The appearance of the equal-arm balance was the result of intuitive thinking. An equal-arm balance has the support (or overhang) point in the center, and the two arms are equal in length; if items of different weights are suspended at the two ends, it will inevitably tilt to the heavier side. This is very intuitive. Huai Nan Zi – Shuo Shan Xun (《淮南子·说山训》) said: "with equal weights, the balance does not tilt," which referred to this phenomenon. Shen Zi (《慎子》) said: "if the balance tilts leftward, the right side is heavier, and vice versa. The light and heavy sides overlap, which is the rule of the nature." It was based on this "rule of the nature" that the ancients made the balance. Evidently, the appearance of equal-arm balances cannot be used as proof that the ancients had understood the principle of leverage. Only when the ancients were able to weigh with unequal-arm balances or steelyards can we conclude that they had mastered the principle of leverage.

Equal-arm balances had been used for such a long time that even Han Shu – Lv Li Zhi (《汉书·律历志》) recorded it as a primary form of weights, namely the "five-weight system": "weight units include zhu, liang, Jin, Jun and shi (stone), used to weigh items . . .The five-weight system uses objects for weighting, and the difference in their size depends on the weight." This five-weight system is a large balance system and the balances used could weigh one stone each.

When the ancients explored new forms of balances, unequal-arm balances emerged as the preliminary product of this exploration. The earliest discussion of the unequal-arm balance appeared in Mo Jing (《墨经》), which said the balance must be leveled upon weighing of an object, because the object and the weight must be of an equal weight. If a heavy object is placed on one side of the balance, the balance will naturally tilt, and the balance can only be balanced when the weight of the object is equal to that of the weight. If the two arms of the weighing instrument are of unequal lengths, and weights and objects of an equal weight are placed on both sides at the same time, the longer arm will naturally be inclined downward due to a greater weight. The second half of this paragraph involves the working principle of the unequal-arm balance, and it is self-evident that to be able to use the unequal-arm balance for weighing, the user must understand the principle of leverage.

It is better if the above interpretation of Mo Jing (《墨经》) can be verified by other literature. Indeed, we also found records of ancients using unequal-arm balances for weighing in other literature. Around the Northern and Southern Song Dynasties period, Wu Zeng's Neng Gai Zhai Man Lu (《能改斋漫录》) quoted a paragraph about weighing with an unequal-arm balance from Fu Zi (《符子》). The original text is "Fu Zi (《符子》) said: barbarian tribes gave a big swine to the king of the Yan Kingdom. . .the king ordered keeping of the swine for butchering. After 15 years, the swine was so big that its feet could hardly support the body. The king, surprised, ordered the weighing official to weigh it, and with ten bridges used, the swine's weight was still beyond measure. So the king ordered the river official to measure with a floating boat." The "weighing official" was an official responsible for weighing. "Bridges" were the so-called shadufs used in the Warring Kingdoms period. Liu Xiang's Shuo Yuan – Fan Zhi (《说苑·反质》) recorded: "a mechanism is made, with a heavy rear and a light front, which is called a bridge." Therefore, "measuring with a bridge" means weighing using the leverage. This record, in combination with Mo Jing (《墨经》), suggests people at that time had known the principle of leverage.

The strongest evidence should come from archaeological discoveries. Among the existing ancient balances in our country, the two Warring Kingdoms bronze balances in the National Museum of China were used as unequal-arm balances. The scales of these two balances were flat, with a length of one chi during the Warring Kingdoms Period and a nose button in the middle, which could be suspended for use. There is groove-like wear in the buttonhole, which shows that it had been used for a long time. On the front of the balance, there is a decimal scale line running through the balance. The word "king" is engraved on the balance, so people call it the "king" balance. The balance's nose button is in the middle, which is consistent with the characteristics of an equal arm balance. However, the equal-arm balance does not mark the scale line on the balance surface, and the decimal scale line running through the balance surface gives it the function of an unequal-arm balance. When in use, the object and the weight are, respectively, suspended on the two arms, and a certain position is found to balance them. From the scale of the suspension position and the weight of the weight, the weight of the object can be calculated (see Fig. 8.1). Liu Dongrui, a cultural relic expert, did research on it and proved that these two weighing instruments can be used as unequal-arm balances.

The "king" copper balance and the above literature prove Chinese ancients had mastered the principle of leverage no later than the Warring Kingdoms period.

Using unequal-arm balances for measurement has certain advantages over equal-arm balances. First of all, this was a breakthrough in principle. It signified that the ancients had mastered the principle of leverage, which laid a scientific foundation for the improvement of weighing instruments; secondly, the measurement data became discrete rather than continuous; finally, the measurement range had been greatly increased and was no longer limited to the magnitude of weights. But this method also had its inconveniences in that calculation was required every time to get the

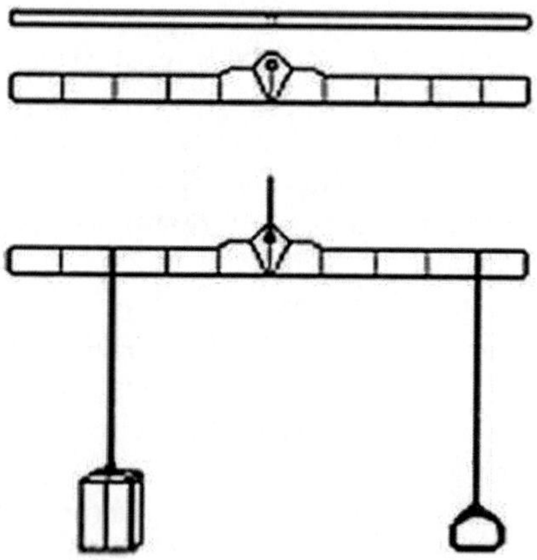

Fig. 8.1 Illustration of the "king" copper balance and weighing with a "king" copper balance

result, which limited its popularization. After a long period of exploration, a new form of weighing instrument, the tie rod scale, appeared. The appearance of this kind of weighing instrument marked that the ancients' use of the principle of leverage had reached the point of perfection. For weighing instruments, the mathematical formula expressing the principle of leverage can be written in layman's terms as:

$$\text{Weight of weight} \times \text{weight arm} = \text{item weight} \times \text{item arm}$$

For the "king" copper balance, the weight of the weight is fixed, and the other three factors are variable. Therefore, calculations are necessary for determining object weights.

In the tethered steelyard, the weight is fixed, and the length of the arm is fixed. In this way, the change in the weight of the object is proportional to the change in the length of the weight arm. Therefore, the weight of the object can be represented by the corresponding length of the weight arm, that is, the measurement of weight is transformed into the measurement of the length of the corresponding weight arm, which brings great advantages. In the Southern Song Dynasty, Chen Chun had a vivid explanation of the use of steel rod scales: "The word weight was derived from the weight of a weight. The weight of a weight was an object, which could be used to determine the weight and was therefore called weight. The weight is about change. If there is any difference in weight, the weight moves back and forth until leveled." (Chen Chun: *Bei Xi Zi Yi – Volume 2 – Jing Quan* (《北溪字义·卷下·经权》)). Moving back and forth is a feature of the approach of measuring the length instead of the weight. This feature makes weighing an easy task.

The measurement data of the balance is discrete, while the tethered steelyard acquires the continually distributed measurement data without the need for conversion, which is another advantage compared to the balance. However, the sensitivity of a steelyard is inferior to that of a well-designed balance, but this is not important for general civilian use. Given the various advantages of the tethered steelyard, it quickly turned popular after emergence and became the most common weighing instrument in ancient China. The steelyard is an example of the ancients' successful use of the principle of leverage and the crystallization of their wisdom.

8.1.3 Buoyancy Experiment and Technology

As early as in the pre-Qin period, the ancients knew about the floating feature of objects and had clever applications in production practice. For example, in Kao Gong Ji – Arrow Maker (《考工记·矢人》), the method adopted by the "arrow maker" to determine the proportions of each part of the arrow shaft is as follows: "Put the arrow into the water to determine the distribution of density based on the floating and sinking patterns of different parts of the arrow shaft, determine the proportion of the different parts of the arrow based on that distribution and install the feature to the arrow tail based on that proportion."

Kao Gong Ji – Wheel Maker (《考工记·轮人》) also mentioned the use of buoyancy in determining the quality of a wooden wheel. To make sure the wheel "has evenly distributed mass and floats and sinks evenly," the maker throws it into the water and checks if it sinks to the same degree. If yes, the wheel's mass is naturally evenly distributed.

In the pre-Qin period, people could also use buoyancy to quantitatively measure the weight of objects. The story of "measuring with a floating boat" recorded in the book "Fu Zi" (《符子》) of the Jin Dynasty cited earlier was about using the buoyancy of water to determine the weight of an incomparably heavy pig given by a barbarian tribe to the king of the Yan Kingdom. If the story in "Fu Zi" (《符子》) is true, this is a wonderful example of the ancients using the buoyancy of water quantitatively. Going further, it would be the famous story of Cao Chong weighing an elephant.

The ancients not only used buoyancy from an empirical point of view, but also theoretically discussed the floating and sinking conditions of objects. For example, Mo Jing (《墨经》) pointed out:

Although an object can be of a large size, it sinks shallowly in the water due to its hollowness. Jing Shuo (《经说》) explained: A hollow object shallowly sinks in the water not because the object itself is shallow, but the buoyancy of the sinking part is equal to the weight of the entire object. This is like exchanging items in the market, where five pieces of item A can be exchanged for one piece of item B. Of course, given the few words in Mo Jing (《墨经》), there are different interpretations of this article, but in any case, it is fair to say that the Mohists had realized that the buoyancy of an object is related to the volume of the submerged part.

Huai Nan Zi – Qi Su Xun (《淮南子·齐俗训》) took bamboos for example: "Bamboos float in nature, but sink when peeled into pieces and bundled, having lost the volume." As the bamboos are hollow, they naturally float in the water, and that is why "bamboos float in nature." If an intact bamboo is peeled into pieces, bundled and thrown into the water, they sink, because "it has lost its volume." In other words, with the bamboo's weight remaining unchanged, its volume is greatly reduced when made into pieces, and it meets greatly reduced buoyancy in water and naturally sinks. Obviously, Huai NanZi (《淮南子》) had realized the close connection between the buoyancy on an object and its drainage volume.

The ancients could also use buoyancy in clever ways. Song Shi – Fang Ji Zhuan Xia (《宋史·方技传下》) recorded an example of the monk Huaibin utilizing buoyancy:

> "A beam floats in the river, tied with eight iron bulls, each of which weigh hundreds of thousands of kilometers. Afterwards, the river rises and overwhelms the beam as well as the iron bulls. People try to find someone who can pull out the bulls. Huaibin used two big boats full of solid earth and tie them to the bulls. He used large lumber as weight to fish out the bulls. After the earth is removed, the boats float, so do the bulls." Huaibin was smart to utilize the buoyancy to lift heavy items.

The ancients also used buoyancy to determine the specific gravity of a liquid. The same object is immersed in different liquids, and its buoyancy is different. Duan Chengshi of the Tang Dynasty mentioned in You Yang Za Zu (《酉阳杂俎》): "lotuses sink in water, while brines can float." Obviously, this is due to the different concentrations of water and brine. Inspired by this, the ancients invented a method to determine the liquid concentration by observing the floating and sinking conditions of lotus seeds. Needham cited an account written by Yao Kuan in the eleventh century when researching the method for measuring saltwater concentration. When Yao Kuan was an official in Taizhou, he used lotus seeds to determine the concentration of bittern every day to investigate salt merchants' fraud. He chose heavier lotus seeds for use. If ten lotus seeds could emerge from the bittern, three or four were thick bitterns. If less than three seeds floated, the bittern was thin. If all the lotus seeds sank to the bottom of the brine, the brine was extremely thin, and even after evaporation, there was not much salt. This method not only made clever use of buoyancy, but also involved certain mathematical statistics.

In Ao Bo Tu Yong (《熬波图咏》), Chen Chun of the Yuan Dynasty recorded an instrument used specifically for testing brines. The lotus seeds used were called "lotus tubes."

> The lotus tube method is to harvest and dip lotus in silt and soak it in four concentrations of brine: the saltiest brine for one spot, three-quarters of brine and one quarter of water for one spot, half water and half brine for one spot and one-third brine and two-thirds of water for one spot. Afterwards, put four lotuses soaked in these four concentrations of brine into four bamboo tubes, and seal the ends so lotuses do not get out. Suck brine with a lotus tube and determine the concentration of the brine based on the floating/sinking of the four lotuses.

The concentration of bittern can be graded with this instrument. This "lotus tube" is similar to a modern float-type hydrometer, in which four lotus seeds are equivalent to the colored balls with different specific gravities, and the specific gravity of the liquid can be judged according to the floating and sinking conditions of these small balls.

At the end of the Ming Dynasty, Fang Yizhi recorded his teacher Wang Xuzhou's observation of the floating and sinking of gold, silver, copper, and iron in the mercury liquid: "Xuzhou said: 'Materia Medica' said that gold, silver, copper and iron float on top of mercury, which is not true. Copper and iron float, while gold and silver sink. Gold and silver will be consumed lightly when taken out and will be eroded." (Fang Yizhi: *A Little Physics Knowledge* (《物理小识》) (Vol.7)). Wang Xuzhou's observation is accurate, which reflects the ancients' in-depth research on the floating and sinking of objects with different specific gravities. This passage also describes the corrosive effect of mercury on gold and silver, which is also valuable in the history of chemistry.

8.1.4 Spray Fish Basin

The spray fish basin is an ancient copper basin similar to the modern washbasin. There are four fish patterns engraved in the basin, and there is a pair of lifting ears on both sides of the upper edge of the basin. The ear is set in such a way that it is easy to lift the basin, while it also serves another function, that is, when the palm of the hand rubs against it, it will make a humming sound, just like the sound that is produced by pulling the string with a bow. Therefore, it is called two strings. This kind of basin has such a feature: when there is water in it, rub the two strings with your hands and the basin can also spray water jets in addition to making a buzzing sound, forming waves on the water surface, which seems very magical. The original water spraying fish basin still can be seen in museums in Hangzhou, Dalian, Chongqing, and other places today. Figure 8.2 shows a schematic diagram of a water-spraying fish basin (according to Needham) in the Chongqing Museum. The bottom of the wash is molded with four fishes. The mouth of the fish is engraved with a streamline, which extends to the basin wall and then along the upper edge of the basin wall. When the two strings of the basin are rubbed by hand, the water column will gush up along the streamline of the fish mouth, so people call it the water spraying fish basin.

In Song Dynasty literature, there are some records that may be related to the water spraying fish basin. For example, Volume 3 of Hui Zhu Qian Lu (《挥麈前录》) by Wang Mingqing mentioned two treasures presented to the Liao Dynasty by Chi Chonggui of the Later Jin Dynasty. Among them was a porcelain basin: "The patterns of two carps are drawn, and if the basin is full of water, the carps are vivid as if they were moving." As we know, when the basin is rubbed, the water in the basin is forced to vibrate and squirt as if it were stirred by the fish in it. This is very similar to the description in Hui Zhu Qian Lu (《挥麈前录》) and it is fair to conclude that the vessel is a water spray fish basin.

In Vol.9 of Chun Zhu Ji Wen (《春渚纪闻》), He Yuan of the Song Dynasty quoted Lu Ting Za Ji (《虏庭杂记》) and also mentioned the fish basin given by Chi Chonggui to the king of Liao, while claiming it was wooden: "the fish basin is a plain wooden basin that is two chis in diameter, with a double-fish pattern within. These fish patterns have both scales and manes and are about five cuns long. If the basin is used to store water, the double-fish pattern slowly appears, and looks very genuine in a moment; when the water is poured, the wooden pattern disappears.

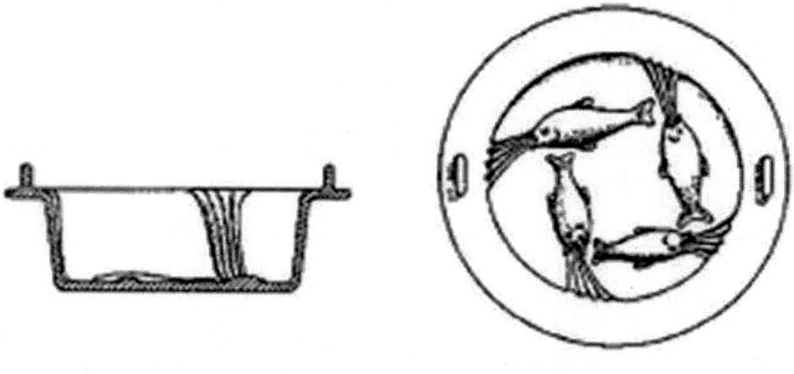

Fig. 8.2 Water-spraying fish basin preserved in a Chongqing museum

Jurong people today still make basins with copper and call these basins double-fish, following its legacy approach." "Slowly appears" seems to be describing how water ripples appear when rubbing has just begun; "looks very genuine" means at a certain point during rubbing, water columns are sprayed as if they were sprayed by real fish in the water. Such a description should be interpreted as a water-spraying fish basin. The quote said it was a wooden basin, which is incorrect, as it would be hard for a wooden basin to vibrate and spray water, and none of the physical items preserved are made of wood.

It should be emphasized here that the quote mentions: "Jurong people today still make basins with copper and call these basins double-fish, following its legacy approach." It articulates the inheritance of the water-spraying copper basin. He Yuan lived in the late Northern Song Dynasty period, and at that time, some people around Jurong, Jiangsu, could already make copper fish basins capable of spraying water. Its origin was the porcelain fish basin of Emperor Chu of Jin Dynasty.

It was originally called a basin, and later it was called a copper wash or double fish copper wash. Later, it evolved from double-fish patterns into quadruple-fish patterns, which suggests people had a deeper understanding of the vibration feature of the fish wash basin. As the water spray fish basin at least sprays four water jets, the four fish patterns correspond to the four water jets, which is an ingenious conception and full of artistic beauty.

It should be clarified that in the descriptions of Hui Zhu Qian Lu (《挥麈前录》) and Chun Zhu Ji Wen (《春渚纪闻》), there is no mention of the need to rub with hands. Is this due to the negligence of the record, or any other reasons? For example, the porcelain basin originally was not a water-spraying fish basin but was made according to a certain optical imaging principle. Here we do some analysis. Also in Volume 9 of Chun Zhu Ji Wen (《春渚纪闻》), He Yuan also described an agate bowl, saying it was "round and clean without carved patterns," and its owner "used it to store water and fill the inkstone. One day he saw a carp in it. Several inches long, the carp was swimming and quite cute. At first, he thought it got into the bowl with water and did not find it interesting. Afterwards, he had doubts and drained the bowl to verify, only to find the carp was gone. He filled the bowl again, and a carp emerged soon. He tried to take it with hands, only to get nothing." As far as the description is concerned, the agate bowl here can only be made based on optical principles as it has nothing to do with friction or vibration. Evidently, the agate bowl has two features. First, there is no fish when there is no water, and the fish can only be seen after water is injected, which mostly adds to its mystery; the other feature is that although the quotation also mentions "swimming" and "emerged soon," but generally speaking, it gives people a static feeling. This is because people see the image of a fish, so naturally they will not feel that the water in the bowl is also fluctuating. On the contrary, the double carp porcelain basin mentioned above has "double carp engravings," originally carved with double fish patterns. After water is injected, "the double-fish patterns emerge" and "leaping like real," as though the bowl were full of waves. This effect is not caused by optical imaging, but can only be caused by mechanical action, so we believe it is the water-spraying fish basin that is described in the paragraph.

What is the reason behind the fish basin's ability to spray water? This of course has something to do with the vibration of the basin caused by hand rubbing. The researchers found that the vibration of the basin is a plate vibration, which is a regular vibration similar to that of a cylindrical plate. When human hands rub the two strings of the basin, the peripheral walls of the basin vibrate. The friction between the palms and the two strings is the excitation source for the vibration of basin, and the energy of vibration is given to the basin surface through friction. Because of the contact between the palm and the two strings, the spot is always at the vibration node position. The two strings are symmetrically distributed with respect to the center of the basin, so the vibration of the basin caused by rubbing the two strings can only be even-numbered pitch lines (such as 4, 6, and 8 pitch lines). The vibration is as shown in Fig. 8.3.

When the peripheral wall of the basin vibrates, it will have a flapping effect on the water in the basin, forcing the water to vibrate correspondingly. At the vibration antinode of the basin, the vibration of the water is also the strongest, and even water columns are sprayed due to the beat of the antinode, and directional waves are formed on the water surface. At the vibrating nodes of the basin, the water does not vibrate, the waves stop on the nodal line, and the bubbles and water droplets on the water surface are also parked on these nonvibrating water surface lines (as shown in Fig. 8.3). In this way, through rubbing of the two strings with hands, a regular corrugated distribution and jets of water are formed on the surface of the basin water.

Water is the carrier of fish basin vibrations, and the amount of water also affects the formation of vibration node lines. Therefore, the amount of water injected must be

Fig. 8.3 Vibration patterns of the water-spraying fish basin

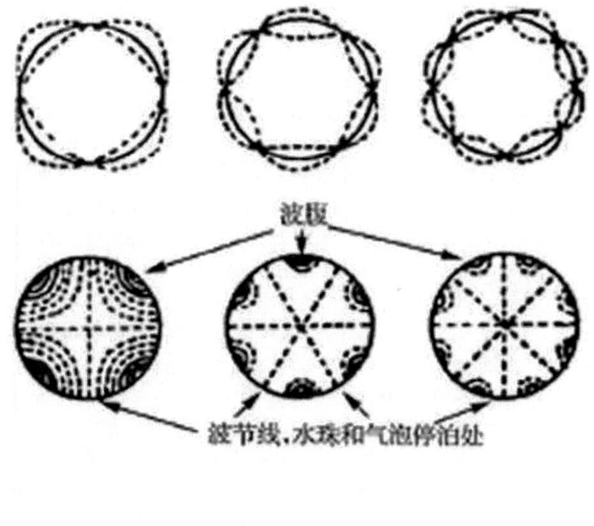

中文	英文
波腹	Antinode
波节线, 水珠和气泡停泊处	Nodal lines, water drops and air bubbles parking spots

properly controlled for the fish basin to produce different numbers of waves and splashes. According to the water-spraying fish basin in the Provincial Museum of Zhejiang, with an amount of water approximately 9/10, 7/10, and 1/2 of the volume of the basin injected, the fish basin produces 4, 6, and 8 standing waves, respectively, namely the fish basin vibrates with 4, 6, and 8 nodal lines, respectively. Among them, the four-node line vibration is the fundamental frequency vibration of the fish basin, triggering the highest water jet and the greatest wave. When the fish basin vibrates, it can be observed that the nodal line of the water surface does not vibrate, and based on this, the position of the nodal line of the basin wall can be determined. In this way, the water spraying fish basin makes the vibration of the curved plate visible. This is indeed a major invention in the history of science. Of course, you can also use dry fine sand instead of water, and the effect is more noticeable.

It should be pointed out that utensils similar to the water-spraying fish basin were also found among utensils made by an ethnic minority of our country. Xu Ke collected a large amount of materials from the Qing Dynasty and compiled a book titled "Qing Pi Lei Chao" (《清稗类钞》), which contains a story called "Li Ziming Collects Ancient Miao King's Copper Pot." The original text is as follows:

> A man named Chen Shunchang bought a copper pot from a Miao peddler on the street by the moat of the ancient city wall. The pot weighed over 5kg. With cold water stored inside, the pot sounded like an organ, a reed, or a bullhorn when its ears were rubbed. The sound was loud and audible and could be heard 1km away. The cold water in the pot foamed like boiling water, splashing very high. The water surface formed an octagon, with the center staying still. It was rumored to be the relic of an ancient Miao king. The pot had a top that was larger than the bottom, and was green all over, with fish-shaped patterns on both ears. Now it was in Li Ziming's possession.

This is probably the earliest clear and comprehensive record of water-spraying fish utensils in history. It describes the shape of the utensil: it looks like a pan, weighs more than 5 kilograms, and is large in the upper part and small in the lower part, with fish-shaped patterns on both ears; it also describes the acoustic performance: If you rub both ears, it will make a loud sound that can be transmitted one km away. The connection between water spray performance and acoustic effect had never been mentioned in previous literature. Whether the claim that the sound "could be heard 1km away" is exaggerative or not, the fact that the copper pot had good acoustic performance is doubtless. The loud sound indicates that it had good vibration performance and good vibration performance means strong water spraying ability. The following description of the water spraying performance of the copper pot proves this point. The line that "the water surface is an octagon, with the center staying still" indicates that the vibration pattern of the copper pot witnessed by the author is an eight-node line vibration. In particular, this paragraph clearly mentions that the copper pot was from the Miao people. "It is rumored to be the relic of the ancient Miao king," which shows that the Miaos cherished the item dearly; "green all over" shows for how long it had been passed down. It is noteworthy that the Miao people had made copper pots with such good water spraying performance a long time ago.

8.2 Acoustics

In the history of Chinese science and technology, acoustics includes two aspects, one of which is the explanation of the nature of sound and related phenomena and technologies, and the other is the knowledge and practice of phonology. In terms of rhythms, we will discuss separately. Here we focus on ancients' exploration of the first aspect.

8.2.1 Nature and Transmission of Sounds

The understanding of sound volatility in ancient China is far behind modern science. From the perspective of physics, the ancients' discussion of the nature of sound remained in the state of describing and contemplating various sound phenomena. But in these descriptions and speculations, there is no lack of insights, and the volatility of sound is also touched.

When the ancients explored the nature of sound, there was a distinctive feature, which its close relation to the doctrine of Qi. This is because the theory of vitality, as the original theory of all things in the universe, had extensively influenced and penetrated into various fields of science. On the other hand, it is also because the ancients realized the production and propagation of sound result from the existence of Qi in long-term observations. The combination of these two aspects helps people to develop a correct understanding of the volatility of sound.

As early as the pre-Qin period, the book *Zhuang Zi* (《庄子》) had already involved the sound phenomenon caused by the air movement in nature. The chapter of Qi Wu Lun (《齐物论》) enumerated the various orifices on the trees, saying that when they are blown by the wind, they will make a variety of different sounds. As the wind varies in magnitude, the sound changes accordingly. When the gale ceases, the sound from these acupoints also stops at the same time, everything is silent, and both are in a state of emptiness and quietness.

This paragraph combines some sound phenomena in nature with air flow and is an explanation of the causes of these sound phenomena, but this explanation has nothing to do with the sound wave, because flow, vibration, and wave are completely different concepts after all.

The chapter of Qi Wu Lun (《齐物论》) also mentioned: "Human speech is different from the sound made when the wind blows. Naturally, speech contains information and can play a role in communication. But what is the nature of language? What is the difference between a man's speech and the chirping of a bird that just came out of the shell?" This paragraph was originally intended to point out that the difference in languages, which are a tool for people to communicate with each other, is relative in nature. However, it unwittingly inspires people to think: Is there a common nature behind various sound phenomena? With today's knowledge, we can definitely answer: All kinds of sounds are all expressed as vibrations of the air. This is their common nature. So what did the ancients know about this?

Wang Chong of the Eastern Han Dynasty realized the sound phenomenon is a special form of air movement, calling it "enveloping":

Living people speak by moving their tongues and opening their mouths, because air is enveloped by their mouths and throats. Playing the reed pipe is also like this. If the reed pipe breaks, air cannot be enveloped, no holes are available for fingers to seal and no music can be made. The pipe is just like a person's mouth and throat; sealing the holes with fingers is just like moving the tongue. (Wang Chong: *Lun Heng – Lun Si Pian* (《论衡·论死篇》)).

Living people refer to people who are alive. Wang Chong believed people's speaking is the same as reed pipes making sounds, and both are the result of air "enveloping." "Enveloping" here should be interpreted as the vibration of air.

Wang Chong also talked about the propagation of sound. He said: "When people sit on the porch and observe the ants on the ground, they can't even see their bodies, so how can they hear their sounds? This is because the ants are tiny – far tinier than humans, and their sound cannot emanate." This means that given the ant's tininess, its sound cannot be heard if it is even a little far from the person, and even if it has a sound, which cannot reach afar. Wang Chong went on to say: "The one-foot long fish moves in the water and shakes the water beside it, with the waves reaching a mere several feet away." Even if the fish is as big as a human, its wave can reach no further than several hundred steps away, while anywhere one kilometer away remains quiet and still. Likewise, human behavior (including speech, of course) changes the state of the surrounding air, and the range of the spread of this change is similar to the outward propagation of water waves caused by fish shaking water. It is worth noting that Wang Chong not only compared the outward propagation of sound with water waves, but also clearly pointed out that sound propagation is a result of the interactions of air, which is consistent with our present understanding of how sound waves propagate.

Since sound is a special form of movement of "air," how to make "air" produce sound through its movement? "Kao Gong Ji – Fu Shi" (《考工记·凫氏》) records the methods of people making musical instruments, which goes:

Different thicknesses lead to vibrations, and vibrations result in sounds.

Seemingly people already knew the vibration of the bell body is the source of the sound of the bell. The tone of the clock is also determined by the vibration of the clock. This kind of knowledge, as far as the ancients is concerned, is a common sense. From the perspective of modern science, it is also undoubtedly correct.

However, Kao Gong Ji (《考工记》) was just a summary of the experience of the ancients. If you delved into it, it would be difficult for them to tell why. In this regard, the Northern Song Dynasty writer Ouyang Xiu once recorded such a question:

A asks B: Casting copper to make a bell, cutting wood to make a stick, knocking the bell will make a sound, but is the sound from the wood? Or copper?

B replies: If you knock the wall with the stick, there is no sound; if you knock the bell, there will be a sound, so the sound is from copper.

Party A says: If you knock the coin with the stick, there is no sound. Is the sound really from copper?

> Party B: The coin is solid, while the bell is hollow, so the sound comes from hollow things
>
> Party A: If the bell is made of wood and clay, there is no sound. Is it really hollow things that can produce sound?

In Ouyang Xiu's article, no answer was given to the question asked by A. As far as modern physics is concerned, these questions are not difficult to answer, but in the context of ancient experience and knowledge, it is difficult to clarify these questions. This example shows that the ancient people's understanding of the sound mechanism of objects is far from modern physics.

Tan Qiao's understanding of sound characteristics in the late Tang Dynasty period is worth mentioning within the scope of ancient knowledge. When evaluating the role of music, he said:

> Harmonious music attracts masculinity, generates breezes and gives birth to everything; disharmonious music transforms air into heavy gales which ruin everything. The music's role depends on the sound, and the sound depends on the air. If the air moves, a sound is generated, and when the sound is generated, the air vibrates, and when the air vibrates, there will be a wind, and everything changes. (Tan Qiao: *Hua Shu – Sheng Qi* (《化书·声气》))

As can be seen, Tan Qiao clearly pointed out: "If the air moves, a sound is generated, and when the sound is generated, the air vibrates," linking the generation and propagation of sound to air vibration and facilitating people's understanding of the nature of sound. Further, Tan Qiao described the dispersion of sound information:

> Shapes and air combine to form sounds. Ears do not hear the sound, which plunges into the ears by itself; the valley does not echo the sound, which fills the valley by itself. The ear is a small orifice; the valley is a big orifice; the mountains and rivers are small valleys; and the sky and the earth are big valleys. With one orifice making a sound, all orifices make sounds; when a sound is heard in one valley, it is heard in all valleys. Sound transmits air. . .air contains sounds. The sound reaches anywhere despite buzzing mosquitoes and flies. (Tan Qiao: *Hua Shu – Da Han* (《化书·大含》))

Shapes refer to an object of a certain shape and certain texture. They interact with air to generate sounds, "when the sound is generated, the air vibrates," and air fills all spaces, and in this way, it will naturally cause the sound to propagate all around in the form of vibrations. Wang Chongzhi's theory involved the limited nature of the range of sound transmission, while Tan Qiao emphasized on the dispersion of sound distribution. Combining the two will lead to more complete understanding of this issue.

In the late Ming Dynasty, scientist Song Yingxing also carried out a more extensive discussion of the nature of sound. Like his predecessors, he also discussed the occurrence and propagation of sounds based on the doctrine of qi (air). He believed that the diffuse distribution of air in space is a necessary condition for sound phenomena, but not a sufficient condition. He said: "Qi is a confusing thing. Even if it can create sounds, but it cannot be generated on its own." (Song Yingxing: *Lun Qi – Qi Sheng Er* (《论气·气声二》)). To form a sound, the diffused air must be subjected to strong influence. This can be "the sound of two gusts colliding," such as the sound of wind; it can also be "the sound of human breath colliding with air," such as a wind instrument; it can also be "the sound of air being broken by a shape." For the latter, Song Yingxing enumerated examples such as flying arrows, jumping whip, strumming, cracking, clapping, and holding an object to hit another object, believing these sounds are caused by the rapid movement of objects hitting air. In particular, he emphasized: "For anything to make a sound by breaking the air, it must be rapid, otherwise it will fail to make a sound; it must be strong, otherwise it will fail." This is because "air does not cause sounds if it becomes static right after being occasionally hit. Only rapid impact can cause air to make sounds." He did the analysis from the perspective of the internal mechanism of air creating sounds and used languages that were far from modern science and his conclusion did not hit the spot. However, he analyzed the internal movement pattern of air, which symbolized the ancients' discussion of the nature of sound was further physicalized.

Song Yingxing also discussed the propagation of sounds. He said:

> When an object plunges into air, it is like plunging into water. Air and water are the same kind of movable matter. When you toss a stone into the water, the water splashes by the length of a fist from the water surface, while the waves spread to meters away and do not stop. Air is likewise, and extremely light sound cannot be heard. (Song Yingxing: *Lun Qi – Qi Sheng Qi* (《论气·气声七》))

Comparing the outward propagation of sound to the propagation of water waves is of course not precise enough, because sound waves are longitudinal waves, while water waves are more complex waves jointly produced by the surface tension and gravity. But in any case, this analogy is a natural expression of the ancients' understanding of the volatility of sound and is also a better explanation in the historical background of the time and therefore has certain historical value.

Finally, let us examine the ancient people's description and understanding of the phenomenon of independent sound propagation. In the early Qing Dynasty, Jie Xuan talked about this phenomenon in the annotations of vol.1 of Wu Li Xiao Shi (《物理小识》). He said:

> Air contains both virtual and physical entities and includes everything that does not constitute a barrier. Everything that has a shape, a color or a sound exists. How could there be any gap between the sky and the earth. . .Sounds from the wind, the water, people, birds and musical instruments mix with one another and are mutually unaffected. Even during late night, these sounds can reach afar.

This means that air is diffusely distributed, and the propagation of sound in the air is of course also diffused, with the sound spreading in all directions and everywhere. But these diffused sounds do not affect each other. Sounds at the same spot coming from all directions can be described as "mixed noises," but are also "distinct and unconcealed." What Jie Xuan described is actually the independent propagation of sound waves. However, it was not until people understood the volatility of sound from the perspective of modern science that this phenomenon was dubbed a "principle" and clarified in terms of mechanism.

8.2.2 Resonance Experiment

In physics, resonance refers to the phenomenon that when one object vibrates, another object vibrates accordingly. Two objects that resonate must have the same natural frequency or frequencies that form a simple integer ratio.

The ancients of our country discovered and recorded the resonance phenomenon incredibly early. In the third and fourth centuries BC, Zhuang Zi – Xu Wu Gui (《庄子·徐无鬼》) recorded the phenomenon of resonance that occurred upon adjustment of the Chinese zither:

> A zither is placed in the same house and same room. When gong is played, the gong string moves, and when jiao is played, the jiao string moves, and the tones are the same. Now if we change one string that is different from all of the five fundamental notes, and when this string is played, all the twenty-five strings move.

This paragraph describes two phenomena. The former is that when fundamental notes such as gong and jiao are played, the strings of other instruments in the same room vibrate, which is the resonance between fundamental notes. The latter is that when a string is adjusted to a tone that is different from any of the five tones of gong, shang, jiao, zhi, and yu, the 25 strings of another instrument resonate once the string is played, which is the resonance between fundamental notes and overtones. The second phenomenon is generally hard to notice, while the ancients could perceive it, which suggests they had carried out incredibly careful observations.

Dong Zhongshu of the Western Han Dynasty also tried to explain the phenomenon of resonance. In Chun Qiu Fan Lu – Tong Lei Xiang Dong Pian (《春秋繁露·同类相动篇》), he wrote:

> Things repel what is different, and echo what is of the same nature. Therefore, things of the same nature interact with each other, and their sounds echo each other, which proves to be true. I have experimented with the instrument. When I play gong, the gong string on another instrument resonates, and when I play shang, the shang string on another instrument resonates. All the five tones resonate, which is nothing magical but a natural occurrence.

Dong Zhongshu believed that objects of the same nature can interact with each other, and the reason why they interact with each other is that they have the same tone, which is an inevitable phenomenon and there is nothing magical. Dong Zhongshu's ability to correctly recognize that this is a natural phenomenon and eliminate the mysterious vibe surrounding it is worthy of recognition. This helped people understand the resonance phenomenon correctly.

It was Zhang Hua of the Jin Dynasty who further explored the resonance phenomenon. He extended the scope of the resonance phenomenon beyond musical instruments. According to legends, there was a big bell in front of the temple at that time. One day the bell suddenly rang for no reason. People were surprised and asked Zhang Hua. Zhang Hua replied that this was because of the collapse of a copper mountain in Shu County, so the bell would ring. Soon reports from the Shu County confirmed it was true. Zhang Hua associated the copper landslide with the bell, which did not necessarily mean that he considered this matter from the perspective of resonance. It can also be explained by Dong Zhongshu's theory "the same kind moves in resonance": The bell is made of copper, and when the copper mountain collapses, and the bell resonates, because "the same kind moves in resonance." However, in Yi Yuan (《异苑》) by Liu Jingshu of the Northern and Southern Dynasties period mentioned another matter about Zhang Hua: "During the Jin Dynasty period, someone used a copper bath basin, which rang in the morning and evening as if it were knocked. So he asked Zhang Hua, who replied: 'This basin echoes the Luoyang bell, which is knocked in the morning and evening. You can make it lighter by filing it, and it will no longer echo the bell'. He did as was told, and the basin ceased to ring." This was clearly an explanation from the perspective of resonance.

Here, Zhang Hua not only identified this as a resonance phenomenon, but also found the source of the resonance and proposed a method to eliminate the resonance. We know that using a file to file a part of the copper basin can change its natural frequency so that it is no longer resonant with the sound, which ends the occurrence of resonance. As can be seen, Zhang Hua's approach to eliminate resonance worked. This method must have been based on his correct understanding of the cause of resonance.

Some experienced musicians had also mastered this method of eliminating resonance. Wei Xun's Liu Bin Ke Jia Hua Lu (《刘宾客嘉话录》) recorded the story of Cao Shaokui of the Tang Dynasty eliminating resonance. During Emperor Kaiyuan's reign, Cao Shaokui was the official responsible for the Court's music affairs. He had a friend, who was a monk and had a chime that often rang without being knocked, which troubled the monk so much that he fell ill. Hearing of it, Cao Shaokui visited him and found the chime also rang when the bell rang. So he told the monk he had an idea and asked the monk to treat him with a banquet the next day. After the lunch the next day, Cao "took out a file and filed the chime at several spots, and the chime ceased to ring. The monk asked why, and Cao replied: 'The chime matches the bell in rhythm and echoes it.' With ecstasy, the monk soon recovered." Cao Shaokui introduced a resonance elimination method that was similar to Zhang Hua's.

In the Song Dynasty period, Shen Kuo began to address the resonance phenomenon with experiments, making further progress in ancients' study of the resonance phenomenon. In Vol.1 of Meng Xi Bi Tan – Bu Bi Tan (《梦溪笔谈·补笔谈》), he said:

> The strings of musical instruments all resonate: when gong is played, the gong string resonates; and when the shang is played, the shang string resonates, and the rest all resonates. This principle should be applied in today's musical instruments. To know which string resonates, first adjust the strings to harmonize the tones, and place a paper man on the strings. When the strings are played, the paper man leaps, while other strings remain still. Even if other instruments are played, the string vibrates, which is called a positive tone.

A "positive tone" refers to the sound that meets the frequency requirements of the rhythm, and here it refers to the tone whose pitch differs by an entire octave, namely the fundamental tone and the overtone. In musical instruments, the two also have a resonance relationship, which however is not easy to find. Shen Kuo's experiment was intended to prove this relationship. To this end, he made a small paper man, placed it on the fundamental string, and plucked the corresponding overtone string and the paper man moved. Then, he played other strings, while the paper man did not move. In this way, he used experimental methods to visualize the resonance phenomenon of the second string when the pitch differed by an octave. Shen Kuo's experiment was centuries earlier than similar paper balance rider experiments in Europe.

Shen Kuo could successfully complete the paper man resonance experiment because he had a thorough understanding of the resonance of musical instruments. In Vol.6 of Meng Xi Bi Tan (《梦溪笔谈》), he talked about the resonance of musical instruments:

> My friend has a lute, which is placed in an empty room, where he plays two tones with a wind instrument, and the lute echoes. When other tones are played, the lute does not echo. My friend thinks this is an unusual phenomenon, which is actually common. The lute resonates with those of the twenty eight tones with the same rhythm, and if it does not resonate with any of the twenty eight tones, it must be a leisure tone...there are many leisure tones, and occasionally some of them are among the twenty eight tones, and people think this is a strange phenomenon, which is actually normal.

Twenty-eight tones refer to the 28 tones commonly used in the folk music of the Yan area at that time. The leisure tones refer to tones other than the common tones. Shen Kuo pointed out: "Those of the twenty-eight tones with the same rhythm echo." The reason for echoing is "the same rhythm," which is common sense. It was precisely because of this understanding that he was able to design and complete the paper man resonance demonstration experiment.

At the end of the Ming Dynasty, scholar Fang Yizhi described and developed Shen Kuo's method in Volume 1 of his Wu Li Xiao Shi (《物理小识》). He said: "When playing the outer string of the zither, the inner string moves along. For example, if you play the Plum Blossom Melody, with a small piece of paper attached to each string, and play a note in the Plum Blossom Melody with a flute, the paper on the string also vibrates. When Cao Shaokui chimes, the bell rings, and Zhang Hua knows a collapse has occurred at Tongshan Mountain. As long as sounds have the same properties, they can resonate even on different things." Here Fang Yizhi even regarded resonance as a symbol of "harmony of sounds from different instruments." Fang Yizhi summarized: As long as sounds have the same properties (of the same frequency or in simple integer ratios), they can resonate even on different things. The practice of Zhang Hua and Cao Shaokui has proved this, Fang Yizhi clearly pointed out that their findings are essentially the same as the resonance on musical instruments, and they are all caused by the "harmony of sounds." These words of Fang Yizhi's marked another step in people's understanding of the nature of resonance.

8.2.3 Sound Transmission Through Solid Items

The transmission of sound can take place in the air or in solids such as the earth. The ancients discovered the phenomenon that sound can propagate in the earth and used it effectively. The ancients' use of the phenomenon of sound transmission through the earth was mainly in the military field. The method of burying the urn into the ground to determine the source of underground sounds recorded in the pre-Qin work Mo Zi (《墨子》) is a typical example. When talking of the method of defending against the enemy digging tunnels to seize cities, the chapter of Bei Xue (《备穴篇》) in this book said:

If enemy forces dig tunnels to attack the city, the defenders should dig wells in the city – digging a well every 5 steps along the wall and next to the bottom of the wall. When running into a high ground, the defenders should dig 487 cm deep, and at low grounds, they should dig 91 cm below the water level. Potters make large jars, each with a volume of more than 414 liters and with a thin layer of leather tightly stretched at the mouth of the jar, which are placed in the well. By making a hearing-sensitive person lie at the mouth of the jar and listen, the defenders can determine the position of the enemy's tunnel, so that they can dig tunnels for counterattacks.

The chapter of Bei Xue (《备穴篇》) introduced another method of determining the position through sounds, which also made use of the earth's ability to transmit sounds. The specific method is:

Bury two jars in a well. The jars' mouths are leveled with the city foundation, with plates placed on them, and people are made to lie on their sides and place their ears on the plates to listen. The density of the wells is also about one every five steps. In this way, the jars are buried at a slightly shallower level, which is susceptible to interference, and the sound transmission effect is not as good as the former. However, as there are two jars buried in the well, there is a slight distance between the two jars, and the orientation of the sound source can be roughly determined based on their loudness difference and then compared with the sound of the jars in the two adjacent wells. In this way, the position of the sound source can be accurately determined. Therefore, this method is superior to the former in terms of orientation accuracy, and it is easy to operate.

Directional devices for determining underground sound sources were often used in ancient warfare. For example, Zong Gongliang of the Song Dynasty introduced two devices of "Urn Listening" and "Earth Listening" with both texts and illustrations in Wu Jing Zong Yao (《武经总要》). Vol.12 of the book (the Chapter of City Defense) said:

> Urn Listening: A 45kg urn is placed in the tunnel, and a man with good ears is chosen to sit beside the urn and listen, in case the enemy is digging a tunnel in preparation for our attack... Earth Listening: Make well-like holes in eight directions on the ground. These holes are 609cm deep each and do not reach the spring. Men with good ears cover new urns with the wells.

These techniques are widely recorded in historical literature, such as Du You's Tong Dian (《通典》) and Mao Yuanyi's Wu Bei Zhi (《武备志》). These are traditional approaches used in ancient warfare.

The ancients also made use of the earth's ability to transmit sounds for monitoring. Vol.6 Defense Chapter of Wu Jing Zong Yao (《武经总要》) recorded the relevant method:

> Those with good ears and sleeping less are selected to lay their heads on hollow barbarian deer pillows, which are made of boar hide. Once there is any activity of troops or horses 15km away, there will be sounds in all four directions. One or two stations are set in each encampment, and three or four stations are set in a large encampment. In isolated towns, one station is set each, and the stations are changed frequently, and a daily rotating schedule is implemented.

The so-called "hollow barbarian deer pillows" are actually hard hollow cylinders made of boar hide, acting as "pickups." When there is activity of troops and horses within a radius of 20–30 km, the sound transmits along the ground, echoes in the hollow cylinders, and is sensed by those laying their heads on them, who then issue the alarm. As sound waves transmit much greater energy when propagating along the earth than in the air, this approach covers a much larger scope than in the air.

In Vol.19 of Meng Xi Bi Tan (《梦溪笔谈》), Shen Kuo said:

The ancient method was to use cattle leather to make quivers, which were meanwhile used as pillows. Hollow quivers were chosen and placed on the ground as pillows, and all activities of troops and horses within miles could be heard. This is because hollowness transmits sounds.

By sleeping on quivers as pillows, soldiers could hear activities of troops and horses in the distance and prepare ahead. Shen Kuo believed this is because the quivers were hollow, as "hollowness transmits sounds." In a sense, this is right. The quivers served as "pickups," and "hollowness" was why they could "pick up sounds." In terms of "picking up sounds," there are two circumstances: First, some sounds of the activities of troops and horses have the same wavelength as the air column in the quiver and resonate with the air volume in the quiver. Besides, for the large amounts of sounds the air column cannot resonate with, the smooth inner wall of the quiver is highly reflective and forms reverberations within the quiver, which can be sensed by human ear. However, all this is based on the precondition of sound transmission. Shen Kuo's explanation to the fact that soldiers slept on quivers was "hollowness transmits sounds" and "on the ground" suggests he realized the earth's ability to transmit sounds was the key.

The phenomenon that "hollowness transmits sounds" was not only used by ancients to receive sounds transmitted by the earth, but also to muffle or insulate the sound. In China, hollow bricks made during the Warring Kingdoms period were unearthed. Buildings made of these bricks could insulate sounds. As for whether people during the Warring Kingdoms period did it on purpose, we do not know for sure. However, Fang Yizhi of the late Ming Dynasty period clearly recorded this technique in Vol.1 of Wu Li Xiao Shi (《物理小识》).

> While illegal minters hide in burrows, people still hear the sound of sawing and filing, so they make walls with urns, with the mouths of the urns facing outward, and the sounds are no longer heard. Why? Because the sounds have been absorbed by the urns.

Fang Yizhi believed the sounds had been absorbed by the urns. The approach recorded in Wu Li Xiao Shi (《物理小识》) is feasible. As the urn wall has a porous structure, the cavities can absorb sound energy, and therefore, this structure can eliminate sounds. While annotating this entry, Fang Yizhi's student Jie Xuan specifically associated this with the practice of laying on one side and placing the ear on the ground to listen, which suggests they might have noticed a certain connection between these two approaches.

8.3 Optics

In ancient China, optical research had been a focus of attention. In these studies, some observed and explained natural light phenomena, some analyzed and explained the nature of light, some explored various types of optical imaging, and some made various kinds of optical equipment to conduct optical experiments. These studies were characterized by an equal focus on theoretical exploration and experimental analysis. Among the many branches of ancient Chinese physics, optics was the most distinctive one.

8.3.1 Observation and Interpretation of Optical Phenomena in Nature

The generation of the ancients' optical knowledge was first based on the observation and interpretation of optical phenomena in nature. Here we mainly discuss the ancients' understanding of the dispersion phenomenon and mirages.

The so-called dispersion refers to the phenomenon in which polychromatic light decomposes into monochromatic light to form a spectrum. The cause of the dispersion may be that when the polychromatic light passes through the medium, the propagation direction of the various colors of light is deflected to different degrees as the medium has different refractive indices for light of different frequencies, so that a spectrum is formed after the light passes through the medium. It is also possible that when the polychromatic light passes through the optical system, the polychromatic light is decomposed due to diffraction and interference. The dispersion phenomenon is common and was recorded and discussed by the ancients. Generally speaking, what they wrote down was mostly in the category of atmospheric light images and crystal dispersion.

Among the dispersion phenomena observed by the ancients, the rainbow was the most common. They recorded the existence of the rainbow very early, and not only used the rainbow to predict weather changes, but also gave the rainbow a strong social meaning. For example, they associated the quality of the social atmosphere with the appearance of the rainbow and at the same time used the rainbow as a divination material to predict the outcome of the war, among others. Although these were somewhat superstition, they prompted the ancients to focus on the observation and research of the rainbow, which objectively promoted the continuous increase of knowledge about the rainbow.

In the process of observing and studying the rainbow, the ancients classified the rainbow according to its color and shape. Cai Yong's Yue Ling Zhang Ju (《月令章句》) said: "Rainbows are generated when yin and yang intersect and are manifested in form and color. The male is called the rainbow, and the female is called the neon." When interpreting Li Ji – Yue Ling (《礼记·月令》), Kong Yingda of the Tang Dynasty provided a detailed explanation to this classification: "That which is bright is male, and that which is dim is female." As can be seen, the ancients divided the rainbow into two categories: one of which is bright, called rainbows and the other is darker, called neon. This classification is, to some extent, equivalent to the modern classification of primary and secondary rainbows. The so-called primary rainbow is the spectral arrangement formed by one reflection and two refractions after sunlight enters water droplets. It is colorful, with a color band where red is on the outer side and purple on the inner side, called a rainbow. The secondary rainbow, also known as neon, is caused by two refractions and two reflections of sunlight entering water droplets, and its color band is contrary to that of the primary rainbow. With one additional reflection, the spectrum is not as bright as that of the primary rainbow. The primary and secondary rainbows can appear separately or simultaneously. Yue Ling Zhang Ju (《月令章句》) said: "The rainbow is to the east, and the neon is always beside." This describes the situation where the two appear simultaneously. As can be seen, the ancients' classification was based on real-world observations.

In addition, there is a kind of rainbow that appears on the fog. Given the small size of the droplets, these rainbows are light in color, typically pale white. The ancients also categorized these rainbows as neon. Shuo Wen Jie Zi (《说文解字》) said: "neon, which is a minor rainbow, is greenish red or white, representing femininity." This definition encompasses the fogbow.

During observations, the ancients found certain conditions are required for rainbows to appear, and rainbows must be in a position opposite to the sun. Li Ji – Yue Ling (《礼记·月令》) mentioned: "Rainbows begin to appear in the month of jichun; and disappear in the month of mengdong." This roughly indicated the season where rainbows appear. Yue Ling Zhang Ju (《月令章句》) went further: Rainbows and neon "often emerge along with clouds in the day and are not seen in the absence of clouds and in gloomy weather." This description is correct.

As for the cause of rainbows, ancients had various theories, which initially were full of speculations and imaginations and far from the actual cause of rainbows. After long periods of discussions, however, people's understanding of the cause of rainbows leapt forward in the Tang Dynasty period. When interpreting Li Ji – Yue Ling (《礼记·月令》), Kong Yingda of the early Tang Dynasty period pointed out: "When the clouds are thin, the sun appears and sheds light upon rain drops to generate rainbows." This description is very accurate: Sunlight seeps through the clouds and shines on raindrops, creating a rainbow. It particularly mentioned raindrops instead of general rain or water vapor, demonstrating a deepened understanding. By the middle of the eighth century, Zhang Zhihe used simulation experiments to verify while pointing out "the sunlight is reflected by rain to generate rainows" in his Xuan Zhen Zi – Tao Zhi Ling (《玄真子·涛之灵》):

> Water is sprayed against the sun and forms rainbows and neon, which are always bent and equally distant from the shadow.

Here, an experiment was conducted to artificially simulate the rainbow phenomenon and directly revealed the cause of generation of rainbows and neon. For the ancients who had no knowledge of light reflection/refraction and the dispersion theory, this is the best way to reveal the cause of rainbow generation.

With "always bent and equally distant from the shadow," Xuan Zhen Zi (《玄真子》) revealed deep scientific knowledge. This is an explanation to why the rainbow is an arc. "Equally distant from the shadow" means each point on the rainbow is equally distant from the observer's shadow, and to achieve this, it must be an arc. This makes sense. When sunlight hits the raindrops to generate a rainbow, the plane of the rainbow is perpendicular to the incident light. In other words, the direction of sunlight is parallel to the axis of the rainbow arc, and the direction of the human shadow is also the direction in which the sunlight projects. Therefore, the description in Xuan Zhen Zi (《玄真子》) makes sense.

Before the introduction of modern western science into China, Zhang Zhihe was the most representative of people's understanding of the rainbow. After that, people simply repeated the same experiment and observed more carefully.

There are many reasons for the dispersion. For example, sunlight can also cause dispersion by irradiating the crystal. This phenomenon was first discovered in the Jin Dynasty period. For example, Volume 11 of Ge Hong's Bao Pu Zi – Nei Pian (《抱朴子·内篇》) recorded five kinds of mica, claiming that they could show various colors under the sun's rays. In the Song Dynasty, there were more discoveries and records. For example, Yang Wen Gong Shuo Yuan (《杨文公说苑》) by Yang Yi in the Northern Song Dynasty period recorded the Bodhisattva stone of Mount Emei: "The color is white as jade, like Shangrao crystals, and five colors are shown in the sun." This actually described natural crystal dispersion. But Yang Yi believed that this phenomenon was caused by the Buddha in Mount Emei, which is wrong. Cheng Dachang of the Southern Song Dynasty corrected this error in Yan Fan Lu (《演繁露》). Cheng Dachang carefully observed the dispersion phenomenon of a single drop of water and realized that the five-color light of the Bodhisattva stone was formed for similar reasons. He said:

> When rain just stopped and sun appears, or when the dews are not dry, the dews stay at the ends of the branches and seem to fall but don't, and all converge into round dots with amazing light. When sunlight gets in, all the five colors show up and keep flashing. This is because the light gives its properties to the water, not because the dews have the five colors…these five colors do not show up on sunless days, which is not because of the Buddha in the Emei Mountain.

The statement that the sunlight gives color to the water implies the colorful lights come from sunlight, which is close to revealing the nature of dispersion. And it makes sense to associate it with crystal association and believe the two are essentially the same.

The ancients had a lot of records about the crystal dispersion phenomenon, but given the limited level of scientific development in the ancient times, their interpretation of its nature had not surpassed Yan Fan Lu (《演繁露》) for a long time.

In the late Ming Dynasty period, China's research on dispersion began to be influenced by Western studies. The missionary Matteo Ricci came to China. Among the objects he carried was a prism, which was used for dispersion performances. This is of course more advanced than the experiment in which the ancients of China "sprayed water against the sun" to simulate the rainbow. As it could stabilize the experimental conditions and control the experimental process, it was helpful for people to explore the nature of dispersion. Even so, people still failed to explain the dispersion phenomenon correctly. For example, the missionary's interpretation of the dispersion phenomenon was based on the thickness of the medium through which light passed, which is undoubtedly wrong.

On the basis of previous explorations, Fang Yizhi, a scientist at the end of the Ming Dynasty, made a summative record of the ancients' dispersion knowledge. He said:

> If the gem is convex, there will be one light beam, and if there are multiple edges, there must be five colors on one side. In the case of Emei gems, there are six sides; when the crystal is placed on the paper, there are three sides; in the case of three-sided crystal, there are also five colors. A valley spring in the sunlight has five colors, and water sprayed under the sun also has five colors. Therefore, we know the colors of the rainbow, the aura of the sun and the moon and the five colors of clouds are subject to the same principle. (Fang Yizhi: *Wu Li Xiao Shi* (《物理小识》) (Vol.8)).

He comprehensively enumerated all kinds of dispersion phenomena, including natural crystals, artificial transparent bodies and rainbows, and sun and moon halos, and believed that they were essentially the same. This statement is reasonable, as the above phenomena are regarded as dispersion of white light today.

Into the Qing Dynasty period, people still discussed the dispersion phenomenon, and knowledge about dispersion continued to accumulate, while there was still no clear interpretation of dispersion. It was not until the middle of the nineteenth century that Zhang Fuxi translated the Theory of Light and used the principles of light refraction and reflection to explain dispersion and the people of China correctly mastered some basic knowledge about dispersion.

Compared with the study of dispersion, more observations and interpretations of the mirage phenomenon were carried out by the ancients.

The mirage is an atmospheric optical phenomenon. When light passes through air layers of different densities and undergoes significant refraction or total reflection, distant scenes are reflected in the air, sea, or ground, thus forming various bizarre and strange scenes. While modern science has provided a thorough explanation to the cause of the mirage, what did people know about it in ancient times?

The mirage is eye-catching with its bizarre landscapes, and as a result, it was discovered and recorded by early humans. Shi Ji – Tian Gong Shu (《史记·天官书》) said: "A mirage appears by the seaside, showing a magnificent scene of palace," Han Shu – Tian Wen Zhi (《汉书·天文志》) had a similar description. More people observed and recorded the mirage in later generations, and relevant ancient books abounded. This leaves us with a credible record to peep into the nature of the past.

The ancients not only recorded the mirages they saw, but also tried to explain their cause. In hindsight, the ancients' insights can be divided into five categories. (Wang Saishi: *Record and Exploration of Mirage in Ancient China, Historical Records of Chinese Science and Technology* (《中国科技史料》), 1988, issue 4).

The first category of explanations believed the mirage was caused by dragons and clams in the sea exhaling air. This theory is often seen in the books of the Han and Jin Dynasties. It was an ancient traditional concept and had many believers. For example, Bo Wu Zhi (《博物志》) said: "There are clams in the sea, which exhale air to form images of palaces," which was quoted by many literati of later generations. The ancients lacked the proper scientific knowledge and understandably came up with this explanation. Moreover, this explanation implied that the mirage is related to water, which is not without any reasonable element. But in any case, it is a fictitious theory after all, which cannot stand the test of time. After long-term observation and research, people gradually became suspicious of this theory. In the Song Dynasty period, Su Shi had a poem "Deng Zhou Hai Shi" (《登州海市》), which read:

> There is an ocean of clouds in the endless space in the east, and mountains emerge out of the hollow space. all kinds of scenes are generated, how would shells and pearls be stored in palaces? Knowing everything I see is mirage, I distrust my eyes and ears.

Here, the poet described a mirage and pointed out it was an illusion. The line "how would shells and pearls be stored in palaces" specially pointed out mirage could not be real palaces. In Meng Xi Bi Tan (《梦溪笔谈》), Shen Kuo also recorded the mirage in detail and finally pointed out: "some say it is caused by the air exhaled by dragons and clams, but I doubt it." It was during constant observations and studies of the mirage that the theory of dragon/clam exhaling air was abandoned.

Another category is the re-emergence of wrecks. According to the theory of a field inevitably turns into the sea, the ancients believed some cities and items would sink into the underground or sea without eroding. Once suitable conditions are available, they will re-emerge at the same spot in their original appearance. In Vol.41 of Qi Xiu Lei Gao (《七修类稿》), Lang Ying of the Ming Dynasty said:

> The mirage in Dengzhou seems strange to people, who cannot find a theory to explain it. With limited knowledge, average people understandably cannot find an explanation. Given the fact that mirage occurs at fixed locations and its content is constant, and mirage only happens in spring and summer, its cause can be determined. The ancient saying goes: oceans turn into fertile fields as time goes by. How do we know the location where mirage occurs was not a city or mountain? In spring and summer, earthly air is generated, and things

that have long been accumulated underwater are reflected by the air. This is why there is no mirage in autumn and winter. The content of today's mirage is nothing but cities and mountains, and what is strange about that?

Lang Ying made this proposition based on the theory that the field inevitably turns into the sea and the observation that mirages mostly occur in spring and summer. He believed that "earth air occurs in spring and summer" and brings out images of objects deposited in underwater cities. In Vol.3 of Lv Yuan Cong Hua (《履园丛话》), Qian Yong of the Qing Dynasty talked of the Lake City in Gaoyou and held a similar view: "The Gaoyou Lake was formed after a city sunk, just like the Hongze Lake, which was formed by the old city of Sizhou, where those near the lake could see cities, terraces, people and horses traveling back and forth. For this reason, how do we know the mirage was not an ancient city? What appears is their essence." Of course, this explanation does not make sense from the perspective of modern science. It is, however, a speculation by the ancients to explore the cause of the mirage, with its own basis for reasoning, and cannot be simply regarded as absurd.

Another category proposed the condensation of winds and air as the cause. This theory believed the mirage was caused by the condensation of natural wind and air over the sea. In Vol.23 of Yu Zhi Tang Tan Hui (《玉芝堂谈荟》), Xu Yingqiu of the Ming Dynasty said: "The mirage is created by the gas in the ocean, not the gas from clams." In Vol.31 of Shui Dong Ri Ji (《水东日记》), Ye Sheng said: "the mirage phenomenon flourishes in the spring when there is a slight southeastern wind...its color is the color of water and slightly greenish, which is probably formed by the cyclone over the water." In Vol.18 of Liang Shan Mo Tan (《两山墨谈》), Chen Toing said: "the mirage of cities, people and horses is suspected to have been formed by the vapor and dust floating in the sunlight and the earth gas when the pond is overwhelmed, and occasionally changes." Associating the formation of the mirage with the effect of gases, these theories got rid of the traditional confinement of mythology and made a step toward the edge of science. As the ancients had no concept of air density, nor did they know light is refracted when passing through air of different densities, it was impossible for them to determine the scientific cause of the mirage. It was commendable that they could associate the mirage with winds and gases.

Another category of theories associated the mirage with the reflection of light and the atmosphere. In Guang Zhi Yi (《广志绎》), Wang Shixing of the Ming Dynasty described his understanding of the mirage: "Up close, there is nothing but sun glow; from a distance, you can see cities and markets." He already began to associate the formation of the mirage with the effect of light. In Vol.9 of Shu Yuan Za Ji (《菽园杂记》), Lu Rong further articulated this theory: "The mirage in Denglai is said to be the illusion caused by divine items, but why is it not caused by the divine gases from the mountains and rivers? By observation, we can find the so-called terraces and mirage are all caused by the gases from the mountains and rivers reflected by the sunlight, and are neither from clam gases nor divine items." This explanation associated gases with light and marked a step further from the theory of wind/air condensation.

The final category ascribed the formation of the mirage to the reflection from vapors. In Vol.2 of Wu Li Xiao Shi (《物理小识》) of Fang Yizhi of the late Ming

Dynasty discussed the mirage, saying: "The mirage over the Taishan Mountain is formed by fog and may be seen once every month...some believe it is caused by clam gases, which is not true." Zhang Yaoxing said: "there is a tower behind the city of Dengzhou, and below it is the sea. The tower faces several islands, and the mirage must have come from there. In the spring or fall of every year, the mirage appears when it gets slightly gloomy and changes instantly. I have personally borne witness to it. White gases arise from below the islands like a tide, and the mirage of pavilions appears accordingly." Based on these descriptions by Fang Yizhi and Zhang Yaoxing, Jie Xuan proposed the vapor reflection theory. When annotating this entry in Wu Li Xiao Shi (《物理小识》), he said:

> The light is reflected by the vapor, objects appear. The fog surges, which means the vapor arises. As the water can mirror things, so can the vapor that arises from it. As the vapor changes, the object it mirrors changes.

This theory by Jie Xuan has much to do with his theory of image dispersion and distribution (see the relevant sections hereof). He believed "people on the ground are mirrored to the sky at all times." Their images are distributed across the sky, and there is a giant mirror in the sky capable of reflecting the images of ground objects distributed across the sky. Like water, vapors can also mirror things, and those mirrored form the mirage. In Tian Jing Huo Wen Hou Ji (《天经或问后集》), Jie Xuan and You Yi provided a clearer explanation to the vapor mirroring theory:

> When the water is still, it mirrors people and objects like a mirror; when water vapor rises, it hangs in the air and also mirrors people objects like a mirror. Some think it is due to the air of the mountain and the sea, but do not know it is due to the moistened air.

The vapor mirroring theory proposed by Jie Xuan remained far from the real cause of the mirage identified by modern science. However, it was the most scientific of all ancient theories. The phenomenon explained by Jie Xuan et al. was the upward mirage. As we know, due to the evaporation of seawater and the passage of cold water, the atmosphere on the sea will be warm in the upper part and colder in the lower part, which aggravates the phenomenon that the upper part of the air layer is denser than the lower part. Just like a mirror, it reflects the distant scenery and forms a mirage. Therefore, the theories of Jie Xuan and others are closer to the understanding of modern science to some extent.

In ancient China, there were some other theories on the cause of the mirage in addition to the above five categories, but none of them reached the level of modern science. In the mid- and late nineteenth century, systematic western optics knowledge spread into our country. According to the western knowledge, the interpretation of the mirage was based on the knowledge of light refraction, which was consistent with the understanding in modern science and was accepted by the Chinese science community. At this point, all that was left of China's traditional theories on the cause of the mirage was eternal historic value.

8.3.2 Arguments over Distance and Size of Sun

The most attractive scene of the ancients' observation and interpretation of light phenomena in the nature was a debate over the distance and size of the sun. This debate originated from the famous story of "youngsters debating about the sun." According to Lie Zi – Tang Wen (《列子·汤问》), the story goes like this:

> Confucius traveled to the east and saw two youngsters debating. When asked, one youngster said: "I think the sun is close to us when it rises, and distant when in the middle of the day." The other youngster believed just the opposite. One youngster said: "The sun is as large as a wheel when it rises and is as small as a pan when in the middle of the day. A distant object is small, and a close object is big, right?" The other youngster said: "The sun is dim and cool when it rises, and scorchy in the middle of the day. It is hot when near, and cold when far, right?" Confucius could not tell who was right, and the two youngsters laughed: "Don't you claim to be knowledgeable?"

It is generally believed the book of Lie Zi (《列子》) was written in the Jin Dynasty period, while its stories were mostly the ancient stories that took place in the Zhou and Qin Dynasties periods. As this story was mentioned in Huan Tan's Xin Lun (《新论》), we have the reason to believe people in the pre-Qin era had already raised this question.

It is understandable for Confucius not to have an opinion on this issue as a humanity scholar. However, his silence did not affect the passion of astronomists, and in the Western Han Dynasty, people began to propose theories. Sui Shu – Tian Wen Zhi (《隋书·天文志》) systematically recorded discussions of this issue by people before the Sui Dynasty period. These discussions started with Guan Ziyang in the Han Dynasty period:

> In Xin Lun (《新论》), Huan Tan said: Han lieutenant Guan Ziyang believed the sun is far when above and close when on the sides. Why? In the morning when the sun rises, the sun looks close. At midnoon, the sun looks far. The more he observed, the more he believed the sun is far above and close when on sides. The sun is the heavenly sun, and fire is the earthly sun. When the earthly sun rises, the heavenly sun falls. Now we put a fire on the ground and measure its heat from aside and above, and the heat is different at different distances. When the sun is right above and covers people, people are right under the sun, and therefore it is hot when the sun just rises. Then, the sun passes through the cool zone, and therefore it is cool when the sun sets to the west.

The debate between the two children was based on different physical principles. One child held the view that a near object is larger, and a distant object is smaller, while the other held the view that the near object is hot and the far object is cool. Guan Ziyang agreed with the former and modified the latter. He said that the sun is indeed farther away from people at noon than in the morning, but "the sun is the heavenly sun," which is different from the common fire on the ground. When the common fire rises, and the "heavenly sun descends" and as a result, "the sun is hotter than when it rises because it is in the middle of the sky and covers people, who are right underneath the heavenly sun." The conclusion following "measuring with accuracy" might be Guan Ziyang's imagination or measurement inaccuracy, as the result of instrument measurement was contradictory to Guan Ziyang's conclusion.

Wang Chong of the Eastern Han Dynasty supported the second youngster's view, claiming "the sun is close in the middle of the day and distant upon sunrise." In terms of astronomy, Wang Chong agreed the sky and the earth are two parallel planes and the sun moves on the sky plane. At noon, the sun is right above the observer like the right angle of a right triangle, while in the morning and dusk, the sun is on the two sides like the hypotenuse of the triangle, and therefore "the sun is close in the middle of the day and distant upon sunrise and sunset." As for changes in the size of the sun, it is because "the sun is bright in the middle of the day and therefore small; and dark upon sunrise and sunset and therefore big. This is because the fire is small when observed in the day, and large when observed in the evening." (Wang Chong: *Lun Heng – Shuo Ri Pian* (《论衡·说日篇》))

Both Guan Ziyang and Wang Chong believed the distance between the sun and the observer changes in the morning and at noon, which is different from the universal sky theory, which suggested "the sun, the moon and the stars are equally distant from the center of the ground regardless of the season and time of the day." (*Jin Shu – Tian Wenzhi* (《晋书·天文志》)). Thus, universal sky theorists' explanation to this issue could only be based on the precondition that the sun's distance to the observer does not change, which is somewhat consistent with modern science.

Universal sky theorist Zhang Heng provided his insight. Sui Shu – Tian Wen Zhi (《隋书·天文志》) quoted his Ling Xian (《灵宪》) as saying:

> When the sun is thin, it looks dim. When you look at the bright place from a dim place, brightness has nowhere to bend and therefore looks big. When the sun is in the middle, the key and the earth are equally bright, and therefore light looks small. Fire appears luminous at night but dim in the day. The moon to the night is similar but slightly different from the sun to the day.

Both Zhang Heng's and Wang Chong's arguments were based on changes in brightness and contrast. They noticed this factor is indeed an important reason for the changes in the sights of the sun and the moon. We know from physics that among two objects of the same size, the brighter one looks larger, and this optical illusion is called light penetration. What Zhang Heng and Wang Chong noticed is the effect of light penetration on this issue. The difference in their conclusions was due to the different theories on the structure of the universe they believed in.

Shu Xi of the Jin Dynasty further explored this issue. He believed that the size of the sun on the side and above the head did not change. The reason why it looked different in size was due to various reasons. He said:

> …… When the celestial body is on the side, you look aside and that is why the sun looks bigger upon sunrise. While the sun remains constant in size, its image may expand or contract. When it contracts, it looks smaller, and when it expands, it looks bigger, which is how it works. Besides, when the sun rises and appears white, it is big but not scorchy; when the sun rises and appears red, it is big and scorchy; it depends on how it appears to people's eyes, and has nothing to do with the distance. If you place a tripod in open air, it will look

> like a pot. Under a roof of 10 rens (16m), even a tall man will appear short; things may look different from the way they are. The size and shape of objects can be confusing, and cannot serve as the basis for drawing conclusions. (*Sui Shu* – *Tian Wen Zhi* (《隋书·天文志》))

Shu Xi's discussion involved the effect of three factors. "When the celestial body is on the side, you look aside and that is why the sun looks bigger upon sunrise." He was talking about the difference between looking aside and looking up, which was a physiological cause; "If the sun is white upon sunrise, it is not hot though big; if the sun is red upon sunrise, it is both big and hot," which talked about the light penetration effect due to the difference in brightness and contrast; "it is that the object is set off against something, not that it has a different shape," which said the object was set off against the landscapes in the background. Relevant modern studies suggest these are basically the three reasons for the changes in the size of the image of the sun between the morning and the noon. As can be seen, Shu Xi's explanation is quite complete, and the conclusion "the object can be confusing and the shape can be deceptive, and therefore cannot serve as the basis for theories" is totally correct.

After Shu Xi, Zu Xuan of the Liang Dynasty also discussed the issue. He said:

> The sun looks big when on sides and small when above. It is difficult to look up at the sun, and easy to look at the sun at eye level. It is up to the perspective, not the distance. If we hang beads at an extremely high spot, or at an extremely distant spot, the sizes of the beads will seem distinctly different.

Zu Xuan's description did not go beyond Shu Xi's theory, but he explained difference between looking aside and looking up. As for the difference between the heat from the sun in the morning and at noon, Zu Xuan's explanation is similar to Guan Ziyang's. This explanation actually involves the change in the incidental angle of the sunlight and potentially makes sense. Zu Xuan's explanation also mentions the accumulation effect of heat, which is also correct.

The explanation by Jiang Ji of the Later Qin Dynasty elevated the ancients' study of this issue to a new height. He said:

When Zihang said the sun descends and it is hot under the sun, I think he means the sun appears bigger in the eyes than other celestial bodies, which is because it is close. All the celestial bodies across the sky and the circumference and peripherals that are measured through shadows are mere what they appear in people's eyes. When the sun just arises, it appears dim because it is aside, and when it reaches the middle of the sky, it appears bright. The luminance is actually the same according to measurement. There is no difference when the sun is aside or above. The essential light of the sun emanates the brightness and dazzles people's eyes, and as a result, it seems small in people's eyes. When the sun just rises, there are roaming gases on the ground, which obscure the sunlight and do not dazzle people's eyes, when the sun is red and large. In the absence of roaming gases, the sun is white and smaller. As the ground gases do not reach the sky, the sun is red in the morning and white at noon.

When ground gases arise and cover the entire surroundings, they connect with the sky, and the sun appears red even in the middle of the day.

Jiang Ji's contribution is twofold: First, he combined theoretical discussions with instrument observations, proving that the angular distance to the sun does not change from morning to noon and correcting the Guan Ziyang's misleading statement on this issue. Second, he explained why "the sun is red in the morning and dusk and white at midday," because "there is roaming air on the ground, which absorbs the sunlight." This involves the phenomena of atmospheric absorption and extinction. We know from physics that gas molecules, dust, and small water droplets contained in the atmosphere have a certain scattering effect on sunlight. In the early morning and evening, the sunlight is slanting on the ground. The atmosphere through which it passes is much thicker than that at noon. At this time, the scattering effect is much stronger. The ones that can reach the ground are mainly those with strong penetrating ability and long wavelengths, namely red and orange lights. This is what Jiang Ji said, "The sun is red in the morning and evening and white at midday." Jiang Ji explained this phenomenon by saying that "there is roaming air on the ground, which absorbs the sunlight." Although he failed to reveal its formation mechanism in detail, his approach undoubtedly built a bridge to the correct understanding of this phenomenon.

After Jiang Ji, the explanation as to why the sun appears distant in the morning and close at noon by Fang Zhongtong, a scholar in the early Qing Dynasty period, is also worth mentioning. As for the ground air said by Jiang Ji, he had a new explanation:

When the sun just rises and is not scorchy, this is because the earthly gas spreads horizontally, and that is why the sun appears big; the gas is thick and separates the heat, and that is why the sun is not scorchy. When the sun is scorchy and not big at noon, this is because the gas is thin on the ground, and that is why the sun does not appear big; the gas is thin and easily penetrable, and that is why it is scorchy. Besides, when the sun just rises, the light cuts through the boundary of the round earth with light force, and that is why it is not scorchy; at noon, the sunlight directly sheds upon the ground with strong force, and that is why it is scorchy. When the sun just rises, humans view it at eye level, and that is why it appears big; at noon, humans look up, and that is why it appears small. (Fang Yizhi: annotations to *Tian Lei – Qi Ying Cha* (《天类·气映差》) by Fang Zhongtong, *Wu Li Xiao Shi* (《物理小识》) (Vol.1)).

Fang Zhongtong's explanation was a new answer by the Chinese to this traditional question after introduction of the western concept of the Earth into China. As this answer was based on the concept of the Earth, there were descriptions such as "the sun just rises...the air is far apart...the realm of the Earth is cut by the light" and so on. Fang Zhongtong introduced the concept of the Earth into the discussion of the traditional issue of the distance and size of the sun in order to point out that the thickness of the air layer in different directions is also different and its effect on the penetration of sunlight, thus advancing the discussion by the Chinese people of the issue to a new level.

8.3.3 Understanding of the Nature and Propagation of Light

The ancient Chinese's understanding of the nature of light had undergone an evolutionary process. In the beginning, the ancients used the traditional theory of vitality to answer the question of what light is.

According to the vitality theory, gases are the essence of everything in the universe, and light is of course no exception. Therefore, in nature, light should be a type of gas. This is exactly what the ancients believed. For example, a famous physician Yihe in the Spring and Autumn period mentioned: "there are six types of gases in the world. . .the six gas is gloominess and sunshine/wind and rain/darkness and brightness." (*Zuo Zhuan – Shao Gong Yuan Nian* (《左传·昭公元年》)). Darkness and brightness are the manifestations of light, and the difference lies in the intensity of light. Yihe juxtaposed darkness and brightness, which suggested he failed to perceive the unity of the two. Anyway, as "brightness" can be regarded as a manifestation of light, he undoubtedly believed the light is gaseous.

Huai Nan Zi – Yuan Dao Xun (《淮南子·原道训》) presented the same explanation from the perspective that materials evolve into creatures:

> That which is intangible is the ancestor of everything; that which is quiet is the ancestor of sound. Is it true that the light and water, which are the son and grandson, are both from that which is intangible?

"That which is intangible" refers to vitality. This paragraph tells us: light is a direct product of vitality and should be a gas itself. Later generations had many theories, which simply said light is a gas. For example, in Vol.1 of Zhu Zi Yu Lei Ji Lue (《朱子语类辑略》), Cai Yuanding was quoted as saying:

> The sun is in the center of the earth and the moon travels in the sky. Therefore, light is the sun gas that spills from the empty spots in the surroundings and as a result, the moon receives the light.

Here Cai Yuanding simply said sunlight is the sun gas. Ming Dynasty medical scientist Zhang Jiebin was clearer about it:

> That which is bright is light, which is a gas from the fire; that which is hot is the substance of the fire. For example, a lamp fills the entire room with light, which is caused by the gas; the coal that fills the furnace has heat but no flames, which is caused by the substance. (Zhang Jiebin: *Lei Jing – Su Wen – Tian Yuan Ji* (《类经·素问·天元纪》))

He not only articulated the idea that light is a gas, but also described the difference between the light source and light.

Based on the notion that light is a gas, the ancients further discussed some issues related to light propagation.

Since light is a gas, the process of light emitting from the light source is the process of the light gas leaving the light source and spreading outward. The ancients called this an exterior scene. By the "scene," they meant light. Corresponding to the exterior scene, there is an interior scene, which means that the object is illuminated and reflected. The expressions of exterior and interior scenes can be seen in the pre-Qin literature. Zeng Zi – Tian Yuan (《曾子·天圆》) said "the fire is the exterior scene, and water is the interior scene," and Xun Zi – Jie Bi (《荀子·解蔽》) said "that which is opaque is the exterior scene, and that which is transparent is the interior scene," both of which mentioned exterior and interior scenes. In the Han Dynasty period, Huai Nan Zi – Tian Wen Xun (《淮南子·天文训》) explained the "exterior scene" and the "interior scene" with the air theory:

> That which is bright exhales air, and that is why the fire is called an exterior scene; that which is dark inhales air, and that is why water is called an interior scene.

"That which is bright" refers to the light source, such as a fire, which emanates light, called exhaling air, which is the exterior scene; "that which is dark" refers to the reflector, such as water, which can receive incoming light rays and reflect them to form images, as if there were an object, called inhaling air, which is the interior scene. Simply put, the difference between the exterior and interior scenes is exhaling or inhaling air.

There are different names for "exterior scenes" and "interior scenes." In Ling Xian (《灵宪》), Zhang Heng said: "The sky is round and the earth is square. That which is square governs darkness, and that which is round governs brightness."

"The sun is just like fire, and the moon is like water. The fire emanates light, and the water mirrors scenes." Here the "exterior scene" was replaced with "light," and the "interior scene" was replaced with "mirrors scenes." Ge Hong said: "Xihe is the exterior scene and hot, and Wangshu is the interior scene and cold" (Ge Hong: *Bao Pu Zi – Nei Pian -Shi Zhi* (《抱朴子·内篇·释滞》)). Xihe refers to the sun and Wangshu refers to the moon. All these descriptions are essentially the same and fall under the scope of Huai Nan Zi (《淮南子》). The ultimate basis for these descriptions is the common belief "light is air."

Parallel to the view that light is air is the theory of limit of light travel, which believed that the propagation of light is subject to a certain range. This kind of theory was quite common in ancient times. For example, Zhou Bi Suan Jing (《周髀算经》) said: "The sunlight propagates over 167,000 km in all directions, and people can see no farther than the sunlight's reach." Huai Nan Zi (《淮南子·地形训》): "the candlelight over the river shines thousands of kilometers away." Kai Yuan Zhan Jing (《开元占经》) of the Tang Dynasty quoted Shi Shen as saying: "the sun shines 162,000km, forming a diameter of 324,000km." Shang Shu Wei – Kao Ling Yao (《尚书纬·考灵曜》) said: "the sun shines 306,000km away." These figures are all different but have one thing in common: there is a limit to the traveling distance of light.

The concept of the limited traveling distance of light was a natural inference from the ancients' belief that light is air. Now that light is air, the light from the light source can only travel within a limited range, as it is inconceivable that a finite amount of air can spread to infinite space. In addition, in the process of light propagation, the phenomenon that it is bright when near and dark when far easily leads to the assumption that it can only spread in a limited range. According to this, the ancients' theory of limited travel distance of light has a certain basis, which is not pure imagination. Even with today's knowledge, that theory is not entirely unreasonable. Imagine a point light source that radiates light energy outward in the form of a spherical wave. The light energy received per unit area at a certain distance from the light source is inversely proportional to the square of the distance. In this way, when the distance increases to a certain extent, the light energy received per unit area will be lower than the perception threshold of the receiver, and this distance can be regarded as the limit of light propagation. In this sense, the propagation of light is indeed subject to a limit.

Another related issue is the speed of light: Is the propagation of light instantaneous or does it take time? The ancient Chinese did not have the concept of action at a distance. They thought a priori that the propagation of light takes a certain time, that is, the speed of light is limited. The ancients did not have a specific discussion on this issue, but we can indirectly get a glimpse of their attitude from some related materials.

Mo Jing – Jing Xia (《墨经·经下》) recorded pinhole imaging. When explaining that experiment, Jing Shuo (《经说》) mentioned: "The light is shed on man as if shot." The word "shot" implied the idea that light has a speed. Yu Di Shu (《与弟书》) by Cai Mo of the Jin Dynasty also clearly mentioned the light speed. The original text is as follows:

> Drums and beacons are used as ears and eyes in the army. Beacons can be seen at a distance, and drums can be heard at a distance. The image and the sound can travel hundreds of kilometers, which man cannot cover, serving to predict.

The above means light and sound are the same, and it takes a process for both to propagate, or in other words, both have a speed. "Travel hundreds of kilometers" indicates a high speed.

In the Tang Dynasty, Gan Zibu wrote Guang Fu (《光赋》), which discussed the nature of light and mentioned:

> While the light fills the space, it is not accumulated or congested. Light travels extremely fast, while people cannot directly feel it.

This paragraph vividly described the characteristics of light and expressed the ancient's notion that light has a speed.

The creation of the concept of the light speed is logically related to the ancient people's understanding of the nature of light: since light is air, the process in which an object is illuminated by light is the process by which the air of light reaches the illuminated object from the light source. Of course, it needs a propagation process. In this way, it is very natural for the concept of speed to come into being.

The last issue to be discussed here is the ancient people's understanding of the way light travels: Does light follow a straight line or a curve? Based on direct observations, it is assumed light travels in straight lines, and it is easy to garner such experience in life. Therefore, it was not difficult for the ancients to master this knowledge. Ge Hong of the Jin Dynasty said: "The sun and the moon cannot shine into curved caves." (Ge Hong: *Bao Pu Zi – Wai Pian – Bei Que* (《抱朴子·外篇·备阙》)). He also said: "The thunder cannot be tuned to resonate with gold and stone, and the sun and moon cannot curve their light." (Ge Hong: *Bao Pu Zi – Wai Pian – Guang Pi* (《抱朴子·外篇·广譬》)). Zhang Zai of the Northern Song Dynasty said: "Light from the burning sun goes straight." (Zhang Zai: *Zheng Meng – Can Liang Pian* (《正蒙·参两篇》)). All these are saying light travels along a straight line.

On the other hand, some ancients believed light travels in curves, which was mostly mentioned in astronomical works. For example, when explaining the solar eclipse, Du Yu of the Jin Dynasty suggested:

> The sun and the moon travel to the same spot, and the moon covers the sun, and ellipse happens...when there is a halo around the sun, sunlight overflows. (*Hou Han Shu* (《后汉书》) (Vol.28), quoted from Liu Zhao's annotations)

The ancients believed the sun and the moon are of equal size, and if light travels along a straight line, the ellipse cannot happen. Du Yu explained: the ellipse occurs because the moon covers the sun, and sometimes when the two get too close, sunlight overflows into the shadow of the moon. His words obviously expressed his belief that light can bend.

Jiang Ji of the Later Qin Dynasty also believed light travels in curves when explaining lunar eclipses:

> The luminance of the sun reaches even if it is dim and radiates in the four directions and all across the universe. The stars and the moon luminate like fire ascends from sparkles, and their light cannot be removed. The sun and celestial bodies are like full-faced drums. (Jiang Ji: *Hun Tian Lun Da Nan* ("浑天论答难"). Sun Xingyan: *Xu Gu Wen Yuan* (《续古文苑》) (Vol.9)).

That is to say, as the sunlight is limited by the celestial sphere when it travels, it spreads along the surface of the celestial sphere against the sun. Only the front side cannot be illuminated, which forms a dark void, and a lunar eclipse occurs when the moon passes there. As light travels along the curved surface of the celestial sphere, it is of course a curve.

The explanation to the cause of shadows on the moon by Zhu Xi of the Southern Dynasty also implied the thought that light travels in curves. He believed the shadows on the moon were projections of the earth on the moon, pointing out:

> The sunlight, accompanied by the moon, cannot penetrate the substantial barrier, and therefore there is some dark halo. (*Zhu Zi Yu Lei Ji Lue* (《朱子语类辑略》) (Vol.1)).

According to the knowledge of the ancients, the sizes of the sun and the moon are much smaller than that of the earth. In this way, the shadow formed by the earth under the sun's rays cannot be cast on the surface of the moon in a straight line and geometrically. To maintain the validity of Zhu Xi's statement, one must think that sunlight travels around the ground along a curve.

In summary, the ancient Chinese did have the knowledge that light travels long distances and can travel along a curve when blocked by objects. This is consistent with the ancients' speculation about the nature of light: since light is air, it certainly cannot be ruled out that it can bend around objects when obstructed. But it needs to be pointed out that the ancient Chinese theory of light travel is fundamentally different from the modern wave optics' view that light travels under the law of waves. The two must not be confused.

8.3.4 Fang Yizhi's Air-Light Wave Theory

Among the ancients' speculations about the nature of light, there was an outlier, which was the light wave theory proposed by Fang Yizhi, a scholar in late Ming Dynasty, in Wu Li Xiao Shi (《物理小识》).

In terms of the understanding of the nature of light, Fang Yizhi modified prior generations' theories. As mentioned in the previous section, when discussing the nature of light in ancient China, it was generally believed that light itself is a special kind of gas, and the process of luminescence is the process by which the gas of light reaches the recipient from the light source. Fang Yizhi believed that light is a manifestation of the gas and a special form of gas movement. He said: "The light decides brightness and darkness, just like the sun which decides brightness and gloominess. The fire has no substance, and the substance is shown by the light that sheds upon an object." His student Jie Xuan commented: "The gas is inherently light, and it is emitted by the sun's fire. It takes the gas as its substance, not the sun's fire. Therefore, the light by the window and in the air, where the sun's fire cannot reach, is residual reflections." That is to say, the function of the light source is to stimulate the light already contained in the gas. The light takes the gas as its substance and is not directly emitted by the light source. This is obviously different from the previous theory that light is the gas.

So, how does as manifest as light? In Vol.1 "On Light" of his Wu Li Xiao Shi (《物理小识》), Fang Yizhi described:

> Gases condense into shapes and manifest in light and sound, and air that has not condensed into shapes roams and blows. Therefore, the shape is limited to the boundaries, while light and sound propagate around. Air is not hollow, but transforms into shapes and vice versa.

Shapes refer to matters with a certain shape and texture; boundaries refer to the space that the "shape" occupies. Fang suggested the light and sound are the result of air being triggered. A tangible object occupies the space equivalent to its volume, and intangible light and sound propagate around from the spot of triggering. According to Fang, "space is filled with gases" (Fang Yizhi: *Qi Lun* (《气论》), *Wu Li Xiao Shi* (《物理小识》) (Vol.1)), namely gases fill the entire space without any gaps. Thus, once stimulated at one spot, the space is actuated everywhere, and the disruption is propagated through layers of air. It is like tossing a stone into the water to generate ripples, which spread constantly. Fang Yizhi's description is obviously a clear image of waves, which we call the air-light wave theory.

Fang Yizhi's air-light wave theory was very primitive, as it neglected cycles and lacked a clear description of vibrations. However, it was indeed a wave theory. There is a universal but incorrect notion that believes that vibrations without a cycle are not a wave. In fact, according to the definition in physics, waves are a disruption that propagates in the medium. Further, a propagation with changing states is a wave. Cyclicality is not the necessary condition for the formation of waves. A pulse wave is a wave. The solution provided in the wave equation section of modern textbooks of mathematical equations is: $f1\ (x + vt) + f2\ (x\text{-}vt)$, which does not mention cyclicality. The exclusion of cyclicality is an accurate reflection of the commonality of waves, while the claim "a propagation without cyclicality is not a wave" is wrong. The simple harmonic vibration propagation in college physics is only an ideal form of wave, not its general form. Neither did early advocates of the western light wave theory, such as Descartes (1596–1650) and Huygens (1629–1695) mention the cyclicality of light, which makes Fang Yizhi's omission justifiable.

The ancient Chinese, such as Wang Chong and Song Yingxing, had a deep understanding of the volatility of sound, which contemporaries do not deny. In Wu Li Xiao Shi (《物理小识》), Fang Yizhi associated light with sound on multiple occasions, believing the two occur and propagate in the same way, which is a natural reflection of his belief in the light-wave theory. Now that the ancients believed sound is a wave, why did they not acknowledge Fang Yizhi's theory that believed light was a wave? Of course, Fang Yizhi's light-wave theory is incomparable to today's wave optics. In Fang's theory, light and air interact with each other, which is a vertical wave, while modern optics believe light propagates as a result of electromagnetic effect and is a horizontal wave. Fang Yizhi's description of light is closer to modern people's understanding of the volatility of sound, which implicitly suggests it is a wave theory.

Starting with the volatility of light, Fang Yizhi also carried out distinctive analysis of the way in which light propagates. According to the air-light wave theory, light propagates by "swinging and absorbing" and "transforming and echoing." Thus, once encountering an obstacle, light naturally spreads toward the shadow zone.

Based on this, Fang Yizhi proposed an extremely important concept: fat light and thin shadows, which mean light always intrudes into the geometrical optics shadow to enlarge the light zone and reduce the shadow zone. In other words, light can propagate along curves. This is a natural conclusion from Fang's air-light wave theory.

Fang Yizhi proposed this concept when commenting on the western method of measuring and calculating the sun's diameter. He somehow misunderstood Matteo Ricci's sun diameter and claimed Ricci had said: "The distance from the sun to the earth is more than three times the sun's diameter," which is nonsensical and certainly could not be accepted by Fang. Through analysis, he believed the missionary had neglected the factor of "fat light and thin shadows" and therefore came up with a result that was longer than the actual diameter of the sun. He said:

> There are different Western schools, which have different methods, and how can their theories be used as basis? Upon scrutiny, the sun is at the end of the sun shadow, and is therefore this big. This is because sunlight is fat and the shadow on the ground is thin, why? The object is constrained by the shape, and its shadow can easily end, while sound and light often overflow from the object. Sound is invisible, while light is visible and measurable, but cannot be accurately measured. (Fang Yizhi: *Theory of Fat Light and Thin Shadow Can Solve Confusion about Sun 116 Times Larger than Earth* (《光肥影瘦之论可以破日大于地百十六余倍之疑》), *Wu Li Xiao Shi* (《物理小识》)(Vol.1)).

Fang Yizhi proposed "sunlight is usually fat, and the shadow on the ground is thin," which means when obstructed by the earth, sunlight bends around its geometric projection and as a result, the shadow on the ground is smaller, as shown in Fig. 8.4. In particular, he mentioned light shares this same nature with sound, which bypasses any obstacle when obstructed. So does light. While sound cannot be seen, light can be seen and measured. However, such measurement of an illuminant's direction based on the straight movement of light "cannot determine an accurate result."

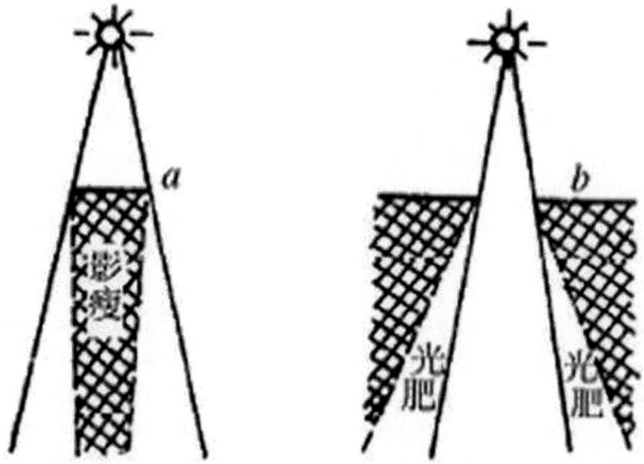

Fig. 8.4 Illustration of "fat light and thin shadow"

中文	英文
日	Sun
地	Earth
影瘦	Thin shadow
光肥	Fat light

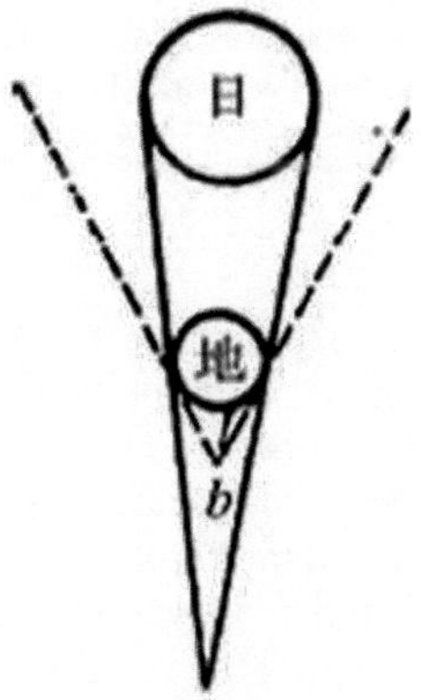

Fang believed the light behind the earth bends into the shadow, and therefore, the end of the sun's shadow is not its geometrical projection (a), but b, and based on b, the angle between the dotted lines can be determined.

Illustration of "fat sunlight and thin shadow on the ground."

A is a screen, and the shadow zone is smaller than its geometric projection; b is a hole, and the light zone is big.

To validate the "fat light, thin shadow" theory, Fang Yizhi further conducted a pinhole imaging experiment. Immediately following the above quote, he said:

> I tried to use paper by making a hole and letting light pass through to shed upon a stone, and the bright spot is as big as the hole. I gradually moved my hand higher, the bright spot is larger than the stone; I made four and five holes and let light shed as usual, and there are four and five shadows; as I moved my hand higher, the bright spots merge into one, while the shadows of the four and five holes can no longer be obtained. This is because light is fat and shadows are thin.

During the experiment, Fang Yizhi made four or five pinholes on the paper to allow sunlight to pass through the pinholes and project four or five bright spots onto the ground; Fang Yizhi believed he could prove the "fat light and thin shadow" is correct if the four or five bright spots merge into one when he holds the paper and "moves it higher." In fact, there are many causes that could lead to this and it is hard to assert the essential cause is "fat light and thin shadow." Anyway, Fang Yizhi tried to verify his theory with experiments, which is a shining spot in the history of Chinese physics.

Jie Xuan's annotations to "fat light and thin shadow" in Wu Li Xiao Shi (《物理小识》 help us correctly understand the intrinsic meaning of this theory. He said:

> The sun emits light thanks to fire. Fire constantly scatters, while in the case of a sphere, it cannot scatter and must circle around the sphere to reach the other pole. It can also circle around the earth or water currents. I observed the sunlight upon sunset, and found the sunlight embraced the earth from west to east as if around a bridge, and I knew its light does

> not travel along a straight line, but can bend along with objects instead of staying on a straight trajectory. . .the same applies to fat light and thin shadow. If light is smaller than the object, light is also fact and does not take a straight line. (Fang Yizhi: *Theory of Fat Light and Thin Shadow Can Solve Confusion About Sun's over One Hundred Times Larger Than Earth* (《光肥影瘦之论可以破日大于地百十六余倍之疑》), annotations by Jie Xuan, *Wu Li Xiao Shi* (《物理小识》)(Vol.1)).

This passage is essentially saying that during the propagation of light, if there is no obstruction, it will proceed along a straight line; if obstructed by an object, it will "be curved along the object and cannot proceed along a straight line." The fact that light propagates along the curve is the primary cause of fat light fat and thin shadow.

Fang Yizhi et al. also explained some common optical phenomena such as mirage with the light wave theory. From today's perspective, these explanations are far-fetched, while they reflect Fang's effort to ensure consistency of his theoretical system, which is noteworthy.

Fang Yizhi's theory of air-light wave theory is generally obliterated except for its influence on the future calendar calculations. This is not only a result of external factors, but also inseparable from its own inherent deficiencies. This theory is basically speculative, coupled with the lack of mathematical means, and the explanations of common optical phenomena are hardly convincing. Compared with the geometric optics that came from the West, it had obvious shortcomings and its oblivion was naturally no surprise.

8.3.5 Imaging Through Reflector

The imaging issue is an important subject of optical research. For ancient Chinese, this is no exception. Ancients' studies of the imaging issue could be classified into three types: reflector imaging, lens imaging, and pinhole imaging. We will cover reflector imaging here.

There are three types of reflector imaging: plane mirror imaging, convex mirror imaging, and concave mirror imaging.

Plane mirror imaging was the earliest, with its origin in "water mirroring the face." Inspired by this phenomenon, people realized any object with a smooth surface can mirror images, which led to the creation of copper mirrors. As early as the Shang Dynasty period, copper mirrors meeting certain quality standards emerged.

For the principle of reflective imaging, Huai Nan Zi – Yuan Dao Xun (《淮南子·原道训》) said: "The mirror reflects the object as it is without any smart design, and the shape, whether it is square, round or bent, is reflected the way it is." "Reflects the object as it is" suggests the image in the mirror is a reflection of the exterior; "without any smart design" suggests such reflection is entirely objective and is therefore credible. The ancients had carried out extensive research on the rules and applications of reflective imaging. The characteristics of flat mirror imaging were also covered in Mo Jing (《墨经》). The Mohists called the flat mirror a "proper mirror," believing the proper mirror generates an image identical to the object rather than the enlarged, reduced, upright, and upturned images generated by curved mirrors. In the case of flat mirror imaging, the object and the image are symmetrical on different sides of the mirror. When the object moves, so does the image, and the two always stay asymmetrical. The object is in front of the mirror, and the image is behind the mirror, and the image and the object are identical. The description of the characteristics of flat mirror imaging in Mo Jing (《墨经》) is plain and brief and is consistent with the actual circumstances of flat mirror imaging.

Given the ability of the flat mirror to reflect light, a complex image can be generated. This was not difficult for the ancients. As long as there were two flat mirrors at hand, it could be easily achieved. Therefore, what we need to understand is how the ancients recorded and explained this matter. When Lu Deming of the Tang Dynasty annotated Zhuang Zi – Tian Xia Pian (《庄子·天下篇》) in Jing Dian Shi Wen (《经典释文》), there was such a sentence: "A mirror can also be mirrored, so if we mirror a mirror with another mirror, there will be endless images." This was a quite intuitive explanation to the complex imaging issue. A Taoist priest by the name of Tang Qiao of the Southern Tang Dynasty went further: "If we mirror an object with a mirror, and mirror the image with other mirrors, the mirrors mirror one another and the image is mirrored once and again, while the patterns on the clothes are unchanged. This object is no different from the image; and the image is no different from the object; so we know the object is not solid, and the image is not virtual." This paragraph pointed out the image in a mirror can generate an image in another mirror, and the image generated is identical to the original object. In other words, multiple flat mirrors can generate complex images because first, the image generated by the flat mirror is identical to the original object; and second, the image can generate another image, and as a result, "mirrors mirror one another and the image is mirrored once and again" – the cycle is endless.

Compared with plane mirror imaging, convex mirror imaging is another matter.

Convex mirrors emerged very early in China. Among existing early bronze mirrors, more than one convex mirror was found to have been made during the Shang and Zhou Dynasties periods. Convex mirror is a divergent mirror, and the image it generates is an upright miniature. The ancients had a clear understanding of this. Mo Jing (《墨经》) said: "The convex mirror generates a miniature," which accurately revealed this characteristic. As the convex mirror generated an upright miniature, it could cover a larger scope than the flat mirror. The ancients leveraged this feature and skillfully selected the curvature of the mirror so that a small convex mirror could also cover the entirety of a person. In this regard, Shen Kuo of the Northern Song Dynasty had a clear description: "When the ancients made mirrors, if the mirrors were large, they must have been flat; if small, they must have been convex. If the mirror was concave, one's face in the mirror was larger than actual; if convex, one's face in the mirror was smaller than actual. As the small mirror could not cover the entirety of one's face, it was made slightly convex so the face appeared small, and that was how a small mirror could cover the entirety of one's face. Then they measured the size of the mirror and adjusted it accordingly to match the size of the mirror to the size of the face. This craftsmanship was so demanding that generations could not master it, and could only smoothen ancient mirrors by scraping them, which was how the ancient craftsmanship was lost." (Shen Kuo: *Meng Xi Bi Tan* (《梦溪笔谈》) (Vol.19)). Shen Kuo correctly described the relationship between how convex the mirror surface was and the size of the image it generated. Although he lamented some mirror makers at that time did not understand the underlying mechanism, his description undoubtedly helped the public understand the imaging mechanism of the convex mirror.

In terms of application, the ancients used the convex mirror (or flat mirror) as a light path conversion device in addition to a mirror. Huai Nan Wan Bi Shu (《淮南万毕术》) said: "With a large mirror hanging high above, one can see neighbors all around while sitting." The "large mirror" here should be a convex mirror, as only a convex mirror allows one to "see neighbors all around while sitting," while a flat mirror can only show the neighbors from a certain angle. However, the annotation by Gao You of the Eastern Han Dynasty undoubtedly involved the reflective function of the flat mirror: "Place a basin full of water under the large mirror high above and you will see neighbors all around." The basin here serves as a flat mirror which reflects the images of neighbors on the convex mirror above to the viewer. In this way, the viewer only needs to look up at the convex to see all the neighbors around. However, as looking up is tiring, the viewer now can look down at the water in the basin. Here the basin serves as a light path conversion device that allows the viewer to observe from a relatively comfortable angle.

Compared with flat mirror imaging and convex mirror imaging, concave mirror imaging is the most complicated: When the object is beyond the center of the sphere, an upturned, reduced image is formed, and the image is between the center of the sphere and the focal point; when the object is between the center of the sphere and the focal point, an upturned, enlarged real image is formed, and the image is beyond the center of the sphere; when the object is within the focal point, an upright, enlarged virtual image is formed, and the image is behind the mirror. While the ancients did not know the difference between a real image and a virtual image, nor did they study the imaging location, they were aware of the effect of the object's position on the imaging result. While observing concave mirror imaging, they paid great attention to how change in the object's position affects the upright/upturned position and size of the image. The Mohists were like that. Mo Jing (《墨经》) recorded Mohists' experiments on concave mirror imaging and their theoretical explanations.

Compared with flat mirror and convex mirror imaging, concave mirror imaging is the most complicated: When the object is outside the center of the sphere, an inverted reduced real image is generated, and the position of the image is between the center of the sphere and the focal point; when the object is located between the center of the sphere and the focal point, an inverted enlarged real image is generated, and the image is outside the center of the sphere; when the object is in the focal point, an upright magnified virtual image is generated, and the image is behind the mirror.

> Mo Jing (《经》): In the case of a concave mirror, one image is small and upturned, and the other is big and upright. The object is located outside of the center.

> Jing Shuo (《说》): Mirror: If within the center, when the object approaches the center, the image grows larger; when the object moves away from the center, the image grows smaller and must be a square: the object starts at the focal point, and the light parallel to the positive axis is reflected back to the mirror and elongated as a very long conjugate point, and the image is far away from the mirror. If outside the center, when the object approaches the center, the object gets more light and the image grows larger; when the object moves away from the center, the object gets less light and the image grows smaller, and must be upturned. The object is concentrated at the center, and the light parallel to the positive axis is reflected as a very long conjugate point, and the image is equal to the object but upturned.

Mo Jing (《经》) failed to record all the three scenarios of imaging, while the description in Jing Shuo (《经说》) covered all the three scenarios. This suggests the Mohists' had carried out delicate observations. The Mohists had already realized when the object is at a different position in front of the concave mirror, the image generated is different. When the object is located at the focal point of the concave mirror, or the so-called "center," an upturned image smaller than the original object will be generated; when the object moves from the distance toward the mirror, the image generated will slowly grow larger; when the object is located between the focal point and the mirror, an enlarged upright image is generated. At this time, when the object moves toward the focal point (namely away from the mirror), the image grows larger and larger. That is to say, whether the object is outside or within the focal point, the image grows larger as long as it approaches the focal point, and vice versa. The Mohists explained it depends on "whether the object is outside or within the center." "Starts at the focal point" and "concentrated at the center" indicate the Mohists' understanding of the mechanisms of generating upright and upturned images. The "center" mentioned by Mohists could only be the focal point, which determines whether the image is upright or upturned. The Mohists' interpretation is essentially consistent with the latticing technique said by Shen Kuo. This will be covered later.

8.3.6 Concave Imaging and Applications

Lens imaging is based on the refraction properties of light, and refraction is an important subject of research in geometric optics. Thus, lens imaging was extremely important in geometric optics, and ancients' discussion of lens imaging became an indispensable part of the history of optics.

In ancient China, the study of lens imaging started late, which was probably due to the failure of the ancients to develop mature glass making techniques. Even today, the earliest time when glass lenses emerged in China remains uncertain. While some of the unearthed glass artifacts of the Han Dynasty are capable of lens magnification, whether they were used as lenses is still being debated in the academic circle. There is also debate over whether they were invented by the Chinese or imported from abroad. While these questions are certainly important, what role these instruments played in the development of ancient optics is more important.

Unfortunately, ancient literature that talked about lens in a clear context of imaging is relatively new. Although some earlier literature did explain from the perspective of lens imaging, this is dubious and it is therefore hard to reach a conclusion. In Hua Shu – Si Jing (《化书·四镜》), Tang Tanqiao mentioned: "I have four mirrors, namely gui, zhu, di and yu." The object appears large in gui, small in zhu, normal in di and upturned in yu. Needham interpreted Tang Tanqiao's "four mirrors" as four kinds of lenses, of which gui is a double concave lens, zhu is a double convex lens, di is a flat concave lens and yu is a flat convex lens. (John Needham, translated by Lu Xueshan: *History of Chinese Science and Technology* (《中国科学技术史》)(Vol.4) *Physics and Relevant Technologies* (《物理学及相关技术》)(Book 1) *Physics* (《物理学》), Science Press, 2003, p.111–112). However, some scholars also believe they are all curved reflectors, which can also produce images through reflection and are more applicable to "large, small, upright and upturned images.". Therefore, we cannot assert the "four mirrors" here are lenses.

However, we cannot assert the Tang people did not know lens imaging, either. On the other hand, the ancient Chinese had knowledge about lenses at an early time. The story of acquiring fire through an icy lens during the Western Han Dynasty period is a convincing example. As more lenses were introduced into China since the Jin and Sui Dynasties periods, there were more records about "fire pearls" in historical books. The so-called "fire pearls" refer to convex lenses, which are capable of converging and acquiring fire from sunlight. The magnification capability of the convex lens is easily noticeable in the practice of acquiring fire from sunlight. Once people leverage its magnification capability to observe an object, this is associated with imaging. It was totally possible for people to have such knowledge during the Tang Dynasty period. Also in Hua Shu (《化书》), Tan Qiao said: "We set a mirror to see what eyes can't see." The "mirror" here should mean a lens. This sentence means using the magnification ability of a lens to see what we are unable to see or unable to see clearly with naked eye.

If one observes an item with the help of a lens, he or she should see a large, normally positioned image, which can only be a virtual image. So, did the ancients know lens imaging produces a real image? In Qu Yi Shuo (《祛疑说》), Chu Yong of the Song Dynasty revealed some charlatans' scams, one of which was related to optics, called the "scene movement trick." He said: "Despite the various kinds of scene movement tricks, the general practice is to hide the image in the mirror, place a lamp aside to let the mirror and the lamp shed light upon each other and project the shadow to the paper. This trick has been used a lot recently and is simply ludicrous." Undoubtedly, this is an optical imaging activity, and the image produced is a real image. The phrase "projecting the shadow to the paper" indicates the existence of an image screen, and such an image must be a real image. However, such a real image could be produced through both the refraction and reflection of a lens. Given the fact that no image screens were used to display the image in the concave mirror imaging experiments recorded in ancient Chinese literature, the trick mentioned here is more likely to be lens imaging.

"Zhou Bida gave a soup bowl to a friend. When soup was poured into it, the pattern of double cranes was seen and disappeared when the soup was drunk up." It was two cranes that were seen here. Things like this can also be seen in operas. There was a widely popular traditional Chinese opera called the butterfly cup, where the hero had a magical "butterfly cup." Once wine was poured into the cup, a butterfly was seen flying in the cup, and when there was no wine in the cup, the butterfly disappeared. The butterfly cup and the agate urn obtained by Chen Gao are the same kind of things. Now that they are related to people's vision, we can only explain from the perspective of optics.

A story recorded in Chun Zhu Ji Wen (《春渚纪闻》) by He Yuan of Song Dynasty is related to lens imaging. He Yuan mentioned a man called Chen Gao, who acquired an agate urn from an ancient tomb and used it to store water and make ink. One day, he accidentally found a carp in it. He emptied the urn and the carp was gone. When he poured water back into it, the carp appeared again. He tried to grab it, only to get nothing, wondering what magical item this was. Magical as this story sounded, there are many similar records in ancient literature. Some saw a flower, and some saw other things. For example, in Vol.5 Yi Jian Zhi (《夷坚志》) of Tao Shuo (《陶说》), Zhu Yan of the Qing Dynasty recorded:

> The butterfly cup was successfully remade in Houma City, Shanxi Province in the late 1970s. The principle is as follows: the cup is divided into upper and lower parts, with upper part serving as the cup body, and the bottom of the cup body is made into the shape of convex lens, which is mounted on the cup leg at the lower half. A butterfly is installed in the leg of the cup with a thin spring (hairspring). Once the cup is slightly disturbed, the butterfly moves. The colorful butterfly is located at the focal point of the lens or close to the focal point outside the focal point, so that when there is no wine in the glass, the colorful butterfly generates a real image on the same side as the human eye but with a large image distance. When one looks into the glass, the image falls behind the viewer and is naturally invisible. When pouring wine into the glass, as the glass wall is moistened by the wine, the surface of the wine is in the shape of concave and is equivalent to a concave lens. The concave lens and the convex lens form a compound lens. The focal length of the compound lens is larger than that of the convex lens. At the same time, the difference in refractive index between wine and glass is also smaller than the difference in refractive index between air and glass, which further increases the focal length of the compound lens, causing the butterfly to fall within the focal length of the compound lens and generating a magnified virtual image. The compound lens functions as a magnifying glass, so the human eye can clearly see the enlarged butterfly (see Fig. 8.5). When held in the hand, the cup is inevitably subject to external disturbance, and the butterfly flies.

The ancients did not necessarily know the scientific principles mentioned above when making vessels like the butterfly cup and agate urn. They figured out the method of making these vessels in practice. Besides, these vessels did not necessarily follow the same principles. However, the above effect can be achieved through lens imaging anyway.

In the late Ming Dynasty period, missionaries entered China and brought western science and technology knowledge, including optics. Under the influence of western science, Chinese scholars not only had a certain understanding of the phenomenon of lens imaging, but also discussed the imaging mechanism. Late-Ming scholar Fang Yizhi quoted his teacher Yang Guanguang as saying: "For that which is convex, the light beams intersect in the front, and for that which is concave, the light beams intersect in the rear." (Fang Yizhi: *Wu Li Xiao Shi* (《物理小识》)(Vol.2)). That which is convex refers to a convex mirror; and that which is concave refers to a concave

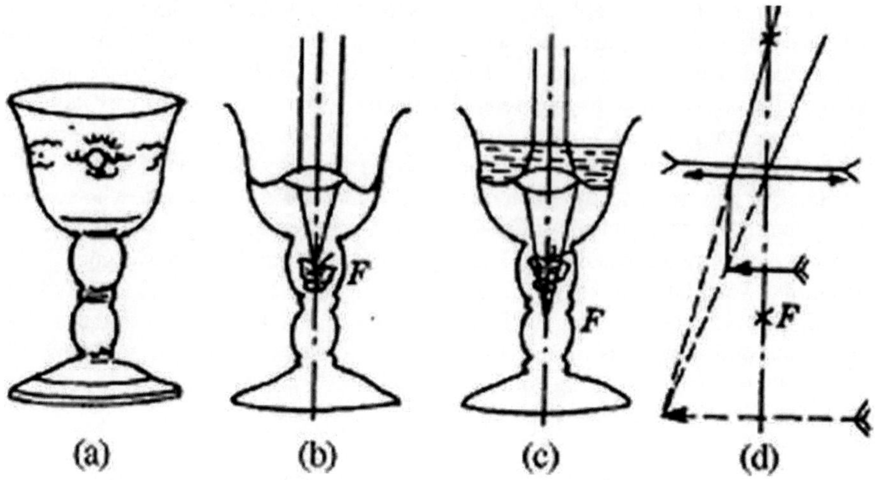

Fig. 8.5 Butterfly cup illustration
(**a**) Appearance (**b**) The butterfly is at the focal point of the lens when the cup is empty of wine (water) (**c**) The butterfly is within the focal point of the lens when the cup contains wine (water) (**d**) Light path illustration

mirror. Here it clearly explained the divergence capability of a convex mirror and the convergence capability of a concave mirror from the perspective of the light path, which symbolized a further step in the study of lens imaging.

In Vol.12 of Wu Li Xiao Shi (《物理小识》), Fang Yizhi also recorded the method of drawing with lens imaging:

> A glass mirror is placed on the window of the darkroom, and the object's shape is minified and projected to the paper and can be drawn from in detail. Drawings like these have a high level of resemblance, including "flowers, trees, insects and objects".

This is not pinhole imaging, which does not require a glass mirror; besides, although pinhole imaging as an adequate depth of focus, its clarity is insufficient, and it is possible to copy-paint "flowers, trees, insects, and objects." Only lens imaging has the characteristics described by Fang Yizhi above. A hole is made on the window of the darkroom, where the lens is embedded, and objects outside of the darkroom are typically twice the focal length of the lens away. Thus, an upturned miniature image is created inside the darkroom, and with paper as the screen, one can copy the image with a brush pen. This is why Fang Yizhi said "the copied painting has a high level of resemblance." By changing the distance between the paper and the lens, one can obtain images of objects at different locations outside of the window, and it is possible to copy-paint "flowers, trees, insects, and objects." In the annotation, Jie Xuan said: "This method of the Far West is drawing objects based on the objects' images." This implies it is lens imaging, as pinhole imaging is by no means "a method of the Far West."

This method described by Fang Yizhi is so creative that it could be tantamount to a camera if the techniques of displaying and fixing the image were available. It was through a similar experiment that Zou Boqi of the late Qing Dynasty invented photography. Fang Yizhi's introduction of lens imaging into drawing was an example of applying science to art. This practice of his was a result of western influence. In Yuan Jing Shuo (《远镜说》), Johann Adam Schall von Bell mentioned that a telescope "can be used to draw in the darkroom," which certainly inspired Fang Yizhi. However, Fang Yizhi used a single piece of lens instead of a telescope, and drawings made in this way have a larger vision.

Zheng Fuguang achieved the greatest results in the in-depth and systemic study of the lens imaging mechanism from the perspective of geometric optics. He created a whole collection of terms and concepts and systematically examined the characteristics and rules of lenses and even lens groups, making a major step in quantitatively revealing the rules of lens imaging. However, his work was completed in the mid-nineteenth century, which was quite close.

8.3.7 Pinhole Imaging

As an important imaging phenomenon, pinhole imaging was thoroughly studied during the development of ancient Chinese optics and was an important of ancients' optics achievements. The earliest work that observed and explained pinhole imaging from the experiment perspective was of course Mo Jing (《墨经》). The Jing Xia (《经下》) chapter of Mo Jing (《墨经》) recorded Mohists' take on this issue:

The image is upturned because of the pinhole on the shading screen.

When light is cast on an object, the shadow is as straight as a bolt if light is reflected. If the light is cast to the downside, it is reflected to the upside, and vice versa. If you block the light with a foot, the projected image is on the upside; if you block the light with your head, the project image is on the downside. With a pinhole far from or near the object, light is cast on the object and projects the shadow of the object to the wall, and therefore, the image on the screen is upturned.

Interpretations of the above texts are mostly based on the assumption that light travels along straight lines. The Mohists were basically saying that in the context of pinhole imaging, due to the existence of the pinhole, the incoming light rays intersect at the pinhole, and the light ray entering from below goes above in the darkroom and vice versa, forming an upturned image in the darkroom.

The "image" here is in a broad sense. It can also be a projection. For example, when a man stands between the sun and a pinhole, there will be the man's upturned shadow, which is a projection, in the evenly distributed sunlight image on the screen behind the pinhole under appropriate conditions. According to Mo Jing (《墨经》), it could be a shadow instead of an image that was observed in the experiment then. Be it a projection or an image, it is an upturned image anyway and gives people the same inspiration, which is light travels along a straight line, and due to the constraint of the darkroom pinhole, the light generates an upturned image after entering the darkroom. According to this description in Mo Jing (《墨经》), the Mohists had a clear concept in this regard.

Due to the decline of Mohism, Mo Jing (《墨经》) had not been known by people for a long time, and Mohists' insights from pinhole imaging experiments had not been well inherited. In the Qin and Han Dynasties periods, the study of pinhole imaging was in a state of rediscovery and starting over in discussing the mechanism. In this respect, much reference was made to the so-called "upturned tower shadow." In You Yang Za Zu (《酉阳杂俎》), Duan Chengshi of the Tang Dynasty said: "As the ocean wave moves, the tower shadow is upturned," which suggests he knew nothing about the imaging mechanism. This phenomenon was also mentioned in Lao Xue An Bi Ji (《老学庵笔记》) by Lu You of the Song Dynasty and Shan Ju Xin Hua (《山居新话》) by Yang Yu of the Yuan Dynasty. In the Ming and Qing Dynasties periods, such records were more commonplace, and some people even specially collected examples of "upturned tower shadows" across the country, which suggested the physical phenomenon of "upturned tower shadows" was widely noticed.

Shen Yue was worth mentioning during the rediscovery of pinhole imaging. In the Ode to Moon, he said:

> I see moonlight at tranquil night, everything is dead silent, square light enters the house, and round shapes come from the holes. (*Yi Wen Lei Ju* (《艺文类聚》) (Vol.1)).

Here he described two phenomena: One is "square light enters the house," namely the moonlight enters the house from the door and projects square lights on the ground in the shape of the door; another is "round shapes come from the holes," namely the full moon outside sheds light through the holes on the wall and projects what is still a round moon onto the ground. In these two circumstances, the moonlight generates the image of the moon inside through the holes, and when there is a full moon, this image is certainly round and irrelevant to the shape of the hole, while for the house door, its size is far larger than what is required for pinhole imaging in comparison to the distance for indoor imaging, and at this time, moonlight can be regarded as parallel light beams, which project the shape of the door instead of the moon's image onto the ground. Shen Yue's comparing these two phenomena suggests he perceived the difference between them. However, it was not until Zhao Youqin of the late Song and early Yuan Dynasties and even Zheng Fuguang of the late Qing Dynasty who managed to reveal the causes of these phenomena.

Zhao Youqin, a.k.a. Qin, alias Jingfu, or Zigong and Yuandu, known as Mr. Yuandu or Yuanduzi (Taoist name), was born in Poyang, Jiangxi. He was the 12th generation grandchild of the imperial family of the Song Dynasty and an important scientist in the late Song and early Yuan Dynasties periods. Zhao Youqin was presumably active in the early Yuan Dynasty period.

In ancient China, Zhao Youqin was the first to discuss in detail the mechanism of pinhole imaging from the perspective of experiments. He first observed sunlight and moonlight forming images through pinholes on the wall and found although the pinholes had various shapes, the images formed by the sunlight through pinholes were always round; and the pinholes could vary in size, while the sun's images generated were always of the same size. The only difference is larger pinholes generated brighter images, and smaller pinholes generated dimmer images. This is always the case with the sun's indoor image in the event of eclipse, when the indoor eclipse image is of the same size as the outdoor one. He also found when the pinhole was large enough to contain the visual diameter of the sun or moon, what appeared indoor was no longer the sun's image, but the pinhole's projection, or "the projection generated by a large hole is necessarily of the same shape of the hole." (Zhao Youqin: *Ge Xiang Xin Shu – Xiao Xia Guang Jing* (《革象新书·小罅光景》)). Zhao Youqin provided a very detailed description. Furthermore, he meticulously designed a massive pinhole imaging experiment and discussed the effects of the various factors during the imaging process.

Using a house as the lab, he dug a round well in each of the left and right rooms. The wells were 133 cm feet in diameter. The left well was 266 cm deep, and the right well was 133 cm deep. Then, he made two wooden plates with a diameter of 133 cm and inserted 1000 candles on each plate. During the experiment, he placed a 266 cm table in the left well and placed a plate with candles in each of the wells, covered the wells with plates, and made a square hole with a side length of 3.3 cm at the center of the plate covering the left well and a square hole with a side length of 5 cm at the center of the plate covering the right well (as shown in Fig. 8.6). The candlelight was shed to the floor slab through the square holes, and when the image distance needed to be changed, a large wooden plate was hung beneath the floor slab as the image screen.

Zhao Youqin's arrangement makes sense. In this way, he could randomly fix or adjust all the imaging factors, thereby clarifying their roles in the imaging process and

Fig. 8.6 Illustration of the device used in Zhao Youqin's pinhole imaging experiment

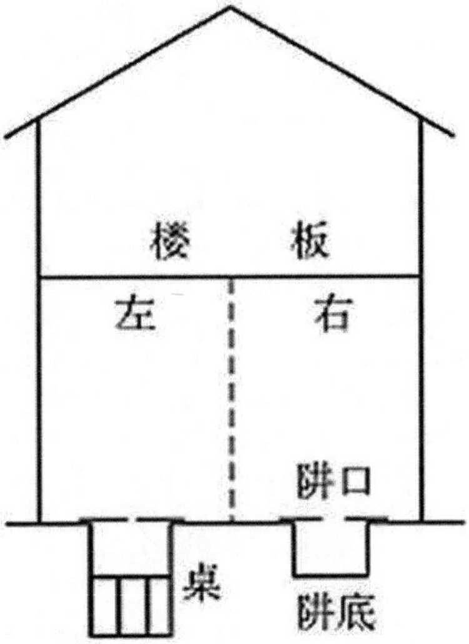

中文	英文
楼板	Floor slab
左	Left
右	Right
桌	Table
阱口	Well top
阱底	Well bottom

revealing the secrets of pinhole imaging. In his own words: “In this way, I can determine how the candle, the light, the hole and the image interact with one another.”

In the experiment, he first kept the distances between the light source, the pinhole, and the image screen unchanged and observed how the size and shape of the pinhole affected the images. He found the two images had similar sizes but different shades of darkness. He tried to explain with pixel overlapping and the straight path of light: “One thousand candles generate one thousand images, which are square dots created by the pinholes.” Each candle projected a bright spot onto the floor slab through a pinhole, and these bright spots were square, just like the pinholes, and their positions corresponded to the candles through the pinholes. Although every bright spot was square, “the one thousand images overlapped with one another and always formed a circle.” In the case of large holes, more light passed through from each candle and the images were naturally brighter as a result of overlapping.

Then, he conducted the simulation experiment of “images eclipsing along with the sun or moon.” He “removed the 500 candles to the east from the right well and noticed the western half of the images on the right floor slab was missing, which was how the images eclipsed along with the sun or moon.” Then, he further adjusted the light sources by extinguishing most candles in the wells, with about two dozen candles left. At this time, the image on the floor slab was a circle consisting of two dozen “square images,” which were not interconnected, and was very dim, which intuitively suggests the round image on the screen was indeed formed by the square bright spots. Finally, he lit only one candle, and there was only one “square image” on the screen. Zhao Youqin explained this is because “when the hole is small, the light shape is especially small, and the hole can fully contain the light.” And this is why “images from a large hole follow the shape of the hole.”

Zhao Youqin also changed the object distance and image distance, conducted the large hole imaging experiment, and finally reached the conclusion: “Therefore the image created by a small hole follows the shape of the light source, and the image by a large hole follows the shape of the hole, which is doubtless.” Wang Jinguang et al. closely examined Zhao Youqin’s pinhole imaging experiment and listed the key elements of Zhao Youqin’s experiment in a table, which is valuable for us to understand this experiment. The table is as follows:

改变的项目		像的大小	像的浓淡（照度）
小方孔	1寸	几乎相同	淡
	1寸半		浓
光源 像距	1千支蜡烛	几乎相同	浓
	二三十支蜡烛 大	大	淡 淡
	小	小	浓
物距	大	小	几乎相同
	小	大	

中文	英文
改变的项目	Changed items
小方孔	Small square hole
1寸	1 cun (3.3 cm)
1寸半	1.5 cuns (5 cm)
光源	Light source
像距	Image distance
1千支蜡烛	One thousand candles
二三十支蜡烛	Twenty or thirty candles
大	Big
小	Small
物距	Item distance
大	Big
小	Small
像的大小	Image size
几乎相同	Almost identical
几乎相同	Almost identical
大	Big
小	Small
小	Small
大	Big
中文	英文
像的浓淡(照度)	Image brightness/dimness (illuminance)
淡	Dim
浓	Bright
浓	Bright
淡	Dim
淡	Dim
浓	Bright
几乎相同	Almost identical

As can be seen from the table, Zhao Youqin discussed almost all the factors involved in pinhole imaging. Moreover, he explained the phenomenon in the experiment from a theoretical perspective based on the assumptions of pixel overlapping and straight path of light, which are correct.

After Zhao Youqin, Zheng Fuguang of the Qing Dynasty further studied pinhole imaging. His Jing Jing Ling Chi (《镜镜詅痴》) and Fei Ying Yu Zhi Lu (《费隐与知录》) contained dedicated descriptions. However, this is something that happened in the mid-nineteenth century.

Guo Shoujing of the Yuan Dynasty, who was active in a time slightly later than Zhao Youqin's, invented an observation instrument that could significantly improve the effect of astronomical observations by successfully applying the principles of pinhole imaging. This is what must be mentioned in the history of physics.

In ancient China, the solar terms, especially the timing of winter and summer, were determined by measuring the length of the shadow of a measurer erected. As the sun's position in the sky changes, the length of the shadow on the ground changes. Therefore, by measuring the length of the ground shadow, one can reversely determine the position of the sun in space. This is how it works. In the pre-Qin dynasty period, the timing of winter and summer was determined by measuring the length of the shadow projected by the midnoon sun, which became an important method of determining the timing.

The measurer is typically 266 cm tall. In order to improve measurement accuracy, Guo Shoujing made major improvements by using a monument-shaped measurer and increasing its length to 1.19 m. At the top of the measurer, a beam with a diameter of 9.9 cm was placed on two dragon statues, and the center of the beam was 1.3 m from the surface. Thus, the distance from the beam shadow to the measurer bottom was 5 times the length of the shadow of the 266 cm measurer. From the perspective of error theory, with the same measurement error, the relative error of a height measurer was a mere one-fifth of the 266 cm measurer. In other words, the measurement accuracy was improved by five times. However, the use of a height measurer aggravated the problem of shadow vagueness. The key to measuring shadows with a measurer was to improve the accuracy of shadow measurement. Due to diffusion of sunlight by air molecules and dust impurities, the edges of shadows were blurred, which was a major obstacle to improvement of measurement accuracy. With the height measurer used, the problem of blurred shadows got worse. Around the solar term of dongzhi, the shadow was large and the problem of blurred edges was more severe, and as a result, the observer could not determine the exact position of beam shadow. Yuan Shi – Tian Wen Zhi (《元史·天文志》) analyzed this as such: "If the measurer is short, the image is short; if the measurer is long, the image is long and the shadow is virtual and dim, and solid shadows are hard to obtain." Basically, it was saying a height measurer can improve measurement accuracy, which is right. The sentence "solid shadows are hard to obtain" is an accurate summary of the height measurer's defects.

Guo Shoujing's solution was to use the shadow definer in measurement. Yuan Shi – Tian Wen Zhi (《元史·天文志》) in detail recorded the structure and usage of the shadow definer, as well as two figures obtained from measurement with the shadow definer:

> A shadow definer is made of copper and 6.66cm wide, with a hole in the center, which is tiny. With the shadow definer as the foot, an axis is set on one end to open and close it and adjust one end of it to make it lean, with the northern side higher than the southern side, and the sunlight travels through the beams. The sunlight is a mere meter long and is looming between the beams. The old approach is to measure the sun's shadow to acquire the image above the sunlight. Now we measure with a beam to acquire the middle image, and no errors are tolerated. The solar shadow in the solar terms of yimao and xiazhi of the 16th year of Emperor Zhiyuan's reign is one zhang, two chi, three cun, six fen, nine li and wu hao long on the 19th day of the 4th lunar month; and the solar shadow in the solar terms of yimao and dongzhi of the 16th year of Emperor Zhiyuan's reign is seven zhang, six chi, seven cun, and four fen on the 24th day of the 10th lunar month.

Fig. 8.7 Shadow definer

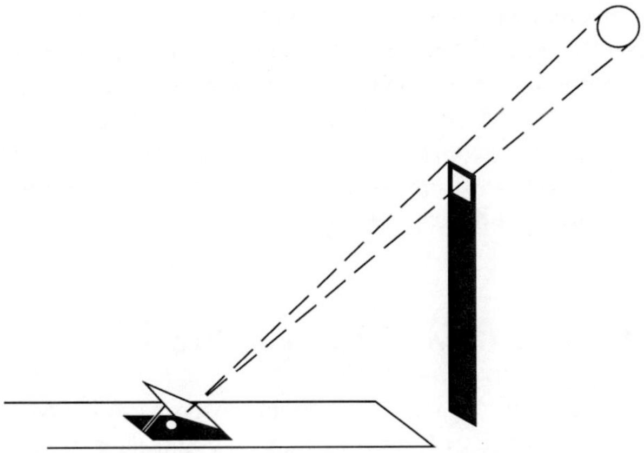

According to this paragraph, the main component of the shadow definer is a thin copper piece, which has a small hole in the center. The copper piece is mounted on a rack, with an axis at the lower end, and the other end can be supported in a slanted direction, with the angle freely adjustable. The rack is moved back and forth on the jade tablet, and when the sun, the beam, and the pinhole are aligned, a rice-sized bright spot can be seen on the jade tablet, with a horizontal line in the middle, as shown in Fig. 8.7. Here the shadow definer serves as a pinhole imager, and the bright spot on the jade tablet is the image generated by sunlight through the pinhole. The horizontal line in the bright spot is the image generated by sunlight shed on the beam at the end of the measurer. Based on the position of the image on the jade tablet, the corresponding shadow length can be accurately measured.

The use of the shadow definer brought about two breakthroughs in the traditional shadow measurement technology: First, it basically solved the problem of "dim shadows and difficulty in obtaining solid shadows" due to the diffusion of sunlight by air molecules and dust and significantly improved the measurement accuracy.

Another breakthrough was described in Yuan Shi – Tian Wen Zhi (《元史·天文志》): "the traditional approach is to measure the shadow cast by the sun with the measurer's end, and the shadow on the upper side of the sunlight is obtained. Now with the measurer replaced with a beam, the shadow in the middle is obtained and even the slightest difference is not tolerated." That is to say, the new approach is not affected by the half shadow of the sunlight. According to the traditional approach, the shadow length obtained is the shadow length on the upper rim of the sun and is shorter than the shadow length in the middle of the sun.

According to the data in Yuan Shi – Tian Wen Zhi (《元史·天文志》), a 1.3 m measurer has a shadow length of 2.558 m, and the solar angle is assumed to be 0.5 degrees. According to general knowledge of trigonometric functions, it is not difficult to determine that the half-shadow projected from the measurer end to the jade tablet is about 53 cm, and the difference between the length of the shadow in the center of the sun and the length of the shadow at the upper rim of the sun is half of that number. Considering Guo Shoujing's reading upon measurement was as precise as 0.016 cm, this error was three orders of magnitude higher than the reading accuracy, which certainly cannot be neglected. Guo Shoujing successfully solved a major problem facing the traditional shadow measurement technology by inventing the shadow definer based on the principles of pinhole imaging, which is an important achievement in observation and measurement technologies in ancient China and should be fully acknowledged.

8.3.8 Latticing in Imaging Theories

Latticing was an important geometrical optics method in ancient China. Based on the light's property of traveling along straight lines, it believes a constraint point exists during pinhole imaging and concave mirror imaging, and this point enables correspondence between an object and its image. This is the ancients' attempt to tackle the imaging issue with equiform geometry. To an extent, it reveals the relevant imaging mechanism.

The term latticing was first coined in Vol.3 of Shen Kuo's Meng Xi Bi Tan (《梦溪笔谈》) of the Northern Song Dynasty:

> Solar flints generate inverted images because of the barrier. Mathematicians call it the "latticing technique." For example, one can paddle because the paddle is subjected to the barrier. A kite flies in the sky and its shadow moves along with the kite, and if the shadow is subjected to the barrier of the window gaps, the shadow moves in the reverse direction of the kite. If the kite moves east, the shadow moves west, and vice versa. Another example is the tower shadow in window gaps, which constrain the tower shadow and make it inverted, just like the solar flint. The solar flint has a depression and produces an upright image if pressed with a finger; if the shadow moves away, nothing is seen; if it passes that point, the image is inverted. Just like window gaps, sculls and bulge, the shadow is subjected to barrier like a paddling. Therefore, when one raises the hand, the shadow points downward, and vice versa. The solar flint has a depression and when it faces the sun, the light converges inward. At a distance of 1 or 2 cuns from the mirror, the light converges into a point in the size of a pea, and burns once meeting inflammables, where the bulge is the thinnest. Not just objects, but people are subjected to the same rule. All are subjected to some barrier. It is even difficult to not reverse the truth, let alone remove the barrier! You Yang Za Zu (《酉阳杂俎》) said "when the wave moves, the tower shadow is inverted", which is nonsense. The truth is the shadow is inverted when entering the window gaps.

The small characters in the quotation are notes added by Shen Kuo himself. Shen Kuo listed two cases of concave mirror imaging and pinhole imaging to explain what is latticing. In the text, yangsui is a concave mirror, and the window gap is equivalent to the pinhole in pinhole imaging. Shen Kuo explained both imaging mechanisms with the latticing technique, believing that the imaging process is like a boatman's operating a paddle, which rotates around its fulcrum, so that the movement of the paddles on both sides of the fulcrum forms a kind of geometry where "the body and the ends lattice with each other." Similarly, in the imaging process of concave mirrors and pinholes, there is also a special point called "obstacle" (which is the focal point in the case of concave mirror imaging and the pinhole in the case of pinhole imaging), which is equivalent to the "fulcrum" upon paddling. The light is constrained by it and converges at the "obstacle," resulting in the formation of an image that has the potential of "latticing." Shen Kuo finally extended this phenomenon to dealing with people and warned people not to be "obstructed" by things, so as not to "exchange good for bad, or reverse right and wrong."

Latticing is a fundamental method in geometric optical imaging. Its scope of applicability is not limited to pinholes and concave mirrors, but also convex mirrors and lenses, and even some wave optics components. The corresponding constraint point cannot only be the pinhole or focal point, but also the center of curvature or light. It can be used to describe the mechanism that leads to upturned images as well as to explain the cause of properly positioned images. The current geometric optical imaging drawing method is essentially latticing. Generally speaking, only one latticing operation is adequate for pinhole imaging, and two consecutive latticing operations are required for sphere mirror and lens imaging to eventually determine the size and position of the image.

The imaging issue can be solved with latticing, which makes sense in its own right. The definition of the image requires dot-to-dot spatial correspondence between the image and the object, and latticing is dot-projection correspondence, the simplest form of spatial correspondence that meets this requirement. Further research indicates latticing of the imaging process also complies with the basic rules of wave optics. Therefore, the concept of latticing has highly profound physical implications. Of course, only people today can acquire this level of insight, which the ancients could not reach.

As a geometric optical imaging method, latticing existed before Shen Kuo. The interpretation of concave mirror imaging in Mo Jing (《墨经》) is essentially latticing. When describing the characteristics of concave mirror imaging, Mo Jing (《墨经》) said: "Concave mirrors generate two types of images, the small and upturned, and the big and upright. This is because the object can be within or beyond the center." Jing Shuo (《经说》) explained: "If the mirror is within the center and near the object, the projection is large and so is the image; if far from the center, the projection is small and so is the image; the images must be upright. This is because the object's projection starts from the center and is projected in the upright position. If the mirror is beyond the center and near the object, the projection is large and so is the image; if far from the center, the projection is small and so is the image. The images must be upturned. This is because of the convergence at the center before projection to the mirror." The "center" here refers to the focal point of the concave mirror, namely the "obstacle" in Shen Kuo's latticing theory. The Mohists were basically saying that when the object is within the focal point and close to the focal point, its projection on the mirror is larger, and the image generated is larger; if far from the focal point, the projection on the mirror is smaller, and the image generated is smaller. However, both are upright. This is because the object's projection on the mirror starts from the focal point and is projected in an upright position. Imaging with the object beyond the focal point is similar, while the image generated is upturned. This is because the light rays from the object first converge at the focal point before being cast onto the mirror. The explanation of the imaging mechanism in Jing Shuo (《经说》) is essentially consistent with the latticing concept. Besides, Mohists proposed the size of the image depends on the distance of the object from the focal point, which is also correct. However, due to the subtlety of the language in Mo Jing (《墨经》) and the decline of Mohism, the Mohist theory explaining concave mirror imaging had been long unknown to people until the Song Dynasty period, when it was promoted by Shen Kuo, who summarized the theory and gave it a special name – the latticing technique.

Neither the Mohists nor Shen Kuo was fully aware of the profound physics implications of latticing. When discussing concave mirror imaging, they used latticing only once, which could explain the size and upright/upturned position of the image, but could not determine its position. Anyway, the emergence of the latticing concept suggests the ancients not only began to analyze imaging from the perspective of the light ray, but also grasped the nature of imaging, which laid a foundation for the further development of geometric optics in ancient China. During the Qing Dynasty period, Zheng Fuguang used the concept of latticing to discuss specific optics issues, while Zou Boqi used Ge Shu Bu (《格术补》) as the title of his geometric optics work. Examples of explaining pinhole imaging with the latticing concept can also be seen in other books, which suggests the latticing concept had certain influence on the development of geometric optics in ancient China. Even today, we can still draw many valuable inspirations from this concept, which is its historical value.

8.3.9 Checking Injuries with Red Light

There is one story that is worth mentioning in the history of optics studies in China, that is, "checking injuries with red light" mentioned in Vol.11 of Shen Kuo's Meng Xi Bi Tan (《梦溪笔谈》). The original text is as follows:

> When Li Chuhou was governor of Luzhou, a man was beaten to death, and Li Chuhou went to check the injuries. When he poured marinated ash water with onto the body, no injuries were visible. An elderly man requested to meet him claiming: "I am an old official. I know how to find invisible injuries. Just place a red umbrella over the body, pour water onto the body, and the injuries will appear." Li Chuhou did as he said, and the injuries appeared instantly. This method has been used by the authorities in the Jianghuai area ever since.

The above is a method of finding injuries the old official had learned through years of experience. Generally speaking, even if there are no outer injuries, there must be inner injuries at the injured spot, and there must be stasis underneath the skin. Such stasis typically appears greenish purple and sometimes is not clearly visible in the sunlight. This is the situation Li Chuhou had found himself in. He adopted the old official's method by pouring water onto the body under a red umbrella, and "the injuries appeared instantly."

Why did this practice work?

From physics we know the wavelength of visible lights ranges from 0.39 μm of the purple light to 0.77 μm of the red light, with orange, yellow, green, blue, and indigo in between. The sunlight is a combination of these lights and appears white. In the context of injury checking, the white light is not contrasting to green and purple lights and the injuries are therefore hard to identify. By placing a red umbrella over the body, the old official actually used it as a light filter to derive red light from the sunlight. As red light has the longest wavelength of visible lights, stasis under skin is typically greenish purple and greenish purple light has the shortest of wavelengths, the difference between the wavelengths of the stasis and the light increases, and the contrast between the stasis and its surroundings increases, the stasis is easily noticeable. Therefore, this story in Meng Xi Bi Tan (《梦溪笔谈》) is an early record of the application of light filtering in China.

As for "pouring water onto the body," the purpose is generally to increase skin transparency and make the injuries underneath more easily noticeable.

The quoted text contains the description of "pouring marinated ash water onto the body," which includes the character zi (胾), whose exact meaning has not been determined by the academia. Zi means a big chunk of meat, while in the context of the story, it was apparently impossible to rub with a chunk of meat. Afterwards, Li Zhichao found the "Wine and Food" chapter of You Yang Za Zu (《酉阳杂俎》) of the Tang Dynasty mentioned "Lao, Zi and Chun all refer to wine pulp." Then, he found Han Shu – Shi Huo Zhi (《汉书·食货志》) mentioned the phrase "糟胾." When annotating "胾," Yan Shigu of the Tang Dynasty believed it was a kind of wine pulp. Thence, Li Zhichao believed the characters of "胾" and "胾" were used interchangeably in the Tang Dynasty period, first by Duan Chengshi, and then by Shen Kuo in the Song Dynasty period. By "糟胾," Shen Huo meant the marinating liquid. Li Zhichao's theory is convincing. So why pour marinating liquid or ash water? Wang Jinguang et al. pointed out: "This is probably because the ash water is alkaline and can remove grease and whiten the skin, thereby making it easier for an abnormal color beneath the skin to show; the marinating liquid is alcoholic and has a similar function."

This story recorded in Meng Xi Bi Tan (《梦溪笔谈》) drew the attention of many in history. Many books mentioned it, and the description in Xi Yuan Ji Lu (《洗冤集录》) by Song Ci of the Southern Song Dynasty went a step further:

> When checking the body and the fractured part, I saw no traces. So I poured marinating liquid and vinegar onto the body and cover the part to be examined with a umbrella made of fresh oily silk in open air, and observed from above the umbrella against the sunlight, and the injury was spotted. On a gloomy or rainy day, hot charcoal can be used to shed light upon it. This is a good method.

The practice of covering with an umbrella made of fresh oily silk marked a step toward light filtering materials, and "using hot charcoal to shed light on a gloomy or rainy day" was illuminating with a man-made red light source, which could achieve the same effect as filtering with a red umbrella. Song Ci was a forensic expert in ancient China, and it was logical for him to make these improvements to "checking injuries with red light."

The ancients' focus on the issue of "checking injuries with red light" was related to the demand for forensic technologies in the ancient society. In terms of the scope of application, it was part of the forensic science; in terms of the principles involved, the concept of light contrast had to be used. There were other examples that involved light contrast in ancient social life. For example, to observe the solar ellipse in ancient times, the ancients "placed a basin of water in the courtyard and observed it from day to dusk." (*Kai Yuan Zhan Jing* (《开元占经》)(Vol.9)). Given the moderate reflectivity of light on the water surface, this could greatly reduce the intensity of sun glare. This method, however, was not good enough, either. Given the high level of transparency of water, if the water is not deep enough, the diffused light from the basin bottom will be strong, which leads to poor image contrast and a poor effect. Therefore, people observed with an oil basin instead. As the oil surface is less reflective of light and less transparent, the diffused light from the basin bottom is weaker, which adds to the image contrast. Meanwhile, as oil is more viscous, the reflective surface is stable and is much easier to observe. As a result, this became an important method of observing the ellipse in ancient China. Like "checking injuries with red light," this was a scientific method found by the ancients during long-term practice and was a major achievement in the history of optics.

Because the transparency of the water is high, if the water is not deep, the diffuse reflection light at the bottom of the basin will be stronger, which makes the contrast of the image worse and the effect is not good. So later people used oil basins to observe. The reflectivity of the oil on the surface of the light is smaller, and the transparency of the oil is small, which makes the diffuse reflection light of the bottom of the basin weaker and increases the contrast of the image. At the same time, the viscosity of the oil is large and the reflecting surface is stable, so the observation effect is much better, so it has become an important method for observing solar eclipses in ancient times. This method, like the "red light inspection," is a scientific method that conforms to the principles of optics and is an important achievement in the history of optics.

8.3.10 Translucent Mirror

The translucent mirror is a special kind of copper mirror. In ancient China, copper mirrors were typically made of copper-tin alloy, with patterns and texts inscribed on the back. The front was the reflective surface, which was polished until capable of mirroring. In appearance, the translucent mirror was no different from a general copper mirror. However, when a light beam was cast on the mirror surface and reflected to the wall, the patterns and inscriptions on the back of the mirror would be magically shown on the wall, as if the light beam reflected the patterns and texts on the back to the wall through the copper mirror, therefore called a translucent mirror.

The origin of the translucent mirror cannot be determined yet. Although many pre-Qin copper mirrors were unearthed, no one has checked them one by one for translucent mirrors. Among the copper mirrors preserved by Shanghai Museum, there is one relic from the Western Dynasty period. On its back are eight characters "reflecting sunlight and brightening the world," and when put in the sunlight, it clearly reflects the patterns and texts on the back to the screen. Undoubtedly, this is a translucent mirror. This suggests translucent mirrors were made no later than the Western Han Dynasty in China. As for whether the translucent mirrors were an accidental finding by the craftsmen in the Western Han Dynasty period or were purposefully made, this remains a mystery.

Wang Du of the Sui Dynasty authored the fantasy novel *Gu Jing Ji* (*the Tale of Ancient Mirror*,《古镜记》), which mentioned an ancient mirror, which was said to "reflect all the patterns and texts on the back when in the sunlight." Albeit a novel text, the description was very consistent with the characteristics of a translucent mirror. It seems Wang Du might have heard of the translucent mirror and found it magical, and that was why he mentioned it in the novel as a magical phenomenon.

After the Sui Dynasty period, translucent mirrors continued to be made. Powerful evidence includes the existence of translucent mirrors made in the Tan Dynasty period and constant records about translucent mirrors after the Song Dynasty period. On the other hand, people developed a strong interest in the "translucence" mechanism of translucent mirrors and many scholars began their explorations. Shen Kuo of the Northern Song Dynasty recorded his and his contemporaries' reflections on this issue. In Vol.19 of Meng Xi Bi Tan (《梦溪笔谈》), he recorded:

> There is a translucent mirror, and on its back are inscriptions, which are a mere twenty characters, which are of far antiquity and are almost invisible. When the mirror is placed in the sunlight, the twenty characters are projected to the eaves and are highly clear. Someone knows the cause, saying the thin part cools first upon casting, while the back is thicker and cools later, and as a result, copper contracts more. Although the characters are on the back, they are vaguely visible on the mirror and therefore appear in light. This is true according to my observation. I have three mirrors, and there are also mirrors collected by others, and all these mirrors are the same. The texts, drawings and inscriptions on the backs of these mirrors are not much different and these mirrors all look very ancient. But only this mirror is translucent, while other mirrors, albeit extremely thin, are not translucent. I reckon the ancients must have had a unique method.

This paragraph is the earliest known text of the ancients discussing the principles of translucent mirrors. This point of view believed the patterns and texts on the back shown through the reflected light are not the result of light passing through the mirror. The cause still lies in the reflective surface of the mirror, "Although the texts are on the back, they are still slightly visible in the mirror, and therefore appear in the light."

So what caused the texts to be "vaguely visible in the mirror"? The opinion quoted by Shen Kuo believed this is because of thermal expansion and contraction. For the same kind of objects, the larger the volume, the greater the thermal expansion or contraction. Therefore, during the cooling process of a finished copper mirror, thin parts contract less and thick parts contract more, and as a result, depressed dents similar to the patterns and texts on the back form on the mirror. Shen Kuo agreed with this explanation, while arguing as thermal expansion or contraction is a universal phenomenon, how come not every mirror is translucent?

Shen Kuo's question was answered by the moderners. In 1975, Fudan University worked with Shanghai Museum to duplicate a copper mirror with "translucence" capability by using the quenching method. This suggests people in the era of the Song Dynasty had a point. However, researchers also found this approach is highly demanding for the mirror thickness and other conditions, and not every mirror thus made can be translucent.

In Xian Ju Lu (《闲居录》), Wu Qiuyan of the Yuan Dynasty presented another explanation:

> If the pattern of a coiling dragon is inscribed on the back of the mirror, and the pattern of a dragon is made on the mirror surface. Then, a different type of copper is embedded. With the mirror surface flattened, lead is added on it, and the mirror is placed in the sunlight, and the light changes along with the different types of copper.

Wu Qiuyan tried to explain with the difference in mirror reflectivity. He believed translucence could be achieved if he embedded into the mirror another kind of copper with patterns identical to those on the back and smoothened it, because different parts of the mirror had different abilities in reflecting light. Wu Qiuyan personally witnessed someone breaking a translucent mirror to verify this assumption, which proved to be true. This assumption was backed by many subsequent scholars, such as Fang Yizhi, a scientist of the late Ming Dynasty.

Wu's explanation makes sense. It is true different reflective abilities can achieve "translucence," and furthermore, there is physical evidence from history. In fact, given the different materials embedded in the mirror, the surface grinding properties are different, which will necessarily make the patterns "slightly visible." Therefore, it is not "the translucence of copper" alone that leads to "translucence."

Zheng Fuguang of the Qing Dynasty opposed Wu Qiuyan's point of view and approved of Shen Kuo's, believing "slightly visible patterns" were the fundamental cause of translucence and made an important addition to Shen Kuo's analysis, pointing out in addition to the cooling process, the scraping of the mirror surface during the processing also led to the existence of the patterns. In "Making Translucent Mirrors" in Jing Jing Ling Chi (《镜镜詅痴》), he pointed out, due to different cooling speeds during forging, the copper has different contraction forces and as a result, the mirror has an uneven surface, and such unevenness is difficult to thoroughly remove during the scraping of the mirror, "the scraping force varies with the unevenness of the mirror, and there is always unevenness." Unevenness leads to "translucence." Zheng Fuguang compared this with water reflection. He said: "When the water is still, it is as flat as a whetstone, and sheds light upon the walls. If it moves, ripples are generated and cause unevenness. . .as the copper mirror is inadequately grinded, it is mostly uneven, which is not noticeable in the case of mirroring, but is very visible in the sunlight." It was appropriate for Zheng Fuguang to explain the magnification effect with the rippling phenomenon in water that reflects light and helped people understand the causal relationship between "slightly visible on the mirror" and "translucence."

In fact, there are many ways to make translucent mirrors. For example, we can simply make pattern grooves on the mirror by manually grinding and polishing. As the grooves are shallow, smooth, and edge-free, they are hardly visible to naked eye, while appearing "translucent" when reflecting light. Of course, it would take a lot of effort to make the patterns in the reflected light identical to the patterns on the back of the mirror.

The issue of translucent mirrors drew attention abroad, too. In the Edo period of Japan, which was contemporary with China's Ming Dynasty, a kind of translucent "magic mirror" was made. In the early Meiji period, "magic mirrors" were already quite commonplace. It was in the 1930s that westerners first saw translucent mirrors and since then, the European science community had discussed why copper mirrors could be "translucent" for nearly a century, making translucent mirrors an issue of worldwide attention.

The ancient Chinese came up with these techniques of making translucent mirrors, including the cooling effect mentioned by Shen Kuo, the embedding method mentioned by Wu Qiuyan and the scraping method added by Zheng Fuguang, all of which could be used to make translucent copper mirrors. The scraping method used by Japan to make "magic mirrors" was similar to that said by Zheng Fuguang. However, all these techniques required sophisticated craftsmanship, and it would not be easy for the ancients to achieve that.

8.3.11 Telescopes

Telescopes originated in the West. In terms of the application of telescopes, China is the benefactor of the cultural exchange between the East and the West.

In the early seventeenth century, Galileo Galilei (1564–1642) made the first astronomical telescope to observe the stars and the moon. He also authored the book Starry Messenger, recording his observation findings. Galileo's work was soon introduced into China. In 1615, Portuguese missionary P. Emmanuel Diaz (1574–1659) wrote the book Essays on Heavenly Questions, which said:

> In recent years, a western scholar adept at calendars tried to know the secrets of the sun, the moon and the stars and found his eyes inadequate, so he created an instrument for assistance. With this instrument, he observed a small item 30km away and found it was right before his eyes. Then, he observed the moon, which appeared a thousand times larger than usual. He observed the Venus, which looked as big as the moon, and its light waned and waxed just like that of the moon. . .he carried this instrument to China and afterwards elaborated on its smart application.

This paragraph was the earliest to have introduced the telescope and its observation findings into China and focused on the findings of Europeans observing the sky with the telescope. "A western scholar" referred to Galileo, and "an instrument" referred to the telescope. The book, however, did not describe the structure of the telescope. The last sentence in the quote also suggests telescopes had not yet been introduced into China.

Telescopes were introduced into China around mid-1720s. In 1626, German missionary Johann Adam Schall von Bell (1591–1666) authored the *Explanation of the Telescope*, which was the earliest treatise on telescopes in China. The book contained an illustration of the entire telescope and articulated: "where is this telescope from? The astronomer from across the ocean." The "astronomer" here referred to Galileo. This book covered the usage and principles of the telescope, enumerated the findings from observation of celestial bodies in the solar system and the Galaxy and other stars with the telescope, further explained the findings from Galileo's observation of the sky, and introduced the method of making the telescope:

> Make a round mirror that seems flat with glass, called tube lens or the so-called "center-protruding mirror" or "front mirror". Make a small concave mirror, which is called the "eye-side mirror" or so-called "center-depressing mirror" or "rear mirror". These mirrors must be adequately reflective and have appropriate matching, lengths and proportions to focus on a small, distant object and mirror the object exactly as it is.

The "center-protruding mirror" is a convex mirror, which is used as an objective lens here, namely the "front mirror" and "canister mirror." The "center-depressing mirror" is a concave mirror, which is used as an eyepiece here, called the "rear mirror" and "eye-side mirror" in the quoted text. A telescope of this structure is a typical Galileo telescope.

The telescope is used in the following way:

> A telescope has two lenses only, while canisters can be added at will, inserted into each other and stretchable. If connected with a screw, they can be moved up and down. With only one mirror used, one can see items 200 steps away, up to 30km away. It can be used to observe the moon, Venus, the sun, Jupiter, Saturn, and the various stars in the night sky. When observing the sun and Venus, one can add a green mirror or place a piece of white paper under the mirror. Besides, it can be used for navigation, drawing in the darkroom and particularly war.

This paragraph provides a detailed description of how to use the telescope. "Canisters can be added at will, inserted into each other and stretchable," which is intended to adjust the distance between the eyepiece and the objective lens to make it easy to observe objects at different distances. In particular, the paragraph introduces two ways of using the telescope: direct viewing and observation of images on a screen. For general objects, one can directly view at the eyepiece. When observing bright celestial bodies such as the sun and Venus, "one can add a green mirror" to protect the eyes. Both are in the category of direct viewing, which sees a virtual image generated by light passing through the telescope. For bright objects such as the sun, one can place a piece of white paper under the eyepiece as the image screen and adjust the telescope to allow the sunlight to pass through the telescope and generate a bright spot on the paper for observation. What is observed in this case is the sun's real image. This approach saves the viewers' fatigue and protects the eyes. However, this does not work for general objects, as the image generated on the screen would be too dim.

Although the book contained a mere 5000 characters and brief descriptions, and the light path illustrations in the book were wrong, it did introduce the telescope to

China. It aroused Chinese people's passion for telescopes and allowed them to grasp the basic method of making telescopes, which is one of its historical achievements.

In Europe, scholars first pointed telescopes to the sky after their invention and achieved amazing results. In China, astronomers embarked on the same path, fully aware of the importance of the telescope to astronomical observations. As early as 1629, or the second year of Emperor Chongzhen's reign, Xu Guangqi submitted a petition to the emperor suggesting the manufacturing of "instruments for urgent use," including "three telescopes," requesting the emperor to set aside workforce and materials for that purpose. Despite the dire situation facing the Ming Dynasty and the tight reserves of the Treasury, Emperor Chongzhen surprisingly granted permission, when telescopes had existed in China for no more than 2 or 3 years. The academic hold different views on whether the telescopes requested by Xu Guangqi were actually complete. Two years later (namely 1631), however, he did observe solar and lunar ellipses with a telescope and found it superior in observing the sky. In subsequent Chinese and western astronomical debates, Xu Guangqi further described the necessity and ways of observing with a telescope:

Upon solar eclipse, the sun is so dazzling that you see the eclipse first before the sun appears; upon lunar eclipse, with roaming gases around, you first see the moon before seeing the eclipse: The difference is moderate. Now we want to see the real difference. With a recently made peeing canister, we acquire the image of solar eclipse in the sealed room, and the image from initial eclipse to round is accurate. For lunar eclipse, we look up, and when the two bodies separate and emerge, they are very clear and distinctly different from observations with naked eye. (*Ming Shi – Li Zhi Yi* (《明史·历志一》)).

What is a peeping canister? Li Tianjing, who succeeded Xu Guangqi as the person in charge of calendar work, said: "The telescope, also known as the peeping canister, consists of canisters that are interconnected and stretchable and glass at both ends, with its length depending on the distance of the object." (*Ming Shi – Tian Wen Zhi Yi* (《明史·天文志一》)) Obviously, the "peeping canister" mentioned by Xu Guangqi here is the telescope. This paragraph was intended to point out the superiority of the telescope in observations, believing it is much more accurate than directly observing with naked eye. Xu Guangqi specially pointed out the telescope should be used in different ways for observing lunar and solar ellipses. For lunar ellipses, direct viewing should be used, while for solar ellipses, the screen imaging method should be used. These are the two methods described in the *Explanation of Telescope*.

The ancients not only realized the importance of the telescope for astronomical observations, but also perceived other purposes of the telescope. Li Tianjing said, the telescope "can not only be used to observe the sky, but also to observe objects several kilometers away, as though they were right before your eyes. This is very useful for observing enemies launch cannons." The fact that telescopes can be used in war was mentioned in the *Explanation of Telescopes* and was further reiterated by Li Tianjing. In fact, before Li Tianjing said this, Chinese scientist Bo Jue had already creatively mounted a telescope on a self-made copper cannon in 1631, which was the first such application in the world. Therefore, famous science historian Needham said: “Whether Bo Jue was an independent inventor of the telescope or not, he deserved the honor of first using it on a cannon.”

After the publication of the Explanation of Telescopes, private production of telescopes in China began. For example, Sun Yunqiu, a maker of optical instruments in Jiangsu, made lenses by himself and produced telescopes with good performance around mid-seventeenth century. Bo Jue was senior to Sun Yunqiu and made telescopes at an earlier time. This suggests the Chinese had quickly accepted the telescope, which was a foreign invention.

Even in literature, descriptions of the telescope could be spotted. In chapter novel Shi Er Lou (《十二楼》) in the early Qing Dynasty period, there was a legendary story of a boy from a noble family peeping the life of an official’s daughter using the telescope (called a “Thousand Kilometer Lens” in the book) and “deceiving” her into marriage. The book described the shape and characteristics of the telescope as such:

> This mirror uses several tubes of different sizes. The thin ones are included in the thick ones, so as to be stretchable or retractable. A so-called thousand-kilometer lens is embedded into the two ends of the tube, and can see objects faraway. Although the word “thousand-kilometer” is exaggerative, and it is impossible to see the Yue Kingdom from the Wu Kingdom, or see the Chu Kingdom from the Qin Kingdom, it is possible to see objects within several hundred kilometers. As for people or objects within tens of kilometers or hundreds of steps, they not only look far, but are clearly distinguishable. Therefore, this is a real treasure.

This paragraph cannot be regarded as a novelist’s imagination. It was impossible for those who had not personally witnessed a telescope to write a paragraph like this. This is effective proof that optical instruments such as the telescope were valued by intellectuals of that time. The earliest known version of Shi Er Lou (《十二楼》) is the version of the 15th year of Emperor Shunzhi’s reign of the Qing Dynasty (1658), which was a mere 30 years from the introduction of the telescope into China.

8.4 Electromagnetics and Thermal Science

8.4.1 Observation and Interpretation of Electromagnetic Phenomena

It was not until the emergence of modern science that people grasped the essence of electromagnetics. The ancients’ relevant achievements were embodied by the records and explanations of the relevant electromagnetic phenomena, which will be described in this chapter.

The ancients’ observations and discussions of electricity phenomena focused on static electricity and lightnings. The phenomenon of static electricity attraction was found early in China. Chun Qiu Wei – Kao Yi You (《春秋纬·考异邮》) completed in the Western Han Dynasty period said: “Magnets attract iron. Tortoiseshells attract.” Tortoiseshells are the shells of a type of marine reptiles and are insulated.

"Tortoiseshells attract" means fractioned tortoiseshells can attract tiny things, which is therefore a static electricity phenomenon.

Although the ancients found "tortoiseshells attract," their explanation of this had nothing to do with electricity. They did not realize at all this phenomenon is essentially the same as the lightning. Wang Chong's comment in Lun Heng – Luan Long Pian (《论衡·乱龙篇》) is typical. He said:

Tortoiseshells attract tiny things and magnets attract needles, all by themselves and without the assistance of other matters. Why can't other similar matters be attracted? Because they have different air properties and cannot be sensed.

Wang Chong believed tortoiseshells can attract tiny things because they have the same "air properties" as the items attracted. Guo Pu of the East Jin Dynasty shared a similar view in his Shan Hai Jing Tu Zan (《山海经图赞》): "Magnets attract iron and tortoiseshells attract tiny things, because Qi has potential senses." Such an explanation is obviously a theory that takes things for granted: As long as two objects attract each other, you can always say they have the same Qi properties. This is not of much help to people's understanding of the essence of such a phenomenon. However, such theories could prevent people from resorting to theology and had certain historical value.

According to Vol.57 of San Guo Zhi (《三国志》), when Yu Fan of the Wu Kingdom was a child, he heard: "Amber cannot attract rotten mustard and magnets cannot attract bent needles." Amber is fossilized tree resin that has good insulation properties and can attract tiny items when subject to frictions. The claim that amber subject to frictions cannot attract rotten mustard is true because "rotten mustard" is conductive because it contains water. However, the assertion "magnets do not attract bent needles" is evidently wrong. We are not sure if the conclusion "amber cannot attract rotten mustard" is based on the ancients' observations or mere speculation.

The ancients not only noticed the phenomenon of "amber attracting tiny things," but also applied it in practice as the standard for identifying genuine and fake amber. In the Southern and Northern Dynasties period, Tao Hongjing said in his Ming Yi Bie Lu (《名医别录》): "genuine amber attracts mustard when rubbed in the palm and becoming heated." The more obvious the property of static electricity, the higher quality the amber. Tao Hongjing's identification method is correct.

Static electricity is mostly subject to frictions, which generate electricity that is accompanied by sparkles and slight sounds, which were also found and recorded by the ancients. Zhang Hua of the Western Jin Dynasty said in Bo Wu Zhi (《博物志》): "When people today comb hair and undress and dress themselves, there is light and sound upon combing and dressing/undressing." What Zhang Hua described is the light and sound caused by the electricity from frictions when people comb and undress themselves. In You Yang Za Zu (《酉阳杂俎》), Duan Chengshi of the Tang Dynasty recorded another kind of light caused by electricity: "when you rub the fur of a black cat in darkness in the opposite direction, you can see sparkles." Here he chose a black cat and rubbed the cat's fur in darkness in the opposite direction in order for the sparkles to be more easily noticeable. Although using a white can could achieve the same electricity effect, the sparkles would not be noticeable. Many similar phenomena were also recorded by subsequent generations, suggesting this is an electricity phenomenon that was often noticed by the ancients.

Electricity discharge at the pointed edge, another static electricity phenomenon, was also recorded in ancient literature. Han Shu – Xi Yu Zhuan (《汉书·西域传》) recorded "sparkles generated at the tip of the spear," which is essentially the electricity discharge at the tip of the spear, which is made of metal, under certain conditions. As the spear stood erect in the open air, there could be dim light from electricity generated if the spear was located at an elevated spot and there happened to be an electricity-containing cloud above, thereby noticed and recorded by people. Of course, the ancients just observed the phenomenon, but did not quite understand what might have caused it. They explained as such: "sparkles at the spearhead indicated warfare, which is a sign of military victories." (*Han Shu – Xi Yu Zhuan* (《汉书·西域传》)). This became the basis for levying wars. While there are some other similar records in ancient literature, we list one example here to give the reader a general picture.

The earliest phenomenon found by the ancients is undoubtedly lightnings. When lightnings occurred, the dazzling light and deafening sound would not be nothing to even the most primitive humans. Therefore, it is pointless to determine when the ancients found lightnings. Our concern is the ancients' discussions of the cause and nature of lightnings. In ancient China, people were yet to have the concept of electricity in modern science and therefore were unlikely to explain lightnings with positive and negative charges. Instead, they explained with the yin-yang theory. Huai Nan Zi – Di Xing Xun (《淮南子·地形训》) mentioned on multiple occasions: "yin and yang approach each other to generate thunder, and fiercely interact with each other to generate lightnings." In Lun Heng – Lei Xu Pian (《论衡·雷虚篇》), Wang Chong of the Eastern Han Dynasty explained with a similar view: "In the midsummer, the sun predominates, and yin rides on it. When yin and yang compete, they become entangled, and when they become entangled, they ignite." In history, there were many such descriptions, which were China's traditional theories over the cause of thunder and lightnings. From a philosophical perspective, this theory is splendid, as lightnings indeed can be regarded as the result of the interactions between opposites. However, there remained a long way ahead for this theory to develop into the modern theory over the cause of lightnings in physics. Of course, superstitious theories that believed thunders are the range of the angry "Thunder God" are ridiculous.

When observing the effect of lightnings on matters, the ancients noticed a strange phenomenon, which was described a lot in ancient literature. The most representative of these was the description by Shen Kuo in Vol.20 of Meng Xi Bi Tan (《梦溪笔谈》), which recorded:

> Chamberlain Li Shun's entire household was once struck by a major thunder. In the west chamber of his house, the thunder fire burst out of the window and hit the eaves. Everyone thought the house had been ruined and ran out to avoid the danger. When the thunder stopped, the house remained the same, and the walls and windowpane paper were all black. In a wooden grid where several utensils were stored, the silver on all the silver buckles on the lacquerware melted down to the ground, while the lacquerware was not burned. A sword, which was made of extremely hard steel, was melted into liquid in its storage chamber, while the chamber remained intact. It is generally believed when a fire burns, grass and wood are

burned first before metals and stones can be burned. In this case, however, all the metals are burned, while grass and wood remain, which cannot be predicted with common sense.

What Shen Kuo recorded was a strange phenomenon after a lightning struck: All the metal objects had melted, while wooden objects remained intact. There were all kinds of utensils on the wooden racks in the house, including silver lacquerware, where all the silver melted and flowed to the ground, while the lacquerware was not burned. A stiff sword melted into liquid steel in the sheath, while the sheath remained unchanged. These phenomena recorded by Shen Kuo can be explained with the principles of electricity known today: As a lightning strike is a high-voltage discharge, and a high-voltage discharge generates a high-frequency alternating magnetic field, where a conductor generates an eddy current due to the effect of the magnetic field. When reaching a certain magnitude, the eddy current melts the conductor, while nonconductors "were not burned." Shen Kuo said this phenomenon "cannot be predicted with common sense," which was naturally due to the limited level of science development at that time. Nonetheless, he recorded details of the fact and provided a practical historical example for us to understand the phenomenon of eddy currents, which was very precious.

Perception of the magnetic phenomenon also started early in China. As early as the Warring Kingdoms period in fourth century B.C., the book of Guan Zi (《管子》) already mentioned the concept of magnets. In the third century B.C., Lv Shi Chun Qiu – Jing Tong (《吕氏春秋·精通》) articulated that magnets could attract iron, saying: "Iron approaches magnets, which may attract them." In fact, it was certain the ancients found the phenomenon of magnets attracting iron earlier than the time of Lv Shi Chun Qiu (《吕氏春秋》), as Lv Shi Chun Qiu (《吕氏春秋》) only mentioned that property of the magnet, but did not specially describe it.

In the Han Dynasty period, people developed a further understanding of the iron-attracting property of magnets. Huai Nan Zi – Lan Min Xun (《淮南子·览冥训》) said: "While the magnet can attract iron, it is difficult for the magnet to attract tiles." Shuo Shan Xun (《说山训》) said: "Magnets can attract iron, but not copper." These suggest people already knew magnets can attract iron, but not some other matters, particularly copper, which cannot be attracted by the magnet though it is metal. The contemporary work of Huai Nan Wan Bi Shu (《淮南万毕术》) mentioned the experiment of "magnets repelling chess pieces," which involved the phenomenon of magnet repulsion.

As to why magnets attract iron, the ancients explained from the perspective of the Primordial Qi, believing magnets and iron have the same Qi. This explanation is rough, as iron items or copper items do not attract each other, but it is impossible that they have different Qis.

Another achievement of ancient Chinese magnet knowledge was the discovery of the polarity of magnets and magnetic declination. In this regard, this book has dedicated a chapter and will not repeat it here.

The discovery of the magnetic shield is yet another achievement of ancient Chinese magnet knowledge. In the early Qing Dynasty period, Liu Xianting wrote in his Guang Yang Za Ji (《广阳杂记》): "someone asked me: 'magnets attract iron, so what can shield them?' My nephew Aru said: 'only iron can shield them.' The person who asked the question tried it and returned to say: 'I tried and it is true.' I said: 'why try it? This is the law of nature'." Liu Xianting believed experiments are not necessary, which is not correct. However, he did the right thing by recording the whole process of the incident and providing future generations with knowledge about magnetic shielding.

The ancients' discovery of electromagnetic phenomena provided another achievement of practical value, which is the observation and recording of aurora. As we know, the sun keeps emitting high-speed charged particles. When these particles approach the earth, they will bend to the north and south poles under the effect of the geomagnetic field, collide with high-level air molecules or atoms, and make them glow in an excited state. This is called aurora. The ancients did not know this theory, but they observed aurora and recorded it. The ancient book of Zhu Shu Ji Nian (《竹书纪年》) recorded an aurora sighting in c.950 B.C.: "In the late years of the reign of King Zhao of Zhou, there was a clear night, and light of five colors could be spotted all around." This record described the time, orientation, and light color of the aurora and was quite in detail. According to statistics, there were over 100 dated auroras recorded in China before the tenth century. Given the fact that auroras generally appear in high-altitude regions and appear only occasionally in mid- and low-altitude regions, this number should not be taken lightly, as it indicates the great effort the ancients made in observing and recording auroras. Around the same time, a mere 30 auroras were recorded in Europe. These precious records from ancient China provided very valuable historical data for studying solar activity and geomagnetic changes.

8.4.2 Methods for Starting a Fire

Fire is essential for human survival and subsistence. Initially, people acquired fire from nature (such as a forest fire caused by lightning) and used fire by keeping the fire burning. The ash layer several meters thick in caves once inhabited by Peking Man suggests this.

As humans evolved, people began to start fires instead of acquiring fire from nature and keeping the fire burning. The time and process of inventing artificial methods to start fires are not known today. Han Fei Zi – Wu Du (《韩非子·五蠹》) said: "In the far antiquity, a saint rubbed wood for fire to cook raw meat, which pleased the people, who made him king, called Suirenshi." This suggests the earliest method used by the ancients to start fires was to generate heat through frictions, which was the so-called "drilling wood for fire."

It took certain skills to rub wood for fire. Although Zhuang Zi – Wai Wu (《庄子·外物》) already mentioned, "wood and wood rub each other to burn," it is difficult to make two logs burn by rubbing them, and that is why Qing Dynasty scholar Yu Yue said when annotating this: "While Huai Nan Yuan Dao Xun (《淮南原道训》) said two logs can burn when rubbed, that didn't actually happen." However, if you do it the right way, it will not be difficult to start a fire by rubbing wood. In the early days of the People's Republic of China, some ethnic minorities of China preserved some fire-starting methods, such as the bamboo-sawing method of the Kucong people, and the wood rubbing method of the Hlai people on the Hainan Island, which proved the assertion that ancients started fires by rubbing wood was credible.

In ancient literature, legends of rubbing wood for fire were not rare, while literature describing the specific fire-starting method was scarce. Vol.2 "Rocks, Bamboos and Fires" of Wu Li Xiao Shi (《物理小识》) by Fang Yizhi of the late Ming Dynasty provided a precious record in this regard:

> If you scrape a rock with steel sickle, sparkles appear, and when you approach them with paper, it burns. You can also acquire fire from the bamboo. Take a dry bamboo and break it, cover it with paper ash, make a hole through the bamboo and cut through the hole with a bamboo sword back and forth, and after you do this three or times, fumes arise. After you do this a dozen times, fire falls into the hole and the paper ash turns red.

Here Fang Yizhi recorded two methods for starting fires: making fires with flints and making fires by rubbing wood. The wood-rubbing method he recorded is unique: He used an inflammable dry bamboo, which was cut into two bamboo tiles, paper ash was put on one of them, and the other was covered. A hole was made on the bamboo tiles, and a bamboo sword was used to repeatedly cut through the holes. The bamboo shavings cut off were very hot, fell from the holes, and piled on the paper ash, which had poor heat conductivity and a low combustion point. It was not easy for the heat of the bamboo shavings piled up to disperse, and when adequate bamboo shavings were piled up and the temperature reached the combustion point of paper ash, the paper ash began to burn, achieving the purpose of starting a fire. The paper ash used here should have been the residual ash pieces after preliminary burning of paper, and these ash pieces contained some carbon, which could sustain brief reignition. We cannot assume paper ash is the ash after paper is fully burned, as ash cannot be burned again.

Fang Yizhi's record is highly valuable. As a minister of the former Ming Dynasty, he went into exile around Lingnan to flee the Qing Dynasty's hunt. This record might have benefitted from his sightseeing in Lingnan and reflects the experience of ethnic minorities in starting fires.

As for the flint fire-starting method mentioned by Fang Yizhi, this was another popular fire-starting method in ancient China. This method was created after iron-ware appeared, around the Spring and Autumn and Warring Kingdoms periods. This approach involved the use of an iron sickle to knock on hard flints, and the iron filings rubbed, knocked, and peeled off were heated. Sparkles were generated on the surface of these filings due to oxidation combustion, and when inflammable fibers such as moxa were used to pick up these sparkles, a fire was started. This approach was easy and practical and became the most commonplace fire-starting method in ancient times.

In ancient China, there was also the record of "making fire with pearls." The earliest record appeared in Guan Zi – Chi Mi (《管子·侈靡》): "Pearls represent both yin and yang, and therefore can generate fires." (Original note: Pearls are from water and reflect light and are therefore both yin and yang. If facing the sun, they look like fires.) The quoted note is said to be from Fang Xuanling, while some believe it was by Yin Zhizhang. Whomever it was, it was another insight by a person in the Tang Dynasty anyway. The original text only said "pearls look like fires," while the note articulated it was making fires with pearls. Pearls are opaque and cannot acquire fire from the sun. However, if we assume the pearls, here are actually quartz or some other transparent items, which became round, transparent, and shiny due to all kind of factors. These actually constituted a convex lens, which could focus sunlight into fire.

During the Western Jin Dynasty period, famous naturalist Zhang Hua said in Bo Wu Zhi (《博物志》): "As for the method for starting a fire, many mentioned the use of pearls, which has not been tried." Evidently, the method of using pearls to start fires had been widely talked about, while few had actually tried it, which was probably because of the lack of glass lenses during that time and pearls suitable for starting fires were hard to find. In Vol.8 of Shi Yi Ji (《拾遗记》), Wang Jia of the Eastern Jin Dynasty recorded a fire accident that occurred to a rich man, who "scattered egg-sized pearls in basins and vases across the courtyard, which was called the treasury courtyard…the next day, a fire started in the warehouse and burned one tenth of the transparent pearls, all of which were dry and inflammable items." Wang Jia believed these pearls attracted fire from the sun and caused the fire accident. In terms of content, Shi Yi Ji (《拾遗记》) can be classified as a fantasy novel, while this assumption by Wang Jia makes sense scientifically. In the Tang Dynasty period, lenses were increasingly used, and new convex lenses kept being imported from abroad. Vol.197 of Jiu Tang Shu (《旧唐书》) recorded: "The Kingdom of Champa, called the Southern Place of Elephant Forests in the Han Dynasty period, sent envoys to donate domesticated rhinos in the early years of Emperor Zhenguan's reign. In the fourth year, its king Kanharpadharma sent an envoy to donate fire pearls, which were of the size of eggs, round, white and smooth, radiating light meters away and looking like crystal. At midnoon, when these pearls faced the sun, moxa was placed beneath and burned." The fire pearls mentioned here are obviously convex lenses, which are capable of focusing and acquiring fire from the sun. This is one of the important reasons why they were donated to the emperor.

The donation of fire pearls was also recorded in Nan Shi (《南史》), Liang Shu (《梁书》), and Wei Shu (《魏书》), which suggests the method of starting fires with pearls was gradually popularized with cultural exchanges between China and foreign countries.

Before the method of starting fires with pearls was popularized, the unavailability of glass lenses probably inspired the ancients to make transparent ice into the shape of lenses to acquire fire from the sun. Huai Nan Wan Bi Shu (《淮南万毕术》) said: "Peel ice to make it round, raise it toward the sun, and place moxa in its shadow, and a fire is started." This record is clear and accurate. "Peel ice to make it round" is about how to make an ice lens; "place moxa in its shadow": moxa is inflammable, and "shadow" is undoubtedly the light beams that converge at the focal point. This paragraph, be it a record from practice of the ancients or their speculation, suggests the people of the Han Dynasty had a clear understanding of how to start a fire with lenses, which can be ascertained.

So, can lenses made of ice be used to start fires? The answer is yes. Zheng Fuguang of the late Qing Dynasty had conducted a simulation experiment, where he used a tin kettle with a slightly depressed bottom and full of hot water. With the kettle, he ironed ice chunks to acquire transparent, crystal-like ice lenses, and tried to use this to get fire from the sun and succeeded. He summarized the key points of acquiring fire with ice lenses as such: "It must be a very sunny day. The ice chunk must be big with a slightly shallow bulge and must be steady and not shaken." (Zheng Fuguang: *Fei Yin Yu Zhi Lu: Xue Bing Qu Huo Tu Jing Tong Li* (《费隐与知录·削冰取火凸镜同理》)). Zheng Fuguang's experience makes sense. As we know, the light collection ability of a lens is the square of the ratio of its diameter to the focal length (relative diameter). Thus, the longer the diameter, the greater the ability in collecting light. On the other hand, if the bulge is shallow, the focal length is great, which hampers light collection. In this regard, Zheng Fuguang explained: "As the fire is from the sun's heat, and the ice is cold and can reduce the heat of the sun, the bulge must be slightly big to keep away the cold." So the great focal length was intended to reduce the coldness of the ice lens, so that it would be easier to start a fire.

In ancient times, another method was to acquire fire by focusing with a concave mirror, which was called a solar flint (yangsui) by the ancients. Kao Gong Ji (《考工记》) mentioned "the instrument of jiansui," to which Zheng Xuan annotated: "jiansui is the instrument used to acquire fire from the sun and moon." Zhou Li – Qiu Guan (《周礼·秋官》) said: "Sixuan gained possession of flints to acquire fire from the sun." Sui here refers to the solar flint, which is capable of acquiring fire from the sun and was valued by the ancients. Li Ji – Nei Ze (《礼记·内则》) also contains the records of "a metal flint to the left" and "a wooden flint to the right," which is proof. The metal flint is the solar flint, and the wooden flint should be a tool used to drill wood for fire on gloomy and rainy days. Literature of the Han Dynasty contained detailed records of the specific method of getting fire with solar flints. Huai Nan Zi – Tian Wen Xun (《淮南子·天文训》) said: "The solar flint burns in the sunlight and starts a fire." Gao You of the Eastern Han Dynasty annotated: "Solar flints are metals. Take a solar flint and rub to heat it, place it in the sunlight and place moxa underneath, and the solar flint burns and a fire is acquired." Shuo Lin Xun (《说林训》) said: "When getting fire with flints, the fire can't be acquired whether too distant or too close. It should be in the right distance." Shuo Lin Xun (《说林训》) emphasizes the fire medium should not be too distant from or too close to the mirror, but should be placed in the right distance, namely at the focal point. Here it implied the concept of the focal length. In Lun Heng – Shuai Xing Pian (《论衡·率性篇》), Wang Chong said: "When you rub a moon-shaped sword and lay it down to face the sun, you can also get a fire. The moon-shaped sword is not a solar flint. It can generate a fire because it has been rubbed." This paragraph suggests even a concave metal reflective surface such as "a moon-shaped sword" can have good reflective properties if "rubbed" and can also be used to acquire fire from the sun, which does not require a special "solar flint." This suggests people developed a further understanding of physical process of acquiring fire with a concave mirror.

After the emergence of the method of acquiring fire by lens focusing, some also called lenses solar flints. But generally speaking, when talking about getting a fire with solar flints, the ancients meant acquiring fire through the reflective focusing of concave mirrors.

The ancients also attached importance to the igniting materials when artificially starting fires. In addition to the moxa and paper previously mentioned, there was an igniting material that was highly worth mentioning – candles. In Qing Yi Lu (《清异录》), Tao Gu of the Song Dynasty said: "When one feels an urge at night, it takes too much time to light a candle. An intelligent man dips a strip of cloth into the sulfur and sets it aside for future use. Once meeting fire, such a strip of cloth burns instantly, and there is light immediately. There are traders of this, renaming it as 'huocun'." In Chuo Geng Lu (《辍耕录》), Tao Zongyi of the Yuan Dynasty also mentioned this igniting material, calling it "fazhu," and said: "In the 6th year of Emperor Jiande's reign during the Later Zhou Dynasty, poor concubines made a living by selling fazhu." According to these records, "fazhu" was a small woodchip with molten sulfur. Sulfur has a low combustion point and is highly inflammable. It burns instantly once meeting fire and is not easy to put out. If what Tao Zongyi said is true, "fazhu" had already been made into commodities for trading in the Northern and Southern Dynasties period. "Fazhu" was an important invention that brought great convenience to people's lives, and therefore had been in use for a long time until the nineteenth century, when matches that could start a fire through frictions invented in Europe were introduced into China. As matches could be used to both start a fire and pass along a fire and were extremely easy to use, the traditional fire-starting tools of China were gradually replaced.

8.4.3 Traditional Temperature Measurement Method

It was not until the establishment of the temperature system and invention of the thermometer that the quantitative measurement of temperatures became possible. Before that, people had no concept of temperature, and thermal measurement in a modern sense was impossible. However, the phenomena of hotness and coldness were an objective existence and ancients inevitably experienced these phenomena in production and life, which brought a question: How to determine the temperature of an object? Long-term explorations of this enabled the ancient Chinese to develop a traditional approach to the qualitative determination of temperatures.

People first rely on their feelings to determine temperatures. In the literature of ancient China, there are abundant terms used to describe the temperatures of objects, such as "icy," "cold," "cool," "warm," "hot," and "scorchy" ranging from low to high temperatures. These terms were closely related to people's subjective feelings. In fact, any nation in the world relies on their intuitive feelings as the key means of determining temperatures when they need to determine the temperatures of objects in everyday life.

Such a method that depends on the intuitive feelings of the human body involved a good practice in ancient China, as recorded by Jia Sixie of Northern Wei in Qi Min Yao Shu (《齐民要术》). In "Cheese Making" of the Sheep Keeping chapter of the book, Jia Sixie mentioned the cheese "must be slightly warmer than the human body." Moreover, "Bean Making Method" mentioned the cheese should be "generally as warm as one's armpit," and "insert the hand into the pile and check if it is as warm as the armpit." These described how the ancients determined whether the temperature of an object is satisfactory based on the temperature of the human body as the benchmark. The temperature of the human body generally does not change much, and the armpit temperature is relatively stable compared to other parts of the human body. With the armpit body temperature as the benchmark, the result is naturally more accurate. Therefore, the method proposed by Jia Sixie based on the body temperature under the armpit is scientifically reasonable.

The ancients also used another method to determine the temperature of an object: by observing changes in the object' state caused by the thermal effect. Lv Shi Chun Qiu – Shen Da Lan – Cha Jin (《吕氏春秋·慎大览·察今》) described one of these methods: "When you examine the shadow at the doorstep, you will know the movement of the sun and moon and changes in the weather; when you observe the ice in the bottle, you will know the temperature and the reserve of fish and turtles." This describes how to roughly determine the temperature range based on changes in the state of water, which makes certain sense scientifically. As we know, with the atmospheric pressure remaining unchanged, the relative temperature of water is constant. Generally, if the atmospheric pressure does not change much, the relevant temperature of water basically remains unchanged. Thus, it is in principle feasible to roughly determine the temperature range by observing whether the water in the bottle turns into ice.

This method recorded by Lv Shi Chun Qiu (《吕氏春秋》) was also mentioned a lot in subsequent literature, such as Huai Nan Zi – Shuo Shan Xun (《淮南子·说山训》), which said: "By observing the water in the bottle, one can know how cold it is." Bing Lue Xun (《兵略训》) said: "By observing the water in the bottle, one can know the temperature outside." If one sees ice in the bottle, the temperature must be low, and if the ice melts into water, it suggests the temperature rises. As can be seen, the ancients relied on this method, which was a step further from subjective feelings, as it was based on objective factors. Subjective feelings are easily affected by people's status and lead to wrong conclusions. For example, Liu Xiang's Xin Xu – Ci She (《新序·刺奢》) told such a story, which went like this: During the Spring and Autumn period, King of the Wei Kingdom wanted to build a garden with pools and trees in freezing weather. A minister persuaded him not to as freezing weather would hurt the workers. The king was doubtful, asking: Is it cold? The minister said: Your majesty surely does not feel cold with the fox-fur robe all over you, the bear-fur pad in your chair, and the furnace indoor. But the people had no clothes or shoes, and they of course felt cold. In this story, the king misjudged the outside temperature because he had always felt warm. However, the method recorded in Lv Shi Chun Qiu (《吕氏春秋》) is not affected by the temperature felt by the observer. If the king of Wei Kingdom had used this method by checking if the water in an outdoor bottle melted into ice, he probably wouldn't have had the idea of building a garden.

Among ancient temperature determination methods, another method was worth mentioning, which was the "hotness method," used to determine high temperatures. The "hotness method" was essentially observing the color of the furnace fire to determine the furnace temperature. This method developed in pace with the ancients' smelting technology and was the product of the ancients' smelting experience. In smelting practice, ancient workers developed an easy approach to determining hotness, which was first recorded by Kao Gong Ji (《考工记》):

> Upon smelting of the metal and tin, the color changes from black to yellowish white; from yellowish white to greenish white; and from greenish white to green, when the smelting can be done.

The "metal" refers to copper. Metal and tin are smelted into a bronze alloy. The paragraph described changes in the color of the flame during the smelting process. The description in Kao Gong Ji (《考工记》) seems scientifically sensible. Upon smelting of metals, the furnace temperature varies with the flame color, which depends on the emission spectrum of gaseous metal atoms generated during the smelting process as well as the radiation background. Metals contain impurities such as carbon and sodium and have different vaporization points. During the heating process, the furnace shows different flame colors depending on the different vaporized substances generated. When heating begins, the hydrocarbons and some impurities attached to the mineral materials burn to produce black gases. As the furnace temperature continues to rise, the flame color turns yellowish white, which is caused by the vaporization of sodium atoms contained in the metal. When heating continues, the flame color turns blue and white. At this time, the vaporized metal atoms are dominated by zinc. The zinc burns at high temperature to produce white zinc oxide. At about 1200 °C, zinc will be completely volatilized. At this time, the furnace is "pure" and the furnace temperature is high enough to be used for casting. As can be seen, this method of roughly determining the furnace temperature by observing the flame color of the furnace is feasible. To this day, the flame color is still used to determine the progress of the chemical reaction in the furnace in some smelting processes, and the operation is performed along with instrument monitoring.

China's traditional temperature measurement method could only roughly determine changes in temperature. In the seventeenth century, some important scientific instruments were invented in Europe, including the thermometer. This invention was introduced into China through missionaries and drew the interest of Chinese scholars, with people trying to make "the thermometer" both officially and privately. Afterwards, through long-term evolution, a new-type thermometer with fixed scales and not subject to the change in atmospheric pressure was gradually popularized in China, leading to more scientific temperature measurement.

8.4.4 Thermometer

Thermometers are important scientific instruments and play a vital role in scientific research and daily lives.

During the Qin and Han Dynasties periods, people roughly sensed the temperature by observing changes in the states of objects, such as the previous quotation from Lv

Shi Chun Qiu (《吕氏春秋》) "you can know the temperature through the ice in the bottle." However, "a bottle of ice" could serve as a primitive thermometer at most, as it could only roughly display changes in the outside temperature within a limited range and did not have any scales and naturally cannot be regarded as thermometers.

Early forms of quantitative thermometers appeared in the 1760s and 1770s in China, first introduced by Belgian Christian missionary Ferdinand Verbiest (1623–1688) through his works Ling Tai Yi Qi Tu (《灵台仪器图》) and Yan Qi Tu Shuo (《验气图说》). The former was complete in 1664, and the latter in 1671. These two books were subsequently included in Xin Zhi Ling Tai Yi Xiang Zhi (《新制灵台仪象志》) by Ferdinand Verbiest, with the former serving as the illustration and the latter becoming part of it, namely Vol.4 Yan Qi Shuo ("验气说").

In Yan Qi Shuo ("验气说") of Xin Zhi Ling Tai Yi Xiang Zhi (《新制灵台仪象志》), Ferdinand Verbiest described the method of making thermometers: "To make a thermometer, use glass vessels, such as A, B, C and D; place a wooden rack, as shown in Fig.108 (note: the numbering in the original book). The ball above (A) connects with the tubes B, C and D below, and their lengths are subject to certain rules. The wooden rack is as long as the tubes and consists of three layers, to correspond to the three domains of the Primordial Qi between the heaven and earth. The smaller parts of the tubes below are aligned with the horizon. The bigger parts above are divided into ten scales on each side. None of the scales are evenly divided and must correspond to the trend of the temperature. Therefore, the scales are at different distances from the horizon, and temperatures vary accordingly. . .the temperature follows an inevitable rule, and the temperature naturally corresponds to it, leading to unequal scales." According to the description in this paragraph and subsequent texts, the thermometer made by Ferdinand Verbiest was a U-shaped tube made of glass, which was connected to a copper ball at one end, and had an open orifice at the other end. The tube and the ball were partially loaded with water. With a horizontal line as the benchmark, the tube was divided into the upper and lower parts, with the upper part longer than the lower part. The tube had uneven scales on both sides for use in temperature measurement (as shown in Fig. 8.8. The number indicated in the illustration is the number in the original book).

The use of unequal scales was due to the understanding of the spatial distribution of temperatures by missionaries at that time. Verbiest said: "There are three domains between the heaven and the earth. The upper domain is close to fire and therefore often hot; the lower domain is close to water and earth, which are often in the sunlight and are therefore warm; the middle domain is distant from both the heaven and the earth, and is therefore cold." This is the Three-Domain Theory popular in the West. Verbiest believed changes in the temperature of the air are caused by the interactions among these "Three Domains," and the scales on the thermometer should correspond to the spatial distribution of the "Three Domains" in the air and reflect them. He said: "The temperature follows an inevitable rule, and the scales should correspond to temperatures in an uneven way." That is to say, miniaturizing the distribution of people's imaginary "Three Domains" onto the thermometer forms such uneven scales. These unequal scales, untrue as they were, provided a scaling method after all.

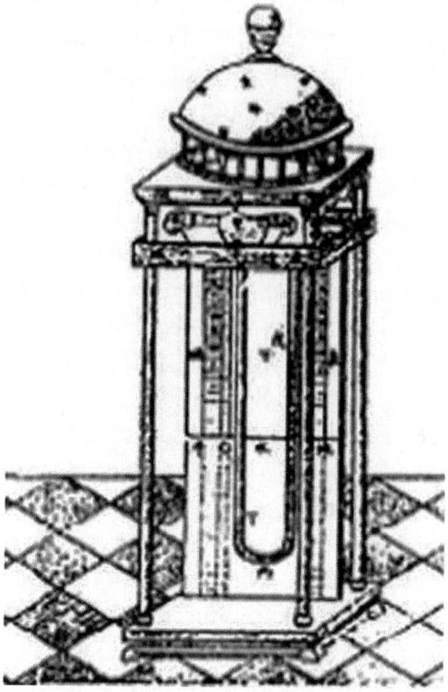

Fig. 8.8 Thermometer in Verbiest's Ling Tai Yi Xiang Zhi (《灵台仪像志》)

As for the operating principles of this thermometer, Verbiest explained: "It is true the rise and fall of the water level are caused by the temperature, but why? Once the upper ball A touches external hot air, the air it contains is released and fills, while there are no gaps to unleash the air, which will inevitably force the water in the left tube to flow down from the horizon to D, and the water in the right tube to flow up from the horizon to E. This is how the thermal principles work. The principles of coldness work in the reverse way. Objects penetrated by cold air contract, such as the ball A. Once it touches external cold air, the air it contains contracts, and the water in the left tube will have to rise." Based on the effect of thermal expansion and contraction, this paragraph basically articulated the operating principles of this thermometer. This, however, is not enough, and this thermometer has an open end, which is connected to the atmosphere.

Verbiest specially explained the reason for doing this. He said: "If we seal the orifice of the tube to disconnect the inside from the outside air, the air in A and D will be forced by the external coldness to contract and condense. A and D are made of copper and iron and will break, and external air will fill the gaps. Besides, as external air is very hot, the internal air will necessarily expand, and without a gap to unleash, and A and D cannot contain, they will necessarily break and air will be unleashed." As can be seen, Verbiest considered sealing the tube orifice, but was worried temperature change might cause the copper balls to break under the influence of ancient Greek theory "nature abhors a vacuum." Therefore, Verbiest allowed the tube orifice to be connected with the outside. He did not have a scientific concept of the atmospheric pressure, which is understandable. However, this kind of thermometer would inevitably be affected by changes in the atmospheric pressure. That is to say, rise or fall of the water column in the tube alone could not reflect the temperature. Besides, the scales on Verbiest's thermometer were also random and there were no fixed points, and therefore could not give accurate temperature values, but only measure changes in the temperature. In conclusion, this was an early primitive air thermometer.

After Verbiest, there were others trying to make thermometers in China. In the early Qing Dynasty period, Huang Lvzhuang invented a "temperature measurer," which could measure the air temperature and body temperature. In the middle of the Qing Dynasty period, Hangzhou locals Huang Chao and Huang Lv also made "thermometers." However, as the original record is too brief, we cannot comment on these private inventions.

8.4.5 Hygrometer

The ancients were well aware of changes in air humidity. As early as in Huai Nan Zi – Shuo Shan Xun (《淮南子·说山训》), the ancients pointed out: "By hanging feather and charcoal, you can know the humidity." Tai Zu Xun (《泰族训》) of the same book said: "When there is wetness, you don't need to see it. Just check if the charcoal gets heavy." Tian Wen Xun (《天文训》) said: "If dry, the charcoal is light; if humid, the charcoal gets heavy." As can be seen, it was known at that time that the weight of certain substances varies with humidity. The ancients leveraged this effect by hanging objects with the same weight but different humidity-absorbing abilities (e.g., feather and charcoal) at both ends of a balance, which became a simple balance-type hygrometer. Upon use, the balance is first made to be balanced, and once the atmospheric humidity changes, the two objects will weigh differently as different amounts of moisture are absorbed (or evaporate), and as a result, the balance becomes imbalanced and turns to display the change in the air humidity. This balance-type hygrometer was not just the ancients' imagination, but had been actually used. According to Hou Han Shu – Lv Li Zhi (《后汉书·律历志》), around the solar terms of dongzhi and xiazhi every year, the emperor would "go to the front palace, gather eight men talented in eight skills, order the playing of eight musical instruments, listen to the music, order the measurement of solar shadows, tune the rhythms and measure the earth and ashes," to determine if winter and summer had really arrived. "Measuring the earth and ashes" here means testing with the balance-type hygrometer. In ancient China, there were no scales on the balance-type humidity measurer, which was never intended to be quantified, and therefore could not be called a hygrometer.

The upper part is tightly clipped, and a length measurer penetrates the lower part, with a flat plate installed under the measurer. Make the center of the measurer, namely the vertical line of the string directed toward the ground center. The measurer is decorated with dragon and fish patterns. The principle is like this: when dry, the dragon meter turns left; when humid, the dragon meter turns right. The addition to or reduction of the gas dryness or dampness will result in the meter turning left or right to the corresponding degree, and the scale of addition or reduction is clearly engraved on the left and right sides of the flat late, which makes the instrument complete. The upper surface of the flat plate is divided into the left and right parts, with 10 scales of different breadths, which indicate the degree of dryness or dampness. The left side is the dryness part, and the right side is the dampness part. The rightness of the string that is stretched or tightened due to the weather corresponds to the breadth of the scales, and therefore, the scales are used to correspond to it.

In China, the hygrometer was also introduced by Verbiest. It was also in the section of Yan Qi Shuo ("验气说") of Xin Zhi Ling Tai Yi Xiang Zhi (《新制灵台仪象志》). Verbiest introduced the method of making and using the hygrometer: "Now we need to detect changes in humidity. Among all the animals, only the tendons of beasts are obvious, and therefore their tendons are used in measurement instruments. See Fig.109 (note: the number in the original book). The approach is to use a new deer-tendon string that is about 66.6cm long and 3.3cm thick, hang two objects with equal weights on both sides and a horizontal bar is placed."

There are various types of hygrometers. In ancient China, the balance-type hygroscopic hygrometer was used. Verbriest's hygrometer was also hygroscopic in principle, but it was a hanging string type in form. He used deer tendons as the string, fixed the upper end, and hung an appropriate weight on the lower end, fixed a pointer on the string, and the pointer was carved into the shape of fish. This string will be twisted after it absorbs moisture. With a different level of moisture absorption, there is a different twist angle. The size of the turning angle is displayed on the dial by the pointer, which serves to measure humidity (as shown in Fig. 8.9, and the number in the figure is the number in the original book number).

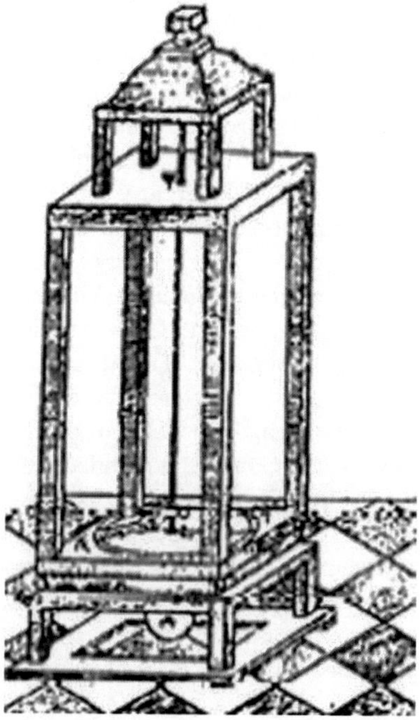

Fig. 8.9 Hygrometer introduced in Verbiest's *Ling Tai Yi Xiang Zhi* (《灵台仪像志》)

The suspended string hygrometer is simple in structure and easy to use and is therefore quite popular. However, there is also room for improvement. For example, Verbiest's unequally divided scales on the chassis of the hygrometer could not accurately reflect the changes in air humidity. But this was the first hygrometer with quantitative scales in China. In addition, this type of hygrometer is also described in western books, which however were later than Verbiest's introduction. Evidently, this type of hygrometer was introduced into China quite early.

After Verbiest, Chinese scholars also tried to make instruments to measure atmospheric humidity. According to Huang Lvzhuang's biography recorded in Vol.6 of Zhang Chao's Yu Chu Xin Zhi (《虞初新志》), Huang Lv Zhuang made a so-called "dampness tester" in 1683: "There is a needle in it, which can rotate left and right. If dry, it can rotate to the left; when wet, it can rotate to the right. It operates in this way without a single error and can predict the weather." However, its structure and principle were not been recorded, and as a result, we know nothing about the specific form of this "dampness tester."

References

1. Xiu, O. (1988). *Liu Yi Bi Ji* (《六一笔记》). *Shuo Fu San Zhong* (《说郛三种》), Shanghai Classic Publishing House.
2. Wang, J. & Yu, S. (1987). "*Optical knowledge in Tan Zi Hua Shu* (《谭子化书》)". Fang Lizhi: *History of Science History Papers* (《科学史论集》), (pp.213–220). University of Science and Technology of China Press.
3. Wang, J. & Zhenhuan, J. (1986). *History of Chinese optics* (《中国光学史》), edition (p.54–56). Hunan Education Press.
4. Wang, J. & Zhenhuan, J. (1986). *History of Chinese optics* (《中国光学史》), Hunan Education Press, p.86.
5. Zhichao, L., & Xu, Q. (1985). "*Latticing Technique in Optics in Ancient China*". *Physics* (《物理》), *14*(12).
6. Zhichao, L. (1995). Textual analysis of "Autopsy with Red Light" in Meng Xi Bi Tan ("《梦溪笔谈》'红光验尸' 的文字考证"). *Tian Ren Gu Yi* (《天人古义》), Henan Education press.
7. Wang, J., & Zhenhuan, H. (1986). *History of Chinese optics* (《中国光学史》), (p.95–96). Hunan Education Press.
8. Wang, J. & Zhenhuan, H. (1990). *Overview of history of physics in ancient China* (《中国古代物理学史略》), Hebei Science and Technology Press, p.160.

Astronomy and Diplomacy: Official Contacts in Astronomy Between China and Korea

9

Yunli Shi

Contents

Abstract

China and Korea are close neighbors and convenient transportation contributed to the close contacts and exchanges between the two countries from ancient times. As astronomy played a vital role in shaping its ideology and bureaucracy, the rulers of Korea traditionally put particular emphasis on learning the knowledge of astronomical calendar of China. As a result, the Korean astronomy has always been under the influence of China and thus belongs to the same academic system. The academic achievements made by China in astronomy were introduced to the Korean Peninsula in a systemic way and in turn promoted the development of its astronomy significantly.

Keywords

Korean Peninsula · Astronomical observation · The Star Atlas · Astronomers · Eclipses · Tough transformation · Christian missionaries

Y. Shi (✉)
Department for the History of Science and Scientific Archaeology, University of Science and Technology of China, Hefei, China
e-mail: ylshi@ustc.edu.cn

X. Jiang (ed.), *The High Tide of Science and Technology Development in China*,
History of Science and Technology in China,
https://doi.org/10.1007/978-981-15-7847-2_9

China and Korea are close neighbors separated only by a strip of water. Convenient transportation contributed to the close contacts and exchanges between the two countries from ancient times in such areas as politics, economy, culture, and scientific technology. Since at least the dynasties of Silla, Baekje, and Goguryeo, the ancient Korean authorities have learned from or imitated China regarding political system, and astronomy played a vital role in shaping its ideology and bureaucracy. For this reason, the rulers of Korea traditionally put particular emphasis on learning the knowledge of astronomical calendar of China. The rulers of China have long regarded the issuance of almanacs to the dependent countries as a symbol of the status and authority of the sovereign state. Therefore, despite that the Koryon astronomers made some breakthroughs of their own in astronomy indeed; they generally copied the Chinese astronomical calendar from disciplinary system to specific knowledge; as a result, the Korean astronomy has always been under the influence of China and thus belongs to the same academic system. Due to the fact that academic exchanges in this area were made generally at the government level, astronomy has become a special part of their diplomatic activities and revealed an obvious interrelationship with their political relations. Such kind of exchanges not only showcased how China's advanced astronomy influenced its neighbors but also reflected its special role in the political relationship between countries. This exchange in astronomy with diplomatic agenda originated in the Yuan dynasty, developed in the Ming dynasty, and prospered in the Qing dynasty. As a result, the academic achievements made by these Chinese dynasties in astronomy were able to be introduced to the Korean Peninsula in a systemic way and in turn promoted the development of its astronomy significantly.

9.1 Prolonged Initial Stage

According to *The History of Koryo·Lizhi*, "There was not a single emperor in history who did not attach lots of weight to the study of astronomical calendar, therefore when the calendar of Zhou dynasty got lost amid its collapse, our country created the calendar of our own" (Zheng Linzhi. *The History of Koryo* (Vol. 50)). Even though this is not necessarily a faithful historical account, it indeed reflects in some way the connection between the astronomy of China and that of the Korean Peninsula.

Since around 57 B.C., Silla (B.C.57–902), Goguryeo (B.C.37–668), and Baekje (B.C.31–660), three independent nations, occupied the Korean Peninsula successively, all of which made contacts with the dynasties of China in areas such as politics, economy, culture, and scientific technology. Influenced by the political system of China, these countries set up the department of astronomical calendar of their own. For instance, back in the sixth century, Baekje set up a "department of calendar" (Linghu Defen. *The History of Zhou Dynasty* (Vol. 49)). In the middle of the seventh century, "Queen Sondok ordered the construction of an astronomical observatory" (Anonymous author. A *Brief History of Koryo* (Vol. 2)). By the eighth century, it started to create "the dictionary of clepsydra. . . . . .and appointed 6 officials as director in charge of academic study and another one for history" and later on "appointed one official as director of astronomical study, and 6 officials as director in charge of clepsydra, and renamed director in astronomical study as celestial manager" (Jin Fushi. *Samguk sagi* (Vol. 8–9; Vol. 38–39)). There is no historical account regarding when the official department of astronomy was established, but it is generally believed that this country had officials in charge of astronomical calendar like "Rizhe."

There are plenty of records on astronomical observation in *Samguk sagi* written by a Korean historian in 1145, a significant body of which was done independently by the Korean astronomers in the three dynasties of Silla, Goguryeo, and Baekje. According to these records, the observed objects they chose for studying abnormal celestial phenomenon (including jiaoshi, yueyanxing, wuxinglingfan, kexing, huixing, liuxing, and so on), terminology used, and how constellations were named were the same as those in China, indicating that the astronomical institutions of the three dynasties adopted pretty much the academic system of China.

With regard to astronomical calendar,. The Baekje dynasty "used *Yuanjia Calendar* of Song dynasty of China and took the month of Jianyin as the beginning of the year" (*Book of Zhou.* (Vol. 49)). The Goguryeo dynasty "sent ambassadors to Tang dynasty to ask for an issuance of calendar" in A.D. 619 (Jin Fushi. *Samguk sagi* (Vol. 20)), and the calendar introduced at that time was probably *Wuyin Calendar* being used in the Tang dynasty. By A.D. 674, the Silla dynasty "sent palace guard Danai madefu to Tang dynasty to learn from its astronomical calendar in an attempt to reform the calendar of its own" (Jin Fushi. *Samguk sagi* (Vol. 7)), and the calendar brought into the Silla dynasty was supposed to be *Linde Calendar.*

In terms of its stellar system, the star atlas (a total of 20 star atlas) from A.D. 357 to the sixth century were Chinese in style including constellations of Three-Legged Crow, Four Symbols, Big Dipper, Sagittarius, and the Lunar Mansions, as found in the ancient tomb of the Goguryeo dynasty. It suggested that the knowledge of stellar system in China had already been popularized in the Korean Peninsula at that time. By the year 629, "an eminent monk Daozheng of Silla dynasty came back from Tang dynasty with a piece of star atlas" (Zheng Linzhi. *The History of Koryo* (Vol. 76)) indicating the introduction of Chinese constellation in the Korean Peninsula.

Shortly after unifying the Korean Peninsula and founding the Koryo dynasty in A.D. 918, Wang Jian set up the departments of Taibujian and Taishiju in charge of affairs in astronomy and calendar. Later on, Taibujian was renamed as Sitiantai, Sitianjian, and Guanhoushu successively until the eventual merging of the departments of Sitianjian and Taishiju into the department of Shuyunguan. By comparison, we can find that the astronomical institution of the Koryo dynasty is pretty much the same as that of the Tang dynasty and Song dynasty, although on a smaller scale.

At the beginning of the Koryo dynasty, it adopted the Chinese reign titles such as post-Tang, post-Jin, post-Han, Post-Zhou, Song, and Liao, which suggested that it had used the astronomical calendar of these Chinese dynasties (Zheng Linzhi. *The History of Koryo* (Vol. 1–3)). Eventually, *Xuanmin Calendar* written by Xu Ang from the Tang dynasty was mastered by officials in charge of astronomy of the Koryo dynasty and established as its official calendar that was to be used as the basis for the compilation of almanacs of ordinary people and for the forecast of eclipses. Nevertheless, officials in charge of astronomy in the Koryo dynasty failed to master the calculation of the movements of five stars.

Shortly after the Mongol conquest of China, the Koryo dynasty became under the tough control of Mongolia as its tributary state. However, this relationship of the two countries resulted in an even closer contact in astronomy. Due to political reasons, the governments of the Yuan dynasty often gave away their almanacs to the kings of the Koryo dynasty as gifts, among which the administration of Kublai Khan granted almanacs as many as 15 times (Song Lian. *History of the Yuan Dynasty* (Vol. 4–17). *A Brief History of Koryo* (Vol. 26–29) also recorded the Yuan's issuance of calendar in Koryo: in 1264, "Envoy Han came back from Mongol with a piece of western brocade and a copy of almanac granted by the Emperor." In February 1268, "Envoy An Qinggong came back from Mongol with a piece of western brocade and a copy of almanac granted by the Emperor." In February 1296, "Envoy Jin Guang came back from the Yuan dynasty......with two bottles of wine and a copy of almanac granted by the mother of the Emperor"). Besides, the government of the Yuan dynasty probably often sent forecasts of eclipses to the Koryo dynasty. For instance, as recorded in *The History of the Koryo Dynasty·Astronomy*, "On lunar January 1st 1320, the 7th year of the reign of Zhongsu Emperor, Yuan government sent officials to inform of an eclipse"; "On lunar April 1st 1351, the first year of the reign of Kongmin Emperor, Yuan government sent officials to inform of possible eclipse that turned out to be true......on lunar September 1st of the next year, Yuan government informed of possible eclipse that turned out to be true"; and so on.

Shoushi Calendar was officially adopted by the Yuan Government in 1281, and in the same year, Kublai sent ambassadors to grant the new calendar to the government of the Koryo dynasty. Wang Tong and other ambassadors who went to Koryo were astronomers who "conducted astronomical survey in the daytime and observed celestial phenomenon at nights" and "asked for the map of our country" during their stay there (Zheng Linzhi. *The History of Koryo* (Vol. 29)). There is no doubt that their work was part of the project of astronomical survey (which was called Sihai Survey) launched by Guo Shoujing, as this survey ranged from "Koryo dynasty to the east, to the Dian Lake to the west" (Song Lian. *The History of Yuan Dynasty* (Vol. 48)). The latitude of the Koryo dynasty (less than 38 degrees) recorded in books such as *The History of Yuan Dynasty·Astronomy* was supposed to be measured by Wang Tong and his fellows. When converted to the modern units of measurement, the latitude was $37°42'$ which was pretty much the same as that of the capital city of the Koryo dynasty.

After the introduction of *Shoushi Calendar* to the Koryo dynasty, the government sought to master its compilation and calculation methods. Nevertheless, the Yuan dynasty forbade any forms of astronomical study outside of the supervision of the ruling government, just as previous dynasties like the Tang dynasty and Song dynasty. Therefore, it was bound to be difficult for the Koryo dynasty, as a tributary state, to learn from the Yuan dynasty on its astronomical calendar. However, a policy made by the Yuan dynasty as part of its move towards greater control over Koryo provided convenience for the Koryo Government in some way.

In an attempt to strengthen its control over Koryo, the Yuan dynasty established a custom at its very beginning, under which the heirs to the throne had to be sent to the capital city of Yuan to grow up there in the culture of Yuan and get married with

Mongol women, only after which were they allowed to go back to Koryo. This custom provided opportunities for Koryo to learn from the astronomical calendar of the Yuan dynasty. In 1278, the Koryo heir Wang Zhang (1275–1325) was sent to the Great Capital of Yuan and grew up there. By 1298 when he failed to ascend the throne back in Koryo, he returned to Yuan where he stayed until 1308 when he went back to his country again to be enthroned. Yet it was before long that he returned to the Great Capital to rule Koryo by giving orders from Yuan, and only until 1313 he came back to Koryo briefly after he had abdicated.

It was in 1298 when he returned to the Great Capital that Wang Zhang "upon seeing how the officials in the department of Taishiyuan knew well about the astronomical calendar, intended to introduce the astronomical knowledge to Koryo. In 1303 and 1304, he ordered Cui Chengzhi of Guangyang city to offer up 50 kilograms of gold in exchange for instructions on astronomical calendar, and by doing so he managed to grasp the essence of the astronomy of Yuan dynasty." As a result, *Shoushi Calendar* was mastered initially by Koryo astronomers and thus used as the guidelines in compiling their own almanac that was officially issued by the ruling period of Emperor Zhongxuan (1309–1313).

Afterwards, Cui Chengzhi imparted what he mastered to Jiang Bao who for his good command of Shoushi Calendar was later on appointed as Sili of the department of Shuyunguan and further promoted as Guanzheng eventually. According to his professional knowledge, Jiang compiled *Calculation Methods of Shoushi Calendar* of two volumes that echoed down the ages. However, as Koryo astronomers failed to master how to extract a root, "the chapter of eclipse followed that of *Xuanming Calendar*" (Zheng Linzhi. *The History of Koryo* (Vol. 50)). Furthermore, what is equally apparent is that they could not grasp the knowledge regarding how to forecast the movements of planets.

Since the establishment of the Ming dynasty, Koryo has maintained a diplomatic relationship with it. It marked the start of exchanges in astronomy between the Ming dynasty and Koryo dynasty that Ming Government sent envoys there with gifts including "a copy of Datong Calendar" in 1369 (Xie jin. *Authentic Records of Emperor Mingtaizu* (Vol. 44)). At that very year, the Koryon Government sent Cheng Zhunde to Ming, hoping to learn about the compiling method of Datong Calendar (Zheng Linzhi. *The History of Koryo* (Vol. 42)), and Cheng returned with a copy of *Datong Calendar* given by Zhu Yuanzhang (Xie jin. *Authentic Records of Emperor Mingtaizu* (Vol. 46)).

Another major masterpiece of Chinese astronomy that was introduced to the Koryo dynasty apart from *Shoushi Calendar* was *Butian Verses.* Since then, the systemic knowledge of Chinese constellations with 28 lunar mansions as its framework in the form of verse has been officially absorbed in the Koryon knowledge system of astronomy.

9.2 Fruitful Development

Shortly after his founding of the Joseon dynasty in 1392, Li Chenggui formed a diplomatic relationship with the Ming dynasty. For political reasons, the Ming Government established it as a custom to present almanacs to the Joseon dynasty

Fig. 9.1 *Ming Star Atlas*

every year, "with one copy for the emperor of Koryo and another 100 copies for the ordinary people" (Shen Shixing. *Records of Laws and Systems of Ming Dynasty* (Vol. 121)). All of the first kings of the Joseon dynasty paid relatively much attention to astronomical calendar and thus laid a special stress on its development. A case in point is the well-known *Ming Star Atlas* that Li Chenggui asked people to carve on stones. This map has been preserved until today (see Fig. 9.1). According to its inscription, the predecessor of this star map had been "kept in the City of Pyeong-yang before it sank in the river in turmoil caused by war." Li was presented a copy of the map after he ascended the throne, which he valued so much that he ordered people to re-carve it on stones. The re-carved map thus became the standard star atlas of Koryo until the seventeenth century when the European star map was introduced through China.

Li Tao was the third king of the Joseon dynasty who reigned from 1418 to 1450. He was an ambitious king, intent on promoting the development of Korean rites, music, and culture, and thus astronomical calendar which served as a symbol of the tributary state of a country became a focus. He also put particular emphasis on the absorption and introduction of the Chinese astronomy. It was under this background that exchanges in the field of astronomy were strengthened unprecedentedly between the Ming dynasty and Joseon dynasty.

The Joseon dynasty continued to use the incomplete version of *Shoushi Calendar* and *Xuanmin Calendar* in the beginning, even though "the emperor of Ming had granted *History of Yuan*" to the second king of the Joseon dynasty Li Fangyuan (*Compendium of Historical Documents* (Vol. 242)). The officials of astronomical calendar failed to take advantage of the full version of *Analysis of Shoushi Calendar* documented on the chapter devoted to the Chinese calendar until 1419, the second year of the reign of King Li Tao, when Liu Tingxian, an official in the department of Lingshuyun, pleaded with King Sejong to order civil ministers to figure out how the Chinese calendar worked. Li Tao adopted his suggestion, holding that "nothing was more important than this in running a country as a king," and thus "ordered the civil ministers in the Ministry of Art and Culture including Zheng Qinzhi to study *Shoushi Calendar* so as to get a broad outline of it, and the heads of this ministry such as Zheng Zhao to dive deeper to grasp the whole system of it" (Anonymous author. *Movements of rahu ketu lilith adam.* Yuantong*: The Calculation Methodology in Datong Calendar*).

Meanwhile, Li Tao sent astronomers to study in Ming. By various means, Li managed to obtain a number of works on the astronomy of the Ming dynasty (Anonymous author. *Movements of rahu ketu lilith adam.* "*On the* Calculation Method of Solar Latitude by Yuan Ton and original block-printed edition of the Huihui Calendar by Beilin" written by Shi Yunli and Wei Tao in 2009 mentioned its detailed discussion on the Chinese astronomical works. *The Chinese Journal for the History of Science and Technology,* 30(1), 31–45). Apart from official history books like *Da Ming Calendar*, *Gengwu Yuan Calendar*, *Analysis of Shoushi Calendar*, and *Discussion on Shoushi Calendar*, several masterpieces written by Chinese astronomers in the beginning of the Ming dynasty were also included:

> *The Calculation Methodology in Datong Calendar,* which consisted of *The Calculation Methodology of the Movements of the Sun, The Calculation Methodology of the Movements of the Moon, The Calculation Methodology of Eclipses, The Calculation Methodology of the Movements of the Five Planets, The Calculation Methodology of Calendar Day,* and *The Calculation Methodology of Siyu* (the Four Residuals). It was compiled by Yuan Tong, an imperial astronomer of the Ming Bureau of Astronomy around 1384 under the order of Emperor Zhu Yuanzhang, and became the official edition of *Datong Calendar* used as reference for the calculation work of calendar and astronomy in the Bureau of Astronomy.

1. *The Islamic Calendar* (called *Huihui Calendar* in Chinese) was an Arabic masterpiece of astronomy that was jointly translated by Chinese officials and Islamic academics in the field of astronomical calendar from 1382 to 1384 under the order of Emperor Zhu Yuanzhang. The re-compiled edition of it currently preserved in Nanjing was finished by Bei Lin in 1477 when he worked as the assistant imperial astronomer of the Ming Bureau of Astronomy. The version consisted of *Explanation of the Islamic Calendar*, *The Islamic Calendar*, *The*

Calculation Method of Longitude, and The Calculation Method of Latitude. These methods can be used to calculate the Islamic Calendar, ecliptic longitude and latitude of the sun, the moon and the five planets, eclipse, and the astronomical phenomenon Lingfan (literal meaning encroachment; the occultation and conjunction among the moon, the five planets, and the fixed stars when the angular distance between these celestial objects was small enough).

2. *The Calculation Method of Solar Latitude* written in 1396 by Yuan Tong under the order of Emperor Zhu Yuanzhang was designed to integrate the calculation methods of the sun in the Huihui Calendar and the Datong Calendar. The "latitude" in the book title referred to the Islamic Calendar (The only existing copy is stored in the Koryon Kuizhangge Library. "*On the* Calculation Method of Solar Latitude by Yuan Ton and original block-printed edition of the Huihui Calendar by Beilin" written by Shi Yunli and Wei Tao discussed on the content and importance of this book).
3. *The Calculation Methodology of Astronomical Calendar in the Western Regions* was compiled by Liu Zhengxin (?~1449), an official in the Bureau of Astronomy who worked as assistant to the emperor of Ming. It was another important work on the Islamic Calendar in the beginning of the Ming dynasty.

Arguably, every single one of the most important masterpieces on astronomy written by the astronomers in the beginning of the Ming dynasty had passed into the hands of the astronomers in the Joseon dynasty. It came as a surprise for us to find out how these astronomers from a tributary state could pull off such feats obtaining a thorough collection of the official astronomical masterpieces of Ming so rapidly, especially when we considered how Ming forbade strictly any forms of astronomical study or printing of astronomical calendar books in private. With Zhu Yuanzang and other emperors of the Ming dynasty being vigilantly on guard against seigniors, there was no way that they offered official help when the Joseon kings went to great lengths to learn the Chinese astronomical calendar, viewed as "embodiment of Heaven's Mandate (Tianming)" and obtained the related works on it. They must have broken the stranglehold of the Ming Government to connect directly with the Ming Bureau of Astronomy, which could be proved by the other two works on astronomy of the Ming dynasty the officials of Li acquired.

1. A volume of *Eclipses of the Past* contained in *The Calculation Methodology of Eclipses* collected in *The Calculation Methodology in Datong Calendar.* As recorded in the book, "Koryo was informed of the eclipses in the Shoushi Calendar by letter from China," and the specific cases of eclipses from the ruling period of Xuande to Zhengtong that were forecasted based on the calculation methods in *The Calculation Methodology of Eclipses* were also documented.
2. A volume of *Lingfan of the Five Planets in the 10th year of Xuande reign* was basically an astronomical log book of the phenomenon of Lingfan among the moon, the five planets, and the stars observed in the 10th year of Xuande reign that were followed through the analysis tools in *The Islamic Calendar.*

In the Ming dynasty, the Bureau of Astronomy was the only institute able to calculate eclipses every year with the official calculation methods of astronomical calendar, and the calculation of the phenomenon of Lingfan that took place among the moon and the five planets was also one of the regular works of the astronomers in the Ming Bureau of Astronomy who were specialized in the Islamic Calendar. Sun Nengchuan in the end of the Ming dynasty documented "a copy of the Lingfan of the Moon and the Five Stars in the 24th year of Hongwu Reign (1391)" (Sun Nengchuan. *Library Catalogue of the Cabinet* (Vol. 7)) in his compilation of *The Catalog of the National Collected Books* (Neige) in 1602. As the phenomenon of Lingfan played a vital role in the Chinese traditional astrology, it was apparent that these calculation projects were designed to facilitate the forecasting of Lingfan by official astrologists. Facts proved that the major reason for the decision of Emperor Zhu Yuanzhang to translate *The Islamic Calendar* was that it could fill this need. Furthermore, according to the historical records, Zhu used the calculation of the Bureau of Astronomy on Lingfan and its significance in astrology to justify the "Heaven's Mandate" in his approach to dealing with the seigniors of Zhou, Qi, Tan, and Lu ("A study of the adoption of Huihui Calendar in the Ming dynasty based on Wuxing Lingfan in the 10th year of Xuande reign" written by Shi Yunli, Li Liang, and Li Huifang discussed in details the motivation behind Zhu Yuanzhang's order to translate and compile *Huihui Calendar* as well as how he used the forecasts of Lingfan for astrological purpose).

Both of these documents were obviously highly confidential matters of Ming, especially the latter one. However, as secret as they were, the Joseon dynasty could still manage to "be informed by letter from the Ming government," which meant that it must have had a special access to the state secrets of Ming. Of course, they could surely achieve their purpose by bribing the officials in the Bureau of Astronomy into "instructing" them, just as what Cui Chengzhi had done to learn *Shoushi Calendar*. But we cannot rule out the existence of "pushing hands" between the Joseon dynasty and Ming dynasty living in the capital city of China long enough to maintain closer contacts with the government officials of Ming like Wang Zhang.

The introduction of these works provided a rich source of materials for the study of astronomy of the Joseon dynasty. By 1442, Li Tao ordered Li Chunzhi of the department of Fengchangsi and Jin Dan of Fengchangzhubu (both departments were in charge of worshipping rituals) to integrate the finer parts of *Shoushi Calendar* and *The Calculation Methodology of Datong Calendar* to compile *Chiljeongsan* of three volumes and later on *Addition to Chiljeongsan* (This book offered approach to a problem that the Chinese edition of *Huihui Calendar* failed to solve, that is, the conversion of the Islamic solar calendar and the Chinese lunar calendar. This approach is of vital importance given that the calculation methods of Huihui Calendar are based on solar calendar, whereas the Chinese astronomical table is based on lunar calendar; it thus is impossible to use Huihui Calendar without converting them to the Chinese equivalents. More detailed discussion on this topic can be found: Shi Yunli (2003). The Koryon adaptation of the Chinese-Islamic Astronomical Tables. *The Koryon Adaptation of the Chinese-Islamic Astronomical Tables. (57). 25–60*) of three volumes based on an assimilation of books including *The Islamic Calendar* and *The*

Calculation Methodology of Astronomical Calendar in the Western Regions. They also re-calculated sunrise and sunset time as well as the day length and night length with the capital city of Joseon as the primary meridian, listed in the appendix of *Chiljeongsan* "as its official calendar to be used permanently." Since then, the Joseon dynasty, for the first time, has its own official astronomical calendar system and used it in the calculation of calendar books and eclipses. In order to provide guidance over the specific calculation of astronomical calendar, astronomers of Joseon compiled another two works on the calculation methods of *Chiljeongsan*, taking the year 1447 as an example to demonstrate how to use the two official astronomical systems to calculate the movements of the sun and the moon as well as eclipses.

Apart from these works, astronomers of Joseon also paid attention to assimilating the astronomical knowledge in other Chinese works. In 1445, Li Chunzhi under the order of the King "collect all of the Chinese works on astronomical calendar to apart from the major official ones, and sort them out into categories to keep the essential parts of them" so as to compile the book of *A Collection of Various Systems of Astronomy*, with four volumes devoted to astronomy, astronomical calendar, astronomical instruments, and sundials, covering a wide range of Chinese astronomical knowledge.

Besides, Li Tao even compiled *On the Calculation of Eclipses* of two volumes in person, taking the year 1447 as an example to demonstrate how to calculate the eclipses of the sun and the moon in two separate volumes. Furthermore, there was a chapter of "Elementary arithmetic" devoted to introducing the basics of arithmetic that would be used in the calculation of astronomical calendar including the extraction of the square root. In the preface Li Chunzhi wrote for this book, he pointed out that "all of the sections in this work were written by King Sejong the Great" including the section of arithmetic. In an effort to mitigate the "difficulty in searching for a specific figure amid a huge amount of ready reference tables," this book listed the formulas to calculate the movements of the sun and the moon to save the bother of table look-up. It would be much easier to apply formulas to work out the figures. In essence, this was basically a summarization of a variety of formulas contained in *Licheng of Shoushi Calendar* (Licheng meant ready reference), yet these formulas had been lost for a long time until the end of the Ming dynasty when astronomers came to figure them out.

Along with an introduction of a number of Chinese astronomical works and a systemic study of them, the Joseon Government also focused on fostering astronomical talents with a command of Chinese astronomy. In 1429 when the government decided to "use Confucion classics as textbook", it stipulated that: "Yin-Yang Theory: *Butian Rhymes of Astronomy*, the calculation of 24 solar terms as well as eclipses of the sun and the moon in *Xuanming Calendar, and* the calculation of 24 solar termsm, eclipses of the sun and the moon and the length of day and night, as well as the movements of the sun, the moon, the five stars and the four faint stars. . .." (*Authentic Records on King Sejong*. (Vol. 47)). These objects had not been changed until the introduction of "Western Learning" to the Koryo Peninsula in the seventeenth century, ensuring a sustained development of the Chinese traditional astronomy in Koryo.

In 1432, Li Tao ordered Zheng Linzhi and Zheng Zhao to "create astronomical instruments to facilitate measurement and calculation of celestial bodies." They thought that among the Chinese astronomical instruments through the ages, "Abridged armilla, scaphe, gnomon among other instruments invented by Guo Shoujing were the most delicate of all time" (Jin Dun. "Abridged armilla. *Authentic Records on King Sejong* (Vol. 77)). Therefore, with copying the measuring instruments of Guo as the priority, they made abridged armilla, small abridged armilla, square table, gnomon, jingfu, and scaphe successively. Later on, astronomers of the Joseon dynasty devised Xuanzhu sundial (literal meaning sundial with hanging beads), Tianping sundial (literal meaning sundial with a balance), Dingnan sundial (literal meaning indicator sundial), timing device (an instrument used to calculate the time based on the movements of stars), small timing device, clepsydra, jade clepsydra, water clock, armillary sphere, and celestial globe (For more details of these instruments, please refer to: J. Needham, Lu Guei-Djen, J.H. Combridge & J.S. Major. (1986). *The Hall of Heavenly Records, Koryon Astronomical Instruments and Clocks 1380~1780.* Cambridge Univ. Press). 1527 and 1602 saw a large-scale revamp of these instruments respectively. Despite this close association with the characteristics of traditional Chinese astronomy, Korean astronomers developed their own unique devices, promoting the development of astronomical observation technologies of the Joseon dynasty in an unprecedented way.

All in all, the exchanges in the field of astronomy with the Ming dynasty in the beginning of the Joseon dynasty gave a boost to the development of the Korean astronomy and, more importantly, led to fruitful academic results. With their achievements even more advanced than that of the Ming dynasty in some specific areas, it ushered in the most splendid chapter of the Korean science history.

It marked the establishment of a comprehensive astronomical research organization with senior talent as well as advanced instruments and knowledge, matched with a conventional management system of talent selection, personnel training, performance assessment, and routine duties (For more details of these institutions, please refer to: *Shuyunguan Journal.* (1979). Chengxin Women's University). This organization was responsible for celestial observation, astrological divination, eclipse forecasting, and the official calendar issuance, just as the Ming Bureau of Astronomy. Needless to say, the academic and political ambition of the kings in the early Joseon dynasty was thus fulfilled. It also implemented a "dual system" in astronomical calendar. On the one hand, they continued to accept the almanacs issued by the Ming Government called *Tang Calendar* and assumed the reign title of Ming dynasty as part of their duties as a tributary state. On the other hand, they compiled and issued their own almanacs using the calculation method of *Chiljeongsan*, called "Xiang calendar" (literal meaning calendar books used in the countryside). Besides, they could also calculate on their own the movements of the sun, the moon, and the five planets and forecast eclipses through *Chiljeong Calendar*, in spite of their tendency to make a comparison among the calculations based on the three calendar books of *Datong Calendar, Chiljeongsan, and Addition to Chiljeongsan* with regard to the forecasts of eclipses; as for *Chiljeong Calendar*, there was only a single copy to be issued for the king every year, "not permissible to publish for other uses." This had not been changed until 1466 when the suggestion of the Bureau of Astronomy that "it is necessary to print two copies of the almanacs for the convenience of comparative analysis, with one for your majesty and the other for this department" was adopted (*Authentic Records on King Sejong*. (October 21st, 2012)).

However, it was apparently intolerable for the Ming Government to see one of its dependencies made such remarkable achievements in astronomical calendar. The dominated upper class knew so well of their suzerain that it took precautions against this risk. For instance, around the Spring Festival of 1469 when the ambassadors of Ming were about to pay a visit to Joseon, the king ordered the administrative organs to make sure that the local governments "do not show the Ming ambassadors Xiang calendar on the pretext of not receiving Tang Calendar yet if they asked to see their almanacs" (*Authentic Records on King Ruizong*. (Vol. 3)). Of course, it was not hard to sweep this whole thing under the rug when several of ambassadors came for a visit; however, it would not be the case if a lot more Chinese officials came over for a longer stay.

In response to Toyotomi Hideyoshi's invasion of Joseon in 1592, the Ming Government sent troops to aid Joseon against the aggression of Japan. What they feared did not come true until 1598 when Ming withdrew troops after its military commander Yang Hao suffered a setback in the War of Daoshan. Ding Yingtai, the head of the Department of War, was thus sent to lead the war. He impeached Yang Hao, which led him to be dismissed from his post, and then Li Yan, the Korean king who spoke out against this unfair dismissal on three charges: First, Joseon deliberately "invited Japan to invade" by trying to resume the territory to the east of the Liao River which it claimed as its lost territory. Second, Joseon overstepped its authority as a tributary state to have its own reign title instead of following that of the Ming dynasty in its compilation of a book that introduced the conditions of Japan titled *A Survey of Haitong* (Haitong referred to the eastern neighboring country). Third, Li Yan "is a brutal and tyrannical ruler overindulged in wine and women" and colluded with Yang Hao to deceive the emperor of the Ming dynasty (For more details this impeachment, please refer to: Sun Weiguo. (2012). On Ding Yingtai's impeachment and the compilation of the historical books in the Ming and Qing dynasty. *Nankai Journal* (3), 74–86. *A Survey of Haitong* offered an introduction of the general situation in Japan including the cultural exchanges between Japan and Koryo, which used as a proof of Li Yan's collusion with Japan. Besides, he also pointed out that the book "adopted and printed the Japanese reign titles of Kangzheng, Kuangzheng, Wenming in large type, whereas the Chinese reign titles of Yongle, Xuande, Jingtai and Chenghua were printed as sub-reign tile in small type. It indicated that Koryo gave fealty to Japan rather than China. Furthermore, Emperor Taizu, Shizu, Liezu and Shengshang were called in this book by their titles, yet another proof of its disrespect for China. With a history of more two centuries of respect and submission to China, it is high time for us confront the Koren kings with interrogations about their loyalty." This was what his second charge referred to).

The impeachment led to another round of political turbulence, and the second charge stoke Korean concerns about the existence of its own calendar "privately issued," and some officials even proposed to stop issuing its own almanacs:

> The calendar promulgated by the emperor of the Middle Kingdom has been used extremely widely that there existed not a single almanac other than this. It is against the norm for our country to create a calendar of our own; therefore once the Kingdom discovered it and condemned us, we do not have a justification for this. All of the almanacs issued by the Kingdom have a stamp on it; without a stamp, it must be a copy privately made. Those who break the rule should be beheaded as stipulated in the law; those who expose such a crime can be rewarded with 50 liang (a unit of weight that equals 50 grams) of silver. Even though we had Tang Calendar printed out in private, once discovered or denounced, it would be convenient to defend ourselves by claiming that we had no choice but printed them in order to promote it across the country. As justifiable as it is, we do not have to be frightened. However, if the case is that we create a calendar of our own, it is bound to be interpreted as a gesture of denying our status of a tributary state of the Middle Kingdom. Even if we adopt the suggestion of the Bureau of Astronomy to amend our calendar according to Tang Calendar, it would still be hard to justify ourselves in the sense that to modify over a thousand almanacs is just impossible for a country as inefficient as we are. What's more, time of the day and night remain unchanged in part of almanacs, therefore once spotted, the almanacs that we created in private are set to be exposed. Given that the distance from the capital city varied from places to places, it is almost impossible to make sure every order is followed through. As a result, if we stick with the decision to issue our own almanacs, the chance is that they will be circulated throughout the country in various means. Ding Yingtai who has a grudge against us is sure to inform the emperor if he happened to come across a copy of our calendar, saying that "it is the definition of deception for Joseon to create its own almanacs when it claims to use the calendar of Ming dynasty. I strongly advised your majesty to ask Joseon about this." If that is the case, the officials of the Bureau of Astronomy would be held in accountability, and probably they will even have to be sent to the capital of Ming to justify themselves. With the possibility of a frame-up plotted by Ding so worrisome and fearful, there is no point in having a new calendar. It is in the interests of our country to abolish our calendar to prevent the potential consequences of having one. Therefore, my suggestion is to stop using the calendar of our own. (*Authentic Records on King Xuanzu.* (Vol. 107))

These words suggest that Ding Yingtai probably had evidence on the Korean almanacs in his hands in his impeachment against Li Yan. It turned out the Joseon Government did not stop issuing their own almanacs even though Li did indeed "ask for advice" about this matter. However, it at least served as evidence that they were fully aware of the political consequences of their actions.

As the political and military contacts between the two countries deepened in the late Ming dynasty, it is an open secret among the government officials of Ming dynasty that Joseon had its own almanacs. This may explain why Mao Wenlong, the vice military governor of Pingliao, asked Joseon for their calendar around the beginning of spring in 1625 according to the Korean historical records:

> The royal court decided to grant the request of Governor Mao for a copy of our almanac. It is the duty of vassal states to use the official calendar of the Middle Kingdom rather than to create their own ones in private. However, with Joseon so far away from the Kingdom, it would take too long for the almanacs issued by the Ming Bureau of Astronomy to be delivered here in our country. Therefore, the government decided to create a calendar of our own but dare not let the emperor know. The request of Governor for our calendar was conveyed by Ambassador Ying Yili to ministers. If we waited for the issuance of almanacs, there would be no ways for us to ensure that they could be delivered in time due to the long distance, but worshipping rituals, astrological work and forecasts in areas vital to our country

> cannot afford any delays. Thus, we imitated the calendar issued by the Middle Kingdom to make a calendar of our own. On this ground, our violation of norms seems justifiable, and thus it is approved to be used to defend ourselves when we convey the calendar to the Governor. (*Authentic Records on King Xuanzu.* (Vol. 8))

It can be seen from these words that the Korean kings and ministers saw no need to hide this fact from the government officials of the Ming dynasty, and they thought their decision to have a calendar of their own was pretty justifiable: with such a long distance from the Middle Kingdom, it would be difficult to ensure the almanacs to be delivered in time, and this uncertainty may lead to a disruption in national affairs such as worshipping rituals, astrological work, and forecasts in areas of vital importance. The official government officials seemed to have no choice but to accept it; after all, the priority of the Ming dynasty at the time was to unite with Joseon against the aggression of post-Jin that was rising as a great power in the east of Liao.

9.3 Tough Transformation in the Period of Revolution

The "dual system" of almanacs turned out to be successful in Joseon, without causing any of the consequences that the Korean astronomers feared. The Joseon dynasty, as a tributary state, maintained a good relation with the Ming dynasty even until the very end of Ming. However, two events that happened in the late Ming dynasty ushered in a new chapter in their astronomical exchanges. The first thing is the introduction of the European astronomy which prompted the Ming Government to transform its astronomical calendar. The other is the rise of the Manchu in the northeastern part of China and its eventual conquest of the capital city Beijing.

In fact, since the European Christian missionaries came to China, the Korean ambassadors in China have heard about them and, more importantly, the European scientific knowledge in various areas such as astronomy that they brought to and spread in China. Li Zuiguang, a well-known Korean academic at the time, recorded that:

> In 1603 when I worked as the deputy minister of education, I once travelled away on official business to the capital city of the Middle Kingdom. Envoys Li Guangting and Quan Xi came over to send me maps of the Europe before I returned. These maps, probably obtained in the capital city, were extremely delicate and intricate especially when it comes to the western regions, covering a wide geographic area from the Middle Kingdom, 8 provinces in the east of our country, to 60 provinces of Japan while retaining a reasonable level of detail. The Europe is the farthest corner of the western regions, miles away from the Kingdom, and thus has not paid tribute until the beginning of the Ming dynasty. The map of Europe was made by its ambassador Feng Baobao, and it is impressive to see how the preface was written in elegant language that is pretty much the same as ours. Li Madou (the Chinese name of the Italian missionary Matteo Ricci) and Li Yingcheng also made the *Map of the Shanhaiyudi* (literal meaning the world map), which has been cited by books including Sancai tuhui (Assembled Pictures of the Three Realms) written by Wang Yi. The Europe spans from the Mediterranean in the south to the Sea of Ice in the north, from the Danai River in the east to the Atlantic. The map took the Mediterranean as the center of the world, hence its Chinese name—Dizhonghai (literal meaning the sea at the center of the earth).

It could well be that Li Zuiguang called Li Madou (Matteo Ricci) by the wrong name "Feng Baobao," for the traditional Chinese characters of these two names looked quite similar. Therefore, the map made by Feng Baobao was probably another version of the *Map of the Shanhaiyudi*, whereas Li did not realize that Feng Baobao was actually Li Madou. As it should be, he also mentioned the European astronomical knowledge conveyed through the captions paired with this map:

> I once saw a map of the sky drawn by Feng Baobao, in which the sky was consisted of nine layers with star layer at the top, the sun below and the moon at the bottom. It seems to be well-founded.

Apparently, this map of the sky that Li saw was that one drawn by "Feng Baobao."

Without doubt, Koryo relied mainly on their envoys to the Middle Kingdom to learn about the scientific knowledge spread by the European missionaries. This can be confirmed in an account of Li that "Envoys Li Guangting and Quan Xi came over to send me maps of the Europe before I returned. These maps were probably obtained in the capital city."

In 1629, the Ming Government officially decided to use the European astronomy as a guide to undertake calendar reform. As the Joseon ambassadors took notice of this move and the reputation of the Christian missionaries in the field of astronomy, they began to make contacts with them. In 1631 when Zheng Douyuan served as an envoy in the Ming dynasty, his translator Li Ronghou got acquainted with Portuguese missionary João Rodrigues (1558, 1561 or 1562–1633 or 1634). In a letter to respond to Li Ronghou's questions on the calculation methods of astronomy, Rodrigues provided brief explanations of the concepts in question and pointed out that "the specific details of the astronomical theories cannot be explained in a few words," hoping to meet up with him so as to discuss further in detail (Quote form a secondary resource "A study of the exchanges between Koryon Envoys to Beijing and the Western missionaries." Yang Yulei. (2006). *The World History,* (5), 126–131).

In the meeting with Zheng Douyuan and Li Ronghou, Rodrigues gave them some western instruments and books as presents which included astronomical devices such as telescopes and clock, and the Chinese version of western astronomical works, all of which were turned over to the higher authorities after they returned to Koryo, and Zheng Douyuan gave a detailed account of this meeting in his "Report on the general situation of the western countries" to the Chengzheng Yuan (the Royal Secretariat):

> The western countries are miles away from the Middle Kingdom and it takes three years to get the capital city of the Ming dynasty. Rodrigues, a friend of Matteo Ricci, had constructed artillery to fend off attacks from other western countries and had an expertise in astronomy. When he arrived in Guangdong province, he offered to use the western artilleries to defend against enemies, which won him the recognition from the emperor who appointed him instructor in the department of military of Dengzhou. He gained special favor at the Ming court, and even respect from the emperor. The Bureau of Astronomy also undertook calendar reform under his guidance.

> I once had a meeting with Rodrigues who, at the age of 97, looked energetic and in high spirits with an imposing appearance. He assented without hesitation to my request to be presented an artillery piece to bring back to Koryo, and gave some other books and instruments as followed:
>
> A copy of *Zhili yuanqi* (origin of the astronomical calendar), a copy of *Tianwen lue* (brief introduction of astronomy), a copy of *Tianwen shu* (a book on astronomy) written by Matteo Ricci, a copy of *Yuanjing shuo* (introduction of telescopes), a copy of *Qianlijing shuo* (introduction of astronomical telescopes), a copy of *Zhifang waiji* (chronicle of foreigh land), a copy of *Xiyangguo fengsuji* (customs of the western countries), a copy of *Shenwei dajing shu* (instruction book of telescopes) presented by the west, two copies of *Tianwentu naneiji* (astronomical charts), two copies of *Tianwen guangshu* (vastness of the universe), and five copies of *Wanli quantu* (map of the world) and a copy of *Hongyipao tiben* (introduction of western artillery).
>
> A pair of telescopes that is worth around 300 or 400 liang of silver can not only be used for astronomical observation but also enabled us to scout the movements of enemy hundreds of miles away. A sundial that can indicate the time of a day, the direction of a place and movements of the sun and the moon; a clock that can ring at 12'o clock; and an artillery piece launched with flints instead of matches can be fired for 4 to 5 times in the time when our fowling piece is fired only twice. Fireworks are made from melted solonetz, and Zimu flower is purple cotton. (Li Rongyuan. *National Treasure* (Vol. 35))

However, King Li Zong (1595–1649) and officials of the Royal Secretariat showed little interest in the advanced western astronomy, the calendar reform undertaken by the Ming Bureau of Astronomy, or the astronomical instruments and books. But he still rewarded Zheng Douyuan, yet not for these astronomical instruments and books:

> Envoy Zheng Douyuan is sharp and capable, and the western artillery piece that he brought back is extraordinarily sophisticated and of great help in military war. It is highly laudable how he has been dedicated to the service of the country. Thus, I decided to give him a raise in salary as a reward. As for his aides and staff worthy of reward, he can report to me about them. (*Journal of the Royal Secretariat* (pp.548))

At the time, the Joseon dynasty was under the great military pressure from post-Jin dynasty in the aftermath of its invasion in 1627 that ended by a surrender treaty signed. In this sense, Li Zong did have reason to praise him. However, it seems that the King and officers in the Royal Secretariat were not on the same page on this decision. These officers even discredited Zheng in front of the King and pleaded for recall of decision.

> The report written by Envoy Zheng Douyuan is extremely preposterous and most of the things that he turned over were showy but not practical. Such a witless envoy should be punished rather than rewarded. It is unreasonable to raise his salary only for an artillery piece that he brought back, thus we ask the majesty to recall the decision. (*Journal of the Royal Secretariat* (pp. 549))

Li Zong emphasized again that:

> The artillery piece he brought back is so sophisticated that it will be impossible to learn about its mechanism without his help. The amount of pieces is not that important. There seems nothing wrong with rewarding Zheng Douyuan for his great achievements after a long and arduous journey abroad. (*Journal of the Royal Secretariat* (pp. 549))

In the next three days, the Royal Secretariat reported to the throne for three times pleading for recall of reward. Eventually, the King unwillingly "took their advice" after turning down for three times, which indicated how powerful the civil officers were in Joseon at the time (*Journal of the Royal Secretariat* (pp. 549–550)).

It suggested that the officials of the Royal Secretariat knew basically nothing about the situations that Zheng Douyuan introduced, nor the significance of the things that he brought back. No doubt they were not interested in learning more of it. Nevertheless, the expansion of the Manchus and its eventual conquest of the Middle Kingdom were bound to change this situation profoundly.

In 1636, Hong Taiji (1592–1643) as the first emperor of the Qing dynasty who had just ascended the throne led the army to invade Joseon Kingdom, forcing it to erase the Ming legacies of reign title and imperial mandate in favor of the reign title of Qing as an official tributary state. The newly founded Manchu Qing dynasty imposed some terms on Joseon including regular financial tribute and demand of two royal family members as hostage. From then on, the Joseon dynasty began to accept the Qing calendar, but it turned out that the whole society of Joseon from the court above to the masses below did not put any faith in the ability of Qing in terms of astronomical calendar, not to mention respect for it. For instance, when the Korean Bureau of Astronomy received the calendar of Qing and spotted differences between this calendar and the calendar of their own in 1639, they recalculated it only to find their calendar was correct. In the meantime, they made an interesting comparison between the calendar of Qing and *Shiyong tongshu* (a calendar written by Ming dynasty people) and found that there was consistency in the calculations of lunar months between the two calendars, whereas when they compared *Shiyong tongshu* with their calendars and Ming calendars many years ago, the conclusion was that the latter two were exactly the same, yet both far different from the former one. Based on these comparisons, they concluded that:

> As far as the Bureau of Astronomy is concerned, the calendar issued by this bureau is meticulously calculated and reliable. Month and year calculated in *Shiyong tongshu* turned out to be inconsistent with those in our calendar. This disparity probably boils down to the extremely limited access to astronomical knowledge under the prohibition against private calendar. Its calculation method proven to cause errors easily is not reliable at all. The Qing dynasty may not master the calculation methodology of the Ming Bureau of Astronomy; instead, it is likely to issue the calendar calculated by *Shiyong tongshu* without making calculations on its own. Thus, we should take the old calendar issued by the Ming Bureau of Astronomy as standard. (*Authentic Records on King Renzu* (Vol. 38))

The fact that King Li Zong took this advice suggested that the Joseon dynasty would rather use the old calendar issued by the Ming Bureau of Astronomy than adopt the new calendar of the Qing dynasty and even went so far as to say that the Qing calendar was compiled based on private calendar in the Ming dynasty "that was created illegally" and thus "not reliable." Apparently, the astronomical calendar and astronomical officials of Qing were held in contempt by the Ming dynasty by the Korean astronomers who still took the Ming astronomy as the prevailing orthodoxy, regretting how "Qing probably failed to master the calculation methodology of the Ming Bureau of Astronomy."

In the meantime, Li Ai (?–1645), the eldest son of Li Zong, and Li Hao (1619–1659), the second son, were sent to Shenyang City as hostage. This was a political event, but as it turned out, this experience happened to turn the two princes into one of the major contributors to a focus of Joseon on the European astronomy introduced to China and the Chinese reform of astronomical calendar, which seems like a repeat of how Wang Zhang created opportunities for the Joseon dynasty to learn about *Shoushi Calendar*.

Both of the princes were sent to the Capital city of Beijing after the conquest of Ming by the Manchu Qing. Li Ai made close contact with the Christian missionary Johann Adam Schall von Bell (1592~1666), the head of the Qing Bureau of Astronomy at the time, and asked to be taught astronomy, which was recorded in the historical literature on the Chinese catholic.

> In the first year of Shunzhi (1644) when Li Ai, the eldest son of Li Zong was sent to the capital city where he knew of the great reputation of Johann Adam and came over to the Catholic church to learn astronomy and other subjects. Johann Adam also invited Li Ai over to discuss further on these topics. They got along well with each other over time. Johann Adam often talked about catholic doctrines which caught the interest of the prince and prompted him to ask for more details. (Huang Feimo. *Introduction of Catholic Doctrines* (pp. 24–25))

In 1645 when the Qing Government decided to send Prince Li Ai, Johann Adam gave him some astronomical instruments and books as gifts.

> By the time when the prince was about to return to Joseon, Johann Adam sent him a lot of books on astronomy, mathematics and Catholic doctrines that he translated, a globe and a portrait of God as presents. The prince was delighted to receive these gifs and wrote in person a letter of thanks. (Huang Feimo. *Introduction of Catholic Doctrines.* (pp. 24–25))

The "letter written by the prince himself" was translated into Latin by Johann Adam and passed down through the ages. In this personal letter, Li Ai expressed his willingness to introduce the western astrological knowledge to Koryo:

> It is not that we do not have these things (refer to globe and astrological book) in our country, it is just the calculation methods we used for several centuries tend to produce errors. Therefore, I am more than happy to receive these valuable presents. When I come back to my country, I intend to spread the knowledge both among the dominated upper class and more importantly, popularize it in the academic circle. I believe the scientific knowledge will not only become a focus of the academic circle, but also help introduce the western science to the masses of Joseon.

Unfortunately, Li Ai "died from disease" shortly after he came back to Joseon accompanied by the Qing envoy around March and April in 1645. It is said that he was poisoned in the palace by his government officials loyal to Ming. Therefore, his astronomy program fizzled out.

After the death of Li Ai, the Qing Government decided to send Li Hao back to Joseon and ordered the Korean minister Han Xingyi to come to the capital city so as to convoy the prince to Joseon. Han saw *Shixian Calendar* in Beijing and purchased a series of works on new calendar. By the end of June in 1645, he came back to Joseon and turned these books over to the court, proposing to carry out a reform on the Korean calendar as "Astronomy and calendar are the first priority of the emperor. It has been more than 400 years since the calendar reform by Guo Shoujing in Yuan dynasty, thus it is high time for us to revise it. Besides, the calendar book written by Johann Adam as I saw in Beijing is good enough to serve as a guide in our calendar reform. I advise that efforts be made by our astronomers to review *Gaijie Map* and *Chiljeong Calendar* in a bid to revise our calendar" (*Authentic Records on King Renzu* (Vol. 46)). Li Zong took his advice and "ordered astronomical officer to make sense of the new calendar" (*Authentic Records on King Renzu* (Vol. 46, 98)) through the books Han Xingyi brought back including Xinli Xiaohuo (meaning a guide to the new calendar), but apparently, it would take time for them to absorb enough knowledge to carry out a calendar reform.

After several months of study of the works brought by Han Xingyi, Jin Yu, head of the Bureau of Astronomy, put forward a proposal to revise calendar in the beginning of February 1646 on the grounds as followed: First, the Middle Kingdom has traditionally revised their calendar on a regular basis. With Shoushi Calendar being used for quite a long time and errors accumulated over time, a calendar reform is inevitable. "Western calendar introduced at this time offers us a great opportunity to revise our own calendar"; second, "the Middle Kingdom had already revised its calendar back in 1639 (This refers to the efforts made by the Ming Government to compile calendar based on the western calculation methods around 1639), which means its calendar issued the next year will definitely be far different from ours." Therefore, "for the parts of new calendar that is largely consistent with the theories in current calendar, we should absorb them in an effort to perfect our calendar." The first reason is basically the same as Huang Xingyi's, whereas the second one is based on the differences between these two calendars. With "the divided opinions over the calendar book Han Xingyi brought back," the only way to make sense of it is to "learn from the author who wrote this book," and given that "the Middle Kingdom forbids tributary states to create their own calendars, he proposed that, "even though it is not permissible to send academics to Qing to learn new calendar, we can appoint one astronomer or two to accompany envoy to the Qing dynasty where interpreters can consult officers in the Bureau of Astronomy in private on the new calendar. If we can get to know about its process of calculation over the recent years, it will probably be not difficult for us to make sense of the methodology and theory of the new calendar" (*Authentic Records on King Renzu* (Vol. 46)).

Around this time, Jin Zidian came back from Beijing with a copy of *Shixian Calendar*. After a study of the book, the Bureau of Astronomy found out major differences in time system and solar terms between the Korean calendar and *Shixian Calendar*. Given that the Korean astronomers "cannot grasp the essence of *Shixian Calendar* only through a copy of it," and it is necessary to get to know how it has been calculated over the past few years so as to master the methodology in the new calendar, they proposed to "send the top talents to Qing's capital city to learn it rather than those who are merely good at calculating" (*Authentic Records on King Renzu* (Vol. 46)).

The Joseon Government thus appointed astronomers to travel to Qing dynasty with envoy Li Jingshi in an attempt to learn about the new calendar and "purchased privately" *Shixian Calendar*. But in a report he sent to the court in Beijing, he said that "we seek widely for *Shixian Calendar*, but to no avail. Our attempts to meet up with Johann Adam also failed." He had not purchased *Shixian Calendar* or met Johann Adam. But "Qi Ying, the son of astronomer Li Yinglin was held hostage in the Middle Kingdom at the very time when Li Jingshi traveled there on official business. Qi Ying has a good command of Chinese and arithmetic." Therefore, Li Jingshi "asked him to learn astronomy and calendar from Johann Adam so as to spread the knowledge to Joseon in the future" (*Authentic Records on King Renzu* (Vol. 47)).

In Feburary 1648 (lunar calendar), "Envoy Hong Zhuyuan came back from Beijing with almanac given by the Qing government, which was *Shixian Calendar*" (*Authentic Records on King Renzu* (Vol. 49)). It marked the official issuance of *Shixian Calendar* to the Korean Joseon dynasty. However, even until that time, the Korean astronomers still used *Chiljeongsan Naepyeon* as the guide to calculate and compile their own calendar. For this reason, it will inevitably cause chaos once the Korean Government adopted *Shixian Calendar* at the time. Just 2 months after Hong Zhuyuan brought back *Shixian Calendar* issued by the Qing Government, the Office of Protocol reported to the throne that:

> Our calendar has been consistent with the Chinese calendar for years until this year which saw inconsistency between them. This must be blamed on the astronomical officers for their errors. If it is true that this year has Run April (meaning there are two Aprils in the lunar calendar of this year) (Note: The Run month of that year was the March based on *Datong Calendar*, whereas based on *Shixian Calendar*, it was the April), instead of Run March, that means official ceremonies, worshipping rituals and other national activities will miss out on the right time, disturbing the social order. I asked your majesty to punish the astronomical officers for bungling.

Since the Protocol Office was a government department in charge of official ceremonies and norms of etiquette, it is naturally unacceptable for it to see such a chaos in the calendar. But they laid the blame wrongly on the Bureau of Astronomy. Fortunately, the ministers had heard of the ins and outs of the matter of *Shixian Calendar*, among whom Jin Zidian and Li Xingyuan came forward to help the astronomical officers out of predicament:

> The Qing dynasty has adopted the new calendar compiled by Johann Adam, whereas our country still continued to use the old calendar. The forecasts of eclipses based on our calendar have been proved to be correct, which means our calculation methods are not altogether wrong. In fact, our calendar is in consistent with calendar issued by the Ming dynasty, and even pretty much the same as the calendar issued by Qing before it moved its capital Shenyang to Beijing. It is only when the western astronomical calendar was introduced to the Qing dynasty that the Chinese government began to use it as a guide to reform its own calendar. The

> western astronomical calendar is different from the Ming's, which is why our astronomers do not grasp the new methodology. (*Authentic Records on King Renzu* (Vol. 9))

They were, nevertheless, defending the Korean calendar in fact, not only by saying "The forecasts of eclipses based on our calendar have been proved to be correct, which means our calculation methods are not altogether wrong," but also by invoking other works on the Ming calendar:

> If we refer to *Shiyong Tongshu* and *Santai Lifa Tongshu* which are compiled in the Ming dynasty, we can find that the Run month of this year also falls on the March. Therefore, it is not for sure that the March is not the Run month of this year. (*Authentic Records on King Renzu* (Vol. 49))

Even though the debate on calendar petered out, it suggested that most of the government officials of the Joseon dynasty would rather stick with the old system than adopt the new calendar issued by the Qing Government. But these was one exception, which is Han Xingyi, the first Korean official that brought Shixian Calendar back to the Joseon dynasty and proposed changes to the Korean calendar. He was the only one who "held family worship rituals based on the calendar issued by the Qing dynasty" but only invited criticism ("people rebuked him strongly for how his ignorance"). The Korean historians who compiled *Renzu Shilu* (Historical records on the reign of King Renzu) even went so as to make a strongly critical and sarcastic comment on this, saying "Xingyi, who knew nothing about astronomy, was so ignorantly assertive as to use the calendar issued by the Qing dynasty when it has not yet been proved to be right. He cannot be put on a par with Chen Xian who adopted the custom of 'Zula' (a form of worship ritual) of Han dynasty. What he did was meaningless!" (*Authentic Records on King Renzu* (Vol. 49)).

The Joseon dynasty's response to *Shixian Calendar* may have a relation with its political attitude towards the Qing dynasty. Even though the Joseon dynasty had surrendered to the Qing dynasty as a vassal state, the people had a sense of cultural superiority, with defiance and hatred against the Qing dynasty which they saw as an uncivilized and barbarous ethnic minority. Before the collapse of the Ming dynasty, they used the reign title of Ming in official documents such as *Renzu Shilu*; after its demise, it did not adopt any Chinese reign title in the internal official documents for a long time. They were more clear-cut in attitude since 1649 when Li Hao succeeded to the throne. The Korean kings and ministers not only held in contempt the Qing dynasty which they described as "full of rude barbarians" but also continued to adopt the reign title of Chongzhen. In this context, it is not hard for us to understand their response to *Shixian Calendar* issued by the Qing dynasty as a symbol of the status of a sovereign state.

However, the stark reality was that the Qing dynasty not only conquered the Middle Kingdom but also became increasingly powerful, which meant to adopt its calendar was an inevitable choice for the Joseon dynasty politically. Therefore, within a month of the Qing Government's first official issuance of *Shixian Calendar*, the Joseon Government wasted no time "sending astronomer Song Renlong to the Qing dynasty in an effort to learn the western astronomical calendar" (*Authentic Records on King Renzu* (Vol. 49). According to the record in the September that year, "sent astronomical official Song Renlong to the Qing dynasty to learn its astronomical calendar." It suggested that Song Renlong went to the Qing only until the September). But Song Renlong recalled that due to the Qing's "strict prohibition of studying astronomical calendar in private," he "only met Johann Adam for once" "exchanging a few words, and Johann Adam sent him 15 volumes on calculation methods and 10 zhang stat map (1 zhang equals 3.33 meters) as gifts in an effort to help him make sense of the western astronomical calendar."

As hard-won as these results, officials in the Bureau of Astronomy were yet dissatisfied with these efforts for two reasons: First, language difficulties meant they had to "express themselves by making gestures while talking, but still failed to convey the idea"; second, in terms of the content he learnt, "it was no more than a glimpse of the whole system, instead of the quintessence of it." But they had already realized that "the western astronomical theories is yet to be verified" and thus proposed to continue to send academics to the Qing dynasty for further study so as to determine whether to adopt the new calendar. In an effort to "communicate effectively in spite of language difficulties," they put forward a good solution, which was to "select the best talents to study the new calendar under proper supervision and send them to Beijing when they comprehend the fundamentals in an effort to help them out with any queries" (*Authentic Records on King Xiaozong* (Vol. 4)).

Astronomical official Jin Shangfan among the group of academics selected "conducted an extremely careful study" on these works, only to get "a rough understanding of it," but still stood out from all other academics, and thus was sent to Beijing in 1651." By "giving the Bureau of Astronomy a large bribe," Jin Shangfan managed to master the calculation methods of *Shixian Calendar* in the next year and returned to Joseon with books including *Richan Biao* (Tables of the motions of the sun) and *Yueli Biao* (Tables of the motions of the moon). Upon returning to Joseon, he "selected several more academics to study under him" while "calculated day and night to revise calendar as soon as possible," and eventually the almanac of the 4th year of Emperor Xiaozong (1653) came out. This almanac, independently compiled by Jin Shangfan, turned out to be in consistent with the *Shixian Calendar* issued by the Qing dynasty that year. Hence, the Bureau of Astronomy officially proposed that "Our country should adopt the new calendar since 1654," which was approved (*Authentic Records on King Xiaozong* (Vol. 8). For more details in Jin Shangfan's trip to China for astronomical study, please refer to: Lin Zongtai. (2013). Koryon scholars' trip to Beijing in the 17–18 century——a case study of Jin Shangfan and Xu Yuan. *Studies In The History of Natural Sciences32 (4), 446–455*). It marked the official adoption of *Shixian Calendar* by the Joseon dynasty. Jin Shangfan was promoted to the rank of Zhengsanpin for its contributions to calendar reform. Unfortunately, he died of disease on his trip to Beijing for further study. Nevertheless, the Joseon Bureau of Astronomy was very determined on the adoption of new calendar at that time and thus proposed to "select and send academics with a good command of astronomy to go to Beijing with envoys," which was approved by Li Hao (*Authentic Records on King Xiaozong* (Vol. 14)).

However, the adoption of *Shixian Calendar* in the Joseon dynasty did not go smoothly. The former head of the Bureau of Astronomy Song Hengjiu presented memorial to the throne, saying that *Shixian Calendar* "has numerous errors." Some officials in the Bureau also believed that "it is better to stick with the old calendar until the new calendar is proven to be accurate." After a round of debates, the government finally decided to officially adopt *Shixian Calendar* while continuing to issue two copies of calendar compiled with the calculation methods of *Datong Calendar* every year, with one for the King and the other for collection." In the next year, Song advised the King once again to "abolish *Shixian Calendar* in favor of *Datong Calendar*" only to be turned down (*Authentic Records on King Xianzong* (Vol. 14)).

Arguably, the adoption of *Shixian Calendar* had been an irreversible trend in the Joseon dynasty by that time. But the problem was that the Qing dynasty remained undecided on whether to adopt the new calendar. In 1666, the Qing Government announced the abolition of *Shixian Calendar* in favor of *Datong Calendar* after Johann Adam was convicted of charges brought by Yang Guangxian. When the almanac issued by the Qing dynasty arrived in Joseon the end of this year, it caused chaos since the Joseon new calendar compiled with the western calculation method had already issued:

> The King asked Zheng Taihe: "why is the Qing's calendar this year different from the former one?"
>
> Taihe responded: "The opinions have been divided on the adoption of the western calendar. Ming dynasty people also commented on how flawed the western methodology was. In this context, the western calendar introduced by Johann Adam was abolished, which is why this calendar was different from the former one."
>
> The King asked: "so which calendar should we use?"
>
> Taihe responded: "the Qing government held worshipping rituals based on Datong Calendar."
>
> The King asked: "Does it mean Shixian Calendar will no longer be used?"
>
> Taihe responded: "It seemingly does."
>
> The King asked: "Have the literati and officialdom been used the new calendar?"
>
> Taihe responded: "As the new calendar just arrived, there is no time for us to print it. Therefore, the worshiping rituals are supposed to be based on the new calendar, whereas the ordinary people will have to use the original one."
>
> Jin Wanji said: "How absurd it is to use two calendars within one country! The new calendar must be printed so as to be issued across the country."
>
> Taihe responded: "You are right, but it is difficult to get it done."
>
> The King said: "Just print the calendar in the form of leaflet to issue them as soon as possible. As for *Shixian Calendar* that we used to use, it should also be issued every year to be referred to in the future even though we do not use them now, just as *Datong Calendar*". (*Authentic Records on King Xianzong* (Vol. 13))

By printing calendar on leaflets, the Joseon Government managed to save itself from a possible predicament of "using two calendars within one country. Yet in the face of an unpredictable Qing government, the Joseon government came up with a plan B, which was to go back to *Datong Calendar* while "issuing *Shixian Calendar* every year."

After redressing the mishandled case of Johann Adam, the Qing Government revived *Shixian Calendar*, and the Joseon dynasty also "restored *Shixian Calendar* in 1669" (*Revised Version of Authentic Records on King Xianzong* (Vol. 11)). Naturally, such a reversal was followed by a series of problems. For example, Song Hengjiu presented to the throne a memorial "on the errors of Shixian Calendar" in the end of this year (*Revised Version of Authentic Records on King Xianzong* (Vol. 11)), which apparently was a manifestation of discontent. In response, Song Yiying, an official in the Bureau of Astronomy, debated with Song publicly on this matter and defeated him eventually. Even two years after the restoration of *Shixian Calendar*, the Bureau of Astronomy still had some work to be done to clean up the mess.

> The Bureau of Astronomy reported that: "According to *Shixian Calendar*, the birth date of the prince is August 15th 1661. However, we went back to *Datong Calendar* in 1666, based on which the Run month should be the October rather than the June, and the August is supposed to be the September. Therefore, the prince was supposed to be born in September. With this restoration of *Shixian Calendar* in 1669, the birth date of the prince should thus be adjusted to the August. Given the significance of this revision, the astronomical officials consulted with the ministers. The King approved this proposal, and the ministers suggested an immediate action to change it, which were again approved by the King.

Of course, the status of Shixian Calendar had remained unchallenged from then on until 1894 when the Joseon dynasty adopted the solar calendar. It was thus reduced to the status of "reference," used only for memorial ceremony, birthday, and picking auspicious days for special occasions (Pak Youngdae. *Compendium of Historical Literature* (Vol. 1)).

These records suggested that even though the Joseon Government had divided opinions on the adoption of *Shixian Calendar* due to its European background, their debate was in essence different from that of the Qing and Ming dynasties on calendar reform, in the sense that the debate in China focused on the legitimacy and political connotation in the adoption of the western astronomy despite some technical problems involved, whereas that in Joseon was centered around technical problem, with the western background and political intonation of Shixian Calendar less remarked on. For Joseon, what it adopted was not a western astronomical system with political connotation, but merely an astronomical calendar system as the symbol of the "legitimacy" of the sovereign state. Thus, the only choice for it was to accept whatever it was asked to no matter whether it was willing to do so or not because it all depended on the "sovereign-tributary" diplomatic relationship between the Qing dynasty and the Joseon dynasty.

9.4 "Secret Front" of the Joseon Dynasty Against the Qing Dynasty

The civil officials of the "Yangban" class led the way in the introduction of Shixian Calendar in Joseon. However, they usually do not have a good command of astronomy, which explained why Jin Yucai proposed to "appoint an astronomical official or two to travel with the envoy on official business to the Middle Kingdom and an interpreter to help consult officials in the Bureau of Astronomy." King Li Zong also took his advice to "select and send the best talents for further study." These academics sent to the Qing's capital city were the astronomical officials and calendar officials in the Joseon Bureau of Astronomy, belonging to the "middle" class that consisted of professionals or experts in various areas. This social class was lower than the class of "Yangban." For this reason, they were not allowed to visit the Middle Kingdom independently; instead they could only travel there as civil officials' entourage on official business. Besides, even though the Qing dynasty did not forbid the ordinary people learning astronomical calendar in private, and even showed a tendency to encourage folks to study astronomy, it was still not proper for Joseon as a tributary state to list astronomy study as one of its official diplomatic activities, especially when the relations between the two countries became intense. Therefore, the best choice for the Joseon Government was to appoint astronomical and calendar officials as entourage of civil officials on their official visit to the Qing dynasty.

During the reign of King Li Zong and his son Li Hao, it basically became a convention to send astronomical officials to Qing when the Joseon Government encountered problems in its astronomical calendar work. At the very beginning, they sent astronomers as needed, and by 1674–1719 there formed a custom to send astronomers for further study every year. In 1741, the Bureau of Astronomy officially proposed to establish it as an institution "making it a convention to send talents every year for study." In 1762, it was adjusted to "send talents for study every three years," and since 1770 "the astronomical officials can be sent as needed without having to report to the court." As stipulated in 1791, the astronomical officials were to be sent every other year (Wu Han. *The Chinese Historical Materials in the Authentic Records of Joseon.* (pp. 4511, 4835~4836、4590)). With the officer responsible for selection of talents referred to as "FuYan Guan," there were three selection methods followed: First, send astronomical officials who were newly promoted to the second level; second, select the talents through examinations; lastly, officials in the Bureau of Astronomy took turns to be in charge of the calendar work. Since 1724, the objects of selection exam included *Qiyao Chou* (calculation of seven illuminators), *Shuli Jingyun* (collected essential principles of Mathematics), and *Lixiang Kaocheng* (thorough investigation of astronomical phenomenon) (*Authentic Records on King Zhengzu.* (Vol. 33)), with the latter two as major mathematical and astronomical treatise composed imperially by the group of Qing astronomers led by Emperor Kangxi.

From a historical perspective, it was necessary for the Joseon Government to establish it as a convention to send astronomical officials to study in Qing, for the reasons as followed: First, by 1651, Jin Shangfan had only managed to "get a rough understanding of the motions of the sun and the moon" that was enough to meet the needs of the compilation of almanac for the ordinary people; however, the solar and lunar eclipse and "the calculation of five planets was still yet to be grasped." Second, apart from the reversal of calendar that took place after "redressing the mishandled case of Johann Adam," the Qing Government continued to make innovative breakthroughs in terms of astronomy, including *Lixiang Kaocheng* that was compiled during the reign of Kangxi and adopted during Yongzheng; *Addition to Lixiang Kaocheng*, compiled and adopted during the reign of Qianlong; and so on. Third, the Joseon astronomers often encounter tough problems to be solved, which boiled down to the fact that: "our astronomical calendar is mainly based on a study of the Chinese works and instruments. With, yet, the contacts and academic exchanges not as frequent as needed, it has been too much of a challenge for us to master all of its essentials even with great intelligence" (*Authentic Records on King Zhengzu.* (Vol. 33)).

These officials that were sent to the Qing dynasty for further study under the "Fuyan Guan" institution seemed to form a "secret front" against Qing in the sense that the establishment of this convention was a one-sided effort of Joseon, rather than an official bilateral agreement between the two countries. Just as in Yuan and Ming dynasties, they faced various difficulties to overcome but generally could manage to attain their goals. Besides, the Qing Government did not prohibit the common people from studying astronomical calendar and phased in a diplomatic policy of "Dehua" (meaning culturally influence) towards the Joseon, under which the Joseon envoys and officials were usually given special treatment. The Qing Government basically lifted the upper limit on the number of delegates of the Joseon diplomatic corps ever since the reign of Emperor Kangxi. Delegates and even their entourage were granted free access to the government departments of Qing and thus were able to make friends with the Chinese upper class. In this sense, it was much easier for these "Fuyan Guan" officials to make contacts with the officials in the Qing Bureau of Astronomy (including the Christians who worked in this department). From the historical records, they seemed to have established a relatively stable "social network" in Qing, and a typical example was the special relationship that He Junxi and his son maintained with the Joseon "Fuyan Guan" officials.

As posts in the Chinese ancient astronomical agencies were hereditary, the sons of astronomical officials often inherited their fathers' title. He Junxi had held senior posts in the Bureau of Astronomy during the reign of Kangxi, and his sons all worked as astronomical officials, responsible for astronomical calendar and calculations. This family had an unexpectedly close relationship with these Joseon "Fuyan Guan" officials.

In 1705, the Joseon Government sent calendar official Xu Yuan to Beijing for further study of the calculation methods of *Shixian Calendar* which Jin Shangfan had not mastered completely. The teacher he got in touch with in Beijing was He Junxi, under the guidance of whom Xu Yuan mastered the calculation methods of the sun, the moon, and the five planets in a more thorough way. But since "it was done in private," Xu Yuan did not finish studying all of the theories. Thus, he came back to Beijing in 1708 once again to study the rest of the calendar (Xu Yuan. Specific analysis of new calculation methods of astronomy. *Specific Analysis of Calculation Methods*).

By 1713, He Guozhu, as calendar official in the Bureau of Astronomy, accompanied the imperial envoy Aqitu and military governor Mukedeng to Joseon to carry out geographical survey. During this survey, he not only discussed on mathematical problems with the Joseon mathematicians including Xia Hongzheng but also taught Xu Yuan for a while. "Even though Xu Yuan has grasped the calculation methods and how to use the astronomical instruments, He Guozhu still let him leave for Yizhou with the corps in order to help him master the fundamental principles more thoroughly. The instructions of these devices were contained in books *Yixiang Zhi*, *Huangchizhengqiu* and so on. Guozhu also said: 'feel free to send a letter to me listing the astronomical works and devices that Joseon do not have, I will manage to find them for you.'" From these words, we can see that the relationship between He Guozhu and Xu Yuan was quite close. Guozhu not only imparted plenty of professional knowledge to Xu Yuan and gave him so many books but also promised to find astronomical works and instruments that Joseon did not have for him. On account of their close relationship, the Joseon minister suggested that "it would be better to appoint Xu Yuan to accompany envoys to Beijing in days to come." And King Li Chun took his advice (*Authentic Records on King Suzong.* (Vol. 54)).

As expected, Xu Yuan was sent to Beijing again as "Fuyan Guan" the next winter and "met with He Guozheng to buy books on astronomical calendar and calculation instruments. He also received 9 copies of book including *Rishi Buyi*, *Jiaoshi Zhengbu* and *Licao Pianzhi*, and 6 kinds of calculation devices from this insightful calendar official, among which there was a western clock that was extremely delicate and sophisticated." Upon turning over all of these books and instruments to the court, he also put forward a plan "to copy" a clock to place in the Bureau of Astronomy (*Authentic Records on King Suzong.* (Vol. 56)), which was approved by the King.

That was not the end of the relationship between He family and the Joseon "Fuyan Guan" astronomical officials. Twenty years later, as Qing adopted *Lixiang Kaocheng*, the Joseon calendar was again inconsistent with the newly issued Qing calendar. In an effort to keep up with the developments, the Joseon Bureau of Astronomy appointed An Chongtai to study in Beijing, only to find that An's teacher was again a member of He family, "Chongtai accompanied the Envoy to Beijing so as to consult with He Guoxun on the calculation methods which he failed to comprehend. He also purchased 3 copies of *Qizhengziyu Wannianli*, a copy of *Shixianxinfa Wugengzhongxingji*, a copy of *Ershisiqi Hunxiaozhongxingji*, a copy of *Riyuejiaoshi Gaoben* and a western sundial with his own money" (*Authentic Records on King Yingzu.* (Vol. 35)). According to the name and post of He Guoqin, he was probably one of He Guoxun's sons who shared the character "guo" in their names.

Over a decade later in 1743, the Joseon Bureau of Astronomy sent An Guobin to Beijing in an effort to learn the calculation method in "Ziqi" program that was newly added to the Qing calendar. "An Guobin and interpreter Bian Zhonghe, Jin Zaixuan got in contact with Dai Jinxian and He Guochen, officials of the Qing Bureau of Astronomy who imparted all of the knowledge with regard to Ziqi calculation methods, position measurement and picking auspicious days as well as the eclipse forecasts in the new calendar to them" (*Authentic Records on King Yingzu.* (Vol. 59)). One of the teachers was Christian missionary Ignaz Kögler (1680–1746) whose Chinese name was Dai Jinxian, and the other one was He Guochen, another astronomical official with "guo" in his name.

Apart from the Chinese astronomical officials of the Qing Bureau of Astronomy, the Joseon envoys and "Fuyan Guan" officials also had opportunities to get acquainted with the western missionaries. The contact between An Guobin and Ignaz Kögler mentioned above is a case in point. Hong Darong (1731–1783), a well-known Joseon scholar, kept a detailed record on this:

> Matteo Ricci travelled to China during the Wanli Reign period of the Ming dynasty in a bid to spread western mathematical knowledge...... by the end of the reign of Kangxi, with a growing number of missionaries coming over to China, the Emperor ordered the compilation of *Shuli Jingyun* (Essential principles of mathematics) which was an encyclopedia of Chinese and Western mathematics for the Bureau of Astronomy to study as the basics of astronomical calculations. Emperor Kangxi also built astronomical observatory in the capital for these foreign missionaries and scholars to conduct surveys. Ever since the reign of Kangxi, some envoys of the eastern states might visit the observatory on their trip to Beijing, and the western scholars were delighted to show them paintings, portraits of God and instruments around the institute while sending western books or devices to them as gifts. Envoys were usually more than happy to receive these rare gifts. Liu Songling and Bao Youguan lived in the southern hall (Note: Liu Songling and Bao Youguan referred to Christian missionaries Augustin von Hallerstein (1703–1774) and Antoine Gogeisl (1701–1771)), with a good command of mathematics and the best of instruments among the other halls. Thus they were especially popular among envoys from the east. (Hong DaeYoung. *Zhanxuan Journal* (Vol. 4))

From these words, we can see that it was easy for the Joseon envoys, especially "Fuyan Guan" officials, to make contact with western missionaries. For instance, in 1766, Hong Darong traveled to Beijing as envoy accompanied by astronomical official Li Dexing whose mission was "under the order of the court to consult Liu and Bao on the movements of the five planets and the details of the new astronomical calendar, and purchase astronomical instruments." Upon their arrival in China, he visited Liu Songling along with Hong for several times, asking some questions on the "calculation methods of the latitude and altitude of the five planets" and "calculations of astronomical calendar," and looked around telescopes, clocks, and various other astronomical devices displayed in the halls.

Of course, the officials of Qing Bureau of Astronomy imparted astronomical knowledge and gave away books and instruments to the Joseon "Fuyan Guan" astronomical officials absolutely in private. These activities, thus, were conducted only under certain conditions, one of which was "huge bribe." The Joseon Government usually granted the "Fuyan Guan" astronomical officials some extra money as "allowances" "in case of special use." For those who came back with technical know-how or purchased important books or instruments, the King would give them some rewards such as "raising salary." These financial incentives prompted astronomical officials to compete intensely for the "lucrative job" of "Fuyan Guan" to a point where the Bureau of Astronomy even had to plead with the King to take some steps to "prevent such a mess" (*Authentic Records on King Zhengzu*. (Vol. 33)). Some interpreting officials also took part in the astronomical study. For example, as recorded in *Beiju Tenglu*: In 1741, "interpreter An Guolin and Bian Zhonghe often went to Catholic churches to approach Dai Jinxian and Xu

Maode for astronomical works, and went to great lengths to obtain *Riyue Jiaoshibiao, Baxian Duishubiao, Duidhu Chanweibiao, Riyue Wuxingbiao, Lvlv Zhengyi, Shuli Jingyun, Rishi Chougao and Yueshi Chougao*." For these interpreting officials, the Joseon Government also "grants rewards" (*Authentic Records on King Yingzu*. (Vol. 59)).

The implementation of the "Fuyan Guan" institution had a huge influence on the development of Joseon astronomy. Through the efforts made by these "Fuyan Guan," the Joseon astronomers could not only solve difficult problems in the astronomical calendar in time but also rapidly absorb and introduce the "new theories" that emerged in China. Since Jin Shangfan's introduction of the calculation methods of *Shixian Calendar*, "Fuyan Guan" astronomical officials continued to introduce the following "new theories" successively.

1. The calculation methods of five stars' motions and eclipses in the *Shixian Calendar*. It was mastered by calendar official Xu Yuan on his two trips to Beijing. In 1710, Xu Yuan finished the compilation of *Xicao Huilei*, which recorded the calculation methods that he acquired.
2. *Lixiang Kaocheng*. This book was officially adopted by the Qing Government in 1724. The Joseon Bureau of Astronomy proposed to send calendar officials to Beijing in 1728 "in an effort to purchase the imperially compiled calendar and master its theories," yet it was not until the next year that they managed to purchase this book and "issued" it under the order of the King by the Bureau of Astronomy. When astronomical officials found it "too difficult to work out even with great efforts," it proposed to send calendar officials for further study in China, which was approved (Wu Han. *The Chinese Historical Materials in the Authentic Records of the Joseon*. (pp. 4429, 4431)). In this sense, the Joseon astronomers only grasped the calculation methods of *Lixiang Kaocheng* after this trip to Beijing.
3. *Lixiang Kaochengbiao* and *Addition to Lixiang Kaocheng*. *Lixiang Kaochengbiao* was a new calendar table compiled by Dai Jinxian and his colleagues based on Newton theories on the sun and the moon. It was attached to the book *Lixiang Kaocheng* (Shi Yunli and Xing Gang. The First Chinese Version of the Newtonian Tables of the Sun and Moon. Chen, K.-Y., Orchiston, W., Soonthornthum, B., and Strom, R. (eds.). *Proceedings of the Fifth International Conference on Oriental Astronomy-Chiang Mai*. (pp. 91–96). University of Chiang Mai Press). When it was officially adopted by the Qing Government in 1734, the Joseon Government noticed the disparities of this calendar with the former one in terms of solar terms and wasted no time in sending astronomical official An Chongtai to Beijing where he found out that it was because "*Lixiang Kaocheng* used the year of 1684 as epoch, whereas the new calendar used the year of 1723 as epoch." Therefore, they "purchased *Richan Biao, Yueli Biao, Qiyao Lifa, and so on*" and the Bureau of Astronomy proposed to "use the year of 1736 as epoch to compile new calendar based on these theories" (*Authentic Records on King Yingzu*. (Vol. 40)). The *Richan Biao* and *Yueli Biao* that they brought back were supposed to be part of *Lixiang Kaochengbiao*.

Addition to Lixiang Kaocheng provided a theoretical explanation and calculation introduction on *Lixiang Kaochengbiao.* It was compiled by the Qing Bureau of Astronomy in 1742. As recorded in the Joseon *Beiju Tenglu*, a year after the book was compiled in 1743, "Envoy's interpreters An Mingyue, Jin Tinghao, Li Jixing purchasd 10 copies of *Addition to Lixiang Kaocheng* and the imperial calendar official Jin Ruitai purchased a copy" (*Compendium of Historical Literature.* (Vol. 1)). According to the record in *Yingzu Shilu*, in the next year, "based on a study of the *Addition to Lixiang Kaocheng* purchased by interpreter An Mingyue and imperial calendar official Jin Ruitai, the astronomical official An Guobin managed to grasp the fundamental theories of the new calendar book" (Wu Han. *The Chinese Historical Materials in the Authentic Records of the Joseon dynasty.* (pp. 4525–4526)). It suggested the Joseon astronomers mastered the theories in *Addition to Lixiang Kaocheng* this year.

The establishment of astronomical "Fuyan Guan" institution improved the ability of the Joseon dynasty to keep up with the developments in the Qing astronomy: It took over half a century to master the calculation methods in *Shixian Calendar* (1644–1708); only 4 or 5 years at most for them to grasp the theories in *Lixiang Kaocheng*; for the *Lixiang Kaochengbiao* and *Addition to Lixiang Kaocheng*, they only spent one year or two in making sense of them roughly. With all the efforts made by "Fuyan Guan" and other envoys to the Qing dynasty, the major astronomical works were introduced to the Joseon. After a period of absorption, the Joseon astronomers got a good command of the western astronomical knowledge that was introduced to China in the Ming and Qing dynasties, and these theories also sparked a lot of further discussions on the astronomy. It thus might well be used to solve the difficulties they encountered in the work of astronomical calendar.

9.5 "Astronomy Diplomacy" of the Qing Dynasty with the Joseon Dynasty

Even though the ancient China is traditionally the absolute center of the Korean astronomy, it by no means implies that the exchanges in astronomical science between the two countries were one-sided efforts of Koryo. Since at least the Yuan dynasty, the ruling governments had taken the initiative to carry out a form of astronomical activity towards the Joseon, which was calendar issuance. What is worth noting is that the Chinese dynasties issue calendar not for the purpose of spreading knowledge and technology to the Koryo, but only to showcase their status as a soveign state. In this sense, calendar issuance was a diplomatic activity with political agenda.

This convention of calendar issuance continued in the Qing dynasty and became more sophisticated and celestial: by the end of every year, the Joseon Government was obliged to send "imperial calendar official" to Beijing to present a "letter of calendar request" to Qing's Office of Protocol in request of the calendar for the next year; as a rule, the Qing Government granted the Joseon a copy of imperial calendar and 100 copies of calendar for ordinary people, along with a "letter of calendar issuance" ("letter of calendar request" and "letter of calendar issuance" during the fifth year of Shunzhi and the 51st year of Qianlong has been compiled and published. Please refer to: Tongwen huikao (Vol. 1). (1978). (pp. 803–814). The Ministry of Culture and Education of the Republic of China). However, unlike the calendars issued by the Yuan and Ming dynasties, those issued by the Qing dynasty indeed contain information regarding the Joseon, that was, to include the Joseon capital city Hanyang in a list of sunrise and sunset time and the day and night length of all the cities in the Qing dynasty. With the adoption of the western astronomy, the astronomers of the Qing were capable of such calculations. Of course, this convention with political agenda contributed to its astronomical development. At least, the calendar issued by the Qing every year served as a guide for the astronomical work of their Bureau of Astronomy, enabling them to check if their own calendar was accurate enough.

The other "astronomical diplomatic" activity towards the Joseon was the forecasts of sun and moon eclipses. Eclipses were regarded as an astronomical phenomenon of ill omen in the ancient China. Therefore, the government would hold "salvage" ceremony when an eclipse took place. In an effort to get prepared for this "salvage" ceremony, the Chinese astronomers (of the government) began to make forecasts of eclipses, probably from the Han dynasty.

As mentioned above, the Yuan Government already informed the ruling Joseon dynasty on the Korean Peninsula of eclipse forecasts back in the fourteenth century. Whether this convention continued in the Ming dynasty was yet to be confirmed. Nevertheless, the historical records suggested that this convention was resumed in the Qing dynasty. Besides, the adoption of the western astronomical calculation methods allowed the Qing Bureau of Astronomy to forecast eclipses that were likely to take place both in the capital city Beijing and in other provinces or neighboring countries (mainly Koryo). As a rule, the Office of Protocol in the Qing dynasty must provide forecasts of eclipses for all the related provinces 5 months before eclipses so as to "concert actions in salvage ceremony." If the Joseon Bureau of Astronomy predicted an eclipse, it would report to the King on the forecasts of eclipses and got prepared for holding "separate salvage ceremonies." Unlike its orders for the provinces, the Qing Government required the Joseon kings to make an official report on their work of "salvage" and observation with eclipse picture attached. The report was usually sent to Qing's Protocol Office which would "in turn report to the Qing emperor after a careful study." Whereas for its own provinces, the Qing Government did not have such requirements, instead it only require the Bureau of Astronomy to make an official report of this sort to the Emperor through the Protocol Office, mainly to confirm whether the observations were consistent with forecasts. Of course, the Qing Bureau of Astronomy had another responsibility, which was to cast a horoscope for eclipse.

The official documents issued by the Qing Protocol Office to the Joseon kings were compiled in *Tongwen Huikao* (*Tongwen Huikao* (Vol. 43–44)). These documents recorded 69 solar eclipses and 157 lunar eclipses from 1721 to 1879 (For more details, please refer to: Shi Yunli, Lv Lingfeng, Zhang Binglun. (2000). Addition to the documents of eclipse forecasts in the Qing dynasty. *The Chinese Journal for The*

History of Science and Technology (Vol. 21). (pp. 270–281)). Each document was attached with a "reply" from the Joseon king on their observations of an eclipse. Regrettably, except for the reply of "failed to observe" due to the bad weather, only four documents were complete, while the other replies only recorded the specific dates without any details of the eclipses. What's more, the four complete documents were shown at the front of "Original compilation" and "A sequel of the original compilation" in *Tongwen Huikao* as a format sample of reply to documents on eclipse forecasts. It can be seen in the replies without observation details that "please see xx for format sample."

According to the research, in the periods without motions of calendar reform or disputes over calendar, eclipse observations were just a copy of the forecasts, with a conclusion of "complete consistency" between observation and forecast. So, how about the Joseon replies? With the Joseon replies largely omitted, the information available was not insufficient for us to make judgments. But a document from the Qing Protocol Office recorded in *Tongwen Huikao* was quite interesting and helpful for us to learn about the details of "eclipse diplomacy" between the Qing dynasty and the Joseon dynasty.

This document was about the lunar eclipse on December 24, 1749. According to the eclipse forecasts of the Qing Bureau of Astronomy in June:

> The Joseon eclipse will last 4 minutes and 9 seconds. The first contact will occur in 3:22 a.m., middle of the eclipse 4:33 a.m., and the last contact 5:45a.m.
>
> However, the observations of the Joseon Bureau of Astronomy were:
>
> The eclipse lasted 3 minutes and 46 seconds. The first contact occurred in 3:26 a.m., middle of the eclipse 4:34 a.m., and the last contact 5:30 a.m.

Thus, the Joseon king reported to the Qing dynasty that "the time when the lunar eclipse occurred was inconsistent with the forecasts" (the original context was lost) and presented the observations in details. Upon reading this reply, King Qianlong attached great importance to this matter and instructed the Bureau of Astronomy to make a thorough investigation. The Bureau of Astronomy wasted no time in responding that this inconsistency was down to the erroneous observations of the Joseon. This reply was later on forwarded to the Joseon king, along with an account of investigation results as followed, apart from a recap of this whole thing:

> A lunar eclipse occurs when the Moon passes through the Earth's shadow, in this sense, the time when it takes place is supposed to be the same no matter where it is observed. But the fact is that the eclipse cannot be observed at the same time in every corner of the earth, which explains the inconsistency in the time when it occurred. In November 15th, 1749, the lunar eclipse lasted for 4 minutes and 9 seconds, according to the forecast I invited officials of the Protocol Office to observe the eclipse with telescopes and other instruments in the observatory, along with the other astronomical officials. Our observations were exactly the same with our forecasts and thus Joseon's observations of the magnitude of eclipse were not supposed to be inconsistent with what we had forecasted. The picture presented by the Joseon suggested that the eclipse lasted for 3 minutes and 46 seconds, but with a careful measurement, the time period was only 3 minutes and 32 seconds, still inconsistent with its observation. Therefore, we cannot be sure if its observations were accurate or not.

Graph 9.1 Forecasts and observations of the eclipse in 1749

中文	英文
食分	Magnitude of eclipse
初亏(小时)	First contact (hour)
食甚(小时)	Middle of an eclipse (hour)
复圆(小时)	Last contact (hour)
礼部预报值	The forecast by the Qing Protocol Office
朝鲜观测值	The observation by Joseon
观测值-预报值	The observation-the forecast

> With regard to the time when this lunar eclipse occurred, the picture presented by the Joseon suggested that the first contact, which was when penumbral eclipse began, was 4 minutes later than the forecast, whereas the last contact, which was when penumbral eclipse ended, was 3 minutes earlier than the forecast. The penumbral eclipse usually begins when the penumbral part of Earth's shadow starts moving over the Moon, while it ends usually when Earth's shadow completely moves away from the Moon. However the edge of the shadow is not easily seen by the naked eye, so the stage when penumbral eclipse begins is usually observed after it has already began and its end tends to be observed even before it ends completely. Thus inaccuracy in observations of these phases may happen. Middle of an eclipse is when the magnitude of eclipse reaches its highest, which usually lasts for a few minutes without any changes. Thus it is not convincing when the picture suggested it was a minute later than the forecast...... This Bureau suggests your majesty issue an imperial order to the Protocol Office to ask the Joseon king to engage in concerted actions of protection by following the instructions of the Office from now on. (*Tongwen Huikao* (Vol. 43).(pp. 25–26))

In fact, the disparity between the Joseon observations and forecasts is not huge (Graph 9.1). The error of magnitude of eclipse was only 4% of the diameter, and in the middle of eclipse, the error was just 0.07 hour (4.2 min). But the Qing Bureau of Astronomy a little bit overreacted, saying "Joseon's observations of the magnitude of eclipse were not supposed to be inconsistent with what we had forecasted," and its observation of the middle of eclipse was "not convincing." This was denying the authority of the Joseon in terms of astronomical observations in favor of theirs. They proposed at last to "engage in concerted actions of protection by following the instructions of the Office from now on," seemingly implying that what the Joseon should do is just to follow their instructions. It suggested that the exchanges with the Joseon in eclipse were more of a ceremonial activity. It was not that the Qing Government genuinely wanted to know about the accuracy of its forecast on Joseon's lunar eclipses. In other words, this kind of exchanges is a form of "diplomacy" rather than a scientific cooperation.

	食分	初亏(小时)	食甚(小时)	复圆(小时)
礼部预报值	0.42	3.37	4.55	5.75
朝鲜观测值	0.38	3.43	4.57	5.70
观测值一预报值	−0.04	0.07	0.02	−0.05

Nevertheless, the "astronomical diplomacy" of the Qing dynasty influenced the Joseon system of eclipse forecasts. They shared the Chinese view on the solar and lunar eclipses. Therefore, they had the same institutions of protection and observation as China. The Joseon Government stipulated that the Bureau of Astronomy must offer forecasts to the King 3 months before the eclipses occurred. The accuracy would be tested in the observation of eclipses, and astronomical officials were to be punished if the forecasts went wrong. But since the forecasts of the Qing dynasty usually arrived 4 or 5 months before, the Joseon Government required its astronomical officials to deliver forecast 5 months in advance, so as to test their professional capability (*Shuyunguan Journal.* (Vol.2)).

References

1. Changbom Park. (2007). *Astronomy, traditional Koryon science* (pp. 51–60). Seoul: Ewha Woman's University Press.
2. Cheng Zhoude. (1979a). *Compilation of works on astronomical phenomenon* (Vol. 77). Chengxin Women's University.
3. Cheng Zhoude. (1979b). *Compilation of works on astronomical phenomenon* (Vol. 1). Chengxin Women's University.
4. Guo Shirong, & Li Di. (2002). The origin of discussion between He Guozhu and Koryon Hong Zhengxia on mathematical problems. *Inner-Mongol Normal College Journal, 33*(2), 209–212.
5. Hong Wansheng. (2000). A dialog on arithmetic between the East and West: Hong Zhengxia vs. He Guozheng. *Chinese Studies, 20*(2), 57–80.
6. Hong Yixie. (1946). *The History of the Koryon Science* (p. 58). Zhengyin Publisher.
7. Li Changqing, Baixin. (2011). The life and achievements of He Guozong. *Xianyang Normal College Journal, 26*(3), 97–101.
8. Li Chunzhi. (1982). *A collection of various systems of astronomy* (A separately published volume of *Compendium of the Koryon Science Classics*). Chengxin Women's University.
9. Li Chunzhi. (1986). *The preface for on the calculation of eclipses* (Compedium of the Koryon scientific works). Lijiang Press.
10. Li Junjie. (1997). *Frescoes in the ancient Goguryeo tomb—Collected works of the International Seminar on Goguryeo Study* (pp. 332–340).
11. Li Liang, Lv Lingfeng, & Shi Yunli. (2010). A analysis of the compilation and adoption of *Datong Calendar* based on the differences of calculation methods. *The Chinese Journal for the History of Science and Technology, 31*(4), 414–431.
12. Li Zuiguang. (1994a). *Zhifeng Leishuo* (the last volume) (p. 515). Yiyou Press.
13. Li Zuiguang. (1994b). *Zhifeng Leishuo* (the last volume) (p. 488). Yiyou Press.
14. Lin Zongtai. (2011). A study of the introduction of the Western astronomy to the late Joseon. *Science & Culture Review, 8*(1), 51–60.
15. Shi Yunli. (1998a). On the ancient Koryon scholars' research into *Shoushi Calendar. Studies in the History of Natural Science, 29*(4), 312–321.
16. Shi Yunli. (1998b). A study of the Koryon version of *Butian Verses. Historical Records of the Chinese Scientific Technology, 19*(3), 69–79.
17. Shi Yunli. (2008). Reforming astronomy and compiling imperial science: Social dimension and the Yuzhi Lixiang kaocheng houbian. *East Asian Science, Technology and Medicine, 28*, 47–73.
18. Shi Yunli, & Lv Lingfeng. (2002). Norms of etiquette, preaching and eclipse forecasts. *Journal of Dialectics of Nature, 24*, 44–50.
19. Shi Yunlli, Li Liang, & Li Huifang. (2013). A study of the adoption of Huihui Calendar in the Ming dynasty based on Wuxing Lingfan in the 10th year of Xuande reign. *Studies in the History of Natural Science, 32*(2), 156–164.

20. Sun Guangsi. (1986). *The preface for calculation methods of Shoushi calendar.* Lijiang Press.
21. Sun Weiguo. (2003). The special treatment of the Qing dynasty for the Joseon Kings and Envoys. *Modern Koryo*, 39–41.
22. Wu Han. (1964a). *The Chinese historical materials in the Authentic Records of the Joseon Dynasty* (pp. 3748, 3753, 3757, 3768, 3770).
23. Wu Han. (1964b). *The Chinese historical materials in the authentic records of the Joseon Dynasty* (pp. 3874–3876).
24. Yang Yulei. (2006). A study of the exchanges between Koryon envoys to Beijing and the Western missionaries. *The World History, 5*, 126–131.

Stacking Techniques and Higher-Degree Interpolation: Summation of Series and Interpolation in Ancient China

10

Zelin Xu

Contents

Abstract

Duo Ji, a branch of traditional Chinese mathematics developed in Song and Yuan Dynasties, is essentially a solution for summation of arithmetic progression of higher degree as we know today. Based on that, ancient mathematicians in China developed a solution for high-degree interpolation called Zhao Cha, a finite-difference time-domain method that is the inverse of Duo Ji.

Keywords

Chinese mathematics · Stacking techniques · Higher-degree interpolation

Duo Ji, a branch of traditional Chinese mathematics developed in Song and Yuan dynasties, is essentially a solution for summation of arithmetic progression of higher degree as we know today. Based on that, ancient mathematicians in China developed a solution for high-degree interpolation called Zhao Cha, a finite-difference time-domain method that is the inverse of Duo Ji.

10.1 Duo Ji (Summation of Series)

The famous deer-dividing and female weaver problems in *Shuai Fen*, a chapter dedicated to decreasing progression in *Jiu Zhang Suan Shu* (*Nine Chapters on the Mathematical Art*), were the earliest examples of arithmetic progression. The deer-dividing problem

Z. Xu (✉)
Institute of History, School of Humanities, Donghua University, Shanghai, China

X. Jiang (ed.), *The High Tide of Science and Technology Development in China*, History of Science and Technology in China,
https://doi.org/10.1007/978-981-15-7847-2_10

involves an arithmetic progression of 5, 4, 3, 2, 1 and the female weaver problem a geometric progression of 1, 2, 4, 8, 16. Simple problems of summation of arithmetic progression can also be found in the chapter dedicated to proportion. Later, formulas for summation of arithmetic progression were introduced in *Zhang Qiujian Suan Jing* (*Zhang Qiujian's Mathematical Manual*), a book published in the fifth century.

Special study on arithmetic progression of higher degree started in Northern Song dynasty (1031~1095) with Shen Kuo, who, in volume 18 of his famous *Meng Xi Bi Tan* (*Dream Pool Essays*) wrote that: "There are plenty of methods for calculating the volume of different shapes, such as chu meng (wedge), chu tong (cuboid frustum), fang chi (cube), min gu (cuboid), qian du (triangular prism), bie nao (pyramid), yuan zhui (circular cone) and yang ma (rectangular pyramid), with the exception of xi ji (summation of arithmetic progression of high degree)." Then, he invented the following method of xi ji: "The word xi ji means gap between stacked objects, like chess pieces on top of one another, layers of platforms or piled up liquor jars, the shape of which resemble an upside-down funnel, with all four sides being oblique. Due to the gap around edges, using the method for calculating the volume of a cuboid frustum would always end up with figures smaller than the actual value. And here's my solution: First, calculate volumes of its top and bottom layers with the cuboid frustum method, then subtract the width of the upper base from that of the lower base, multiply the difference by its height, divide the result by 6, and then combine it with the previous numbers."

As shown in Picture 10.1, the chess pieces are an analogy for stacked cubes, so are layers of platform for a trapezoid and the liquor jars for stacked spheres, as shown in Picture 10.2. All three examples are similar to a cuboid frustum in shape, only with gaps in between. So, to work out their volume using the cuboid frustum method, as introduced in the chapter dedicated to calculating volumes (沈括: *Dream Pool Essays (梦溪笔谈)*, Volume 18, the *Imperial Collection of Four* (四库全书) version) in *Jiu Zhang Suan Shu*, results will be under-calculated.

By setting the width of the upper base of either of the three shapes as a, length as b, width of the lower base as c, length as d, and height as n (in the liquor jars' case, length is the number of jars), Shen Kuo's formula for the volume (or total number in the liquor jars' case) is as follows:

$$V = \frac{n}{6}[(2b + d)a + (2d + b)c] + \frac{n}{6}(c - a)$$

Picture 10.1 Lei Qi (stack of chess pieces)

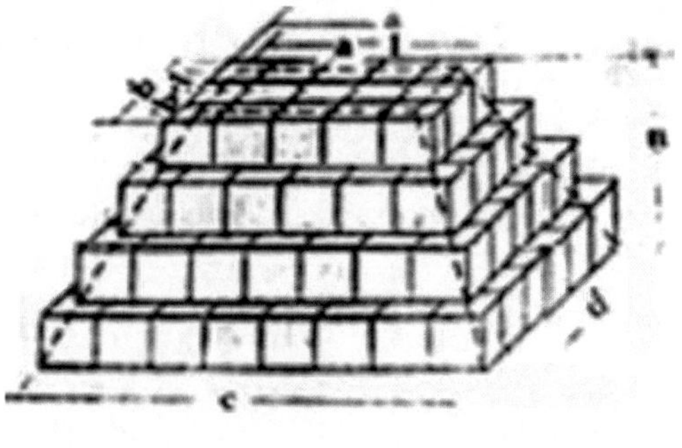

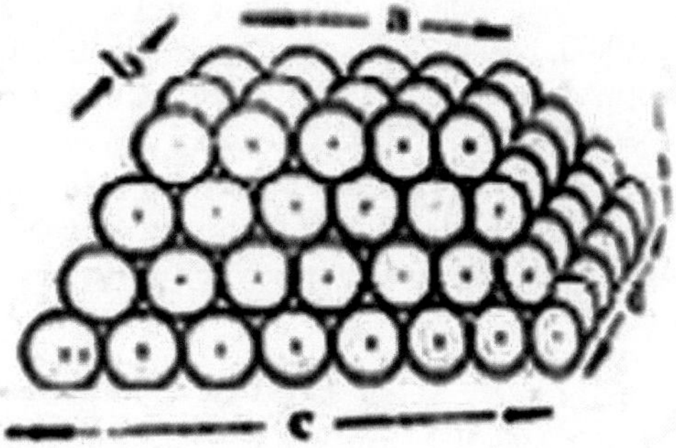

Picture 10.2 Ji Ying (stack of liquor jars)

The xi ji method is a solution for summation of second-order arithmetic progression, which marked the beginning studies on arithmetic progression of high degree over the next few centuries till the end of the Qing dynasty. After Shen Kuo, Yang Hui in Southern Song dynasty introduced the duo ji method after discussing the calculation for the volume of polyhedrons. By comparing fang duo (cubic stack) to fang ting (cubic platform), he set the lower area of the cube as a, the upper area as b, height as h, and volume as s, and the formula goes

$$S = a^2 + (a+1)^2 + \cdots + b^2 = \frac{n}{3}\left(a^2 + b^2 + ab + \frac{a-b}{2}\right)$$

Then, by comparing guo zi duo (fruit stack) to fang zhui (trapezoid), he set the lower area as n and volume as s, and the formula goes

$$S = 1^1 + 2^2 + \cdots + n^2 = \frac{n}{3}(n+1)\left(n + \frac{1}{2}\right)$$

By comparing san jiao duo (triangular stack) to bie nao (pyramid), he set the lower area as n and volume as s, and the formula goes

$$S = 1 + 3 + \cdots + \frac{n(n+1)}{2} = \frac{n}{6}(n+1)(n+2)$$

Then, by comparing another fruit stack to chu tong (cuboid frustum), he set the lower width as a, length as b, the upper width as c, length as d, height as n, and the volume as s, and the formula goes,

$$\begin{aligned} S =& ab + (a+1)(b+1) + \cdots + cd \\ =& \frac{n}{6}[(2b+d)a + (2b+d)c] + \frac{n}{6}(c-a) \end{aligned}$$

The first three examples were first mentioned in Yang Hui's book, and they should be seen as variations of the fourth example, which uses the same formula as Shen Kuo's. However, Yang Hui made no comment as to where the formulas were from.

In Yuan dynasty, Zhu Shijie provided a more systematic discussion on the duo ji method, basing on Jia Xian's triangle theory. Various types of duo ji formulas were applied in several chapters of *Si Yuan Yu Jian* (*Jade Mirror of the Four Unknowns*) (1303), such as *Jiao Cao Xing Duan* (*wild rice stacks*), *Ru Xiang Zhao Shu* (*graphical presentations*), *and Guo Duo Die Cang* (*fruit stacks*).

In *Jiao Cao Xing Duan*, he introduced a series of formulas for calculating the volume of triangular stacks while using the Tian Yuan method to establish equations.

Problem 1:

A total of 680 bundles of jiao cao (wild rice) are to be stacked into the luo yi (lit. "minus-one", which means the number of bundles reduces by one as it goes up one layer, naming of different stacks follows the same rule throughout the article, therefore will not be further explained—Translator's note) shape. Question: how many are in the base layer?

The answer is: 150.

Solution: set number of bundles at the base layer of the stack as unknown number one, then work out the result based on the given product. Now we determine the constant term as 4080, coefficient of the linear term as 2, coefficient of the quadratic term as 3 and coefficient of the cubic term as 1, then calculate the cube root. Thus, the result matches the answer.

$$S = \sum_{i=1}^{n} \frac{1}{2} i(i+1) = \frac{1}{3!} n(n+1)(n+2)$$

The problem involves a stack of wild rice in the jiao cao luo yi shape (a.k.a. Yang Hui's triangle). With a known volume of 680, using the formula for calculating the product of stacks in this particular shape to establish a cubic equation and work out the number of bundles at the base layer of the stack as 15.

Problem 2:

A total of 1820 bundles of wild rice are to be stacked in the sa xing shape (lit. scattered stars). Question: How many are in the base layer?

The answer is: 13.

Solution: set number of bundles at the base layer of the stack as unknown number one then work out the result based on the given product. Now we determine the constant term as 43680, coefficient of the linear term as 6, coefficients of the quadratic term as 11 and 6 and coefficient of the cubic term as 1, then calculate the cube root. Thus, the result matches the answer.

The problem involves a stack in the sa xing shape (or minus-one triangular stack). With 1820 as the given product, by using the formula for calculating the product of stacks in this particular shape

$$S = \sum_{i=1}^{n} \frac{1}{3!} i(i+1)(i+2) = \frac{1}{4!} n(n+1)(n+2)(n+3)$$

to establish a quartic equation and work out the number of bundles at the base layer of the stack as 13.

Problem 4:

A total of 8568 bundles of wild rice are to be stacked into minus-one sa shape. Question: how many bundles are in the base layer?

The answer is: 14.

Solution: set number of bundles at the base layer of the stack as unknown number one then work out the result based on the given product. Now we determine the constant term as 1028160, coefficient of the linear term as 14, coefficients (朱世杰: *Detailed Analysis on Jade Mirror of the Four Unknowns (四元玉鉴细草). Complete Collection of Historical Documents of Science and Technologies in China: Mathematics (《中国科技典籍通汇·数学卷》),* book 1, edited by 郭书春, Henan Education Press, 1993, page 1241) of the quadratic term as 50 and 35 and 10, and coefficient of the cubic term as 1, then calculate the fourth root. Thus, the result matches the answer.

The problem involves a stack in the minus-one sa xing shape (or triangular sa xing shape). With 8568 as the given product, by using the formula for calculating the product of stacks in this particular shape

$$\sum_{i=1}^{n} \frac{1}{4!} i(i+1)(i+2)(i+3) = \frac{1}{5!} n(n+1)(n+2)(n+3)(n+4)$$

to establish a quintic equation and work out the number of bundles at the base layer of the stack as 14.

Problem 6 under *Guo Duo Die Zang Men* (*Stacks of Fruit*) in *Si Yuan Yu Jian, Book III.*

A total of 924 fruits are to be stacked into the triangular minus-one sa xing shape. Question: How many fruits are in the base layer?

The answer is: 7.

Solution: set number of fruits at the base layer of the stack as unknown number one then work out the result based on the given product. Now we determine the constant term as 665280, coefficient of the linear term as 120, coefficients of the quadratic term as 274 and 225 and 85 and 15, and coefficient of the cubic term as 1, then calculate the fifth root. Thus, the result matches the answer.

The problem involves a stack in the triangular minus-one sa xing shape. With 924 as the given product, by using the formula for calculating the product of stacks in this particular shape

$$\sum_{i=1}^{n} \frac{1}{5!} i(i+1)(i+2)(i+3)(i+4) = \frac{1}{6!} n(n+1)(n+2)(n+3)(n+4)(n+5)$$

to establish a sextic equation and work out the number of fruits at the base layer of the stack as 7.

By summarizing the formulas listed above, we can see that Zhu Shijie used a series of formulas for san jiao duo (triangular stacks) problems.

$$\sum_{r=1}^{n} \frac{1}{p!} r(r+1)(r+2)\cdots(r+p-1) = \frac{1}{(p+1)!} n(n+1)(n+2)\cdots(n+p) \tag{10.1}$$

When $p = 1, 2, 3, \ldots$ it's respectively the solution for jiao cao duo (hay stack), jiao cao luo yi xing duo (triangular stack), sa xing duo (minus-one triangular stack), etc., which were later referred to as yi cheng (multiplied by one) triangular stack, er cheng (multiplied by two) triangular stack, and san cheng (multiplied by three) by Li Shanlan in the Qing dynasty.

Problem 5 under *Dui Ji Huan Yuan Meng (Source-returning in Stack Problems)* in *Suan Xue Qi Meng* (*Introductions to Mathematic Studies*).

> There is a square-shaped stack of fruits, with 44 at each layer. Question: How many fruits are there totally?
>
> The answer is: 29370.
>
> Solution: Count the number of fruits at the base layer, plus one and a half, then multiply the newer number by the previous one. Plus half to the result, then again, multiply the newer number by the previous one to obtain the result as 88110. Divide it by three and the result matches the answer.

This problem involves calculating the product of a si jiao duo (square-shaped stack). Set the number of fruits at each layer as n, and the solution can be interpreted as

$$S_n = \sum_{r=1}^{n} i \times i = \frac{1}{3}\left[\left(n + 1\frac{1}{2}\right)n + \frac{1}{2}\right]n = \frac{1}{3!} n(n+1)(2n+1).$$

Problem 12 under the same chapter is the source-returning calculation of problem 5. The text goes

> A total of 29870 fruits are stacked in the si jiao duo shape. How many fruits are in each layer?
>
> The answer is: 44.
>
> Solution: Count the total number of fruits and multiply the number by three to obtain 88110. Then, set coefficient of the linear term as 0.5, coefficient of the quadratic term as 1.5, and

coefficient of the cubic term as 1, then calculate the cube root. Thus, the result matches the answer.

With the given product of the stack as 29370, use the formula for calculating the product of stacks in this particular shape to work out the number of fruits in each layer as 44.

Problem 3 *under Guo Duo Die Zang Men* in *Si Yuan Yu Jian*

A total of 540 fruits are stacked in the minus-one si jiao duo shape. Question: How many fruits are in the base layer?

The answer is: 8.

Solution: set number of fruits in the base layer of the stack as unknown number one then work out the result based on the given product. Now we determine the constant term as 6480, coefficient of the linear term as 2, coefficients of the quadratic term as 5 and 4, and coefficient of the cubic term as 1, then calculate the cube root. Thus, the result matches the answer.

The problem involves a stack in the minus-one square shape. With 540 as the given product, by using the formula for calculating the product of stacks in this particular shape

$$S_n = \sum_{r=1}^{n} \frac{1}{3!} n(n+1)(2n+1) = \frac{1}{3!}(n+1)n(n+1)(n+|2)$$

to establish a quartic equation and work out the number of fruits at each layer of the stack as 8.

From the examples above, we can see that Zhu Shijie used a series of formulas for square-shaped stacks problems.

$$\begin{aligned} &\sum_{r=1}^{n} \frac{1}{p!} r(r+1)(r+2)\cdots(r+p-2)(2r+p-2) \\ &= \frac{1}{(p+1)!} n(n+1)(n+2)\cdots(n+p-1)(2n+p-1) \end{aligned} \tag{10.2}$$

When $p = 2, 3$ it's respectively called si jiao duo (square-shaped stack) and si jiao luo yi xing duo (minus-one square-shaped stack).

Zhu also discussed lan feng xing duo (stacks in the shape of a mountain crest), with the general term being $\frac{1}{p!}i(i+1)(i+2)\ldots(i+p-1)i$. The chapter titled *Jiao Cao Xing Duan* in *Si Yuan Yu Jian* involves a lan feng xing duo problem as the following:

A total of 3367 bundles of wild rice are to be stacked in the lan feng shape. Question: how many are in each layer?

The answer is 12.

Solution: set number of bundles in each layer of the stack as unknown number one then work out the result based on the given product. Now we determine the constant term as 80888, coefficient of the linear term as 2, coefficients of the quadratic term as 9 and 10, and coefficient of the cubic term as 3, then calculate the cube root. Thus, the result matches the answer.

With the given product of the stack as 3367, use the formula for calculating the product of stacks in this particular shape

$$S = \sum_{r=1}^{n} \frac{1}{2!} i(i+1)i = \frac{1}{4!} n(n+1)(n+2)(3n+1)$$

to establish a quartic equation and work out the number of fruits at each layer of the stack as 12.

Problem 4 under *Guo Duo Die Zang* in *Si Yuan Yu Jian* involves a san jiao (triangular) lan feng xing duo (or minus-one lan feng xing duo under the chapter *Jiao Cao Xing Duan*)

A total of 630 fruits are to be stacked in the san jiao lan feng shape. Question: how many are in each layer?

The answer is 6.

Solution: set number of fruits in each layer of the stack as unknown number one then work out the result based on the given product. Now we determine the constant term as 75600, coefficient of the linear term as 6, coefficients of the quadratic term as 35 and 50 and 25, and coefficient of the cubic term as 4, then calculate the fourth root. Thus, the result matches the answer.

With the given product of the stack as 630, use the formula for calculating the product of stacks in this particular shape

$$\sum_{r=1}^{n} \frac{1}{3!} i(i+1)(i+2) \times i = \frac{1}{5!} n(n+1)(n+2)(n+3)(4n+1)$$

to establish a quintic equation and work out the number of fruits at each layer of the stack as 6.

In fact, Zhu Shijie used a formula of summing the first n terms in lan feng xing duo problems

$$\begin{aligned} S_n &= \sum_{r=1}^{n} \frac{1}{p!} r(r+1)(r+2)\cdots(r+p-1)r \\ &= \frac{1}{(p+2)!} n(n+1)(n+2)\cdots(n+p)[(p+1)n+1] \end{aligned} \tag{10.3}$$

When $p = 1, 2, 3 \ldots$ it's respectively the solution for si jiao duo (square-shaped stack), lan feng xing duo (mountain-crest-shaped stack), san jiao lan feng xing duo or lan feng geng luo yi xing duo (triangular or minus-one lan feng xing duo).

Other than that, Zhu Shijie listed a stack problem of the following type:

$$\sum_{r=1}^{n} \frac{1}{p!} r(r+1)(r+2)\cdots(r+p-2)(2r+p-2)\cdot r$$
$$= \frac{1}{(p+2)!} n(n+1)(n+2)\cdots[n+(p-1)]\left[2(p+1)n^2+\left(p^2+2\right)n+(p-2)\right]$$

When $p = 3$ it's called a si jiao lan feng duo (square-shaped mountain-crest stack) (Picture 10.3).

Zhu Shijie's stack-problem formulas were, to a certain extent, connected to the old method of seven powers recorded in the beginning of *Si Yuan Yu Jian*. The sum of the first n terms in the p row of the table equals the nth term in the $p+1$ row or duo ji formula (10.1); the rth term in the p row is

$$\frac{1}{p!} r(r+1)(r+2)\cdots(r+p-1)$$

And the rth term in the $p+1$ row is

$$\frac{1}{(p+1)!} r(r+1)(r+2)\cdots(r+p)$$

Thus, formula (10.1) is obtained. *Si Yuan Yi Juan* also involves the following duo ji formulas:

Picture 10.3 The old method of seven powers (Jia Xian's triangle), *Si Yuan Yu Jian*

$$\sum_{r=1}^{n} r[a+(r-1)b] = \frac{1}{3!}n(n+1)[2bn+(3a-2b)]$$

$$\sum_{r=1}^{n} r[a+(n-r)b] = \frac{1}{3!}n(n+1)[bn+(3a-b)]$$

$$\sum_{r=1}^{n} \frac{1}{2!}r(r+1)[a+(r-1)b] = \frac{1}{4!}n(n+1)(n+2)[3bn+(4a-3b)]$$

$$\sum_{r=1}^{n} r^2[a+(n-r)b] = \frac{1}{3!}n(n+1)\frac{bn^2+(4a-b)n+2a}{2}$$

Among the above, formulas (10.3) and (10.4) are respectively the bian duo (variations) of formula (10.1) and (10.2). In other words, the general term in formulas (10.1) and (10.2), when multiplied by r, equals the general term in formulas (10.3) and (10.4). Of these equations, formula (10.1) is the basis, upon which all the other formulas can be derived. The method for derivation was very likely zhao cha (in wasan, the traditional Japanese math system, duo ji formulas were derived using the zhao cha method).

10.2 Zhao Cha (High-Degree Interpolation)

In ancient China, in order to determine exact time of the shuo (new moon) and jiao shi (eclipses) in calendrical calculation, the true position of the sun and the moon's movement in the zodiac and moon's path must be accurately calculated. Early calendars, such as *Gu Si Fen Li* (*lit. Old Quarter Calendar*), *Tai Chu Calendar* (104 B.C.) of Western Han dynasty, and *San Tong Calendar*, all adopted the ping shuo (lit. constant syzygy) calculation method, in which the time of new moon was determined according to the moon's mean motion during a synodic month. However, the moon's revolution is a variable motion, so is the sun's ecliptic motion, which means it's impossible to calculate the exact time of new moon using the ping shuo method. In Eastern Han dynasty, Liu Hong (130?~196) became the first to suggest, *in Qian Xiang Calendar*, using linear interpolation to determine the time of new moon, a process called ding shuo (lit. determining new moon). And Liu Hong successfully worked out the degrees of the moon's daily movement during one anomalistic period. By setting the number of days as n, total degrees the moon moves in n days is $f(n)$. As for degrees the moon moves in $n + s(n < n + s < n + 1)$ days, he used a formula of linear interpolation $f(n + s) = f(n) + s\Delta$ for calculation, in which $\Delta = f(n + 1) - f(n)$, as first difference. This calculation method was then inherited by mathematicians in the Three Kingdoms Period and the Northern and Southern dynasties. Due to the ever-changing speed of the moon's movement, $f(n)$ is not a linear function, so second-order difference $\Delta^2 f(x) \neq 0$. That's why Liu Hong's linear interpolation calculation didn't arrive at accurate values.

After Zhang Zixin (dates of birth and death unknown), an astronomer of the Northern Qi dynasty, discovered that the sun's movement through the zodiac was uneven; calendarists started to consider establishing a mathematic method to accurately calculate the position of celestial bodies in motion. In Sui dynasty, Liu Zhuo introduced, in *Huang Ji Calendar*, a formula of quadratic interpolation at equal intervals for calculating degrees of apparent motion of the sun, the moon, and the five planets.

$$f(kl+m) = f(kl) + \left(\frac{\Delta_k + \Delta_{k+1}}{2}\right)\frac{m}{l} + (\Delta_k - \Delta_{k+1})\frac{m}{l} - \frac{\Delta_k - \Delta_{k+1}}{2}\left(\frac{m}{l}\right)^2 \tag{10.4}$$

In the formula listed above, l is the length of a solar term (in days), $f(kl + m)$ is ying suo fen (difference between the actual degrees and the average degrees of the sun's movement) of the sun's motion $x = kl + m$ days (wherein $0 < m < l$) after winter solstice. Obviously, $f(0) = 0$, $\Delta_k = f(kl + l) - f(kl)$ are the sun yi lv (difference in shadow lengths as measured by a sundial) within each solar term.

Liu Zhuo's formula, though much more precise than former ones, still had its flaws due to the fact that the I, representing days of a solar term, does not change at equal intervals, nor are the sun, the moon, and the five planets move in a uniform accelerated motion (which means third-order difference $\neq 0$). These two problems were later solved, respectively, by Yi Xing (673~727) in Tang dynasty and Guo Shoujing (1231~1316) in Yuan dynasty.

In the 15th year of Kaiyuan Period of Emperor Xuanzong's reign in Tang dynasty, or 727 A.D., Yi Xing introduced a formula of quadratic interpolation at unequal intervals in *Da Yan Calendar.*

$$f(t+x) = f(t) + \left(\frac{\Delta_1 + \Delta_2}{l_1 + l_2}\right)x + \left(\frac{\Delta_1}{l_1} - \frac{\Delta_2}{l_2}\right)x - \left(\frac{\Delta_1}{l_1} - \frac{\Delta_2}{l_2}\right)\frac{x^2}{l_1 + l_2} \tag{10.5}$$

In which ($l_1 \neq l_2$, $x < l$), and when $l_1 = l_2$, it becomes the same formula as Liu Zhuo's.

In a chapter titled *Chuang Fa Wu Duan* (lit. five innovative methods) in the *Shou Shi* (lit. time-telling) *Calendar* (1281) of Yuan dynasty, essential innovation was achieved in calculating the sun's ying suo, the changing of speed in the moon's motion and the degrees of the five planets' movement. The author of the calendar, having considered variable-speed motion of the sun, the moon, and the five planets, believed that distance is a cubic function of time and, therefore, should be calculated using the zhao cha method. However, the calendar didn't determine the formula of third-order interpolation. Instead, the problem was solved with a table of differences. In *Da Tong Calendar*, included in *History of Ming*, the passage about the calculation of the sun's apparent motion is as follows:

> Set ri ping cha (lit. per-day difference) in stage one, which is 476 minutes 25 seconds, as the fan ping ji (lit. average total). Then, subtract the er cha(difference between stages) of stage two, which is 1 minute 38 seconds, from the yi cha (difference between days) of stage one, which is 18 minutes 45 seconds, to obtain 37 minutes 7 seconds as the remainder, or fan ping ji cha (lit. average total difference). Then, break the er cha of stage one, which is one minute 38 seconds, in half and get 69 seconds as fan li ji cha (lit. average vertical difference). Add

> fan ping ji cha, which is 37 minutes 7 seconds, to fan ping ji, which is 76 minutes 25 seconds, to obtain 513 minutes 32 minutes, which is called ding cha (lit. constant difference). Then, subtract fan ping ji cha, which is 37 minutes 7 seconds, from fan li ji cha, which is 69 seconds, to obtain 36 minutes 38 seconds, as shi (lit. dividend). Then, divide time length per stage, which is 14 days 82 quarters, by shi, to obtain 2 minutes 46 seconds, as ping cha. Then, divide time length per stage by fan li ji cha (69 seconds) twice to obtain 31 wei as li cha (lit. vertical difference).
>
> To calculate ying suo, multiply length of the first and last days on the calendar by li cha then add the result to ping cha, and then multiply it again by length of the first and last days on the calendar, subtract ding cha from the result, then multiply the remainder by length of the first and last days to obtain ying suo ji (lit. product).

Provided the earth covers 365.25° revolving once around the sun, then each quadrant contains 91.31°. According to *Shou Shi Calendar*, quadrants measured before and after winter and summer solstices are, respectively, 88.91 days and 93.71 days. Provided the sun moves one degree per day, then increment of the former and decrement of the latter are both 2.40°. So, the formula for calculating ji cha (lit. accumulated difference) before and after winter solstice is as follows:

$$f(x) = 513.32x - 2.46x^2 - 0.003\,1x^3 \tag{10.6}$$

in which $0 < x \leq 88.91$. Coefficient of x's linear, quadratic, and cubic terms (all taking positive values) are respectively called ding cha, ping cha, and li cha. In Shou Shi Calendar, the quadrant from winter solstice to spring equinox (totally 88.91 days) is divided into six stages (length of each being $l = 88.91 \div 6 = 14.82$ days), and the actual degrees of the sun's movement at $l, 2l, \ldots, 6l$ were measured and then subtracted from which the average degrees to obtain ji cha $f(kl)$, $i = 1, 2, \ldots, 6$. Divide ji ri (lit. accumulated days) by ji cha to obtain ri ping cha $\frac{f(il)}{il}$, and then calculate its yi cha and er cha, constituting a difference table with equal intervals of stages. According to *History of Ming*, the results above can be listed as in the following table (张廷玉 and others: *History of Ming: Records Chapter Nine* (《明史·志第九》), *Complete Collection of Astronomical and Calendrical Works Throughout History* (历代天文律历等志汇编), Zhong Hua Book Company, 1976).

段数 i	积日 x_i	积差 $F(x_i)$	日平差 $F(x_i)/x_i$	一差 Δ_i^1	二差 Δ_i^2
第1段	l	0.705 80	0.047 625	−0.003 845	−0.000 138
第2段	$2l$	1.297 64	0.043 780	−0.003 983	−0.000 138
第3段	$3l$	1.769 37	0.039 797	−0.004 121	−0.000 138
第4段	$4l$	2.114 87	0.035 676	−0.004 259	−0.000 138
第5段	$5l$	2.328 00	0.031 417	−0.004 397	
第6段	$6l$	2.402 62	0.027 020		

中文	英文	中文	英文
段数	Stage	积日	Accumulated days
第1段	Stage 1	积差	Accumulated difference
第2段	Stage 2	日平差	Per-day difference
第3段	Stage 3	一差	Difference between days
第4段	Stage 4	二差	Difference between stages
第5段	Stage 5		
第6段	stage 6		

According to the definition of cha fen (difference), we can determine $f(0)$, $\Delta^1 i$, $\Delta^2 i$, $(\Delta^3 i = 0)$. Furthermore, a difference tale with equal intervals of days can be obtained simply by adding and subtracting. By analyzing the table, we can see that all third-order differences are equal, which means fourth-order differences are equal to 0, so it's a cubic interpolation polynomial.

$$f(x) = d + ax + bx^2 + cx^3$$

And, because $f(0) = 0$ (degrees moved in stage 0 is 0), which means $d = 0$, so $f(x) = ax + bx^2 + cx^3$

Therefore, the original cubic interpolation function is now transformed into a quadratic interpolation function.

$$F(x) = \frac{f(x)}{x} = a + bx + cx^2 \tag{10.7}$$

Using the formula of quadratic interpolation, we can work out the expression of $F(x)$ and then obtain

$$f(x) = xF(x) = 513.32x - 2.46x^2 - 0.003\,1x^3.$$

In Shou Shi Calendar, cubic interpolation was also used for calculating motion of the moon and the five planets.

Interpolation is essentially functional approximation calculated using the finite difference method. Another type of mathematic problems associated with the finite difference method is summation of series. The introduction text of problem 5, as well as its footnote, provides a summation formula for third-order arithmetic series. The problem is as follows:

> The government was recruiting soldiers using an exponential method. 3^3 of soldiers were recruited on day one and 4^3 on day two. Each solder would cost 250 wen (monetary unit of ancient China) and a total of 23400 soldiers had been recruited, costing 23462 guan (monetary unit of ancient China, containing 1,000 wen). Question: How many days did the recruitment last?

In this problem, first difference is established as $\Delta f(x) = (2 + x)^3 (x = 1, 2\ldots 15)$, and $f(n)(n = 15)$ is to be calculated. The formula Zhu Shijie provided is this

$$
\begin{aligned}
f(n) =& n\Delta + \frac{1}{2!}n(n-1)\Delta^2 + \frac{1}{3!}n(n-1)(n-2)\Delta^3 \\
&+ \frac{1}{4!}n(n-1)(n-2)(n-3)\Delta^4
\end{aligned}
\tag{10.8}
$$

in which, Δ, Δ2, Δ3, Δ4, respectively, represent the first difference of each order.

Zhu Shijie wrote in the footnote for this problem:

> Set the number of soldiers recruited on day one as shang ji (lit.first product), shang ji minus one as jiao cao di zi ji (lit. hay stack product); or second product, shang ji minus 2 as san jiao di zi ji (lit. triangular stack product); or third product, shang ji minus 3 as san jiao luo yi ji (lit. minus-one triangular stack); or fourth product. Multiply differences by products, and combine the four numbers to obtain the total number of soldiers recruited.

Set the days of recruitment as n, then (朱世杰: *Jade Mirror of the Four Unknowns (四元玉鉴). Complete Collection of Historical Documents of Science and Technologies in China: Mathematics (《中国科技典籍通汇·数学卷》)*, book 1, edited by 郭书春, Henan Education Press, 1993, page 1250)

With "shang ji minus one," or $n-1$, as a hay stack model of the general term, or di zi, and it arrives at

$$\sum_{k=2}^{n}(k-1) = \frac{1}{2!}n(n-1)$$

with "shangji minus two," or $n-2$, as a triangular stack model of the general term, or di zi, and it arrives at

$$\sum_{k=3}^{n}\frac{1}{2!}(k-2)(k-1) = \frac{1}{3!}n(n-1)(n-2)$$

With "shangji minus three", or $n-3$, as a minus-one triangular-stack model of the general term, or di zi, and it arrives at

$$\sum_{k=4}^{n}\frac{1}{3!}(k-3)(k-2)(k-1) = \frac{1}{4!}n(n-1)(n-2)(n-3);$$

Plus, Zhu Shijie "obtained first difference as 27, second difference as 37, third difference as 24 and fourth difference as 6."

Which means, by setting $f(n)$ as the total number of soldiers recruited in n days, or $f(n) = \sum_{k=1}^{n}(2+k)^3$, and establish that $f(0) = 0$, then the first difference $\Delta f(0) = 27$, second difference $\Delta^2 f(0) = 37$, third difference $\Delta^3 f(0) = 24$, and fourth difference $\Delta^4 f(0) = 6$. To sum up, Zhu Shijie's formula of quartic-difference high-degree interpolation is as follows:

$$
\begin{aligned}
f(n) =& 27n + 37 \cdot \frac{1}{2!}n(n-1) + 24 \cdot \frac{1}{3!}n(n-1)(n-2) \\
&+ 6 \cdot \frac{1}{4!}n(n-1)(n-2)(n-3) \\
=& n\Delta f(0) + \frac{1}{2!}n(n-1)\Delta^2 f(0) + \frac{1}{3!}n(n-1)(n-2)\Delta^3 f(0) \\
&+ \frac{1}{4!}n(n-1)(n-2)(n-3)\Delta^4 f(0)
\end{aligned}
\tag{10.9}
$$

Zhu Shijie explicitly pointed out that coefficients of the second, third, and fourth terms in the formula are exactly the sum of series in the "triangular stack system." Therefore, it has been a popular opinion among the scholars studying history of math in China that Zhu Shijie had obtained Newton's interpolation formula.

Not only does formula (10.10) provide the result of $\sum^{n}(r+2)^3$, but it also displays the duo ji method in general. The structure of the formula shows that a given summation of progression can be, in proper order, broken down into the sums of all triangular stacks, coefficients of which, or differences, are determined by the given summation, as in

$$
\begin{aligned}
f(n) =& \sum (r+2)^3 \\
=& 27\times(\underbrace{1+1+1+\cdots+1}_{\text{共}n\text{项}}) + 37\times[\underbrace{1+2+3+\cdots+(n-1)}_{\text{共}(n-1)\text{项}}] \\
&+ 24\times[\underbrace{1+3+6+\cdots+\frac{1}{2!}(n-2)(n-1)}_{\text{共}(n-2)\text{项}}] \\
&+ 6\times[\underbrace{1+4+10+\cdots+\frac{1}{3!}(n-3)(n-2)(n-1)}_{\text{共}(n-3)\text{项}}]
\end{aligned}
$$

or

$$
f(n) = 27n + 37\sum^{n-1} r + 24\sum^{n-2}\frac{1}{2!}r(r+1) + 6\sum^{n-3}\frac{1}{3!}r(r+1)(r+2)
$$

which is the same as formula (10.10).

In the same way, summation formulas for all problems under this chapter can be reckoned. For example, the total number of workers in problem one (Same as previous footnote, page 1251)

$$
f(n) = \sum^{n}[64 + 7(r-1)] = 64n + 7 \cdot \frac{1}{2!}(n-1)n
$$

and the total number of recruited soldiers in problem four

$$f(n) = \sum^{n} (r+4)^2$$
$$= 25n + 11 \cdot \frac{1}{2!}(n-1)n + 2 \cdot \frac{1}{3!}(n-2)(n-1)n \qquad (10.10)$$

Using the method demonstrated in formula (10.10), along with incomplete induction, Zhu Shijie's summation formulas for square-shaped stacks, mountain-crest stacks, and square-shaped mountain-crest stacks can be reckoned.

The calculation method mentioned above is the same method for calculating the sun's ying suo ji in *Shou Shi Calendar*. Zhao cha method works by breaking the given duo ji down into the sum of all triangular stacks ($p = 0, 1, 2, \ldots$). It is a general method for the deduction of formulas of high-degree arithmetic progression and establishment of interpolation polynomials.

Generally speaking, a p-order polynomial can be shown as a p-order interpolation formula. Obviously, sum of a p-order arithmetic progression with a general term in the form of p-order polynomial is a $p+1$-order polynomial, which can be shown as a $p+1$-order interpolation formula. So, zhao cha is a method for establishing interpolation formulas for polynomial functions.

Guo Shoujing and Wang Xun (1235–1281) didn't adopt the term "zhao cha." It was up till early Qing dynasty that Huang Ding, in his book *Tian Wen Da Cheng Guan Kui Ji Yao* (*Digests of Astronomical Achievements*), called the method in *Shou Shi Calendar* the "zhao cha method for stack problems," indicating that he noticed the connection between duo ji and zhao cha. In *Si Yuan Yu Jian*, Zhu Shijie didn't really point out the connection between zhao cha and the third-order interpolation method in *Shou Shi Calendar*. Rather, the invention of zhao cha was based on similar studies in times past.

The Spread of Traditional Chinese Mathematics in the Sinosphere and Its Influence

11

Zelin Xu

Contents

Abstract

The Sinosphere refers to China, the birthplace of Chinese characters, and its surrounding areas, including Vietnam, Korean Peninsula, Japan, and the ancient Ryukyu Kingdom. With the language of Chinese as the medium, China's neighboring countries and ethnic groups learned not only national systems, political thoughts, and various kinds of knowledge from China's different dynasties and developed similar cultures and values, but more importantly, traditional Chinese mathematics had great influence on these neighboring countries.

Keywords

Sinosphere · Traditional Chinese mathematics · Chinese astronomical calendar · Korean mathematics

The Sinosphere refers to China, the birthplace of Chinese characters, and its surrounding areas, including Vietnam, Korean Peninsula, Japan, and the ancient Ryukyu Kingdom. In ancient times, people with farming culture were mainly distributed in these areas, and their leaders were crowned by ancient Chinese emperor who enjoyed tribute presented to them. In history, these countries used Chinese characters in writing or mixed it with their own characters. Ancient officials

Z. Xu (✉)
Institute of History, School of Humanities, Donghua University, Shanghai, China

X. Jiang (ed.), *The High Tide of Science and Technology Development in China*, History of Science and Technology in China,
https://doi.org/10.1007/978-981-15-7847-2_11

and intellectuals mostly used classical style of writing (known as Han Wen or Chinese written language in Japan, Vietnam, and Korea) as their written language. With the language of Chinese as the medium, China's neighboring countries and ethnic groups learned national systems, political thoughts, and various kinds of knowledge from China's different dynasties and developed similar cultures and values. The influence of Chinese science and technology culture on neighboring countries is one demonstration.

11.1 Mathematical Exchanges Between China and Korea

As a close neighbor of China, Korea paid tributes to ancient China and received conferred titles in different dynasties. Its culture, science and technology are deeply influenced by Chinese culture. In the history of mathematics exchanges in Sinosphere, Dongsuan (literally means eastern calculation, Korea's traditional mathematics), derived from Zhongsuan, or Chinese calculation, has inherited Chinese mathematics culture successively, preserving many related ancient literature and systems of Chinese calculation, providing many academic resources for Hesuan (literally means Japanese calculation, Japanese traditional mathematics) as well.

Chinese astronomical calendar was introduced to Korea in the Qin and Han Dynasties (221–220 AD). According to Houhanshu-Xunli liezhuan (Book of the Later Han-biographies of benevolent officials), during the early years of the west Han Dynasty (206 BC-25 AD), the descendant of the noble Langya Family, Wang Zhong, from Buqi (in Shandong Province) moved to Korean Peninsula to avoid chaos and rebellion of empress Lv's family. Korea was in the early days of Wei's reign back then and had frequent communications with the Han Dynasty. In 108 BC, Emperor Wu of the Han Dynasty destroyed the Wei Family of Korea and set up four prefectures, namely, Zhenfan, Lintun, Lelang, and Xuantu, bringing the connections between Korea and China closer. Wang Jing, the eighth grandson of Wang Zhong, lived in Nehan County, Lelang Prefecture. He studied the Book of Changes when he was still young, and he read widely, "has good commands of astronomy and many other skills." From Wang Zhong to Wang Jing, generations of immigrants spread the astronomy and mathematics of the Han Dynasty to the Korean Peninsula. According to the ancient Japanese documents, such as the Kojiki or records of ancient matters and *Nihon Shoki*, or the Chronicles of Japan, in the third century, Wang Ren from Baekje brought Chinese culture from mainland to Japan. In the sixth century, Wang Daoliang from Baekje, Wang Baosun, Guan Le, and other so-called naturalized people brought many Chinese science knowledge including calendars into Japan. Therefore, during the Qin and Han Dynasties, Chinese immigrants introduced Chinese calculation along with other cultures into the Korean Peninsula.

From the fourth century to the seventh century, tripartite confrontation among Koguryo, Baekje, and Silla was formed on the Korean Peninsula, leading to closer connections with Chinese culture. In 372 AD, Koguryo began to set up Taixue, or imperial knowledge schools, with Sinology as its main teaching content. The calendar calculation of Yuan-jia li, created by Liu Song of the Southern Dynasty of China, was adopted by Baekje at that time. According to *Nihon Shoki*, in the fifth century, some Chinese descendants who immigrated to Lelang and Daifang county and immigrants from Baekje of Koreans were called "people who cross the ocean" or "naturalized people." They traveled to Japan from Baekje located in the south of Korea, mastering all kinds of handicrafts and having good commands of Chinese, and brought all kinds of techniques to Japan. Among them, those who were called "doctors" or "teachers" were employed by the Japanese court in ministry of appointment and ministry of finance in charge of document record, foreign affair document, accounting, tax collection, etc. As a result, Chinese culture was introduced to Japan, including knowledge of arithmetic, computing skills, and calendar. According to *Nihon Shoki*, Japanese imperial court sent envoys to Baekje in 553 AD for doctor of medicine, geomancy, and calendar, bringing back "divination books, almanacs and different medicines" to Japan. So, the next year, Wang Daoliang, or Shide, the doctor of geomancy; Wang Baosun, or Gude, the doctor of calendar; and others came to Japan from Baekje.

During the Sui and Tang Dynasties (581–907), the exchanges of astronomy and mathematics between China and Korea saw important development. In 676, Silla unified Korea and in 682 established imperial college imitating the system of Tang Dynasty, bringing Chinese calculation books and mathematics systems to Korea. The records of the history of the Three Kingdoms of Korea documented literature of Silla which mentioned the Nine Chapters, or *The Nine Chapters on the Mathematical Art*, the Classic of Zhui, or Zhui Shu (Methods for Interpolation). However, these two books, San Kai and Liu Zhang, have not been recorded in Book of Sui, Old Book of Tang, and New Book of Tang.

Korea directly adopted the Chinese calendar for a long time. In the second year of Linde during the reign of Emperor Gaozong in the Tang Dynasty (665), Linde calendar created by Li Chunfeng was issued, which was used until 728 AD. According to the Records of the Three Kingdoms, in the first year of Yongchun during the reign of Emperor Gaozong of the Tang Dynasty (692), monks from Silla returned to their home country after studying the calendar in China and "presented the astronomical map" to the king of Silla. In 718, Silla established the system of clepsydra imitating the Tang Dynasty. In the second year of Tanbao of Tang (749), Silla followed China, "appointed one doctor of astronomy and six doctors of clepsydra," gradually forming a complete astronomical organization system. From 766 to 779, Jin Yan from Silla went to Tang to study astronomy and calendar and returned to serve as doctor of heaven in Silla. Goryeosa or History of Goryeo records that "the calendar was changed in the previous Tang Dynasty and it was then changed for twenty-two times. However, Goryeo used the same calendar until Emperor Zhongxuan of Goryeo announced to change the calendar of Shoushi in the Yuan Dynasty." From the time of Silla to the time of Emperor Zhongxuan of Goryeo (1309–1313), the Xuanming calendar of Tang xu'ang was adopted in Korea for 400 years. Compared with the calendar of the Han Dynasty, the calculation method of the calendar of Sui and Tang Dynasty was improved in both the astronomical accuracy and the mathematical method. For example, the method of adjusting the sun, the method of second differences interpolation, the difference table, and the table containing the tangent function value were all advanced mathematical methods in the world at that time. These mathematical knowledge along with Chinese calendar were introduced into the Korean Peninsula.

In 918, Wang Jian established the Goryeo Dynasty and inherited the former dynasty's mathematics system due to the tradition of Confucianism and imperial examination. According to volume 188 of Supplementary Documents for Reference, "election examination," the arithmetic examination in the Korean Dynasty was still dominated by the Chinese calculation. The latest arithmetic books popular in China at that time were also introduced to Korea at that time. Xiejia could be the arithmetic book Xiecha micro arithmetic book in the Song Dynasty.

The reign of Li Family for over 500 years in Korea witnessed a golden age of Korea culture. During its peak time, the fourth generation of the Shizong Dynasty (1419–1450), copper movable-type printing was applied to compile and print a large number of Chinese books. Hangul, or the Korean alphabet, was also created at this time. Calendar computation was one of the seven studies advocated by the government and was charged by Ministry of Revenue. The imperial court set up the posts of mathematics including professor, master, scholar, and tutor and retained the traditional mathematics examination system. The middle class entered the court through examination of music (played in bamboo pitch pipes), medicine, mathematics, and calendar. After the end of the Yuan Dynasty, many Chinese arithmetic books were lost in China but was found preserved in Korea. In the eighth year of Xuande (1433), the Yang Hui Suan Fa (Yang Hui's Methods of Computation) was published in 1274 in the Ming Dynasty; Xiang Ming Suan Fa (the Detailed Methods of Computation) of An Zhizhai and the Introduction to Computational Studies of Zhu Shijie were copied at this time. Later these literature were introduced to Japan and back to China, playing an important role in Sinosphere's inheritance and development of mathematics in Song and Yuan Dynasties. The earliest existing works of Korean mathematics, Nine Figure Strategy, the Origin of Calculation, and the Detailed Explanation of the Calculation in the seventeenth century were all based on the abovementioned books. Although abacus was introduced into Korea at this time, it was not accepted by the traditional Korean calculators who use rods.

The reign of Yingzu (1725–1776) and Zhengzu (1777–1800) in Korea experienced another Korean culture boom, when Kangxi and Qianlong of the Qing Dynasty were ruling China at the same period. According to Supplementary Documents for Reference, Shu Li Jing Yun (Collected Basic Principles of Mathematics) was introduced to Korea and greatly influenced Korea's most representative calculation book, Books for Rod Calculus, which also listed many other Chinese computation books along with Shu Li Jing Yun, including Introduction to Computational Studies, Suan Fa Tong Zong (Systematic Treatise on Arithmetic), Yang Hui Suan Fa, Complete Description of Calculation Methods (written by Jiang Yuancheng of the Qing Dynasty), and Hun Gai Tong Xian (an abridged Chinese translation of Astrolabium). The Li Dynasty also sent officials to the Qing Dynasty to study western astronomy.

Korean mathematics reached its peak in the nineteenth century. Nan Bingzhe, Nan bingji, Li Shanghe, and Zhao Yichun, four outstanding mathematicians in Korea during Zhezong's (1849–1863) and Li Taiwang's (1864–1895) reign, wrote a lot of literature on mathematics. The mathematical exchanges between China and Korea were also very frequent during this period. Scholars in the Qing Dynasty collected and studied many Chinese traditional arithmetic books and translations of western mathematics which were translated into Korea and went popular, including *Jiu Zhang Suan Shu* (*The Nine Chapters on the Mathematical Art*), *Shu Shu Jiu Zhang* (*Mathematical Treatise in Nine Sections*), *Ce Yuan Hai Jing* (*Sea Mirror of Circle Measurements*), *Yi Gu Yan Duan* (*Old Mathematics in Expanded Sections*), *Introduction to Computational Studies*, *Jade Mirror of the Four Unknowns*, *Tong Wen Suan Zhi*, and *Chi Shui Yi Zhen*. Suan Xue Zheng Yi (the true methods for calculation) of Nan Bingji and Yi Suan of Li Shanghe absorbed and learned from mathematics of the Song and Yuan Dynasties. At this time, China was the only channel for Korea to access western mathematics.

11.2 Traditional Chinese Mathematics in Japan

From 300 BC to 300 AD, Japan entered the Yayoi period when iron culture and pottery culture coexist. Chinese rice cultivation technology, bronze ware, iron ware, and other metal culture were introduced into Japan through Korea. Japan established a tributary relation with China and formed increasingly closer connection with the mainland. Records of Japan appeared in Chinese history books since the Han Dynasty. According to *Han Shu*, or *Book of Han*, there were Japanese across the Sea of Lelang regularly sending delegates to offer tributes to Lelang, a county set up by Emperor Wu of the Han Dynasty on the Korean Peninsula. According to *Hou Hanshu*, *Book of the Later Han*, and *Records of the Three Kingdoms* – Records of Wei-The Japanese people of Wa – Himiko, the queen of Yamatai of Japan, sent envoys to Luoyang, and Emperor Ming of the Wei Dynasty crowned her as "Pro Wei Japanese King." Later it also sent envoys to pay tribute to the Western Jin Dynasty.

During the Qin and Han Dynasties, a large number of Chinese immigrants came to Japan due to the natural disasters and wars. According to the Japanese historical books *Kojiki* and *Nihon Shoki*, the "naturalized people" of Han nationality from Korean Peninsula spread Chinese culture and agriculture technology to Japan. Chinese characters, Buddhism, and Chinese classics such as The Analects, Book of Poetry, Book of Documents, Book of Change, The Rites of Zhou, Spring and Autumn Annals, and Thousand Character Classic were introduced into Japan at that time. At the beginning of the sixth century, Japan recruited Doctor of classics Duan Yang'er from Baekje to teach these classics. Three years later, Dr. Han Gao An Mao was also recruited. With the introduction of Chinese characters, China's number system, simple arithmetic knowledge, and measurement system were systematically introduced to Japan during this period. China's astronomical calendar system was also used for the purpose of agriculture production. According to *Nihon Shoki*, in the 15th year of the emperor's reign (554), Doctor of calendar, Gu De, or Wang Baosun and Doctor of Yi, Shi De, or Wang Daoliang and Doctor of medicine, Nai Shuai Wang Youling tuo, were invited to Japan. In 602, Guanle from Baekje also went to Japan with books on calendar, astronomy, geography, magic skills, and alchemy. It can be seen that Chinese mathematics, along with Confucianism, astronomy, and calendars, was introduced to Japan indirectly from mainland immigrants through the Korean Peninsula.

From the fourth to the seventh century, Japan entered the Kofun period with slavery system or the era of "Yamato period" in Japanese history when the Japanese Imperial court ruled from "Yamato Province." In 592, Empress Suiko ascended the throne, entering Asuka period (592–710), followed by Nara period (710–794). For nearly 400 years from 794 to 1192, Minamoto no Yoritomo established Kamakura shogunate, historically known as the Heian period. The seventh to the ninth century is an important period when Chinese mathematics systematically spread to Japan on a large scale.

Prince Shōtoku, appointed by Empress Suiko as the regent, carried out the reform at the beginning of the seventh century, sending envoys to China for four times (in 600, 607, 608, and 614) to learn Chinese culture. Emperor Kōtoku succeeded in 645 and rolled out "Taika Reforms," implementing centralized land-equalization system and Zu Yong Diao taxation system like the Tang Dynasty of China, and their clothing, dressing, and cultural relics were also modeled in China. For 264 years from 630 to 894, 19 groups of envoys were dispatched to China successively. During this period, Chinese culture was systematically adapted to Japan. In 710, Thousand Character Classic, The Analects, Erya (the first surviving Chinese dictionary), and Qimin Yaoshu (Essential Techniques for the Welfare of the People) were introduced to Japan. In 717, Abe no Nakamaro and Kibi no Makibi came to Chang'an of China with the Japanese mission of 557 people, learning astronomy, mathematics, music, and calligraphy and taught those knowledge in Japan when they came back. Several books taught by Kibi no Makibi include Taien calendar, *The Nine Chapters on the Mathematical Art*, *Zhoubi Suanjing* (*The Arithmetical Classic of the Gnomon and the Circular Paths of Heaven*), Ding Tian Lun, Records of the Grand Historian-Tiangongshu, Book of Han-Astronomy, and Book of the Jin-Astronomy. He also brought back to Japan 130 volumes of Tang Li (the ritual of Tang), 1 volume of Taien calendar, 12 volumes of Dayanli Licheng, 10 volumes of Yueshu yaolu (Important records about writings on music), and some astronomical instruments.

During the Taika Reforms, Japan issued gakuryou, or orders of education, based on the imperial examination and education system of the Tang dynasty. According to the orders Yōrō Code issued in 718 and Yōrō Code-explanation in 733, Japan imitated the arithmetic system of Tang, setting 2 doctors of calculation and enrolled 30 students in the imperial university. Their mathematics textbooks included *The Nine Chapters on the Mathematical Art*, *The Arithmetical Classic of the Gnomon and the Circular Paths of Heaven*, *Haidao Suanjing* (*Sea Island Mathematical Manual*), *Sunzi Suanjing* (*Master Sun's Mathematical Manual*), *Wucao Suanjing* (*Mathematical Manual of the Five Government Departments*), *Zhui Shu* (*Method of Interpolation*), *Chong Cha* (*Commentary on Double Differences*), *Liu Zhang* (six chapters), *San Kai*, and *Jiu Si*. The first seven books were from China, and the last three books were not recorded in Chinese historical literature. It could be Chinese calculation books written by Korean people and introduced to Japan.

Yōrō Code was drafted based on Taihō Code issued by Emperor Wenwu in the second year of Dabao (702), indicating that the arithmetic official system of the Sui and Tang Dynasties and Chinese mathematics books were introduced to Japan before 702. Apart from these mathematics books mentioned above, there were many others introduced to Japan in Sui and Tang Dynasties. In the era of Kampyō (889–897), Fujiwara no Sukeyo (?–897) compiled List of Writings Currently Held in the Nation of Japan at the imperial order, which records Chinese classics in Japan at that time, including 85 astronomical books with 461 volumes and 54 calendar books with 167 volumes. Among them, mathematics books include:

Nine Chapters (9 volumes, commented by Liu Hui), Nine Chapters (9 volumes, commented by Zu Chongzhi), Nine Chapters (9 volumes, written by Xu), Nine Chapters and Meanings of the Methods (commented by Zu Chongzhi), Nine Chapters and Their Meanings, Nine Chapters and Their Figures, Nine Chapters and Records of Multiplication and Division, Nine Methods of Arithmetic, Six Chapters (6 volumes, written by Gao), Six chapters and Their Figures, Six Chapters and Personal Records, Nine Divisions, Nine Divisions and Arithmetic, San Kai, Figures of San Kai, Sea Island Mathematical Manual, Sea Island Mathematical Manual (commented by Xu), Sea Island Mathematical Manual (commented by Zu Chongzhi), Figures of Sea Island Mathematical Manual, Method of Interpolation, Xiahou Yang's Mathematical Manual, New Collection of Mathematical Cases, Wu Jing Suan, Zhang Qiujian's Mathematical Manual, Yuan Jia Mathematics, Sun Zi's Mathematical Manual, Wucao Suanjing (Mathematical Manual of the Five Government Departments) (written by Zhen Luan), Yao Yong Suan Li, Brahman Yin and Yang Calculation, Memoir on some Traditions of Mathematical Art, Arithmetic in Five Classics, etc.

Historical documents in Japan of this period recorded some information about the positions of doctors of mathematics and Taoist. In the fifth year of Yanxi, during Emperor Daigo's reign (905), Fujiwara no Tokihira and others began to compile *Engishiki* (*Procedures of the Engi Era*) (927) as requested by the emperor, which recorded all kinds of rules and regulations at that time. Most of the rules remained the same as Yōrō Code. The regulations of Daigaku-ryō, or the imperial university of Japan, were recorded in volume 20:

For teachers, "*Li Ji* (*The Book of Rites*)" and "*Zuo Zhuan* (*Commentary of Zuo*)" should be taught in 770 days. "*Zhou Li* (*Rites of the Zhou*)," "*Yi Li* (*Etiquette and Rites*)," "*Mao Shi* (Mao's (version of the) Book of Songs)," and "Lv (law)" are to be taught in 480 days for each book, "*Zhou Yi* (the book of changes)" for 310 days, "*Shang Shu* (*Book of Documents*)" and "*The Analects of Confucius*" and "Ling (order)" for 200 days respectively, and "*Xiao Jing* (*The Book on Filial Piety*)" for 60 days. Major classics include "San Shi (Three history)" and "*Wen Xuan* (*Selected Literature*)"; minor classics include "*Gong Yang* (commentary on the Confucian Classic Chunqiu), "*Gu Liang* (commentary on the Confucian Classic Chunqiu)," "*Sun Zi* (The Methods of War by Master Sun)," "*Wucao Suanjing* (*Mathematical Manual of the Five Government Departments*)," "*Jiu Zhang* (*Nine Chapters*)," "*Liu Zhang* (*Six Chapters*)" and "*Zhui Shu* (*Method of Interpolation*)," "*San Kai*," "*Chong Cha* (*Commentary on Double Differences*)" and "*Zhou Bi* (*The Arithmetical Classic of the Gnomon and the Circular Paths of Heaven*)," "*Hai Dao* (*Sea Island Mathematical Manual*)," and "*Jiu Si*."

For the graduates, there will be four majoring in classicist, two in literature, two in law, and two in arithmetic (Taoist). They will be given gowns for summer and winter, and the Central Secretariat will give them cloth, rough silk, winter cloth, and cotton in winter and summer.

The graduates majoring in classicist should be able to give lectures on "San Jing (Three classics)" and San Shi (Three history)," and for law students they should be able to teach "law and orders"; for arithmetic students they should be able to teach "Treatise on the calendar of the Han and Jin Dynasty," "Da Yan Calendar," "Nine Chapters," "Six Chapters," "Zhou Bi," and "Ding Tian Lun (Theory of the heaven)."

Ruijū fusen shou and volume two in Erzhong calendar in the late Kamakura period, "Confucian official calendar," also recorded information about arithmetic, and the latter one also documented names of 37 doctors of mathematics from the Yanxi period during the reign of emperor Daigo (901–922) to the Kamakura period (1185–1333). There are also records of graduates in Ryō no Shūge:

There are ten graduates, four studying classicist, two in literature, two in law, and two in arithmetic. For those who are intelligent and excellent in their study, they were given winter and summer cloth, cottons, rough silk, everyday rice, fish, seaweed, small fish, and salt.

Although the systems of Onmyōdō (Yin and Yang) and sando were established, Japan still use Chinese calendar directly, because people who were in charge of Onmyōdō and sando in the court were hereditary families with limited science knowledge, so they were unable to observe and compile the calendar independently. According to Japanese historical records, before Jōkyō calendar (Shibukawa Harumi,1639-1715) was adopted (1685), the Chinese calendars issued by Japan include Yuanjia calendar, Yifeng calendar, Dayan calendar, Wuji calendar, and Xuanming calendar. With the introduction of the system of arithmetic, law and regulation, and astronomical calendar, Japanese mathematics has incorporated Chinese mathematics in its culture.

Due to the political chaos in Japan between the tenth century and the fifteenth century, the law and regulation system of Ritsuryō was declined and changed. Japan's diplomatic relations with China began to weaken, and the economic and cultural exchanges of its people gradually became the mainstream. Kuge (aristocratic class) culture in the Ritsuryō era went through many changes as buke (warriors' house) rise in the Heian and Kamakura period. The imperial court was no longer able to maintain the education system similar to the Tang Dynasty. Mathematics, astronomy, and Onmyōdō completely transformed from imperial education to family education. Doctors of mathematics and calendar were controlled and inherited by several families. Such system of doctor of mathematics was retained to Muromachi period (1392–1573). Up to now, we have not found any mathematics works in the Heian period. Some literature works contains arithmetic-related knowledge. The numbers in the lines of *Man'yōshū* (*Collection of Ten Thousand Leaves*) (around 759) were often in nine-nine song. Kou You and Shi Jie Chao also recorded nine-nine table, the content of which was exactly like China, beginning with "nine-nine eighty-one." Kou You also documented some mathematical games originated from China, including "bamboo problem," "pregnant woman problem," "patient life and death problem," etc.

By the end of Kamakura period, *The Nine Chapters on the Mathematical Art* of China was still popular in Japanese society. Monk Chūgan Engetsu wrote in his autobiography of Volume five of Toukaiichiousyu (Literature collection of the East China Sea) that "he read Classic of Filial Piety and the Analects with monk Daohui in the spring, and also learned Nine Chapters on the Mathematical Method before went back to Daci Temple in the autumn." It describes his experience when he was 12 years old in the first year of Yingchang (1311). The Nine Chapters on the Mathematical Method mentioned by him should be *The Nine Chapters on the Mathematical Art*. The Muromachi period (1338–1573) left us with no mathematics books. Some historical documents in this period recorded some mathematics games, which later appeared in the early mathematics works of the Edo period. For example, Rhymeprose on a Miniature Landscape Garden written by Kokan Shiren recorded the following popular mathematical games at that time: Shibuzu, Baiwujian, Daorenyin, Langdengda, Jizili, Shizhua, Rujin, Yoajin, Chongdan, Xiaotongfu, Poluomenshuangliu, Yijuli, Daoli, Zuozuoli, Youzaili, etc. Arithmetic books of Japanese mathematics in the early years of the Edo period also contained mathematical games, such as Jizili, Baiwujian, Fangzhen, Mufuzi , Daorenyin, and Zuozuoli, which can be found in the literature of the Muromachi period. It's not difficult to tell that some of the games are originated from China.

Due to the prohibition of Western Christian culture in the Edo period (1603–1867), Tokugawa bakufu cut the country from the outside world. Han culture was the only foreign culture introduced to Japan. From the second half of the sixteenth century to the first half of the seventeenth century, the mathematics books of the Yuan Dynasty and the Ming Dynasty were mainly brought to Japan through sea smuggling and trade and through its invasion to Korea. The abacus of the Ming Dynasty was also introduced into Japan through the sea trade around 1570. Tengaku shokan and many other Chinese translations of Western astronomy and mathematics could be also introduced to Japan, but due to the lack of evidence and information further study is required. Japan obtained a large number of Chinese books through sea trades or indirectly obtained a large number of Chinese books through North Korea during Toyotomi (no) Hideyoshi's invasion of North Korea. At least the following Chinese mathematics books were introduced to Japan in this period according to the study:

(1) *Introduction to Computational Studies* (Zhu Shijie, 1299). The time when the book was introduced to Japan is unknown. In 1658, Hisada Gentetsu reprinted it in Japan. In 1672, Hoshino Sanenobu (1638?–1699) wrote three volumes of Annotation of Introduction to Computational Studies. In 1690, Takebe Katahiro (1664–1739) wrote seven volumes of Explanations of Introduction to Computational Studies, which left great influence on Japanese mathematics.
(2) Yang Hui Suan Fa (Yang Hui's Methods of Computation) (Yang Hui, 1378). The time when the book was introduced to Japan is unknown. Ishikuro Nobuyoshi (1760–1836) in the late Bakumatsu period included Yang Hui Suan Fa compiled in the Song Dynasty (3 volumes, compiled by Yang Hui) in his book the Catalogue of Computation and mentioned that at the end of this

book it indicated that the book "was copied by Seki Takakazu (?~1708)." According to this catalogue, Yoshio Mikami found the copy of Ishikuro Nobuyoshi based on the Seki Takakazu's copy of Yang Hui Suan Fa. Seki Takakazu copied the book based on a manuscript from Korea in 1661. The University of Tsukuba, libraries of the Ministry of the Imperial Household, and the Cabinet of Japan still each preserved one copy of the Korean version of Yang Hui Suan Fa.

(3) Suan Fa Tong Zong (Systematic Treatise on Arithmetic) (Cheng Dawei, 1593). The time when the book was introduced to Japan is unknown. Yoshida Mitsuyoshi (1598–1673) wrote Jinkōki based on this book in 1627. Yuasa Tokushi added Kanbun and published it in the name of New compiled Suan Fa Tong Zong with Kanbun (12 volumes), which has huge influence on Japanese mathematics in the early Edo period.
(4) When the book Shu Xue Tong Gui (Rules of Mathematics) (Ke Shangqian, 1578) was introduced to Japan is unknown. It was copied in Japan in 1672. The Chinese version and its Japanese copy were preserved in Sonkeikaku Bunko Library (former Maeda Library). Another copy has been preserved in Jingu Bunko. In 1653, Shimada Sadatsugu cited this book in his Nine Numbers Computation. Miyahi Kiyoyuki also referred to this book many times in his Japanese computation method published in 1695.
(5) The Detailed Account of Computation (Li Changmao, 1659). The time when the book was introduced to Japan is unknown. Yuasa Tokushi mentioned this book in New compiled Suan Fa Tong Zong with Kanbun published in 1675. The book is also mentioned in the Chronicle records of mathematics. The book now is in the Library of the Cabinet national archives.
(6) The author and the compilation time of Suan Xue Qun Qi are not clear. The time when the book was introduced to Japan is also unknown. Yuasa Tokushi mentioned this book in his New compiled Suan Fa Tong Zong with Kanbun. Hirayama Chisato also mentioned the rhymes of this book in volume 5 of his Suan Shu (Computation) in 1789.
(7) Tongling Nine Chapters Shortcut Computation (the author and the time the book was written are unknown). The time when the book was introduced to Japan is unknown. In 1673, the preface of Murase Yoshimasu's Sanpoufutsutan-kai wrote that: "Tongling nine chapters Shortcut computation, Introduction to Computational Studies and Systematic Treatise on Arithmetic are books of different dynasty. Even for researchers and inventors, they can't be read without literary talents." Takebe Katahiro mentioned the rhymes of this book in Explanations of Introduction to Computational Studies. Seki Takakazu recorded the method of Tongling 63/20 on calculating pi in the volume 4 of his Katsuyo Sanpō, which is now preserved in the library of Tohoku University.
(8) Method of Calculating on an Abacus revised by Xu Xinlu in 1573 is preserved in the Library of the Cabinet.
(9) New Methods of Mathematics (Zhu Zaiyu, 1603) was an attached volume of Complete Description of Musical Tone. The time when the book was

introduced to Japan is unknown, and it is now preserved in Sonkeikaku Bunko Library.

(10) One volume of the Xinjuan Jiulong Methods of Calculating (the author and the time the book was written are unknown), the Lianjietang edition, was preserved in Asakusa Library (now transferred to the Library of the Cabinet). The time when the book was introduced to Japan is unknown.

(11) Xijuan Rectified Methods of Calculating (the author is unknown or could be written by Xia Yuanze, 1439?). The time when the book was introduced to Japan is unknown. The copies of Jixintang of Fuzhou are preserved in the library of Tohoku University, Kokura Library of Waseda University, and the headquarters of the National Diet Library.

In addition to the mathematics book, calendar books such as Shoushi calendar were also introduced to Japan at this time. With the development of Kuge (aristocratic class) and Chōnin culture, Japanese in Edo era had great enthusiasm to study mathematics and astronomy. Japanese mathematics developed independently based on Chinese mathematics, of which Suan Fa Tong Zong (Systematic Treatise on Arithmetic), Introduction to Computational Studies and Yang Hui Suan Fa (Yang Hui's Methods of Computation), and Shoushi calendar have the greatest influence on Japanese traditional mathematics. The book of Warizan written by Mori Shigetoshi and Jinkōki written by Yoshida Mitsuyoshi and many other Japanese mathematic books were based on Suan Fa Tong Zong. After Introduction to Computational Studies was copied in Japan, many important algebra achievements such as Tianyuan segment technique, stacking plot and moving difference technique, square cutting technique, pipe cutting technique, and vertical and horizontal diagram were accepted by Japanese mathematicians. With the spread of Introduction to Computational Studies, Yang Hui Suan Fa, and Shoushi calendar in Japan, the mathematical knowledge and methods of the Song and Yuan Dynasties achieved huge progress in Japan and become the most developed mathematics in East Asia.

After Tokugawa Yoshimune announced to lift the ban in 1726, many Chinese translations of Western books related to astronomical calendar were introduced to Japan gradually, including Calendar book of Chongzhen, Shu li jing yun (Collected Basic Principles of Mathematics), Li Xiang Kao Cheng (Study of Calendar calculation), Mei's Complete Description of Calendar Calculation, Elements of Algebra, Algebra, etc. contributing greatly to the promotion of western math in the Japanese community.

11.3 Traditional Chinese Mathematics in Vietnam

Vietnam was called as Jiaozhi in ancient China. It was set as the Xiang county in the Qin Dynasty. Three couties, Jiaozhi, Jiuzhen, and Rinan, were set in the north part of Vietnam in 111 BC during the reign of Emperor Wu in the Han Dynasty. In the eighth year of Jian'an (203), Emperor Xian of the Han Dynasty changed Jiaozhi county to Jiaozhou. In the first year of Taolu during Emperor Gaozong's reign in the

Tang Dynasty (679), Jiaozhou was changed to An'nan Prefecture. In the first year of Kaibao in the Northern Song Dynasty (968), Vietnam broke away from the rule of China and established "Daqu Yue State (Đại Cồ Việt)." In June of the 5th year of Yongle (1407) in the Ming Dynasty, the Ming army entered Vietnam, defeated the Li regime of An'nan, and changed An'nan into Jiaozhi, dividing it into 15 prefectures, 36 cities, and 181 counties, setting up 3 divisions to take charge of its politics. Another 5 prefectures and 29 counties were under the direct control of Chief Secretary. An'nan was then included in the territory of the Ming Dynasty. After five years of suppressing, Vietnam finally was controlled by the Ming Dynasty (1407–1427). In the second year of Xuande (1427), Chen Hao was crowned as "King An'nan" by the Ming court, and Vietnam regained its independence. After that, Vietnam maintained the suzerain vassal relationship with China until ruled by the French colonial in the nineteenth century.

When Vietnam was a part of China, its traditional science and technology culture naturally was the same as Chinese traditional science and technology culture. After its independence, Vietnam followed China to set up its laws and regulations and develop its own culture and science and technology. Therefore, in the Sinosphere, Vietnamese culture was most influenced by Chinese culture. Vietnam imitated China's imperial examination system, including setting mathematic examinations to select talents in the past dynasties. In the fourth year of Zhengfu during Emperor Li Gaozong' s reign (1179), the examination included mathematic examinations. In the fourth year of Shaolong during Emperor Chen Shengzong' s reign (1261), officials were selected through written mathematic examinations. In the second year of Kaida during Emperor Hu Hancang's reign, county examinations were conducted, with the first four examinations testing literature and the last one for calculation. In the fourth year of Shaoping during Emperor Li Taizong's reign (1437), 690 people were selected through written mathematic examination as officials in the court. In the eighth year of Hongde during Emperor Li Shengzong' s reign (1472), officials were tested in forms of written mathematic test. In the second year of Duanqing during Emperor Liweimu's reign (1505), more than 30,000 candidates were tested through written mathematic tests, and among them 1,590 were selected, including Ruan Ziqi. It's not clear whether Vietnam set doctor of mathematics or "ten books of calculation" were appointed as the guide book for examination. It's recorded that Wuyou, a Vietnamese mathematician in the fifteenth century, wrote Dacheng computation based on the Chinese mathematics book, which was later republished by Liang Shirong.

Around the Ming and Qing Dynasties, Chinese abacus was introduced into Vietnam, which is still widely used in Vietnam. Suan Fa Tong Zong (Systematic Treatise on Arithmetic) of Cheng Dawei republished in China in 1716 (55th year of Kangxi in the Qing Dynasty) was widely spread in East Asia including China and Vietnam, which greatly affected the development of abacus in Vietnam. In the 52nd year of Kangxi (1713), the Qing Dynasty set up a school of mathematics to send descendants of privileged families under the banner system to study mathematics. There was a small test for every season and a big test at the end of the year, studying for 5 years. In 1761, Zheng Ying, the general of the government of the An'nan Li

Dynasty, ordered the government to follow the practice of the Qing Dynasty in China, holding math tests of equal division and difference division every 12 years and select 120 people every time.

Due to the havoc of wars and the plunder of colonists in history, the existing ancient books in Vietnam are less than that of other countries in the Sinosphere. Ancient books of mathematics preserved are even fewer. The Institute of Natural Science History of the Chinese Academy of Sciences has collected part of the eighteenth- and nineteenth-century Vietnamese mathematics books, all of which are copies of Mr. Li Yan's eight selective copies, including Suan Shu Di Yun (the Inside of Mathematics Book) (writer unknown); Suan Fa Da Cheng (To Achieve Success in Mathematics) (by Liang Shirong, fifteenth century); Yi Zhai Suan Fa Yi De Lu (by Ruan Youshen, 1829); Shortcut to *The Nine Chapters on the Mathematical Art* (writer and time unknown); Da Cheng Suan Xue Zhi Ming (by Fan Jiaji, time unknown); Zhi Ming Li Cheng Suan Fa (by Pan Huikuang,1820); Li Cheng Suan Fa (by Fan You Zhong, 1705–1719); and Bi Suan Zhi Nan (The Guide to Written Computation) (by Ruan Jin, 1909). In addition, Vietnam's popular arithmetical works include The Guide to Written Computation by Fan Wenyu (time unknown), the wonder of Mathematics (writer and time unknown), Specifying Computation (writer and time unknown), Zong Ju Zhu Jia Suan Fa Da Quan (writer and time unknown), Yi Zhai Suan Fa (by Ruan Youshen), and Jiu Zhang Li Cheng Suan Fa (by Fan Youzhong). These Vietnamese calculation books inherit the traditional Chinese mathematic knowledge in *The Nine chapters on the Mathematical Art* and the Systematic Treatment on Arithmetic.

Chinese computational knowledge also spread in Vietnam along with astronomy and calendar studies. Shang Shu-Yao Dian (Book of Documents-Canon of Yao) records that "He further commanded the third brother Xi to reside at Nan-jiao, (in what was called the Brilliant Capital). to adjust and arrange the transformations of the summer, and respectfully-to observe the exact limit (of the shadow)." It indicates that at the time of Yao and Shun, China's astronomical measurement was already conducted in Vietnam. Therefore, during the period when Vietnam was a part of China, China not only issued the order but also carried out astronomical calculation in Vietnam. For example, the national measurement of land area in the Tang Dynasty reached An'nan.

From the beginning of the Northern Song Dynasty to the early fifteenth century, Vietnam was still an independent vassal state that presented tribute to the emperor of China; China in turn bestowed calendar system to Vietnam. Such system lasted till the Qing Dynasty. In the second year of Yuantong during Huizong's reign in the Yuan Dynasty (1334), the Minister of the Ministry of Appointments was sent to Vietnam, and the officials of the ministry of rites Zhi Xishan was sent as an envoy to Vietnam and bestowed Shoushi calendar to the Chen Dynasty of Vietnam. In the 11th year of Kaiyou (1339) of the Chen Dynasty, Xieji calendar was compiled and promulgated based on Shoushi calendar. Hu Jiyu usurped the Chen Dynasty and abolished the Xieji calendar in 1401 and issued Shuntian calendar. During the Yongle period of the Ming Dynasty, An'nan was governed by China, and Datong calendar was practiced. After its independence, Wanquan calendar was compiled according to Datong calendar.

In September 1790, the 55th year of the Qianlong's reign of the Qing Dynasty, Ruan Guangping earnestly requested the issue of the calendar. Emperor Gaozong of the Qing Dynasty decided to send 20 copies of Shixian calendar of that year to An'nan to follow the time. In 1809, the eighth year of the Ruan Dynasty of Vietnam, Ruan Youshun, the envoy of Vietnam, came to Beijing and bought a copy of *Li Xiang Kao Cheng* (*Study of Calendar Calculation*). After he came back, he asked the emperor to issue the Xieji calendar according to this book.

In 1810 (the 15th year of Jiaqing of the Qing Dynasty), *Li Xiang Kao Cheng* (*Study of Calendar Calculation*) (1713) was introduced to Vietnam. People of An'nan changed the Wanquan calendar into Xieji calendar according to calculation methods of *Li Xiang Kao Cheng*.

The settings and appointments of the Board of Astronomy were basically the same as the Qing Dynasty. In 1809, Ruan Fuying appointed Hou Dengde from the Ministry of Rites to be in charge of astronomic calculation and 12 other people including Ruan Yulin as officials of astronomic calculation. In 1813, Zheng Huaide, a high official from the Minister of Rites, was appointed to be in charge of the Board of Astronomy. Apart from Li Xiang Kao Cheng, there were many Chinese astronomical computation works of the Qing Dynasty preserved in the Board of Astronomy, such as Zhi Zhi Yuan Zhen (Interpretation of the Originality), Yue Ling Cui Bian (Collection of Climate and phenology in a lunar month), Qin Ding Yi Xiang Kao Cheng, Gao Hou Meng Qiu, Guan Kui Ji Yao, Shu li jing yun (Collected Basic Principles of Mathematics), Xin Zhi Ling Tai Yi Xiang Zhi, Wu Lei Mi Qiao, Wu Li Xiao Shi, Ge Zhi Jing Yuan, and Di Qiu Shuo Shu. Most of the astronomical instruments used by the Board of Astronomy followed Western standards, and they were basically bestowed by the Qing Dynasty.

Zhu Zaiyu and the Equal Temperament

12

Fei Xu

Contents

Abstract

Zhu Zaiyu's remarkable encyclopedic contributions in many fields of science and art were rare in ancient China and even in the history of world civilization. He solved the theoretical problem of transition and modulation that had plagued the music field for more than 2000 years. He created the mathematical formula of the equal temperament he called "the new law for density rate."

F. Xu (✉)
Department of Philosophy of Science and Technology, University of Science and Technology of China, Hefei, China
e-mail: xufei@ustc.edu.cn

X. Jiang (ed.), *The High Tide of Science and Technology Development in China*, History of Science and Technology in China,
https://doi.org/10.1007/978-981-15-7847-2_12

Keywords

Zhu Zaiyu · The equal temperament · Arithmetic reasoning · Algorithm analysis · Pitch pipe · Chinese musicology

Zhu Zaiyu was an outstanding scientist and artist in the Ming Dynasty in China. His remarkable encyclopedic contributions in many fields of science and art were rare in ancient China and even in the history of world civilization. One of his most significant contributions in the history of science can be summed up like this: for the first time, he solved the theoretical problem of transition and modulation that had plagued the music field for more than 2000 years. He created the mathematical formula of the equal temperament he called "the new law for density rate." He offered the musical score that could be transited and modulated freely. It is particularly worth mentioning that Zhu Zaiyu also experimented and made physical tuning instruments that strictly conform to the equal temperament – New Chun (newly made tuning instrument) and pitch pipes of different diameters which could achieve the new tuning criteria of the equal temperament. The new tuning instrument had been made more than a century earlier than the similar musical instrument piano in the West. The original pitch pipes of different diameters he made helped to solve the pitch pipe diameter correction for the tuning of the equal temperament once and for all, thus representing the most tremendous achievements in the development of the field of experimental acoustics. Zhu Zaiyu's contributions involved many fields of art, such as temperament, music, and dance, as well as many fields of natural science such as physical acoustics, astronomical calculation, history of measuring systems, and mathematical abacus. The scientific miracle he had created exerted tremendous influence to such a degree that "all could be regarded as breakthroughs, in other words, they were done in a way of reaching the peak of perfection." "No one could have ever overthrown it and refuted it until now; the algorithm he used has been stuck to till now, the figures of his calculation are still used today without having to worry about the validity of the figures." (Liu Fu. Zhu Zaiyu: The Discoverer of the Equal Temperament (十二等律的发明者朱载堉) Collection of Papers to Celebrate the 65th Birthday of Mr. Cai Yuanpei (《庆祝蔡元培先生六十五岁论文集》)(Vol.1):pp.279–310. The first edition of the external edition of the collection of the Institute of History and Language of Academia Sinica, 1933. All the quotes made by Liu Fu are retrieved from the article. Hereinafter, no separate notes will be made according to such quotations). Dr. Joseph Needham, the famous historian of science, believed that Zhu Zaiyu's contributions "may justly be regarded as the crowning achievement of China's two millennia of acoustic experiment and research." (Dai Nianzu, Zhu Zaiyu-The Super Star of Science and Arts in the Ming Dynasty (《朱载堉——明代的科学和艺术巨星》). The People's Publishing House: pp.303, ed. 2011. Tian Huang Zhen Ren Zhu Zaiyu (《天潢真人朱载堉》) Elephant Press. April 2008, the first edition).

In the past hundred years, based on many studies done by many scholars such as Liu Fu, Yang Yinliu, Joseph Needham, Chen Wannai, Cheng Zhenyi, Dai Nianzu, etc., we can now confirm that the theory and practice of Zhu Zaiyu's creation of the equal temperament is one of the great contributions made in ancient China to the world civilization.

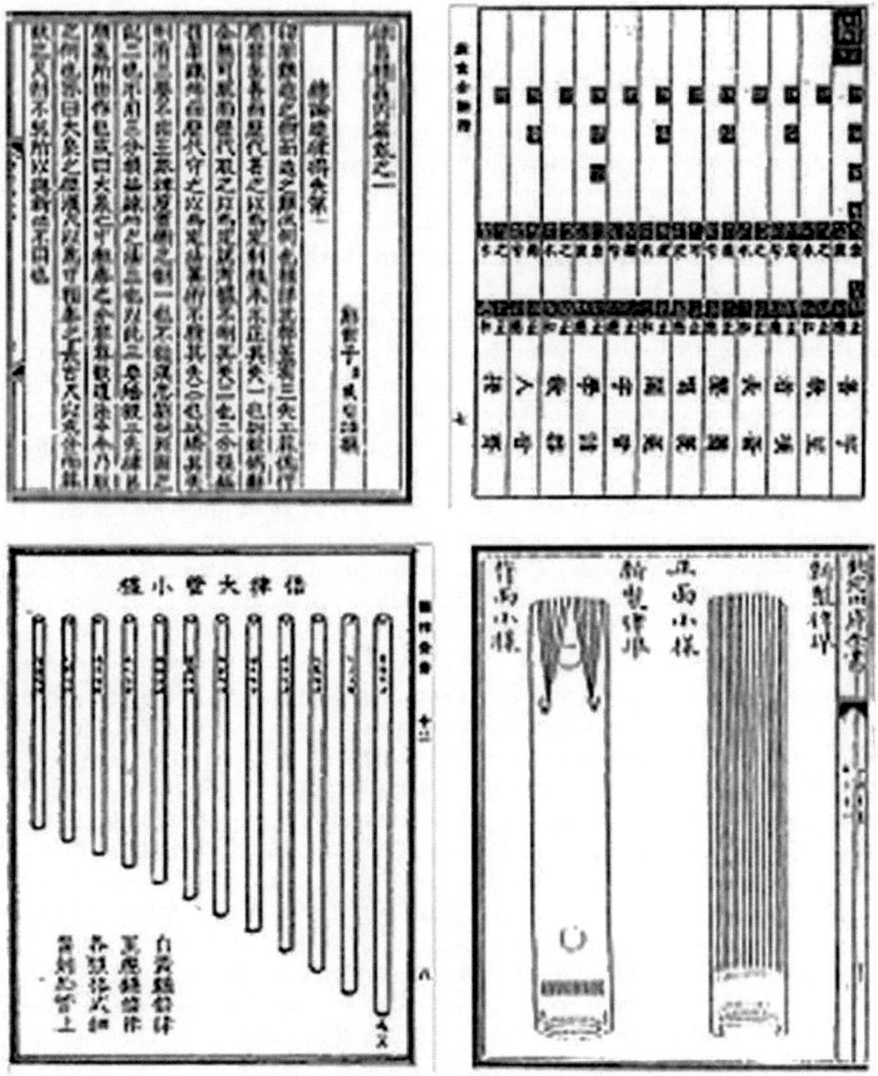

Picture 12.1 Brief of achievements of Zhu Zaiyu's equal temperament

On top of that, through further research, it is not difficult to find that this contribution not only contains the calculation data of the equal temperament that Zhu Zaiyu had left us with which were accurate with more than 20 significant digits but also lies in his academic analysis of theories and calculation formulas, as well as complete experimental verification with proper theoretical guidance. It can be said that Zhu Zaiyu's pitch pipes of different diameters have not only been a great creation in the field of musicology but also the highest achievement in the field of ancient Chinese physical acoustics (Picture 12.1).

Photograph of *Yue Lv Quan Shu* (《乐律全书》) *The Complete Book of the Tuning System*

Photograph of Xuan Gong He Yue Pu (《旋宫合乐谱》) Transition and Modulation Scores

The pitch pipes of different diameters following the equal temperament

The new tuning system following the equal temperament

12.1 Brief of Zhu Zaiyu's Equal Temperament

Zhu Zaiyu (1536–1611), with a style name of Boqin, was an encyclopedic scholar and scientist in the Ming Dynasty in China. His outstanding contribution included the exact description of the equal temperament which he called the "new law for density rate," the production of the earliest physical tuner in the world that conformed to the equal temperament – setting the criteria for chord alignment and tuning; especially in the manufacturing process of tuning instrument, he adopted the pitch pipes of different diameters and originally created the unified mouth correction which solved the physical problem of the pitch pipe mouth correction once and for all and ended the history that the bamboo pipes had not been tuned since ancient times. (《后汉书·律历志》, 中华书局校点本, 第11册, 第3000页。History of the Later Han Dynasty Musicology and Calendrical Science) (《后汉书·律历志》) Vol. 11, P.3000. China Book Company). Since then, people have finally found a tuning method that can arbitrarily achieve the transition and modulation between the twelve semitones. This method is still in use today, which had helped bring a rich and colorful musical life to the mankind. In addition, Zhu Zaiyu had made many achievements in several fields of science and art, such as astronomy, physics, mathematics, and dance. Unfortunately, for a long time after Zhu Zaiyu, his great invention had hardly been recognized. Zhu Zaiyu's new law for density rate and the law of pitch pipes of different diameters were not mentioned in *Ming Shi* (《明史》) *The History of the Ming Dynasty*. In the 120-volume *Yu Zhi Lv Lv Zheng Yi Hou Bian* (《御制律吕正义后编》) "*Sequel to the Imperial Pitch-Pipes Annotations*" compiled in the eleventh year during the reign of Emperor Qianlong in the Qing Dynasty, Zhu Zaiyu's new law for density rate was even denounced as a wild speculation made by Zhu Zaiyu. In the *executive summary of Si Ku Quan Shu* (《四库全书》) *The Complete Works of Chinese Literature*, Zhu Zaiyu was accused because his methods of calculating the equal temperature by adopting the Pythagorean theorem were too farfetched to be trusted.

In the second half of the eighteenth century emerged Jiang Yong (1681 ~ 1762), the first scholar who fully accepted Zhu Zaiyu's equal temperament theory in China. Since then, although there have been many people who studied Zhu Zaiyu among scholars at home and abroad, they still had mixed opinions. Until 1933, Liu Fu (1891–1934) wrote an important article to commemorate the 65th birthday of educator Mr. Cai Yuanpei (1868 ~ 1940), which was entitled "Zhu Zaiyu: the Discoverer of the Equal Temperament." The article was historic and had made people more generally come to understand that such a great invention as equal temperament originally came from ancient China. This article by Liu Fu had been of great significance. Many of its contents and ideas have been quoted repeatedly by later scholars. Miao Tianrui

(1908 ~ 2009), a famous scholar of musicology, praised Liu Fu's article as it "has ended the silence in the field of musicology and played a certain role in introducing and popularizing knowledge and achievements in the ancient musicology." (Miao Tianrui. In Honoring *Lv Xue Xin Shuo* (《律学新说》) *A New Account of the Science of the Pitch-pipes by Zhu Zaiyu on its 400th Anniversary. November 1984* Commemorative Meeting of Zhu Zaiyu's Works and Papers and the Symposium on Musicological Studies). Dai Nianzu (1942-), the contemporary scholar specializing in the studies of Zhu Zaiyu, said, "since then, the musicological theories of Zhu Zaiyu has gradually restored its position in the history of science, culture and musicology which should have been recognized overdue for a long time. (Dai Nianzu, Zhu Zaiyu-The Super Star of Science and Arts in the Ming Dynasty (《朱载堉——明代的科学和艺术巨星》). The People's Publishing House: pp.125, ed. 2011. Tian Huang Zhen Ren Zhu Zaiyu, (《天潢真人朱载堉》) Elephant Press. April 2008, the first edition).

In fact, the study of the history of modern Chinese music also started late, one of the more representative of which was the book entitled *History of Chinese Music* published by Wang Guangqi (1892–1936) in 1934. The book discussed more issues of musicology and offered a concise introduction to Zhu Zaiyu's equal temperament. Wang Guangqi paid special attention to Zhu Zaiyu's pitch pipes of different diameters and believed that "whether such an algorithm is reasonable cannot be evaluated without physical experiments." (Wang Guangqi. History of Chinese Music (《中国音乐史》), first edition in 1934, reprinted by Music Press in 1957, p. 94). Since then, Yang Yinliu (1899 ~ 1984), a musicologist, published *Calculation of the Equal Temperament* (《平均律算解》) in the 21st issue of the *Yanjing Journal* (《燕京学报》) in 1937 and began an in-depth study of Zhu Zaiyu's achievements in musicology. In 1952, Yang Yinliu published another book entitled *History of Chinese Music*, which gave him a new understanding of Zhu Zaiyu's new law for the density rate and the pitch pipes of different diameters. Yang believed that there were minor errors in Zhu Zaiyu's pitch pipes of different diameters. Therefore, Yang Yinliu proposed an amendment, which was widely circulated and accepted or cited by quite a few scholars.

Among foreign scholars, some scholars had noticed Zhu Zaiyu's work since the last century. The famous physicist Hermann von Helmholtz (1821–1894) and Hugo Riemann (1849–1919), a German music theorist, had mixed comments regarding Zhu Zaiyu and his works with proper and mistaken conclusions due to language and cultural barriers. They even regarded Zhu Zaiyu as a mythical Fig. 1500 years before Jesus' birth. (Dai Nianzu, Zhu Zaiyu-The Super Star of Science and Arts in the Ming Dynasty (《朱载堉——明代的科学和艺术巨星》). The People's Publishing House: pp.138, ed. 2011. Tian Huang Zhen Ren Zhu Zaiyu, (《天潢真人朱载堉》) Elephant Press. April 2008, the first edition). In 1948, Kenneth Robinson (1917–2006) published a professional article on the studies of Zhu Zaiyu's equal temperament, (Kenneth Robinson, "A Critical Study of Ju Dzai Yu's Account of the System of the Lu-Lu or Twelve Musical Tubes in Ancient China". Inaug. Diss. Oxford, 1948) which made everyone in the Western world know more about the great achievements and contributions of Zhu Zaiyu. In addition, Joseph Needham (1900 ~ 1995), the famous historian of science, believed that "Zhu Zaiyu's formulation of equal temperament may justly be regarded as the crowning achievement of China's two millennia of acoustic experiment and research" and "To China must certainly be accorded the honor of first mathematically formulating equal temperament." (Joseph Needham, Science and Civilization in China. Cambridge Univ. Press, Vol. IV, I:220–228). Fritz A. Kuttner (1903 ~ 1991), an American scholar, also published an article in 1975 with a view that Zhu Zaiyu should share the invention right of the equal temperament with Western scholars; Robinson published an article entitled "On Zhu Zaiyu's Contribution to the Equal Temperament in Chinese Music" in 1980.

Among the contemporary works that studied Zhu Zaiyu with monographs, *Zhu Zaiyu—the Superstar in Science and Arts in the Ming Dynasty* (《朱载堉——明代的科学和艺术巨星》) by Dai Nianzu and *Studies of Zhu Zaiyu* (《朱载堉研究》) by Chen Wannai in Taiwan could be regarded as the most comprehensive. Both have their own features. Dai's work is comprehensive covering the fields of musicology, mathematics, dance, culture, science, and philosophy, while Chen's work focuses on the studies of musicology with experiments on strings and pitch pipes, which can be regarded as perfect combination of theories and practices. These two monographs are the books of must for anyone who is interested in understanding Zhu Zaiyu. Recently, Zhuo Renxiang published *Zhu Zaiyu and His Academic Achievements in the Context of Oriental and Occidental Cultures* (《东西方文化视野中的朱载堉及其学术成就》). It discussed in detail how Zhu's formulation of the equal temperament had been possibly spread to the West with meticulous examination and exploration. (Zhuo Renxiang: "Zhu Zaiyu and Academic Achievements in the Perspective of Eastern and Western Culture (《东西方文化视野中的朱载堉及其学术成就》)", Central Conservatory of Music Press, 2009). In addition, Cheng Zhenyi also has a special study on Zhu Zaiyu's calculation of the equal temperament. (Cheng Zhenyi: Huangzhong Dalv-Acoustic Achievements in Ancient China and the Sixteenth Century (《黄钟大吕——中国古代和十六世纪声学成就》), Shanghai Science and Technology Education Press, 2007).

In addition to the above research, researchers in the musicology community who have focused on Zhu Zaiyu's new law for density rate and the law of pitch pipes of different diameters have also achieved great results, taking the restoration experiment of Zhu Zaiyu's pitch pipe of different diameters as an example. Victor C Mahillon (1841–1924), a Belgian scholar, conducted a measuring experiment in 1890. In China, Yang Yinliu had an analytical calculation in 1937; in 1956, the glass tube simulation experiment was conducted by Zhuang Benli at Taipei Cultural University. Besides, there were Chen Quanfang's test tube simulation experiment and the comprehensive restoration of acoustic experiments conducted by Chen Wannai, Liu Yong, and so on. As for the theoretical discussion and controversies in this regard, there are more to name.

The study of Zhu Zaiyu's equal temperament initiated by Liu Fu's article has been deeply rooted in the hearts of the people. Zhu Zaiyu and his achievements have begun to be included in the annals of Chinese science, technology, and culture in the true sense of history.

However, upon reviewing the history, in order to fully understand Zhu Zaiyu's major achievement in creating the equal temperament, we must have a deeper understanding of the following issues:

1. What is the theoretical approach of Zhu Zaiyu to formulating the equal temperament?
2. What is the specific algorithm Zhu Zaiyu adopted in creating the law of equal temperament?

3. Does Zhu Zaiyu's law of pitch pipes of different diameters meet the theoretical requirements of the equal temperament law?
4. Are the several different views that question Zhu Zaiyu's achievements in the equal temperaments solid-grounded?

It is worth mentioning that even now, in addition to the Complete Works of Chinese Literature, it is not easy to find the full text of Le Yue Quan Shu (《乐律全书》), the Complete Book on Tuning System. The more popular version is the *All-Encompassing Library* (《万有文库》) *Wan You Wen Ku* published by the Commercial Press in 1931, totally 36 volumes. The core idea of Zhu Zaiyu's creation of the new law for density rate and the law of the pitch pipes of different diameters is basically reflected in his book *Lv Xue Xin Shuo* (《律学新说》) *A New Account of the Science of the Pitch-pipes* published between 1567 and 1581, i.e., *Suan Xue Xin Shuo* (《算学新说) *The New Theory of Mathematics* and *Lv Lv Jing Yi* (《律吕精义》) *The Essence of Pitch-pipe Temperament*. Among them, *Lv Xue Xin Shuo* (《律学新说》) *A New Account of the Science of the Pitch-pipes* is the most concise and easy to understand. *Suan Xue Xin Shuo* (《算学新说》) *The New Theory of Mathematics* and *Lv Lv Jing Yi* (律吕精义》) *The Essence of Pitch-pipe Temperament* are very simple and well explained, which fully reflects that Zhu Zaiyu's scientific thoughts and methods had reached a level comparable to modern Western science.

12.2 Theoretical Exploration of How Zhu Zaiyu Formulated the Equal Temperament

For a long time, there have been questions in the academic circles concerning how Zhu Zaiyu formulated the equal temperament. Some scholars believed that Zhu Zaiyu only gave specific data regarding the equal temperament among the 12 semitones, but did not explain the theory. A typical example was Victor C Mahillon (1841 ~ 1924) from Belgium. He said, "Zaiyu didn't explain his theory, he only gave us numbers . . .." (Liu Fu. Zhu Zaiyu: The Discoverer of the Equal Temperament, Collection of Papers to Celebrate the 65th Birthday of Mr. Cai Yuanpei (Vol.1): pp.279–310. The first edition of the external edition of the collection of the Institute of History and Language of Academia Sinica, 1933). The *Executive Summary of the Complete Works of Chinese Literature* (《四库全书总目提要》) (*Executive Summary of the Complete Works of Chinese Literature* (《四库全书总目提要》) *Entry of Yue Lv Quan Shu* (《乐律全书》) *The Complete Book of the Tuning System*) and the *Sequel to the Imperial Pitch-Pipes Annotations* compiled in the 11th year during the reign of Emperor Qianlong in the Qing Dynasty (in 1746) also contained disagreement regarding the issue (Question No. 2. Sequel to the Imperial Pitch Pipes Annotations (《御制律吕正义后编》). (Vol. 180)). Liu Fu the scholar, while highly appraising Zhu Zaiyu's great achievements, also severely criticized Zhu Zaiyu's use of *Zhou Li: Li Shi Wei Liang* (《周礼·栗氏为量》), thinking that Zhu Zaiyu's calculation of pi by adopting the methods shown in Zhou Li: Li Shi Wei Liang (《周礼·栗氏为量》) was in essence of stealing from others while claiming to be the conventional. Dai Nianzu put forward a tolerant view. He believed that "Zhu Zaiyu's fault was sort of plagiarism. He should not have stolen from Zhou Li (《周礼》) the Rites of Zhou to justify his own thoughts in formulating the new law for density rate just because there was the number 2." In the meantime, Dai Nianzu also expressed his most considerate understanding of Zhu Zaiyu, by pointing out that "in the history of China, many scholars and inventors had tried to seek legitimacy of their own works by claiming those discoveries and inventions in the name of the Yellow Emperor or the Duke of Zhou (周公), which was quite expedient and not uncommon. Therefore, Zhu Zaiyu just carried on such a tradition." (Dai Nianzu. Zhu Zaiyu-The Super Star of Science and Arts in the Ming Dynasty (《朱载堉——明代的科学和艺术巨星》). The People's Publishing House: pp.116–117, ed. 2011. Tian Huang Zhen Ren Zhu Zaiyu, (《天潢真人朱载堉》) Elephant Press. April 2008, the first edition).

However, through in-depth literature research, it is not difficult to find that Zhu Zaiyu used the inner square and the outer circle as illustrated in "Zhou Li Li Shi Wei Liang (《周礼·栗氏为量》)" as its theoretical basis, which is not redundant or unnecessary. In addition to profound cultural reasons, such a way of thinking had indeed inspired the formulation of the equal temperament among the 12 semitones. It is on top of this theoretical basis that Zhu Zaiyu had worked hard to establish a theoretical system that could uniformly explain all the difficulties in the law of the pitch pipes of different diameters with tremendous brilliant achievements in the calculation of the new law for density rate. In the comparative study of theory and experiment, Zhu Zaiyu was one of the ancient Chinese scholars who adopted an approach that was the closest to the modern scientific way of thinking.

12.2.1 Science Wonder from the Heluo Studies

As we all know, the studies of Heluo culture is one of the essences of Chinese traditional culture. It profoundly influenced the way of thinking and theoretical construction of ancient Chinese scholars for generations. No wonder that Zhu Zaiyu, who made major innovations like the formulation of the equal temperament, as an outstanding representative of Chinese culture, inherited the essence of Heluo culture. In *Lv Xue Xin Shuo* (《律学新说》) *A New Account of the Science of the Pitch-pipes*, Zhu pointed out unambiguously that the theoretical basis of his thoughts was the study of *He Tu* (《河图》) and *Luo Shu* (《洛书》). The numbers in *Luo Shu* follow the rule that when random combinations of any set of numbers are multiplied with each other, the sum of the results is 9. The numbers in *He Tu* follow the rule that random combinations of any set of numbers are multiplied with each other, and the sum of the results is 6. Zhu Zaiyu traced the number of tones back to the magic numbers listed in *Luo Shu* and the number of scales to the magic numbers listed in *He Tu*. On top of that, Zhu introduced the analytical methods of the ancient measurement units, such as the vertical, horizontal, and slanting lengths of Shu (the grain of Shu, a crop known as millet in English). Such efforts put all the disputes regarding the scale of Huangzhong (黄钟) into the great framework of the studies of Heluo. In addition, Zhu Zaiyu believed that the controversies regarding the length of Huangzhong arose from the ignorance or understanding of the studies of Heluo. (-Origins of Toning System No. 1 (律吕本源第一) Wind Instruments Volume 1, Lv Xue Xin Shuo (《律学新说》) A New Account of the Science of the Pitch-pipes Vol. 1).

Such an analytical method, in the sense of modern science, seemed to fail to comply with reasoning in nature. However, as far as the tradition of Chinese culture is concerned, it was unique. Liu Fu believed:

According to the ancients' arithmetic, the length of Huangzhong was recorded to be ten cuns, nine cuns, or eight cuns one fen. However, the standard was changed a little for the convenience of calculation. Zhu believed that ten cuns should equal nine cuns or eight cuns one fen, thus presenting the vertical, horizontal, and slanting lengths of a grain of Shu (a kind of crop in China). Therefore, the vertical lengths of 81 grains of could be equal to the slanting lengths of 90 grains of Shu or the horizontal lengths of 100 grains of Shu. It seemed a little bit farfetched. However, such a theory was adopted in *Lv Lv Zheng Yi* (《律吕正义》) *Pitch-Pipes Annotations*.

Mr. Liu Fu's comments are not unreasonable. We also need to look at the thoughts of the ancients against the context of the era back then. From a theoretical point of view, the pursuit of unity and simplicity in the understanding of nature has been an instinctive requirement of almost every natural scientist, and so was Zhu Zaiyu. He first elaborated on the unified explanation regarding the length of Huangzhong by resorting to the thoughts of ancient scholars in a hypothetical way based on the study of Heluo. On top of that, a complete set of formal new systems on the theory of musicology was established – the new law for the density rate and the theory of the equal temperament. In this way, it would be unfair to accuse Zhu Zaiyu of adopting ideas from Heluo studies. It cannot be imagined that Zhu Zaiyu would evolve the concepts of intervals and melodic pitches from the sequence of equal numbers directly, to achieve the equal temperament among the 12 semitones. What's more, Zhu Zaiyu took this hypothesis after doing the experiment of arranging millets in person, and Zhu Zaiyu also warned people: "If you don't bother to conduct experiments, you can't experience how wonderful it is to test your theories." (Jia Liang Pian No. 2 (嘉量篇第二), Lv Xue Xin Shuo (《律学新说》) A New Account of the Science of the Pitch-pipes (Vol. 4)). This shows that Zhu Zaiyu Kyu's formulation of the equal temperament, like many major scientific discoveries and inventions in human history, was gradually obtained in a complex process full of conjectures, analysis, verification, and experiments. There were even many coincidences and opportunities (Picture 12.2).

Taking today's natural science as a frame of reference, it seems that the studies of Heluo are just a simple mathematical game. However, in ancient China, the strict mathematical structure of the study of Heluo was developed by the two generations of scholars in the Song and Yuan Dynasties. It had evolved into systematic theories by the Ming Dynasty. It had intersected with the Yin and Yang Five Elements and Taiji Eight Diagrams theory, which constituted the theoretical basis for Chinese scholars to explain everything in the world. Most of what we see today is the negative comments regarding their academic effect, which accused it of having suppressed Chinese scholars' in-depth exploration of the principles of nature to

Picture 12.2 The density rate origin source diagram (《密率源流图》)

such a degree that ancient people often over-relied on one technique to control everything and gradually fell behind the world scientific trend in the shadow of the superstitious practice of The Duke of Zhou's telling fortunes by using the eight trigrams.

In fact, every coin has two sides. The fact that Zhu Zaiyu could have formulated the equal temperament resulted from Chinese scholars applying the studies of Heluo. Although Zhu Zaiyu may not be perfect in the analytical process, the result was the equal temperament theory that has been completely consistent with modern scientific understanding. According to this theory, the equal temperament upon the correction of the pitch pipes' diameters had amazed scholars of mathematics and musicology for generations. It was essentially awe-inspiring. In this sense, the studies of Heluo culture and other related traditional academic theories had indeed played a role in enlightening Zhu Zaiyu's exploration of the equal temperament. Therefore, it is not advisable to think that Zhu's studies were farfetched or not grounded. We should make further analyses in this regard.

12.2.2 Arithmetic Reasoning Carried Down from *Li Shi Wei Liang* (《栗氏为量》)

How did the studies of Heluo and the Duke of Zhou's techniques inspired Zhu Zaiyu with scientific associations and enlightenment so that Zhu Zaiyu could have made creative results?

As is known to us all, ancient scholars often changed the unit of the length of Huangzhong for the convenience of the arithmetic calculations, i.e., the length of

Huangzhong could be 10 cuns, 9 cuns, or 8 cuns 1 fen. The first problem Zhu Zaiyu faced was how to interpret these different lengths of Huangzhong. In order to unify all these figures found in the classics into the theoretical framework of Heluo's theory, Zhu Zaiyu did something unique in that he proposed the explanations of vertical length, horizontal length, and slanting length, holding to the view that 81 vertical lengths of Shu was equivalent to 90 slanting lengths of Shu and 100 horizontal lengths of Shu. (*Lv Lv Jing Yi* the Internal Chapter (律吕精义》) *the Essence of Pitch-pipe Temperament Vol. 4*). In this way, the contradictions and conflicts between different scales of Huangzhong in ancient books were resolved in one attempt, and this interpretation was related to the Heluo theory as a result of Heluo studies. The second problem was that when calculating the diameters of pitch pipes, it was inevitable to use the pi. In order to ensure the completeness of the theoretical system, Zhu Zaiyu must also have used the same theoretical premise. He then proposed that "to calculate the density rate, pi and the diameters were used to make the product." "Such a formula was made by Zu Chongzhi, who got inspired from the Duke of Zhou's techniques." (Jia Liang Suan Jing (《嘉量算经》) Vol.1). Viewed today, such an inference was made quite indiscreetly or unreasonably, as criticized by Liu Fu. However, the method used by Zhu Zaiyu in constructing his theoretical framework was an axiomatic method rarely adopted by ancient Chinese scholars. Starting from the least primitive proposition, all the theoretical issues related to the studies of pitches and tuning work were included in the same theoretical system. Certain logical principles were therefore uniformly explained and narrated, which was the only way for natural science to shift from empirical induction to logical deduction. This was also the weak link in ancient Chinese natural sciences. Since Zhu Zaiyu had made a breakthrough in this respect, he was able to obtain the great achievement of the equal temperament, which could be regarded as inevitable by chance.

In Zhu Zaiyu's view, all mysteries in the world could be revealed with the knowledge of Heluo. He interpreted the 9-cun length of Huangzhong as a ruler and connected it with the numbers listed in Luo Shu. He believed that the length of Huangzhong could be regarded as the foundation of the temperament theory. Suppose Huangzhong was 1 Chi long, Zhu used the measuring unit of the horizontal length of Shu. As it was related to the figures in He Tu, He then drew the theoretical basis of the original standard for measuring Huangzhong. When Zhu Zaiyu perfectly explained these theoretical problems that had been entangled in history, he was quite excited. It followed that he would feel "how came that for centuries no one had ever discovered such a rule," which was a great shame.

Even today, among Zhu Zaiyu's interpretation, except that the vertical and horizontal length rulers were an assumption, the rest of the interpretation was still not unreasonable, at least logically consistent. In particular, Zhu Zaiyu cleverly used the picture of "Zhou Li Li Shi Wei Liang (《周礼·栗氏为量)," i.e., the inner square and the outer circle, to calculate the main data of the equal temperament among the 12 semitones. Here was one of the many examples to show the originality and creativity of Zhu Zaiyu's calculations (Jiang Yong Preface to Lv Lv Chan Wei (《律吕阐微》)) and how he calculated the length and internal and external diameters of the pitch pipes.

Zhou Li Li Shiwei Liang (《周礼·栗氏为量》) had enjoyed a high academic status in the eyes of ancient Chinese scholars. The perfect formulation of the Pythagorean Theorem had led many ancient scholars to simply believe that such an academic outcome should be immortal while universally applicable. Some scholars still asserted that these academic achievements had existed in ancient China since ancient times, by the time that the Western thoughts got spread eastward and Euclidean geometry and trigonometry were introduced into China. Under such a cultural background, Zhu Zaiyu couldn't go far beyond the historical limit. Naturally, he would also choose the principle diagram of the inner square and the outer circle from Zhou Li Li Shi Wei Liang (《周礼·栗氏为量》) and put it as a geometric model that fully explained the principle of temperament. Zhu Zaiyu proposed several key concepts: One was that the sky is round, and nine represents the earth and ten represents the sky, which constituted the basic principle of Heluo's studies. Such an idea was taken as a natural truth in the non-artificial sense. The second was that there were four sides in the sky, each side measured 10 cuns and distributed in a circle, so the circumference should be 40 cuns. The third was that there were four sides on the ground, which were distributed in a square shape, and each side should be 9 cuns. Then Zhu used the Pythagorean to find the string selection criteria which could go from the ground to the health with the diagonal line calculated, i.e., the diameter of the circle. According to such a reasoning logic, the Pi could be inferred as $40/\sqrt{9^2+9^2}$. In his sense, the measurement of circles had been achieved. (*Solve the Circle Power with the Density Rate No. 5* (密率求圆幂第五) *Lv Xue Xin Shuo* (《律学新说》) A New Account of the Science of the Pitch-pipes (vol. 1)) (Picture 12.3).

If we do not put aside the errors of the theoretical model of the square ground and round sky temporarily, such a clever theoretical "derivation" with strong logical

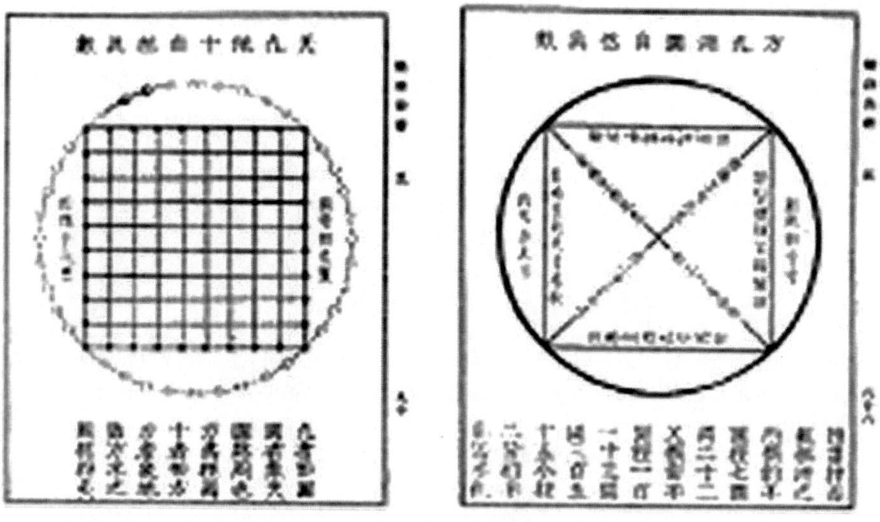

Picture 12.3 Illustrations of Zhu Zaiyu's formulating the theory of temperament

argumentation can be well compared with Western modern science except that the former did not use formalized diction in derivation from the mathematical rule found in Heluo's studies to the illustration of Li Shi Wei Liang in *Zhou Li* (《周礼》) *The Rites of Zhou*; Zhu's formulation fully revealed the magical subtleness of mathematical operations. Such a theory, despite it was against today's science, was also self-contained and self-explanatory, which fully demonstrates the charms of theoretical speculation. It was precisely because of the completeness of this theoretical system that Zhu Zaiyu was convinced of the pi algorithm derived from it to such a degree that he even dismissed the pi calculated by Zu Chongzhi. That well explained why Mr. Liu Fu was dissatisfied with Zhu Zaiyu and claimed that "Zhu Zaiyu's fault was sort of plagiarism and he should not have stolen from *Zhou Li* (《周礼》) *the Rites of Zhou* to justify his own thoughts in formulating the new law for density rate."

What Zhu Zaiyu did was not just theoretical evolution, he also made experiments. He used cards to make a square of 9-cun sides. He also used a square of 1-chi-and-3-cun sides as formed with the diagonal line of that 9-cun square. He then used a paper stripe with marks of measuring units to measure the diameter of the circumcircle of the 9-cun square. What Zhu Zaiyu obtained was that the circumference of the circumcircle of the 9-cun square was equal to 40 cuns. (*Solve the Circle Power with the Density Rate No. 5* (密率求圆幂第五) *Lv Xue Xin Shuo* (《律学新说》) A New Account of the Science of the Pitch-pipes (vol. 1))Therefore, Zhu Zaiyu so firmly believed in the illustration taken from Zhou Li Li Shi Wei Liang (《周礼·栗氏为量》)!

From the perspective of science today, Zhu Zaiyu had made a historic mistake here. That is, with such a test method, what was obtained could not a "natural true number," but an approximate number that could be estimated with considerably errors. If the side of inscribed square of a circle is 9 cuns long, the slanting length of the square should be $\sqrt{9^2 + 9^2} = 12.727922$ inches, and the circumference is 39.985946 ... cuns, instead of 40 cuns. The difference was about 0.01405 ... cuns. But this error was undetectable in Zhu Zaiyu's experiment of measuring the circumference with a piece of paper.

We should also proceed from what things were like in history when discussing issues. Based on Zhu Zaiyu's knowledge level and feasible operation methods at that time, it was still of positive significance for being able to carry out such scientific explorations that were realistic and tangible. Zhu Zaiyu personally conducted a lot of experiments, the data he obtained, although including certain errors due to the empirical misunderstanding of the pi measurement, these errors happened to be within the range allowed by the actual production of the pitch pipes. As a result, Zhu Zaiyu's pitch pipes of different diameters reached the highest level in history, and for the first time he helped realize the beautiful wish of transition and modulation by pipes. Reviewing the development of human civilization, no scientific fields could have reached the summit in one step and exhausted the mysteries of nature. Newtonian mechanics has also been gradually discovered as not absolutely correct after the human observation ability had developed to the microscopic and high-speed physics field, thus allowing the inheritance and development of physics through the proposition of Einstein's theory of relativity. Even embarrassing situations could happen after the Nobel Prize was conferred when people found out that the conclusion of the prize Laurette was unreliable several years later. Therefore, we should not blame Zhu Zaiyu for his mistake regarding the issue of pi.

As Zhu Zaiyu found the logic of the theoretical narrative, his calculations proceeded extremely smoothly. At the same time, he also repeatedly warned: “The natural truth rate in the world and the overall rate of The Duke of Zhou’s Circle could never be ignored in formulating the law of tones.” (*Solve the Circle Power with the Density Rate No. 5* (密率求圆幂第五) *Lv Xue Xin Shuo* (《律学新说》) A New Account of the Science of the Pitch-pipes (vol. 1)). When calculating the data such as the length and inner and outer diameters of the different-diameter pitch pipes, he calculated them as follows:

1. The calculation of the length of the pitch pipes can be directly determined by the “new law for density rate.” Of course, when calculating the “new law for density rate,” what inspired Zhu Zaiyu to discover for the first time that the law of Ruibin was the figure $\sqrt{2}$, the method of which was taken from the “inner square and the outer circle” as shown in the Zhou Li Li Shi Wei Liang. “The length of the diagonal line of the inscribed square is equal to the length of the diameter. Therefore, know the length of the diagonal line, you can infer the diameter of the circle.” (Don’t use the three- point Losing and Gaining law, No.3 Lv Lv Jing Yi (《律吕精义》) the Essence of Pitch-pipe Temperament. (vol. 1)). Based on such a reasoning, we can infer that $\sqrt{2}$ was the length of lower Ruibin string. In this regard, we will specifically discuss the issue in the next chapter.
2. When calculating data of the equal temperament, besides the string length of middle Ruibin could be calculated based on the Pythagorean method exerted over the inner square and outer circle as shown in Zhou Li · Li Shi Wei Liang (《周礼·栗氏为量》). Other tones needed to be proportional upon solved. Thus, when calculating the inner and outer circumferences and inner and outer diameters of the different-diameter pipes, Zhu Zaiyu established his theoretical system rigorously. As mentioned earlier, through theoretical analysis and experimental results, Zhu Zaiyu had come up with a “new method of calculating the new law for density rate regarding the arithmetic circumference and diameters that could be uniformly interpreted according to traditional Chinese culture.” This method cleverly avoided the calculation of pi and thus obtained all the data regarding the 36 different diameter pipes. For example, when determining the values of the outer diameter and circumference of the largest pitch pipe, through the theoretical evolution and the actual measurements, Zhu obtained the result that the outer circumference was 40 cuns and the inscribed square had a side length of 9 cuns.

According to this logic, Zhu defined the outer circumference of lower Huangzhong as one-ninth of its length, and the inner diameter of ow octave Huangzhong as one-fortieth of its length. This definition method conformed to the tradition of ancient tuning and was in the same vein as the algorithm of the new law for density rate. It could also uniformly explain the various controversies about the scaled of the Huangzhong pitch pipe in ancient times. For example: According to

Zhu Zaiyu's unified interpretation, the lower Huangzhong (Bei Lv), middle Huangzhong (Single Lv), and higher Huangzhong (half Lv) (the traditional three Chinese octaves include the lower string, the middle string, and the higher string) could be nested in each other seamlessly. The circumference and inner and outer diameters of the remaining pitch pipes can be calculated one by one if they are calculated according to a predetermined ratio (No adoption of the equatorial circumference No. 5, Lv Lv Jing Yi Internal Chapter (《律吕精义》) the Essence of Pitch-pipe Temperament. (vol. 2)). It can be seen from this that if the Heluo theory was to be used consistently, the definition of the inner and outer diameters of the law must be taken as 1/9 and 1/40 to be in line with the logic of reasoning.

Although Zhu Zaiyu circumvented the use of pi by resorting to the Pythagorean method, he still had to use a pi, which is $40/\sqrt{9^2+9^2}=3.142696805$.

From the perspective of today's science, Zhu Zaiyu's use of pi was not rigorous enough. However, because his calculation errors were mostly between 5 and 9 out of a thousand. Viewed from the perspective of processing arts, the impact of such errors on accuracy could be basically negligible. Therefore, Zhu Zaiyu's theory and practice not only had the universal significance of scientific methodology and epistemology, but the series of calculation results he had obtained have remained amazing to date. This is not only because Zhu Zaiyu had personally conducted experimental verification of blowing and acoustic tests regarding the pipes back then. The following theoretical analyses will also prove that Zhu Zaiyu's law of pitch pipes of different diameters completely met the requirements of the equal temperament concerning transition and modulation.

12.2.3 Complete System with Unified Interpretation as the Pursuit

From a historical point of view, Zhu Zaiyu adopted Li's measuring system and the numbers in *He Tu* and *Luo Shu* as the logical starting point for his description of the new law for density rate and deduced it through constant experiments. The correct conclusion was drawn at the forefront of the times with the final completion of the great theory and practice of the equal temperament. This process was also noteworthy in the history of Chinese science and technology.

Of course, Zhu Zaiyu also paid a certain price for this attempt to unify the theory. In addition to the suspected farfetched nature in pursuing the perfection of theory, he also believed too much in truth hidden in ancient literature: in particular, the Duke of Zhou's density rate and He Tu and Luo Shu. These theories are still considered to be one of the representatives of Chinese culture. We do not deny their meanings and value because they are completely different from Western learning. However, Zhu Zaiyu's formulation did lack analytical accuracy as the insufficient experimental accuracy led to the imprecise understanding of pi. But if compared with the narratives of many musicologists' ideas and classics in history, Zhu Zaiyu's scientific practice had become extremely precious.

In the Ming and Qing Dynasties, "many works of music ...which had escaped from reality ... with mysterious narratives" could not solve any problems (Yang Yinliu: "A

History of Ancient Chinese Music (《中国古代音乐史稿》), People's Music Publishing House, 1981, p. 1012). There were numerous errors regarding the set of 160 chimes in the Temple of Heaven made under the imperial order of Emperor Kangxi in the Qing Dynasty. (Yang Yinliu: "A History of Ancient Chinese Music (《中国古代音乐史稿》), People's Music Publishing House, 1981, p. 1014). When we compare Zhu's works with other scientific masterpieces in the same historical period as *Compendium of Materia Medica, The Exploitation of the Works of Nature, Complete Treatise on Agriculture, Famine Herbal, and Diary of the Travels of Xu Xia-Ke*, we cannot help wondering how came that Zhu Zaiyu was in the same historical period of authors of these works. Zhu made his own formulations while using the analytical science independently with employment of axiomatic methods in which ancient Chinese science and technology were still accumulating experience and simplicity prevailed at that time. All his explorations were on his own and spontaneously. After all, what Zhu did was unparallel in the history of several thousand years. Therefore, we need to sing high for and attach importance to the full set of scientific research methods combining the theoretical foundation of the studies of Luo He, the deduction method of using the sides of square to solve the diagonal line together with obtaining data through experiments.

Zhu Zaiyu's achievements in scientific epistemology and methodology showed that the development of human civilization did not have a clear division between eastern and western sciences. Once the conditions are met in China in the east, a modern scientific thought may also germinate.

Due to the historical advancement of the scientific method used by Zhu Zaiyu, for the next few hundred years, there were few people, except Jiang Yong, who could understand what Zhu had achieved. On the other hand, among all the scientific achievements in ancient China, Zhu Zaiyu's equal temperament was almost as good as the modern science. How came that? We finally have the answer: adopted by Zhu, the theoretical method has the same principles as the modern scientific method. Guided by such a scientific method, it was most likely to reach a conclusion comparable to that in science today. Although this method was not understood at that time and for quite some time in the future, once the science and technology of the entire society have progressed to a comparable level, people would be surprised to discover the scientific miracle hidden behind the dust of history.

It is true that we cannot simply equate Zhu Zaiyu's pursuit of theoretical completeness with the axiomatized system of Western learning. However, if we want to discuss the earliest and first cases in scientific methodology and scientific epistemology, Zhu Zaiyu's work may also be another spectacle of ancient Chinese science for having comprehensively combined the theoretical system analysis with the precise testing of experiments. Zhu himself could also be regarded as the ancient Chinese scholar whose thinking method and research method had been the closest to those of western modern science.

In the nearly 300 years after Zhu Zaiyu, the only person who understood and accepted Zhu Zaiyu's equal temperament theory was Jiang Yong, a scholar in the Qing Dynasty. What Jiang Yong admired was the method of axiom deduction and argumentation adopted by Zhu Zaiyu and the true numbers of temperament that are consistent with the Chinese cultural tradition and the truth of the world's natural rationality.

12.3 Algorithm Analysis of Zhu Zaiyu's Equal Temperament

The equal temperament law has been a great contribution of the Chinese civilization to world civilization. This is an indisputable historical fact. However, for a long time, there have been doubts about the calculation method of Zhu Zaiyu's formulation of the equal temperament. Many people repeatedly quoted a section regarding the new law for density rate in *Internal Chapter of Lv Lv Jing Yi* (《律吕精义·内篇》) *The Essence of Pitch-pipe Temperament*, "the scale originated from the length of Huangzhong..." as the evidence for Zhu Zaiyu to formulate the equal temperament law. Liu Fu was among those who verified these words. Subsequent research has mostly referred to Mr. Liu Fu's research results. In Mr. Miao Tianrui's book *Studies of Tones*, which has been reprinted several times, also has the same quotation. He believed that: "Zhu Zaiyu's calculation method, according to today's algorithm, is equivalent to the square root of the octave, $\sqrt{2}$, 1.414213 ..., which suggested the half of the octave, which is #f at the six semitones in the twelve equal temperament. Then find the square root of 1.414213, i.e., $\sqrt{1.414213}$, to get 1.189207..., which is one fourth of an octave, i.e., at the third semitones $^{\#}$d in the equal temperament. Then find the cube root of 1.189207, i.e., $\sqrt[3]{1.189207}$ is 1.059463... which is one twelfth of the octave, at the semitone of $^{\#}$c, also the higher octave of any tones." (Miao Tianrui: Musicology (《律学》) p. 145., People's Music Publishing House, 1983). However, such words could not fully demonstrate the formulation of the equal temperament. From the perspective of mathematical calculation, it is also illogical. Why did Zhu Zaiyu think of extracting the square root of the octave and then the square root of that number, till extracting the cube root of the resulting number. It is unimaginable to find the 12th root of 2, $\sqrt[12]{2}$ to serve as the pitch interval of semitones in the equal temperament. Therefore, just quoting the phrase "the scale originated from the length of Huangzhong ..." is indeed not enough to prove that the equal temperament law is the product of rational thinking. In this selection, although Zhu Zaiyu gave a series of accurate calculation results, he did not give a detailed account of the calculation process of the "new law for density rate." As a result, many scholars had to avoid the discussions in this regard or simply skipped Zhu Zaiyu's original idea of formulating the equal temperament while making a simplified summary of its production process. Indeed, it was inevitable that people had to doubt its theory: Zhu Zaiyu's data regarding the equal temperament were generated too suddenly! It seemed that its inventor was unreasonable about it. Since the compilation of the Completed Works of Chinese Literature during the reign of Emperor Qianlong in the Qing Dynasty, people had been questioning it. Scholars abroad also believed that Zhu Zaiyu had only listed the calculation results, but did not explain the theory of how to deduct the equal temperament law. However, as far as Zhu Zaiyu was concerned, by claiming "the creation of a new method is to set one scale as the criteria, divide by the density rate, which is to cover the 12 semitones,"(Solve Pitch and Interval with Law of Density Rate No. 3 Lv Xue Xin Shuo (《律学新说》) A New Account of the Science of the Pitch-pipes (vol. 1)) he had indeed unmistakably recognized the proportional relationship between the 12 temperaments.

The key to the question is: How the law for new density rate came? How did those magical 2, 122 stand out? If the selection in *Internal Chapter of Lv Lv Jing Yi* (《律吕精义·内篇》) *The Essence of Pitch-pipe Temperament* could not serve as the solid evidence to prove that Zhu Zaiyu did formulate the equal temperament law, did Zhu leave behind some more specific formulas in this regard?

Dai Nianzu once made a systematic analysis of Zhu Zaiyu's immortal works entitled *Yue Lv Quan Shu* (《乐律全书》) *The Complete Book of the Tuning System.* He developed Liu Fu and Miao Tianrui's textual research and fully explained the elaborations regarding the new law for density rate in *Internal Chapter of Lv Lv Jing Yi* (《律吕精义·内篇》) *The Essence of Pitch-pipe Temperament*. He also offered simple explanation in modern language (Dai Nianzu. Zhu Zaiyu-The Super Star of Science and Arts in the Ming Dynasty (《朱载堉——明代的科学和艺术巨星》). The People's Publishing House: pp.66, ed. 2011. Tian Huang Zhen Ren Zhu Zaiyu (《天潢真人朱载堉》) Elephant Press. April 2008, the first edition). However, we haven't seen more analysis and research on how Zhu Zaiyu got the results of up to 25 figures. Regarding the literature involved in the author's research, no scholars have yet found an in-depth explanation on this. Most of them reiterated Zhu Zaiyu's text on the new law for density rate in modern language, aiming to dispel the confusion regarding the specific formulation with the correct calculation results. The confusion in the calculation process makes it unavoidable that this great achievement of ancient Chinese phonology often leads to misunderstanding and criticism, not only the peers, the scholars in ancient China, and even the contemporary scholars.

The problems we face are: on the one hand, Zhu Zaiyu had indeed offered a solution to figure out the geometric progression; on the other hand, in the quotations of a large number of contemporary studies of Zhu Zaiyu's works, it seems confusing to approach how he could have calculated these numbers of the toning system.

In view of this, it is necessary to take the historical conditions against which Zhu Zaiyu lived as the framework to thoroughly examine his thought process of formulating the equal temperament to help clarify Zhu Zaiyu's specific algorithm for the equal temperament.

To be sure, the selection regarding the new law for density rate in *Internal Chapter* of *Lv Lv Jing Yi* (《律吕精义·内篇》) *The Essence of Pitch-pipe Temperament* was insufficient to demonstrate how the equal temperament among the 12 semitones was formulated. In this regard, the more representative texts should be taken from the questions listed in *Suan Xue Xin Shuo* (《算学新说》) *The New Theory of Mathematics*.

12.3.1 Scientific Associations Evolved from the Illustration in *Li Shi Wei Liang* (《栗氏为量》)

In theory, to formulate the 12 equal temperament law, the following prerequisites must be met:

1. Establish an octave interval or string length ratio of 2:1.
2. Establish a proportional relationship between pitch change and string length.
3. Deny the possibility of assigning temperaments according to the equal difference from 2 to 1.

The above items can be derived from the existing pitch tuning playing experiments, which had also been the knowledge already possessed by Zhu Zaiyu in his era. The next question is: Why did Zhu Zaiyu start with the key figure of 2 and first determine the string length of the lower of Ruibin. Isn't it too arbitrary to describe this creative breakthrough as decorative, or in the name of Pythagoras to deceive people in the name of Pythagoras? Similar views appeared in the *Executive Summary of the Complete works of Chinese Literature* (《四库全书总目提要》) and *the Sequel to the Imperial Pitch-Pipes Annotations* (《御制律吕正义后编》) and even today's works. Today's works are sometimes seen. Even though it is believed that Zhu Zaiyu "wrongly used the classic record in *Zhou Li Li Shi Wei Liang* (《周礼·栗氏为量》) to justify the legitimacy and validity of his theories (Dai Nianzu. *Zhu Zaiyu-The Super Star of Science and Arts in the Ming Dynasty* (《朱载堉——明代的科学和艺术巨星》). The People's Publishing House: pp.75, ed. 2011. *Tian Huang Zhen Ren Zhu Zaiyu* (《天潢真人朱载堉》) Elephant Press. April 2008, the first edition), it still requires further analysis and demonstration in that all of similar comments have ignored the actual role the number 2 played in the formulation of the equal temperament.

We might as well restore Zhu Zaiyu's original idea of formulating the equal temperament law from the lower of Huangzhong to Huangzhong, the string length is reduced by half, and there are twelve semitones in between. There must be some equivalence relationship between the temperaments to achieve the transition and modulation. It is easy to judge by the playing test that the equivalence relationship cannot be regarded as equal to transition and modulation. Since it cannot be the equivalence relationship between the semitones, it is most likely to be some equal proportion relationship. If the length of the lower Huangzhong is set to 2, then the length of middle Huangzhong is 1. The lower Ruibin is located in the middle of the twelve semitones. In order to achieve the transition and modulation, the length rate of the lower Huangzhong by the lower Ruibin should be equal to that of the lower Ruibin to middle Huangzhong. According to the ancient musical laws like three-point losing and gaining method (三分损益律), ancient wind and stringed instruments were based on the inference algorithm that the standard notes have a fixed length and pitch position. The three-point losing 1 referred to the tonal difference of pure fifth treble, and the three-point gaining 1 referred to the tonal difference of pure fifth treble. The development from three-point losing and gaining method to Zhu's formulation was easy to comprehend. The continued proportional thoughts as mentioned above could be represented mathematically as follows:

Lower Huangzhong/Lower Ruibin = lower Ruibin/Huangzhong
The known figure of the string length for lower Ruibin is $\sqrt{2}$
That is how Zhu Zaiyu used the figure the square root of 2 to formulate the equal temp

Surprisingly, what inspired Zhu Zaiyu to discover for the first time that the string length for the lower Ruibin equals to $\sqrt{2}$ was another way, which was more intuitive and simpler than the above solution. It was the law based on the illustration of "the inner square and outer square" listed in *Zhou Li Li Shi Wei Liang* (《周礼·栗氏为量》). The diagonal line of the inscribed square is equal to the diameter of the circumcircle. Therefore, we can know the diameter of the circle based on the length of the diagonal line of the inscribed square. If the inner square ruler is round and the outside is round, the diameter of the circle is the same as that of the square oblique. According to the idea of "inner square ruler and rounder outer circle," the conclusion that $\sqrt{2}$ is the length of the Ruibin pitch pipe law can also be derived. The method is as follows: ("No use of the Three -point Losing and Gaining Law No. 3", *Lv Lv Jing Yi* (《律吕精义》) *the Essence of Pitch-pipe Temperament.* (Vol. 1)). Similarly, such a method derived from the inner square and outer circle can also be applied to calculate the string length of lower Ruibin as the square root of 2. The formulation was listed as follows:

As shown in Picture 12.4, according to the thoughts of Zhu, the side length of the inscribed square is 2, the string length of the lower Huangzhong. According to what was mentioned in *Zhou Li Li Shi Wei Liang* (《周礼·栗氏为量》), we can make the inscribed circle of the inscribed square twice. It can be easily deduced from the Pythagorean theorem that the side length of the inscribed square in the middle circle is 2, and the side length of the inscribed square in the inner circle is exactly 1, which is the string length of the middle Huangzhong. Since the string length of the lower Huangzhong can be derived from the string length of the middle Huangzhong, $\sqrt{2}$ as an intermediate transition could of course be guessed as the string length of lower Ruibin. That was

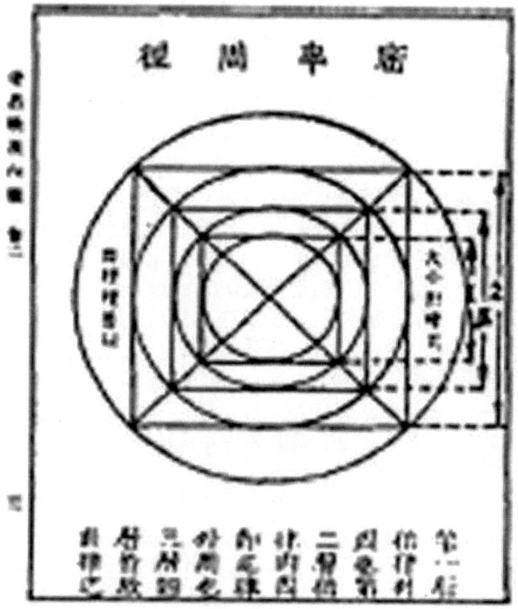

Picture 12.4 How Zhu Zaiyu formulated the semitones of equal temperament: thought diagram no. 1

how Zhu Zaiyu had taken the string length of lower Ruibin as $\sqrt{2} = \sqrt{1^2 + 1^2}$ in the most original manner. Based on this we have reason to believe that Zhou Li Li Shi Wei Liang ("Zhou Li Li Shiquan") did play a role in inspiring Zhu Zaiyu and assisting him in the calculation of Zhu Zaiyu's equal temperament. When solving the calculation problem of "diameter correction for the pitch pipes," the model of the inner square and the outer circle had become the calculation tool directly. In an era where feudal laws and traditional rituals were emphasized, it was a social fashion to cite articles based on citations. However, Zhu Zaiyu did not cite scriptures solely for the purpose of citations. Inspired and proceeded from the scripture of the classical works in the Western Zhou Dynasty, Zhu drew a series of correct conclusions, which fully revealed the scientific charm of abstract thinking. Thus, his methods and achievements were highly recommendable. Even from the perspective of scientific sociology, Zhu Zaiyu's interpretation was not unreasonable. As we all know, computing the square root of 2 has been one of the most exciting parts of Chinese mathematics. Using the description in *Zhou Li Li Shi Wei Liang* (《周礼·栗氏为量》) as the explanation, it is easier for the public to understand and accept it. At the same time, if the octave transition is regarded as a circle that repeats from one point to another, then the circle diameter that divides the circle into two can be compared to the string Ruibin. According to the consistent thinking of Chinese toning system, all the tones originated from Huangzhong. Therefore, the 1-Chi Huangzhong was adopted as cathetus in the Pythagorean theorem; the string length of lower Ruibin could be regarded as the hypotenuse. On the other way, if the string length of the middle Ruibin was taken as the cathethus in the Pythagorean theorem, then the string length of the middle Huangzhong would be the hypotenuse (see Picture 12.5).

Picture 12.5 How Zhu Zaiyu formulated the semitones of equal temperament: way no. 2

This method of fitting mathematical projections to traditional ideas was not surprising in ancient Chinese academic works. On the one hand, *Zhou Li Li Shi Wei Liang* (《周礼·栗氏为量》) inspired Zhu Zaiyu to discover the string length of the lower Ruibin is equal to the square root of 2, i.e., $\sqrt{2}$. On the other hand, it also provided a visual explanation that was in line with the traditional theories of the toning system. In today's words, it gave a geometric interpretation of the equal temperament. It is true that this interpretation was obviously wrong if we take the acoustic vibration theory today as a reference. But at the time, it was an effective theoretical explanation, just as the geocentric theory could also predict part of the lunar eclipse. Science is advanced in the process of constantly discovering errors and flaws in progress. In fact, it is impossible for Zhu Zaiyu not to verify the equation that the string length of the lower Ruibin equals the square root of 2 from the perspective of equal proportion. Therefore, from Si Ku Quan Shu (《四库全书》) on, we should review the accusations to Zhu Zaiyu's citation of *Zhou Li Li Shi Wei Liang* (《周礼·栗氏为量》) in the records in the Complete Works of Chinese Literature.

12.3.2 Mathematical Formulation of the Tones One by One

Once the string length of Ruibin was settled, it would not be too difficult to infer the lengths of other strings according to the equal proportion thoughts. We would like to list the formulas as listed in *Suan Xue Xin Shuo* (《算学新说》) *The New Theory of Mathematics*.

The string length of the lower Ruibin $= \sqrt{\text{Huangzhong}^2 + \text{Huangzhong}^2}$
The string length of the lower Ruibin/2 = the string length of the middle Ruibin *Suan Xue Xin Shuo* (《算学新说》) *The New Theory of Mathematics* Q2

The string length of Jiazhong $= \sqrt{\text{Huangzhong} \times \text{Ruibin}}$ *Suan Xue Xin Shuo* (《算学新说》) *The New Theory of Mathematics* Q4

The string length of the lower Nangong $= \sqrt{\text{Huangzhong} \times \text{the lower Ruibin}}$ *Suan Xue Xin Shuo* (《算学新说》) *The New Theory of Mathematics* Q5

The string length of Dalv $= \sqrt[3]{\text{Jiazhong} \times \text{Huangzhong} \times \text{Huangzhong}}$ *Suan Xue Xin Shuo* (《算学新说》) *The New Theory of Mathematics* Q6

The string length of the lower Yingzhong $= \sqrt[3]{\text{the lower Nanlv} \times \text{Huangzhong} \times \text{Huangzhong}}$ *Suan Xue Xin Shuo* (《算学新说》) *The New Theory of Mathematics* Q7

These calculation formulas clearly showed Zhu Zaiyu's idea of invention. Obviously, these formulas could no longer be derived from the square circle diagram illustrated in Zhou Li (《周礼》). How did Zhu Zaiyu get these formulas? According to our analysis, he still calculated according to equal proportions. Take the length of

Jiazhong as an example. Since Jiazhong is located in the middle between Huangzhong and Ruibin, it follows:

Jiazhong/Huangzhong = Ruibin/Jiazhong.
Also, Jiazhong = $\sqrt{\text{Huangzhong} \times \text{Ruibin}}$.

It was exactly Question No. 4 in *Suan Xue Xin Shuo* (《算学新说》) *The New Theory of Mathematics*.

Extract the square root of the product obtained from multiplying the string length of the middle Ruibin by that of the middle Ruibin; you can get the string length of the middle Jiazhong.

Similarly, the string length of the lower Nanlv = $\sqrt{\text{Huangzhong} \times \text{the lower Rubin}}$

It was exactly Question No. 5 in *Suan Xue Xin Shuo* (《算学新说》) *The New Theory of Mathematics*.

Extract the square root of the product obtained from multiplying the string length of the middle Huangzhong by that of the middle Ruibin; you can get the string length of the lower Nanlv.

At this point, there are still four equal proportions, with item 1 or item 4 known and item 2 and 3 to be solved. For example, among the four consecutive strings, namely, Huangzhong, Dalv, Taicu, and Jiazhong, the string lengths of Huangzhong and Jiazhong are known; please solve the lengths of Dalv and Taicu. Or in the other case, among the four consecutive strings, namely, Nanlv, Wushe, Yingzhong, and Qinghuangzhong, the lengths of Nanlv and Qinghuangzhong are known; please solve the lengths of Wushe and Yinghzong and so on.

Based on the analysis of equal proportions, the calculation of these two items can be transformed into the simultaneous solution of binary equations and the problem of extracting the cubic root.

Take solving the length of Dalv as an example.

We know that Huangzhong/Dalv = Dalv/Taicu.
Dalv/Taicu = Taicu/Jiazhong.
Combining the above two equations,
we know
Dalv = $\sqrt[3]{\text{Jiazhong} \times \text{Huangzhong} \times \text{Huangzhong}}$

It was exactly Question No. 6 in *Suan Xue Xin Shuo* (《算学新说》) *The New Theory of Mathematics*.

Find the product of multiplying Jiazhong, Huangzhong, and Huangzhong. Then extract the cube root of the product, which is the string length of the middle Dalv.

Similarly, we can find the length of the lower Yingzhong. All of these mathematical operations are easy to do. Zhu Zaiyu also calculated the string lengths of Ruibin, Jiazhong, Nanlv, Dalv, and Yingzhong according to the above ideas and gave specific formulas in turn. In *Internal Chapter of Lv Lv Jing Yi* (《律吕精义·内篇》) *The Essence of Pitch-pipe Temperament*, Zhu Zaiyu only gave the algorithm of the length of Lvlv as mentioned above. Therefore, according to the analytical results of *Internal Chapter of Lv Lv Jing Yi* (《律吕精义·内篇》) *The Essence of Pitch-pipe Temperament*, afterwards people argued that Zhu Zaiyu did not give all the formulas of the equal temperament one by one. For instance, F.A. Kuttner believed "Chronologically, here is no doubt that Prince Chu was the first to offer, in 1584, a nine-digit monochord of the 12 pitches of equal temperament. … But Chu's 1584 presentation does not contain a mathematical or theoretical definition of the temperament; it is strictly a numerical or figuring exercise the procedure of which is only partly indicated by the prince." (F.A. Kuttner, Prince Chu Tsai Yu's Life and Work. Ethnomusicology. Vol.XIX, No.2, 1975:168—169). Mr. Dai Nianzu refuted this view. But Mr. Dai's argument can be further enriched. He believed that although Zhu Zaiyu did not enumerate the semitones of the equal temperament by solving the geometric progression, from the point of view of the method for solving arbitrary terms of the geometric progression, the examples he cited were complete. It is superfluous to seek the solutions to other tones. (Dai Nianzu. Zhu Zaiyu-The Super Star of Science and Arts in the Ming Dynasty (《朱载堉——明代的科学和艺术巨星》). The People's Publishing House: pp.202, ed. 2011. Tian Huang Zhen Ren Zhu Zaiyu (《天潢真人朱载堉》) Elephant Press. April 2008, the first edition).

In fact, Zhu Zaiyu didn't get the concept of geometric progression from the perspective of consciousness. He still tirelessly listed the formulas for solving all the semitones of equal temperament, the formulas that have not been paid attention to by future generations.

Let's take our time and continue to read *Suan Xue Xin Shuo* (《算学新说》) *The New Theory of Mathematics*, an important book that contains the essence of the equal temperament calculations; you could see all the calculation formulas of Zhu Zaiyu. In the tenth question, Zhu Zaiyu gave a long list of formulas:

Among them, the lengths of the long strings can be used to determine the lengths of the shorter strings.

The lower $\text{Dalv}^2/2 =$ the lower Taicu.
The lower$\text{Jiazhong}^2/2 =$ the lower Guxi.
The lower $\text{Guxi}^2/2 =$ the lower Yizhe.
The lower $\text{Zhonglv}^2/2 =$ the lower Wushe.
The middle $\text{Ruibin}^2/2 =$ the middle Huangzhong.

Among them, the lengths of the short strings can be used to determine the lengths of the longer strings.

The lower $\text{Yingzhong}^2 =$ the lower Wushe.
The lower $\text{Wushe}^2 =$ the lower Yizhe.

The lower Nanlv ^{2} = the lower Ruivin.
The lower Yizhe^2 = the lower Guxi.
The lower Linzhong^2 = the lower Taicu.
The lower Ruibin ^{2} = the lower Huangzhong.

This set of formulas expresses Zhu Zaiyu's original idea of calculating the 12 semitones of equal temperament more clearly. It was nothing more than continued proportional solutions. He exhausted all the formulas that took the length of the lower Huangzhong and Huangzhong as the first proportional terms, with lengths of Ruibin Jiazhong, Nanlu, Dalu, and Yingzhong known. He then derived the formulas for the other proportional terms. Before people understood the concept of geometric progression, it had been undoubtedly the most effective way to express the calculation by listing the solving formulas of the items of the geometric progression according to the above methods one by one!

The above analysis was about how Zhu Zaiyu discovered the equal temperament, which does not require advanced mathematics, as long as he could calculate the square root. To this end, we can also get confirmation from the passage below by Zhu Zaiyu. In Question No. 1 shown in *Suan Xue Xin Shuo* (《算学新说》) *The New Theory of Mathematics*, Zhu Zaiyu clearly pointed out that the order of solving the 12 semitones of equal temperament is as follows: solve Huangzhong first just as a calendrical scientist would seek winter solstice first. Next solve Ruibin as a calendrical scientist would seek summer solstice. Then solve Jiazhong as a calendrical scientist would seek spring equinox. Afterwards, solve Nanlv as a calendrical scientist would seek the autumn equinox. Next, solve Dalv, which ranks next to Huangzhong. Next, solve Yingzhong, the ending tone of all the 12 semitones. Calendrical scientists would reckon the first date of the winter solace month, the regular months in the months and the leap months for the rest. Huangzhong is the beginning semitone; Ruibin is the middle semitone; and Yinghzong is the ending semitone. Based on the solution to Question No. 1, we can clearly infer in what order Zhu Zaiyu formulated the 12 semitones of equal temperament (Table 12.1).

律名	黄钟	大吕	太簇	夹钟	姑洗	仲吕	蕤宾
次序	1	5	7	3	8	11	2
比喻	冬至			春分			夏至
律名	林钟	夷则	南吕	无射	应钟	清黄	
次序	12	10	4	9	6	1	
比喻			秋分			冬至	

Table 12.1 Zhu Zaiyu's solution order table for the 12 semitones of equal temperament

Chinese	English	Chinese	English
律名	Tone	南吕	Nanlv
次序	Order	秋分	Fall equinox
比喻	Analogue	夹钟	Jiazhong
律名	Tone	春分	Spring equinox
次序	Order	无射	Wushe
比喻	Analogue	姑洗	Guxi
黄钟	Huangzhong	应钟	Yingzhong
冬至	Winter solace	仲吕	Zhonglv
林钟	Linzhong	清黄	Qinghuang
大吕	Dalv	冬至	Winter solace
夷则	Yize	蕤宾	Ruibin
太簇	Taicu	夏至	Winter solace

12.3.3 Great Creations Using Associations and Analogies

According to the mathematical formulas presented by Zhu Zaiyu in *Suan Xue Xin Shuo* (《算学新说》) *The New Theory of Mathematics*, given the contents shown in *Internal Chapter of Lv Lv Jing Yi* (《律吕精义·内篇》) *The Essence of Pitch-pipe Temperament*, we can infer the solving process of the semitones of equal temperament employed by Zhu as follows: Starting from the proportional relationship, inspired by the inner square and outer circle shown in *Zhou Li Li Shi Wei Liang* (《周礼·栗氏为量》), the string length of the middle semitone, i.e., the lower Ruibin, could be calculated with the string lengths of the lower Huangzhong and the middle Huangzhong. Next, with the string lengths of the lower Huangzhong, the lower Ruibin, and the middle Huangzhong, we can solve the string lengths of the lower Dalv, the lower Yingzhong, and so on. Once the string lengths of all the 12 semitones could be calculated, it would be easy to test the equal proportional relationship and the least common multiple 122. What Zhu Zaiyu focused on was not to figure out the mathematical formulations. Instead, he aimed to solve the issues of transition and modulation. Therefore, although he had figured out the string lengths of all the 12 semitones, he did not make further summarization or demonstration regarding the issue of the geometric progression. He tried to comply with the theories proposed by the predecessors, by suggesting the new law for density rate was consistent with the previous theories and the role of Huangzhong's scale. To this end, he meticulously borrowed the thought of the inner square and the outer circle as shown in *Zhou Li Li Shi Wei Liang* (《周礼·栗氏为量》). First, he found the string length of the lower Ruibin is the square root of 2. On top of that, he continued his formulation and discovered the 12 semitones of equal temperament.

From the above textual research, Kuttner's statement that Zhu Zaiyu "would not give a mathematical or theoretical explanation to the equal temperament law" was wrong. Judging from Kuttner's thesis, he made this judgment based on "how Zhu Zaiyu calculated his nine-digit length of the semitones." (Dai Nianzu. *Zhu Zaiyu-The Super Star of Science and Arts in the Ming Dynasty* (《朱载堉——明代的科学和艺

术巨星》). The People's Publishing House: pp.314, ed. 2011. *Tian Huang Zhen Ren Zhu Zaiyu* (《天潢真人朱载堉》) Elephant Press. April 2008, the first edition). Foreign scholars have had doubts about the ancient Chinese literature, which should be understandable. The generation of such negative opinions is partially related to the little research on the algorithm of "new law for density rate" and insufficient publicity regarding Zhu and his achievements. However, the more reasons lie in the mistakes of traditional Chinese academics in the Western modern scientific framework. Chinese mathematics has always emphasized "there are techniques in numbers while there are numbers in techniques." All the truths are contained in the wonderful example of solving and calculations. It was rare to give intermediate steps or make too many abstract proofs. There were mysteries that required the readers to carefully think over. Zhu Zaiyu's calculation was no exception. He gave the calculation formulas of each middle semitone one by one, instead of explaining the calculation process one by one with a kind reminder to readers at the end of the book "scholars should do their best to think over the process by themselves." We could not therefore reach a conclusion that Zhu himself could not understand it either. It is not difficult to reveal the specific process of Zhu Zaiyu's calculation of the 12 semitones of equal temperament when the simulation is repeated according to the historical background at that time. It once again shows the unique style of Chinese academics with integration between theories and practices.

In summary, the following conclusions can be drawn:

1. Zhu Zaiyu accomplished the calculation of the twelve semitones of equal temperament independently. When looking for the calculation methods, he was inspired by the inner square and the outer circle as shown in *Zhou Li Li Shi Wei Liang* (《周礼·栗氏为量》) and the Pythagorean theorem. Over the past over 200 years, the various criticisms of Zhu Zaiyu's citation of *Zhou Li* (《周礼》) *The Rites of Zhou* have been inappropriate.
2. In *Suan Xue Xin Shuo* (《算学新说》) *The New Theory of Mathematics*, Zhu gave the formulas to calculate all the semitones of equal temperament. The view that Zhu Zaiyu just offered some of the calculation formulas was inappropriate. The view that it was unnecessary for Zhu to offer all the details regarding the formulation has not been verified according to the relevant research so far.
3. It is conceivable that Zhu Zaiyu had completed the calculation of the 12 semitones of equal temperament and obtained the highest level of joy. Therefore, he did not tire himself of repeatedly checking the proportional relationship in various ways and came to brilliant conclusions. These conclusions were based on a rigorous calculation process and are completely consistent with the modern expression of the equal temperament. Zhu's achievements were so great that it suggested that the ancient Chinese musicology had reached a level that could well match their western counterparts.

So far, we have every reason to believe that the equal temperament is a wonderful flower of "modern science" born in the ancient Chinese traditional academic soil.

4. For a long time, the highly quoted words from Internal Chapter of *Lv Lv Jing Yi the Internal Chapter* (《律吕精义·内篇》) *The Essence of Pitch-pipe Temperament* could not serve as the sufficient proof to demonstrate the discovery of the equal temperament law. The questions listed in *Suan Xue Xin Shuo* (《算学新说》) *The New Theory of Mathematics* can be regarded as the relevant historical literature in detail, which showed the calculation formulas of the 12 semitones of equal temperament one by one and the original ideas for deriving these formulas. Combining these materials, we can reverse the specific calculation process of the 12 semitones of equal temperament.
5. Since Zhu Zaiyu's focus was to solve the problem of transition and modulation, he completed the calculation of the geometric sequence without subjective discussions correspondingly. Zhu had understood the geometric sequence. But we could not overstress too much its significance. Chen Feng, a scholar in the Qing Dynasty, offered a more systematic way to solve all items in the geometric progression. (Chen Feng. *Sheng Lv Tong Kao* (《声律通考》) Vol. 2).

In fact, in order to ensure the completeness and unity of the theoretical system, Zhu Zaiyu devoted considerable attention to how to resolve the compatibility of the new law for density rate with the studies of Heluo, the techniques and arts of The Duke of Zhou, and Chinese traditional musicology. In his writings, there are quite a lot of space and illustrations to carry out such explanatory expositions. The historical limitations of scholars' perspectives in the Ming Dynasty should be understood tolerantly rather than be particularly demanded.

In any case, the particularity of the highly integrated theory and experiment of ancient Chinese musicology suggested there remains much in-depth research of theoretical issues regarding the subject. The clarification of the formulation of the 12 semitones of equal temperament can help explore the source and brilliant achievements in Chinese musicology.

12.4 Ancient and Modern Scale Conversion and Its Effects on Pitch Pipes' Acoustics with Mouth Correction (for Related Content, Please Refer to Xu Fei. "Effects of the Conversion of Ancient and Modern Scales on the Pitch Frequencies of Pitch Pipes (古今尺度换算对律管发音的影响)". "Chinese Musicology (《中国音乐学》)", 1996, No. 4)

The conversion of ancient and modern scales is an important issue that must be involved in the study of ancient Chinese toning system. The data regarding the pitch pipes recorded in ancient Chinese classics were very rich. According to these historical data, deduction was conducted according to modern physics theory, or simulation restoration experiments were conducted. Then you can get a direct experience regarding the acoustic tests of ancient pitch pipes. However, in this research, the conversion of ancient and modern scales has always been an important reason for the difference in the results of various studies. Take the acoustic studies of

Zhu Zaiyu's different-diameter pitch pipes as an example. There have been many researchers in the past 100 years. Due to the different methods used, especially the different ancient and modern scale conversion formulas adopted by each person, the research results were different. No comparative analyses between them could be conducted with many questions outstanding. This chapter takes Zhu Zaiyu's scale conversion problem as an example to draw a general research conclusion.

In *Yue Lv Quan Shu* (《乐律全书》) *The Complete Book of the Tuning System*, not only did Zhu Zaiyu offered the criteria of pitch pipes made according to the equal temperament law, including the sizes, modes, and sample sketches of pitch pipes, but he also recorded the methods to produce these pitch pipes. According to Zhu Zaiyu's statement, the "new laws are created in accordance with the data given by pitch-pipes for the three octaves, i.e. the lower octave, the middle octave and the higher octave, totally 36 semitones." (The New and Old Pitch Pipes Experiment No. 7. *Lv Lv Jing Yi the Internal Chapter* (律吕精义》) *the Essence of Pitch-pipe Temperament Vol. 5*). All of them are consistent with the requirements of transition and modulation. What we are most concerned about today is whether this conclusion is true. Since the pitch pipes produced by Zhu Zaiyu himself had not been passed down, it is meaningful work to conduct theoretical calculations or restoration experiments in accordance with the documents left by Zhu Zaiyu.

Since the nineteenth century, there have been continuous studies on Zhu Zaiyu's pitch pipe restoration, acoustic studies, or theoretical calculation. Here are some examples of the most representative outcomes.

1. The work of V. Ch. Mahillon (1841–1924), a Belgian acoustician and curator of the Royal Belgian Museum of Musical Instruments. According to the data of Zhu Zaiyu's pitch pipes, he reproduced the three pitch pipes which stood for the semitones of the lower Huangzhong, the middle Huangzhong, and the higher Huangzhong and also conducted the acoustic tests accordingly. He took the measuring unit of the horizontal length of millet (Shu), i.e., 1 Chi equals 23.28 centimeters.
2. The work conducted by Yang Yinliu, a famous musicologist in 1937. He reproduced the pitch pipes according to the reproduced illustration shown in *Yue Lv Quan Shu* (《乐律全书》) *The Complete Book of the Tuning System* and conducted tests in reality before he drew the conclusion that the horizontal length of Shu (millet), also known as the Xia Chi (夏尺), 1 Chi equals 25.48 centimeters.

However, when Mr. Yang calculated the data of Zhu Zaiyu's pitch pipes according to the sizes of the textual research, the results obtained were very unsatisfactory. The intervals between two adjacent semitones are slightly smaller. Therefore, he proposed another amendment to deny the correction parameters for the diameters of Zhu Zaiyu's pitch pipes. He then suggested that to fully comply with the equal temperament law; Zhu's statement that the inner diameters of the various pitch pipes should be subsequently divided by 1.029302236($\sqrt[24]{2}$) should be changed to "sequentially divided by 1.059463094($\sqrt[12]{2}$)." (Yang Yinliu: *Outline of Chinese Music History* (《中国音乐史纲》), pp.:301–304. Shanghai Wanye Bookstore, 1952).

Yang Yinliu's conclusion has a great influence on the evaluation of Zhu Zaiyu's achievements in musicology. Since then, many other scholars adopted the Xia Chi used by Yang Yinliu as a scale conversion scheme to study the actual analog restoration and acoustic measurement including Liu Yong, a 1991 master student of the China Conservatory of Music, and Chen Wannai, a Taiwan scholar, etc. Using the conversion scheme of Yang Yinliu, Liu Yong did a restoration and acoustic measurement experiment of Zhu Zaiyu's different-diameter pitch pipes and concluded that "it proves that what Zhu Zaiyu proposed was exactly the equal temperament without compromise." (Liu Yong: *Study on the Acoustic Measurement of Zhu Zaiyu's Different-Diameter Pitch Pipes* (朱载堉异径管律的测音研究). "Chinese Musicology (《中国音乐学》), 1992, No. 4).

The Taiwan scholar Chen Wannai also conducted a restoration experiment based on the same conversion standard, only find that the actual measurement results were "too much different." Therefore, Chen concluded that "Zhu's so-called equal temperament had not been 100% accurate." (Chen Wannai: *Research on Zhu Zaiyu* (朱载堉研究). *Taipei Palace Museum Series* (《台北故宫博物院丛刊》), 1992, p. 93).

These research conclusions have affected people's correct evaluation and in-depth understanding of Zhu Zaiyu's formulation of the equal temperament, one of the greatest discoveries in ancient China.

In fact, many studies in the past mostly used the theoretical formulas of close-ended pitch pipes for analysis, which inevitably misread the forms and modes of Zhu's pitch pipes and his equal temperament law. From the beginning, the view that Zhu Zaiyu's different-diameter pitch pipes that contain minor errors had caused certain negative impact in the field of musicology was misleading as those scholars could not completely understand the pitch pipes made and used by Zhu Zaiyu. Therefore, this kind of opinion cannot be used as a theoretical standard for evaluating Zhu Zaiyu's achievements in pitch pipe production and musicology.

Our research has proved that if we consider the acoustic effects of the open-ended pipes first and then take into account the correction factors of Zhu Zaiyu's special pitch pipes of different diameters, we could find out that the pitch pipes of different diameters proposed and made by Zhu Zaiyu were completely in line with the formulation of equal temperament and could achieve ideal transition and modulation.

In the study of music theory, there has been a relatively common misconception about the frequency of Huangzhongluo pitch pipe, which went on like this: "For a pipe of the same length, the pitch of the open pipe is one octave higher than that of the closed pipe." In fact, such a view is an idealized conclusion that failed to consider mouth correction. To investigate the pitches of the pipes, the calculation and experiments should be conducted separately over the opening and closing conditions of the pitch pipes. For the Huangzhong pitch pipes, of the same length, the sound frequency of the open pipe is not equal to twice the sound frequency of the closed pipe.

In ancient China, there were many pitch pipes, all of which were generally called Huangzhong pitch pipes. According to the records in the relevant literature, as early as in the Zhou Dynasty, there were jade pitch pipes with both ends open. A set of bamboo pipes unearthed at No. 1 Han dynasty tomb of Mawangdui in Changsha could be regarded as one of the most representative pitch pipes. Twelve pitch pipes are inserted in an embroidered bag with the longest 17.65 centimeters long and the shortest 10.2 centimeters long. The internal diameter of the pitch pipes is 0.65 centimeters. This set of bamboo pipes served as physical reference to study the ancient pitch pipes in China.

However, when specifically studying the pitch frequency of the Huangzhong pitch pipes, many scholars simply thought that the open and closed pipes in the set of Huangzhong pitch pipes differed by only one octave according to the idealized formula of physics. Typical views are as follows:

1. Miao Tianrui: "for the pipes of the same length, the open-ended pipe has a pitch one octave higher than that of a close-ended pipe." (Miao Tianrui Musicology (《律学》) p.9, People's Music Publishing House, 1983).
2. Chen Yingshi: "the pipes of the same length, the open-ended pipe can sound one octave higher than the close-ended pipe does." (Xue Liang ed.: Handbook of Music Knowledge (《音乐知识手册》), China Literature and Art United Publishing Company, 1984, p. 310).
3. Wang Guangqi: "the pipe with both ends opened has an-octave pitch higher than the pipe with one open end of the same length." (Wang Guangqi. History of Chinese Music (《中国音乐史》) p.50, first edition in 1934, reprinted by Music Press in 1957).

In addition, the famous musicologist Yang Yinliu holds a similar view. The experts mentioned the mouth correction of the pitch pipes to varying degrees. However, they did not clearly clarify the relationship between the mouth correction and the pitch frequency of the pitch pipes. Instead, they would not consider the issue of mouth correction unless the difference between the air column vibration and the string vibration is stressed. Therefore, it is believed that the pitch of the open pitch pipe is one octave higher than that of the closed pitch pipes. Due to the influence of this concept, in the general books of musicology, music history, and even the history of weights and measures, most of them only give the mouth correction formula of the closed tube as if the pitch for the high-ended pipes can be self-evident.

In addition, the reason why people generally use the correction formula of the close-ended pipe is that some scholars believed that the ancient Chinese Huangzhong pipes are closed rather than open-ended. A typical view is, for example, as Zeng Wuxiu said: From the relationship between Huangzhong pipes and Huangzhong flutes, they were likely to be closed (Collection of Ancient Chinese Weights and Measures (《中国古代度量衡论文集》) pp.154–164, Zhongzhou Ancient Books Press, 1990). In Examination of Weights and Measures in Chinese Dynasties (《中国历代度量衡考》) compiled by Qiu Guangming, there was a similar view. The book only gives the theoretical frequency calculation formula of the close-ended pitch pipes, the source of which, the author did not state (Qiu Guangming: "Examination of Weights and Measures in Chinese Dynasties (《中国历代度量衡考》), Science Press. p.2. 1992); Wang Guangqi, a scholar in the earlier period, believed that "the ancient pitch-pipes in China should be those with one end closed for sure." (Wang Guangqi. History of Chinese Music (《中国音乐史》) p.51, first edition in 1934, reprinted by Music Press in 1957). In A History of Music in China (《中国音乐史纲》) by Yang Yinliu, the data regarding pitch pipes are calculated as they were close-ended. The frequency of the Huangzhong pitch pipe is 346.743 Hz. (Yang Yinliu: A History of Ancient Chinese Music (《中国古代音乐史稿》) pp.78–79. People's Music Publishing House, 1981, p. 1012).

Miao Tianrui was the only musicologist who adopted the open-ended pitch pipes to conduct the calculation and reached the conclusion that the frequency of Huangzhong was 693.5 Hz. He set that in the late Zhou Dynasty, 1 Chi (foot) was equal to 23.0886 cm today. Take a 9-cun pipe with an internal diameter of 3 fen with one end closed. For a pipe like that, it was workable to conduct the acoustic test or make the mathematical calculations. However, Miao Tianrui did not explain how he came to a conclusion that the pitch frequency of the pipe was 693.5 Hz or what method he used to reach such a conclusion. (Miao Tianrui: *Musicology* (《律学》) p. 100., People's Music Publishing House, 1983). When the data are compared with Yang Yinliu's calculated frequency of the close-ended pipe's pitch frequency at 346.743 Hz, it can be found that: $693.5 = 346.743 \times 2$. Miao Tianrui's calculation of the pitch frequency of Huangzhong's open pipe is simply based on the product of multiplying the pitch frequency of Huangzhong's closed pipe by 2. In *History of Ancient Chinese Music* (《中国古代音乐史稿》) by Yang Yinliu (page 87) and *History of Chinese Acoustics* (《中国声学史》) by Dai Nianzu (page 149) and other books, there are references to Miao Tianrui's Huangzhong frequency of 693.5 Hz without any controversies. Dai Nianzu was aware that something must have been wrong with the algorithm of Miao Tianrui's pitch pipe pitch frequency. In turn he added: "Although the calculations of Mr. Miao and Yang differ by an octave, the results are consistent."

At a first glance, it seemed that the difference between the open and closed pitch pipes was only one octave. The conclusion that the pitch frequency of Yang Yinliu's closed pipe at 346.743 Hz and the pitch frequency of Miao Tianrui's open pipe at 693.5 Hz was exactly one octave different. In fact, that was an approximate estimation that had ignored the different correction factors for the open and closed pipes. In fact, the calculation of the pitch frequencies of the open pipes is not so simple as imagined.

In view of the abovementioned misunderstandings regarding the pitch frequencies of open and closed rhythms in the study of law, it is necessary to discuss the mouth correction of the pitch pipes more carefully from the aspects of physical analysis.

No matter what the toning system is involved, the pitches of strings and pipes are not equivalent. The string system is relatively simple with relatively unified conclusion, which is not described here. Due to the particularity of the acoustic system in wind instruments, the pitch frequency of a pipe of a certain length is not only related to the length, but also related to various factors such as the diameter, texture, and temperature of the pipe when blown.

The so-called pitch pipe is actually a hollow cylinder made of various materials. The air column in the cylinder is like other elastic materials which will vibrate when subjected to external forces. The excited vibration of the air column will produce waves in both directions. As air is a fluid elastic substance, the vibration of the air column mainly takes the form of longitudinal waves, which determines the fundamental frequency of vibrations and a series of overtones, while a small number of transverse waves can produce high-order overtones. The longitudinal waves and transverse waves are superimposed to form the unique acoustic effect of the pitch pipes.

The characteristic of air column vibration is that the longer the pipe is, the bigger the diameter is, the lower the frequency is, and vice versa. By changing the length and diameter of the pipe, the purpose of emitting sounds with different frequencies can be achieved. For example, Zhu Zaiyu's different-diameter pipes are to create 36 pipes with lengths that conform to the equal temperament law and the diameter changes according to a specific ratio to achieve the purpose of pitching.

The shapes of the ancient pitch pipes could fall into two types, open-ended and close-ended. They were also different in pitching accordingly (Picture 12.6).

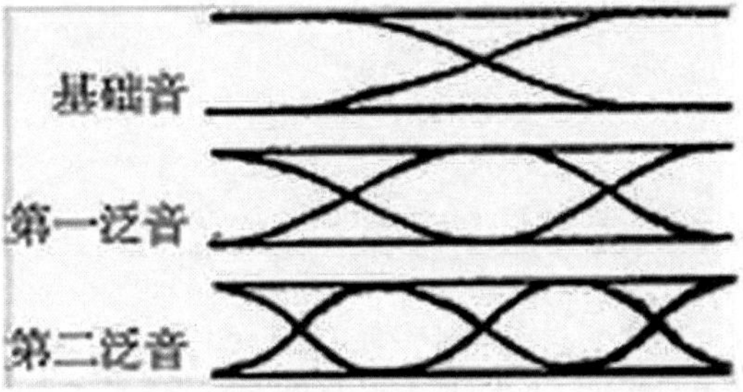

For an open pipe with both ends open, the air column vibration between them is shown on the left. The two ends are the antinodes of the vibration, and the center of the pipe is the node of the fundamental frequency. The first harmonic node is formed at the one-fourth and third-fourth points of the pipe. The second harmonic node is formed at the one sixth, five sixths, and one-half points of the pipe. The air column vibration wave of the open pipe starts from a section of the tube and reaches the other end of the pipe before it is reflected. When the reflected wave returns to the original position, it coincides completely with the phase of the next vibration wave to form a standing wave, the amplitude is strengthened, and the volume is also the largest.

Picture 12.6 Vibration of the open-ended air column

中文	English
基础音	Fundamental note
第一泛音	The first overtone
第二泛音	The first overtone

Picture 12.7 Vibration of the close-ended air column

中文	English
基础音	Fundamental note
第二泛音	The first overtone
第四泛音	The first overtone

Therefore, the fundamental frequency of the open pipe is 2 times as long as the pipe length. The wavelength of the first overtone is one time as long as the pipe. The wavelength of the second overtone is 2/3 of the pipe length. Others can be inferred in a similar way.

In this way, we can get the pitch frequency of the open-ended pipe. (Picture 12.7)

$$F = V/\lambda = V/2L \tag{12.1}$$

The Nth overtone frequency: $F{=}NV/2\,L$ ($N = 1, 2, 3\ldots$)

Where L = length of the pipe, V = sound velocity in the pipe, λ = wave length

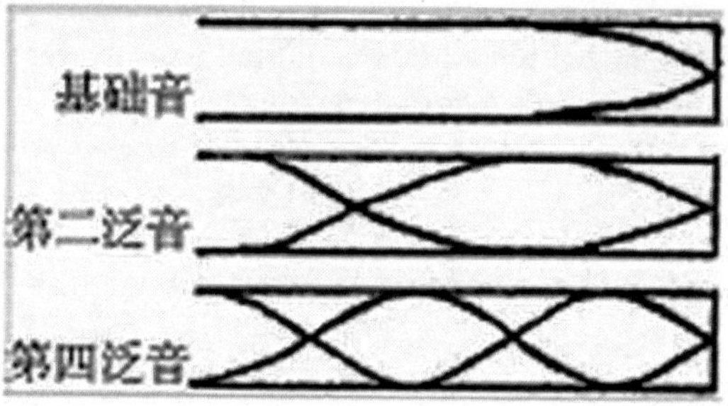

For a closed pipe with one end closed and one end open, the open end is where the antinode is. Since the air at the close end is always not vibrating, it is always at a node. The specific air column vibration is as shown in the right picture. Since the vibration wave of the air column in the closed pipe needs to go back and forth twice to make the reflected wave and the reoccurring wave coincide in phase, its period is $T = 4\ L\ /\ V$. In addition, the vibration in the closed pipe can only generate even-numbered overtones such as the second, fourth, sixth, etc., but cannot produce odd-numbered overtones such as the first, third, fifth, and so on.

In this way, we can get the pitch frequency of the close-ended pipe.

$$F = V/\lambda = V/4L \tag{12.2}$$

The Nth overtone frequency: $F = (2\ \text{N-}1)V/4\ \text{L}$ ($N = 1, 2, 3\ldots$)

where L = length of the pipe, V = sound velocity in the pipe, λ = wave length.

From this we can also know if the lengths of the pipes are the same. By comparing (12.1) and (12.2), we infer that the fundamental frequency of the open pipe is twice that of the closed pipe, which is 1 octave higher. This is the theoretical basis of the previous toning theory that "the pitch for a pipe of the same length is one octave higher than that of a closed pipe." However, the above calculation is idealized, without considering the mouth correction of the pitch pipes.

From the above two simple formulas, we can see that the pitches of the pipes are related to the length of pitch pipes. The pitch frequency and overtone frequency are inversely proportional to the pipe length. The longer the pipe is, the lower the frequency is.

However, further physical experiments have found that the actual measured frequency is inconsistent with the calculation result of the above theoretical formula, which is generally lower than the frequency value given by the formula. The reason for this phenomenon is that when the air column in the pipe vibrates, due to the inertia of air in motion, the wave reflection does not occur exactly at the opening, but is reflected from a point outside the opening surface, which increases the effective length of the tube by ΔL, and the sound source that is located at the ΔL outside the opening propagates the sound waves in the pipe to the surrounding space. From this point, the longitudinal wave vibration changes from the plane wave form in the pipe to the spherical wave form outside the pipe. In addition, at the opening of the pipe because the lips cannot touch the opening all the time, there is a gap between where the maximum pressure is, i.e., the antinode position of the vibration and opening of the pipe. These additional effective pipe lengths are the "open end correction" and "mouth correction" required for consideration by wind instrument producers. Only when the mouth correction number is considered can pitch pipes be made according to the equal temperament and can the toning system of pipes be consistent with that of strings.

The determination of the mouth correction number needs to be combined with complex factors such as the shape of the opening of the specific pipe and the comparison of experimental sound measurements. The currently used mouth correction numbers are based on empirical formulas. In the history of modern physics, many scholars have studied the mouth correction parameters. The more famous ones included those done by the famous physicist The Duke of Rayleigh, (1842–1919) and Xu Shou (1818–1884) a Qing Dynasty scholar. Based on the existing research results, for the open-ended pipe with length L and inner diameter D, the correction value of the orifice can be:

$$\Delta L = 0.306D \tag{12.3}$$

When calculating the fundamental frequency (pitch frequency), the effective length of the pitch pipe is (L + 2ΔL), because the two ends of the open tube are calibrated.

For the closed pipe with length L and inner diameter D, the correction value of the orifice can be taken as

$$\Delta L = 5D/3 \tag{12.4}$$

Strictly speaking, there are other factors that can also affect the pitches of the pipes, including temperature, (Tang Lin et al.: "Music Physics (《音乐物理学》) p. 108. China University of Science and Technology Press, 1991) wall thickness, (Tang Lin et al.: "Music Physics (《音乐物理学》) p. 107. China University of Science and Technology Press, 1991) and wall material. However, according to many experiments conducted by relevant scholars, if the pipe wall is not very thin and soft or very hard or flexible, "Generally speaking, the pipe wall material has little effect on the tone." Accuracy can be ignored in theoretical analysis

(E.G. Richardson. ed. Technical Acoustics (《声学技术概要》) pp.449. Science Press. 1961). All these factors can be ignored in non-high-precision theoretical analysis.

Although the velocity of sound is a function of temperature, in general calculations, it is enough to take $V = 340$ m / s, which is approximately equivalent to the velocity of sound waves when the air temperature is 14 degrees Celsius. According to the above analysis, at normal temperature, when $V = 340$ m/s, take into account formulas (12.1) (12.2) (12.3) (12.4). The formula for calculating the pitch frequency of the open-ended pipe is:

$$\begin{aligned} F_{(open-ended\ pipe)} &= V/2(L + 2 \times 0.306D) = 34000/2(L + 0.612D) \\ &= 17000/(L + 0.612D) \end{aligned} \tag{12.5}$$

For close-ended pipe, given the mouth correction, the formula for frequency goes on like this:

$$\begin{aligned} F_{(close-ended)} &= V/4(L + 5D/3) = 34000/4(L + 1.6667D) \\ &= 8500/(L + 1.6667D) \end{aligned} \tag{12.6}$$

where L is the length of the pipe and D the internal diameter of the pipe. The unit for length is centimeters, and the unit for frequency is Hertz.

The above theoretical formula of the pitch frequency of pitch pipe is suitable for the theoretical calculation of the frequencies of ancient Chinese pitch pipes. When calculating the close-ended pipes, the formula (12.6) is the same as the formulas adopted by Yang Yinliu and Chen Wannai. But for the calculation of the pitch frequency of the open pipe, compare formula (12.5) and formula (12.6) and we will know: you can't simply think that the pitch frequency of the open pipe is 1 octave higher than that of the closed pipe, because the formula (15) $\neq 2 \times$ Formula (12.6).

According to different research purposes, more accurate calculation formulas in acoustic research can also be used, but the formulas (12.5) and (12.6) currently used are more accurate than the idealized formulas that ignore the mouth correction. They are used to calculate the frequency relationship of a set of pipes. As for more precise mathematical research on the pitch frequencies of pitch pipes, we must also refer to professional acoustic works and adopt more rigorous formulas.

Through the analysis of the mouth correction principle, we learned that in the study of toning system, it is not possible to continue to use the overly crude statement that "the sound of an open-ended pipe of the same length is one octave higher than that of a closed pipe."

Using the abovementioned mouth correction method to analyze some historically controversial issues regarding toning system, such as whether Zhu Zaiyu's different-diameter pipes conform to the equal temperament law etc., new results can be obtained to clarify the dispute and distinguish differences. When analyzing Mengkang pitch pipes of different diameters in the Western Jin Dynasty and the pitch pipes of different diameters made by Ruan Yi and Hu Yuan in the Song

Dynasty, the above theories also helped us to confirm that the two sets of pitch pipes earlier than those of Zhu Zaiyu had already achieve a three-point losing and gaining relationship. (Xu Fei. Examination of the Mengkang Different-Diameter Pitch Pipes in Western Jin Dynasty"(西晋孟康异径管律考证). Historical Materials of Chinese Science and Technology (《中国科技史料》), 2000(3). Xu Fei: "Mathematical Verification of the Acoustic Achievement of Different-diameter Pitch Pipes of Ruan Yi and Hu Yuan in Song Dynasty (宋代阮逸胡瑗异径管律声学成就的数理验证). Research on the History of Natural Science (《自然科学史研究》) 2001(3)).

Using the abovementioned relatively precise formula, one can also make physical verification of Zhu Zaiyu's extraordinary achievements in formulating the equal temperament.

12.5 Historic Achievements of Zhu Zaiyu's Formulation of the Equal Temperament

As we all know, Zhu Zaiyu not only theoretically invented the equal temperament law as he called the "new law for the density rate" but also personally experimented and gave the criteria of strings and pitch pipes according to the "new law for the density rate." Regarding Zhu Zaiyu's research, the conclusions are unanimous and affirmative, and I will not repeat them here. As the pitch frequency of pipes is not related to the length of pipe, the factor of the diameter correction should also be considered. For the past hundred years, the question of whether the pitches of the different-diameter pitch pipes of Zhu Zaiyu met the equal law temperament had been discussed with controversies. Judging from the text of Zhu Zaiyu's original book, he not only produced these pitch pipes, but performed acoustic tests. Therefore, he had reached a conclusion that "the pitch pipes made according to the three-point loss and profit rule were not harmonious." The pitch pipes made according to the equal temperament law Zhu proposed were consistent. (The New and Old Pitch Pipes Experiment No. 7 (新旧律实验第七). *Lv Lv Jing Yi* the Internal Chapter (律吕精义》) *the Essence of Pitch-pipe Temperament Vol. 5)*

Based on literature review, Zhu Zaiyu not only found the problem of mouth correction of the pitch pipes but also solved this problem uniquely through experiments and theoretical calculations with provision of specific data, production technology, and small samples of the different-diameter pitch pipes. His plan was to make a full set of pitch pipes with different internal and external diameters, through the adjustment of which the mouth correction of pitch pipes could be achieved. Was the data of mouth correction adopted by Zhu Zaiyu reasonable and verified? For the past hundred years, many scholars both at home and abroad had conducted the restoration experiments and acoustic tests. However, their results were quite different with both positive and negative response toward Zhu's data. Such phenomena made it hard to decide whether Zhu Zaiyu's different-diameter pitch pipes worked or whether the mouth correction methods he proposed were workable. It had been a historical mystery.

Table 12.2 Conclusions of V. Ch. Mahillon drawn from his research

Chinese	English	Chinese	English
通长	Overall length	闭口	Close-ended
内径	Internal diameter	黄钟倍律	The lower Huangzhong
外径	External diameter	黄钟正律	The middle Huangzhong
音高	Pitch	黄钟半律	The higher Huangzhong
开口	Opening		

Since the last century, there have been continuous studies through the restoration experiments and acoustic tests of Zhu Zaiyu's pitch pipes. Here are some of the most representative outcomes.

1. The work of the Belgian scholar V. Ch. Mahillon. The main research conclusions are shown in Table 12.2 below:

		黄钟倍律	黄钟正律	黄钟半律
通长		2	1	0. 5
内径		0. 05	0. 035 35	0. 025
外径		0. 070 71	0. 05	0. 035 35
音高	开口	$E^{\flat}_{3}$	$E^{\flat}_{4}$	$E^{\flat}_{5}$
	闭口	E_{2}	E_{3}	E_{4}

The results were very satisfactory. The pitches of the three pitch pipes fully met the expectation of one octave away. We know little about the details regarding the restoration experiments done by V. Ch. Mahillon; we can only refer to his conclusions.

2. The work of Yang Yinliu, the musicology in 1937. He denied the mouth correction parameters adopted by Zhu Zaiyu and proposed that the statement that the internal diameters of all pitch pipes shall be divided by 1.029302236($\sqrt[24]{2}$) as suggested by Zhu Zaiyu shall be changed into divided by 1.059463094($\sqrt[12]{2}$) (Yang Yinliu: "*A History of Ancient Chinese Music* (《中国古代音乐史稿》), p.301–304.People's Music Publishing House, 1981).

This proposal of amendment was undoubtedly a fatal denial of the accuracy of Zhu Zaiyu's different-diameter pitch pipes. Yang Yinliu was a famous musicologist in China. His views had always had a great negative impact on people's full recognition of Zhu Zaiyu's great achievements in the equal temperament, especially preventing people from believing in the correctness of Zhu Zaiyu's different-diameter pitch pipes.

3. In the 1990s, Liu Yong, a 1991 master student of the China Conservatory of Music, conducted experiments using Yang Yinliu's Xia Chi conversion scheme and concluded that "it has proven that Zhu's pitch pipes are uncompromisingly in line with the equal temperament." Liu Yong's work was later published in the fourth issue of "Chinese Musicology (《中国音乐学》)" in 1992. But in this issue of the journal, Liu Cunxia held to an opposite view. He still believed that "the data of Yang (Yang Yinliu) were more accurate than the data of Zhu (Zhu Zaiyu)".
4. Chinese Taiwanese scholar Chen Wannai also conducted a restoration experiment according to Yang Yinliu's specifications and found that the measured results were not satisfactory. From this, Mr. Chen concluded: "The equal temperament law discovered by Zhu Zaiyu had not yet reached 100% accuracy." Mr. Chen then referred to Yang Yinliu's ideas and proposed the same modification to revise the pitch pipes of Zhu Zaiyu. He then made calculations and reached a conclusion, saying "the result is completely satisfactory." (Chen Wannai: *Research on Zhu Zaiyu (朱载堉研究)*. Taipei Palace Museum Series (《台北故宫博物院丛刊》) p. 93–94, 1992). This is a relatively recent negative conclusion on Zhu Zaiyu's different-diameter pitch pipes and their advances.

The inconsistency of such large theoretical calculations or the restoration of experimental results inspired us to review our conclusions. Modern physics has provided us with a more reliable theoretical analysis tool. According to this idea, we can get a more convincing new conclusion.

The theoretical verification of Zhu Zaiyu's different-diameter pitch pipes goes on as follows:

According to the previous analysis, when V = 340 m / s, at normal temperature, for the open-ended pitch pipes, the frequency calculation formula upon considering the mouth correction is as follows:

$$\begin{aligned} F_{open-ended} &= V/2(L + 2 \times 0.306D) = 34{,}000/2(L + 0.612D) \\ &= 17{,}000/(L + 0.612D) \end{aligned} \tag{12.7}$$

For the close-ended pitch pipes, the frequency calculation formula upon considering the mouth correction is as follows:

$$\begin{aligned} F_{close-ended} &= V/4(L + 1.6667D) = 34{,}000/4(L + 1.6667D) \\ &= 8500/(L + 1.6667D) \end{aligned} \tag{12.8}$$

where L is the length of the pitch pipe, D is the inner diameter of the pitch pipe, the unit of length is centimeter, and the unit of frequency is Hertz.

There were both open-ended and close-ended pitch pipes in ancient China, which varied from time to time and from place to place. Therefore, no easy generation can be made. As far as the pitch pipes of Zhu Zaiyu are concerned, they were open-ended obviously. In the first paragraph of Wind Instruments Volume 1, *Lv Xue Xin Shuo* (《律学新说》) *A New Account of the Science of the Pitch-pipes*, there was a statement saying that Zhu Zaiyu's pitch pipes must be open-ended.

Next, let's discuss how to convert Xia Chi used by Zhu Zaiyu into a modern unit of measurement.

Zhu Zaiyu used the horizontal Shu ruler also known as Xia Chi to make the pitch pipes. This was the scale that Zhu Zaiyu had worked hard to study. Zhu Zaiyu believed that only by making the Huangzhong according to the scale of the ancient times could the pitches maintained. When V. Ch. Mahillon made the restoration experiments, he used a horizontal Shu Chi, i.e., 1 Chi (foot) = 23.28 cm. According to the sizes of pitch pipes drawn on *Yue Lv Quan Shu* (《乐律全书》) *The Complete Book of the Tuning System*, based on measurements, Yang Yinliu reached the conclusion that Zhu Zaiyu used the horizontal Shu Chi, i.e., 1 Chi (foot). Zhu Zaiyu's horizontal ruler, or summer ruler, was obtained through actual measurement. Its 1 foot = 25.48 cm. Liu Yong and Chen Wannai adopted the Xia Chi as the standard which was calculated by Yang Yinliu during the restoration experiments without any adjustments.

We found that there are better conversion methods. According to Zhu Zaiyu's article in Volume Two "Three Rules of the Present System Lv Xue Xin Shuo (《律学新说》) A New Account of the Science of the Pitch-pipes: "the scale for producing the pitch pipes was 12 cuns and a half of the Xia Chi." Combining the results of modern archaeological research, 1 Chi in the Ming Dynasty could be 32 centimeters.

Table 12.3 Table of parameters of the 36 pitch pipes listed in *Yue Lv Quan Shu* (《乐律全书》) *The Complete Book of the Tuning System*

Chinese	English	Chinese	English
律管名称	Pipe name	倍律夷则	The lower Yize
通长	Overall length	倍律南吕	The lower Nanlv
内径	Internal diameter	倍律无射	The lower Wushe
外径	External diameter	倍律应钟	The lower Yingzhong
倍律黄钟	The lower Huangzhong	正律黄钟	The middle Huangzhong
倍律大吕	The lower Dalv	正律大吕	The middle Dalv
倍律太簇	The lower Taicu	正律太簇	The middle Taicu
倍律夹钟	The lower Jiazhong	正律夹钟	The middle Jiazhong
倍律姑洗	The lower Guxi	正律姑洗	The middle Guxi
倍律仲吕	The lower Zhonglv	正律仲吕	The middle Zhonglv
倍律蕤宾	The lower Ruibin	正律蕤宾	The middle Ruibin
倍律林钟	The lower Linzhong	正律林钟	The middle Linzhong

It was a relatively credible conclusion. Not only was the ruler uniform in the whole country in the Ming Dynasty, but the results of modern scholars' research are basically the same. Therefore, we can immediately calculate Zhu Zaiyu's horizontal Shu Chi (Xia Chi), 1 foot (Chi) = 32 × 10 / 12.5 = 25.6 cm.

Now, let us review the original data of the complete and intact equal temperament given by Zhu Zaiyu. In *Yue Lv Quan Shu* (《乐律全书》) *The Complete Book of the Tuning System*, all the data of the 36 pitch pipes were listed (Table 12.3).

律管名称	通 长	内 径	外 径
倍律黄钟	2.000 000	0.050 000	0.070 710
倍律大吕	1.887 748	0.048 576	0.068 697
倍律太簇	1.781 797	0.047 193	0.066 741
倍律夹钟	1.681 792	0.045 850	0.064 841
倍律姑洗	1.587 401	0.044 544	0.062 996
倍律仲吕	1.498 307	0.043 276	0.061 202
倍律蕤宾	1.414 213	0.042 044	0.059 460
倍律林钟	1.334 839	0.040 847	0.057 767
倍律夷则	1.259 921	0.039 685	0.056 123
倍律南吕	1.189 207	0.038 555	0.054 525
倍律无射	1.122 462	0.037 457	0.052 973
倍律应钟	1.059 463	0.036 391	0.051 465
正律黄钟	1.000 000	0.035 355	0.050 000
正律大吕	0.943 874	0.034 348	0.048 576
正律太簇	0.890 898	0.033 370	0.047 193
正律夹钟	0.840 896	0.032 420	0.045 850
正律姑洗	0.793 700	0.031 498	0.044 544
正律仲吕	0.749 153	0.030 601	0.043 276
正律蕤宾	0.707 106	0.029 730	0.042 044
正律林钟	0.667 419	0.028 883	0.040 847

律管名称	通　长	内　径	外　径
正律夷则	0.629 960	0.039 061	0.039 685
正律南吕	0.594 603	0.027 262	0.038 555
正律无射	0.561 231	0.026 486	0.037 457
正律应钟	0.529 731	0.025 732	0.036 391
半律黄钟	0.500 000	0.025 000	0.035 355
半律大吕	0.471 937	0.024 288	0.034 348
半律太簇	0.445 449	0.023 596	0.033 370
半律夹钟	0.420 448	0.022 925	0.032 420
半律姑洗	0.396 850	0.022 272	0.031 498
半律仲吕	0.374 576	0.021 638	0.030 601
半律蕤宾	0.353 553	0.021 022	0.029 730
半律林钟	0.333 709	0.020 423	0.028 883
半律夷则	0.314 980	0.019 842	0.039 061
半律南吕	0.297 301	0.019 277	0.027 262
半律无射	0.280 615	0.018 728	0.026 486
半律应钟	0.264 865	0.018 195	0.025 732

Chinese	**English**	**Chinese**	**English**
律管名称	Pipe name	半律太簇	The higher Taicu
通长	Overall length	半律夹钟	The higher Jiazhong
内径	Internal diameter	半律姑洗	The higher Guxi
外径	External diameter	半律仲吕	The higher Zhonglv
正律夷则	The middle Yize	半律蕤宾	The higher Ruibin
正律南吕	The middle Nanlv	半律林钟	The higher Linzhong
正律无射	The middle Wushe	半律夷则	The higher Yize
正律应钟	The middle Yingzhong	半律南吕	The higher Nanlv
半律黄钟	The higher Huangzhong	半律无射	The higher Wushe
半律大吕	The higher Dalv	半律应钟	The higher Yingzhong

(The figures in the book have 17 significant digits, here only the former 7 digits are listed. Measuring Unit: Xia Chi, Chi)

Once there were some scholars who tried to check whether the pitch pipes of Zhu Zaiyu met the equal temperament but did not take the original data of Zhu Zaiyu's experiments. Instead, they just assume the internal diameter of Huangzhong pipe was 2 centimeters with an overall length of 64.82581 cm. (Liu Cunxia: Theoretical Test of Zhu Zaiyu's Different-Diameter Pitch Pipes (朱载堉异径管律的理论检验). "Chinese Musicology (《中国音乐学》), No. 4. 1992). Such a testing method that deviates from the original intent of ancient books is not grounded. Zhu Zaiyu's different-diameter pitch pipes resulted from systematic experiments with a systematic concept. It is not possible to set and check Zhu Zaiyu's pitch pipes at will based on the common ratios such as $\sqrt[12]{2}$ and $\sqrt[24]{2}$. In addition, it is not appropriate to use the closed-pipe formula to estimate the pitch frequencies without following Zhu Zaiyu's original intent.

To check whether Zhu Zaiyu's pitch pipes meet the theoretical requirements of the equal temperament, you can use the data in the table above to convert them to a modern metric scale according to the Xia Chi conversion formula and then conduct theoretical or experimental verification.

If we convert the pipe length and diameters of Zhu's pitch pipes according to the formula 1 Chi (foot) = 25.6 centimeter to the modern metric unit, then based on the formula for the open-ended pipes (21), we can calculate the theoretical frequencies, cent values, and pitch differences of Zhu's pitch pipes that were in line with the equal temperament law (Table 12.4).

律 名	管 长	管内径	管外径	理论频率	音分值	音分差
正律黄钟	25.600 0	0.905 0	1.280 0	650.00	0	
正律大吕	24.163 1	0.879 1	1.243 4	688.37	99	99
正律太簇	22.806 8	0.854 3	1.208 1	728.69	197	98
正律夹钟	21.526 8	0.830 0	1.173 8	771.51	296	99
正律姑洗	20.318 7	0.806 1	1.140 2	816.84	395	99
正律仲吕	19.178 2	0.783 4	1.107 7	864.80	494	99
正律蕤宾	18.101 8	0.761 1	1.076 2	915.57	593	99
正律林钟	17.085 7	0.739 3	1.045 5	969.32	691	98
正律夷则	16.127 0	0.718 3	1.015 8	1 026.16	790	99
正律南吕	15.221 8	0.697 9	0.986 9	1 086.34	889	99
正律无射	14.367 5	0.677 9	0.958 7	1 150.02	987	98
正律应钟	13.561 1	0.658 7	0.931 6	1 217.40	1 086	99
半律黄钟	12.800 0	0.640 0	0.905 0	1 288.69	1 184	98

According to the analysis of modern physical theory, Zhu Zaiyu's systematic mouth correction method basically met the theoretical requirements of the equal temperament, and the pitch differences between each two pipes were up to 2 cents

Table 12.4 Theoretical frequency, cent value, and cent difference data of Zhu Zaiyu's equal temperament of 12 semitones

Chinese	English	Chinese	English
律名	Pitch name	正律夹钟	The middle Jiazhong
管长	Pipe length	正律姑洗	The middle Guxi
管内径	Internal diameter	正律仲吕	The middle Zhonglv
管外径	External diameter	正律蕤宾	The middle Ruibin
理论频率	Theoretical frequency	正律林钟	The middle Linzhong
音分值	Cent value	正律夷则	The middle Yize
音分差	Cent difference	正律南吕	The middle Nanlv
正律黄钟	The middle Huangzhong	正律无射	The middle Wushe
正律大吕	The middle Dalv	正律应钟	The middle Yingzhong
正律太簇	The middle Taicu	半律黄钟	The middle Huangzhong

Unit: centimeters

at most, which was undetectable by human ears. Even the accumulated pitch difference over the 12 semitones was only 16 cents. This could already be adjusted through the technical treatment of the blower. However, Zhu Zaiyu had a better way.

As can be seen from the table above, Zhu Zaiyu's pitch pipes had a cent difference of less than 100 cents. Therefore, there was the pitch difference of less than 1200 cents between the low Huangzhong and the high Huangzhong. To make up for this deficiency, Zhu Zaiyu cleverly added the same size of mouthpieces to all the pitch pipes. On the one hand, it was helpful for the pipe blower to blow out the pitch of each pipe. On the other hand, it could systematically increase the pitch frequencies to achieve the perfect octave relationship between the lower, the middle, and the higher pipes.

Zhu Zaiyu stipulated the norm of such mouthpieces in detail. (*Lv Lv Jing Yi* the Internal Chapter (律吕精义》) *the Essence of Pitch-pipe Temperament Vol. 5).* Unfortunately, this exquisite design had been neglected by many scholars for a long period of time. Except for Liu Yong, almost no one was testing Zhu Zaiyu's different-diameter pitch pipes and his equal temperament together with the impact of this mouthpiece. Therefore, we may wish to do a little more work to further consider the impact of Zhu Zaiyu's special mouthpieces.

Suppose the mouthpiece diameter is Δc; after considering the influence of the mouthpiece, the air pressure of pipes is calculated from the center point of the mouthpiece, and the length of the pipe is reduced by 0.5Δc. The frequency is:

$$F_{open-ended} = V/2(L + 2\Delta L - 0.5\Delta c) = V/2(L + 2 \times 0.306D - 0.5\Delta c)$$

As analyzed above, take Xia Chi, the scale, 1 Chi (foot) = 25.6 cm, and Δc = 25.6 × 0.0176 = 0.4506 cm; take V = 340 m / s = 34,000 cm / s, so that the calculation results are listed in Table 12.5:

Table 12.5 Theoretical frequency, pitch value, and pitch difference of Zhu Zaiyu's pitch pipes of equal temperament

Chinese	English	Chinese	English
律名	Pitch name	音分值	Cent value
管长	Pipe length	音分差	Cent difference
管内径	The internal diameter	正律黄钟	The middle Huangzhong
管外径	The external diameter	正律大吕	The middle Dalv
理论频率	Theoretical frequency		

律 名	管 长	管内径	管外径	理论频率	音分值	音分差
正律黄钟	25.6000	0.9050	1.2800	655.65	0	
正律大吕	24.1631	0.8791	1.2434	694.56	100	100

律 名	管 长	管内径	管外径	理论频率	音分值	音分差
正律太簇	22.806 8	0.854 3	1.208 1	735.79	200	100
正律夹钟	21.526 8	0.830 0	1.173 8	779.48	300	100
正律姑洗	20.318 7	0.806 1	1.140 2	825.77	400	100
正律仲吕	19.178 2	0.783 4	1.107 7	874.83	500	100
正律蕤宾	18.101 8	0.761 1	1.076 2	926.82	600	100
正律林钟	17.085 7	0.739 3	1.045 5	981.93	700	100
正律夷则	16.127 0	0.718 3	1.015 8	1 040.31	800	100
正律南吕	15.221 8	0.697 9	0.986 9	1 102.21	900	100
正律无射	14.367 5	0.677 9	0.958 7	1 167.82	1 000	100
正律应钟	13.561 1	0.658 7	0.931 6	1 237.36	1 100	100
半律黄钟	12.800 0	0.640 0	0.905 0	1 311.08	1 200	100

Chinese	English	Chinese	English
律名	Pitch name	正律姑洗	The middle Guxi
管长	Pitch length	正律仲吕	The middle Zhonglv
管内径	Internal diameter	正律蕤宾	The middle Ruibin
管外径	External diameter	正律林钟	The middle Linzhong

(continued)

Chinese	English	Chinese	English
理论频率	Theoretical frequency	正律夷则	The middle Yize
音分值	Cent value	正律南吕	The middle Nanlv
音分差	Center difference	正律无射	The middle Wushe
正律太簇	The middle Taicu	正律应钟	The middle Yingzhong
正律夹钟	The middle Jiazhong	半律黄钟	The middle Huangzhong

Length unit: centimeters

From this, we have reached an exciting conclusion: after considering the system correction factors of the special mouthpieces, Zhu Zaiyu's different-diameter pitch pipes could achieve transition and modulation without any difference. In other words, Zhu Zaiyu's different-diameter pitch pipes fully met the theoretical requirements of the equal temperament.

The data of Zhu Zaiyu's different-diameter pitch pipes were very precise, and subtle changes in the order of millimeters would also be reflected in the pitch frequency. No wonder when talking about the mouthpiece shaping system, Zhu Zaiyu emphasized that they should be exactly proper, without overestimation or underestimation. This exquisite design showed that Zhu Zaiyu not only had a profound training in arithmetic theory but also had originality in the practice of pitch pipe making. With its unique design, Zhu Zaiyu's unique different-diameter pitch pipes had solved the age-old dilemma of transition and modulation both from the theoretical and legal aspects, which is truly amazing!

According to the theoretical analysis and precise calculations of modern physics, Zhu Zaiyu's brilliant achievements in establishing the equal temperament theory and practice could be more comprehensively confirmed. It can be said that Zhu Zaiyu had also made a major contribution to the mouth correction of the pitch pipes with the use of mouth pieces, rather than despite the existence of subtle errors as mentioned above.

As for why there are so many discrepancies in the experimental restoration work done by many Chinese and foreign scholars, especially why Mr. Yang Yinliu and Mr. Chen Wannai had reached a conclusion that denied the accuracy of Zhu Zaiyu's different-diameter pitch pipes, we have also found the cause of the problem. Mr. Yang Yinliu did not strictly abide by Zhu Zaiyu's original methods of using the mouthpieces by neglecting the influence of the special mouthpiece and mistakenly treated Zhu Zaiyu's pitch pipes as close-ended pitch pipes, thus giving wrong conclusion regarding the accuracy of Zhu Zaiyu's pitch pipes. Interestingly, Mr. Yang denied Zhu Zaiyu's pitch pipes. However, his amendment was correct as an independent research result. This showed that the production plan for the pitch pipes that comply with the equal temperament was not exclusively unique. As open-ended pitch pipes, Zhu Zaiyu's production conformed to the equal temperament law. Yang Yinliu's amendment should no longer be used to negate the accuracy of Zhu Zaiyu's different-diameter pitch pipes.

For a long time, Yang Yinliu's misunderstanding of Zhu Zaiyu's pitch pipes was not clearly pointed out, which had a great negative impact on later scholars. For example, although the Taiwanese scholar Chen Wannai also knew that "in Zhu Zaiyu's experiments, open-ended pitch pipes had been used," he still insisted that "in fact, the open pipes and close-ended pipes were both pitch pipes. The open-ended pipes had pitches one octave higher than those of the close-ended pitch pipes." (Chen Wannai: Research on Zhu Zaiyu (朱载堉研究). Taipei Palace Museum Series (《台北故宫博物院丛刊》) p. 86. 1992). Therefore, until the 1990s, Mr. Chen Wannai also neglected the difference between the close-ended pitch pipes and the open-ended pitch pipes as well as Zhu Zaiyu's special mouthpiece. In the acoustic tests of Zhu Zaiyu's pitch pipes, he made a mistake by claiming "use the finger to cover the end and reached a conclusion that Zhu's pitch pipes were not accurate as Yang Yinliu did in proposing the so-called modification law." The results obtained in violation of Zhu Zaiyu's regulation regarding the production of pitch pipes and pitch requirements could no longer be used to verify the correctness of Zhu Zaiyu's equal temperament law. While fully affirming the academic contributions of senior scholars such as Yang Yinliu and Chen Wannai, it is also necessary to clarify their false and negative influence on the future research track regarding Zhu Zaiyu and his contributions to musicology.

In the past period, due to various reasons, the physical laws of mouth correction have not been well described and expressed in certain fields of musicology to such a degree that many scholars simply believed that the tones of some open-ended pitch pipes are just one octave higher than those of the close-ended pitch pipes. Even in some of the current musicology works, such arbitrary statements exist. Due to the vague understanding of the different mouth correction parameters between the open-ended and the close-ended pitch pipes, despite Yang Yinliu and Chen Wannai had had theoretical calculation and restoration experiments, they misunderstood Zhu Zaiyu's pitch pipes as close-ended ones and made the wrong judgment on the accuracy of the pitch pipes.

In modern times, many scholars including Yang Yinliu and Chen Wannai, when studying Zhu Zaiyu's pitch pipes or conducting restoration experiments, did not strictly observe the original requirements of Zhu Zaiyu regarding the parameters of the pitch pipes, ignoring the influence of its special mouthpiece, especially. They treated Zhu Zaiyu's pitch pipes as close-ended pipes in the acoustic measurements, which would inevitably lead to an error of about 3 cents between each semitone, resulting in a cumulative error of 30–50 cents among all the 12 semitones. Such errors did not result from the errors of Zhu Zaiyu's pitch pipes or his equal temperament law. Instead they were caused by the failure to strictly abide by the regulations of Zhu Zaiyu's different-diameter pitch pipes during the restoration experiments or calculation process.

Yang Yinliu's modification which regarded Zhu Zaiyu's pitch pipes as close-ended ones had far-reaching negative influence, which shall be corrected soon without further delay. The analysis of the gains and losses of Yang Yinliu's modification plan also made us understand the great achievements of Zhu Zaiyu's creation of the equal temperament theory and practice from another perspective.

Zhu Zaiyu not only formulated the equal temperament in theory but also gave the criteria for strings and pitch pipes according to the equal temperament law through the hands-on experiment. Research on the restoration of Zhu Zaiyu's string criteria in line with the equal temperament was relatively easy to do, and some scholars have done such research, like Chen Yingshi's work (Chen Yingshi. Tuning Standard and Pitch Pipes (均准'和律琴). Musical Instruments (《乐器》). 1987.(4)). The precise data of the equal temperament calculated by Zhu Zaiyu could be verified from the string criteria through the "new tuning system," a process recorded in Zhu's book entitled *Lv Xue Xin Shuo* (《律学新说》) *A New Account of the Science of the Pitch-pipes*. Therefore, "Zhu's new tuning system was not only a tuning instrument, but also the world's first stringed instrument made in accordance with the theory of equal temperament. The first string instrument in line with the equal temperament law in the West was the keyboard instrument created in about 1711, the piano made by the Italian Bartolomeo Cristofori (1655–1731). The tuning method of the piano is to strike a steel string set with a twelve-average temperament, which is the same as Zhu's hand-plucked strings made in line with the equal temperament, completely identical in nature." (Wang Hongyun. The development of the String Criteria Chinese history (中国历史上弦准的发展)." Research on the History of Natural Sciences (《自然科学史研究》) 1991, No. 4).

Zhu Zaiyu's major achievements in the history of physical acoustics are more revealed in a sense that he discovered through experiments that the mouth correction problem existed in pitch pipes and he successfully solved this problem, giving a set of complete data of pipe lengths and pitches that were totally consistent with the equal temperament law. In the meantime, it also ensured that the acoustic tests of all the pitch pipes data met the physical requirements of the physical tuners made in line with the equal temperament law. Thus, they constituted Zhu Zaiyu's different-diameter pitch pipes (Picture 12.8).

In addition to giving these data, Zhu Zaiyu also gave detailed production processes and small sample drawings. Judging from Zhu Zaiyu's original book, he not only produced these pitch pipes but also performed acoustic measurements,and concluded that the pitch pipes made in line with the three-point losing and gaining

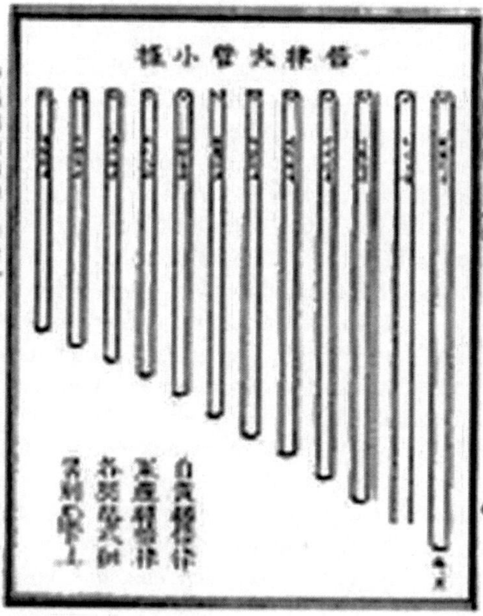

Picture 12.8 Sketch of Zhu's different-diameter pitch pipes

law were not harmonious. The pitch pipes made according to his parameters were harmonic and workable. (The New and Old Pitch Pipes Experiment No. 7 (新旧律实验第七). *Lv Lv Jing Yi* the Internal Chapter (律吕精义》) *the Essence of Pitch-pipe Temperament Vol. 5)*.

However, the acoustic textures of strings and pitch pipes were different. The pitches of pipes were not simply proportional to the pipe length. Other factors included the mouth correction, and other acoustic factors would influence the acoustic performance of pitch pipes. Based on the literature review and the above analysis, we know that Zhu Zaiyu not only found the mouth correction phenomenon of pitch pipes. He also creatively solved the long-lasting problem through experiments and theoretical calculations and gave a set of pitch pipe manufacturing data and small sample artwork that he tested fully to conform to the equal temperament. However, were the mouth correction data used by Zhu Zaiyu reasonable? The final judgment on this question depends largely on the results of the restoration experiments.

More than half a century ago, when famous scholar Liu Fu introduced Zhu Zaiyu's great achievements, he put forward his views on Zhu Zaiyu's data regarding the different-diameter pitch pipes. He believed: "This kind of calculation method, regardless of flaws, it was advisable as a theory with systematic structure. I did not reach such a conclusion out of whim. However, we Chinese have never studied it well, experimented with it, but just treated it indifferently!" Liu also aimed to call for more attention by claiming that "if more scholars had done more experiments on this again ... Zhu's calculation of the different-diameter pitch pipes and production methods would have been recognized and taken as more seriously in the history of musicology in the world."

Over the past hundred years, some Chinese and foreign scholars have successively conducted some restoration acoustic measurement experiments on Zhu Zaiyu's different-diameter pitch pipes. The conclusions of these experiments, however, were inconsistent and even opposing to one another. Some scholars have found that any little technical treatment carried out during the experiments and acoustic tests may have affected the conclusion of the experiment. This has made the question of whether the systematic mouth correction method adopted by Zhu Zaiyu was effective to remain unresolved.

The pitch pipes and the mouth correction among other issues involve physics. In particular, mouth correction was quite professional based on the physical acoustic experiments; many scholars have reached different conclusions due to their different research approaches or the differences in theoretic ground when conducting the pitch pipe studies. Therefore, although some scholars had begun to carry out restoration and acoustic measurement experiments on Zhu Zaiyu's different-diameter pitch pipes since the nineteenth century, quite a few scholars did not strictly abide by Zhu Zaiyu's theoretical requirements and production specifications. No wonder they would obtain different conclusions. In these restoration experiments, a common mistake was to treat Zhu Zaiyu's pitch pipes as close-end pipes and ignore the uniform size of the mouthpiece that Zhu Zaiyu opened at the upper end of each pitch pipe, resulting in the final experimental results with reliability greatly reduced and even conflicting results. If the sketch of the pitch pipe samples as shown in the original book of Zhu Zaiyu had been taken, it would be easy to find that most of the restoration acoustic tests were done in a rather rough way. Therefore, their results were not solid enough for people to believe in.

We can summarize the existing research work on the restoration experiments and others regarding Zhu Zaiyu's different-diameter pitch pipes.

研究者	实验或分析方法	实验结果	发表年份
马　容	仅作三只律管，开、闭口管都做了测音	合律	1890 年
杨荫浏	用闭口管发音公式计算模拟	基本合律，有细微误差	1952 年
庄本立	制作玻璃律管，测定闭口管发音	同上	1956 年
陈权芳	用试管模拟律管，测定闭口管发音	同上	1987 年
陈万鼐	用玻璃及金属材料复原，测定闭口管发音	同上	1992 年
刘　勇	用金属材料复原，测定开口管发音	合律	1992 年
徐　飞	用开口管发音公式并考虑管口校正后计算模拟	合律	1996 年

It can be seen from Table 12.6 above that in the past 100 years, among the restoration and acoustic measurement experiments of Zhu Zaiyu's pitch pipes, Liu Yong's restoration work was relatively in line with the requirements of Zhu Zaiyu's original work. His conclusions also provide a reference close to the truth of history regarding whether we could be certain about Zhu's pitch pipes' conformity of the equal temperament. However, the work of other scholars was based on the adoption of closed-end pipes as a sound measurement standard, which had been wrong in form and specification. Therefore, their conclusions could not naturally be used as a basis for judging whether Zhu Zaiyu's pitch pipes were in line with the equal temperament law.

Although we use modern physics theory analysis to prove that under the premise of strictly following Zhu Zaiyu's production rules, the restored pitch pipes of different diameters as proposed by Zhu Zaiyu were completely conformed to the equal temperament law. But such an analysis needs confirming through experiments. We look forward to that experts will complete this experiment strictly at an early date. The following will focus on the discussion of several principles related to the completion of this experiment.

12.5.1 Basic Norms of Restoring Zhu Zaiyu's Different-Diameter Pitch Pipes

When conducting research on the restoration of Zhu Zaiyu's pitch pipes, we should strictly abide by Zhu Zaiyu's rule of forms and production requirements regarding

Table 12.6 Summary of the restoration and acoustic measurements of Zhu Zaiyu's different-diameter pitch pipes

Chinese	English
研究者	Researcher
马容	Victor C Mahillon
杨荫浏	Yang Yinliu
庄本立	Zhuang Benli
陈权芳	Chen Quanfang
陈万鼐	Chen Wannai
刘勇	Liu Yong
徐飞	Xu Fei
实验或分析方法	Experimental or analytical methods
仅作三只律管, 开、闭口管都做了测音	Three pipes were made with acoustic tests over open-ended and close-ended pipes
用闭口管发音公式计算模拟	The formula for the open-ended pipe of pitch frequency was taken for calculational simulation
制作玻璃律管, 测定闭口管发音	Making glass pitch pipes, measuring pitch frequencies of close-ended pipes
用试管模拟律管, 测定闭口管发音	Using testing tube as the simulator for pitch pipes, measuring the pitch frequencies of close-ended pipes
用玻璃及金属材料复原, 测定闭口管发音	Using the materials of glass or metal to measure the pitch frequencies of the close-ended pipe
用金属材料复原, 测定开口管发音	Using the metal materials to conduct restoration, measuring the pitch frequencies of the open-ended pipes
用开口管发音公式并考虑管口校正后计算模拟	The formula for the open-ended pipe of pitch frequency with the mouth correction was taken for calculational simulation
实验结果	Experimental results
合律	Harmonic
基本合律, 有细微误差	Fundamentally harmonic, with minor differences
同上	Ditto
Chinese	**English**
同上	Ditto
同上	Ditto
合律	Harmonic
合律	Harmonic
发表年份	Year of publication

the pitch pipes. It serves as the basic criterion for the study of the history of science. According to the records in *Yue Lv Quan Shu* (《乐律全书》) *The Complete Book of the Tuning System*, Zhu Zaiyu first proved through experiments that the pipes of the same diameter made according to the ancient three-point losing and gaining law cannot reproduce all the tones or achieve the transition and modulation (*Lv Lv Jing Yi* the Internal Chapter (律吕精义》) *the Essence of Pitch-pipe Temperament Vol. 2)*. Then he analyzed the cause of the problem and found two ways to adjust the pitches of the pipes: change the pipe lengths or adjust the internal diameter. The pipe lengths were supposed to conform to the equal temperament law; Zhu Zaiyu chose to change the diameters of the pipes to conduct the mouth correction. He used 36 pitch pipes to tune the tones in conformity with the equal temperament, representing the 36 tones from the lower octave through the middle octave to the higher octave, equivalent to the 36 notes ranging from the octave lower to the octave higher than middle C note. He set the length of the lower Huangzhong to 2 feet (Xia Chi), the middle Huangzhong to 1 foot, and the higher Huangzhong to a half foot. The remaining lengths of the other pipes can be calculated in the same way as those for the string lengths made according to the new law for density rate. For the internal diameters and external diameters of pipes, follow the methods of the inner square and the outer circle as shown in *Zhou Li Li Shi Wei Liang* (《周礼·栗氏为量》), and set the internal diameter of the lower Huangzhong as 9 cun and 1 fen. (*Lv Lv Jing Yi* the Internal Chapter (律吕精义》) *the Essence of Pitch-pipe Temperament Vol. 2)*. In this way, the two ends of the pipe are like the concentric circles with internal and external diameters. The inscribed square of the external circle has the side equal to the length of the internal diameter. The diagonal line of the external square is taken as the external diameter. Therefore, the internal diameter of the lower octave pipe is equal to the external diameter of the middle octave pipe. The internal diameter of the middle octave pipe is equal to the external diameter of the higher octave pipe (see Picture 12.9).

As can be seen from the picture, the idea of "inner square and outer circle" as illustrated in *Li Shi Wei Liang* (《栗氏为量》) did inspire Zhu Zaiyu in formulating the equal temperament. Geometrically speaking, the internal diameters and external

Picture 12.9 Diagram for the diameter algorithm of Zhu Zaiyu's pitch pipes

diameters that conform to the above graphs are undoubtedly intoxicating in mathematics. Scientific aesthetic functions may sometimes lead scientists to an unexpected harvest. Zhu Zaiyu adopted the inner and outer diameters of the rhythm tubes inlaid with each other, which is in line with the principle of scientific simplicity, and it is easy to draw from the design. Secondly, as most people still do not understand the calculation relationship of the geometric progression, using the geometric diagram of inner square and outer circle, the internal diameters and lengths among other parameters of the pitch pipes could be calculated one by one with understanding of the Pythagorean theorem and the conversion method of diameter and circumference This method of calculation and interpretation were similar to the traditional academic ideas, which were easily to be accepted by the public and the court.

This is why Zhu Zaiyu often gave the calculation data of each pitch one by one in *Yue Lv Quan Shu* (《乐律全书》) *The Complete Book of the Tuning System*. According to him: these data are for correction arithmetic. Because at that time people had almost no understanding of the geometric progression, it is impossible to express the equal temperament with the proportional relationship like modern people. Taking more common scientific understanding ability in the era of Zhu Zaiyu, to accept the equal temperament, he would have to first encounter mathematical obstacles and calculate the numerical values of the geometric progression indirectly. Most people would have difficulty understanding the data regarding Zhu Zaiyu's pitch pipes, the origin of which was not as easy to understand as the three-point losing and gaining law. This might have been an important reason why Zhu Zaiyu's new law for density rate and related series of achievements were not widely accepted for a long time. Zhu Zaiyu had actually foreseen that. Therefore, he repeatedly listed the data of each pitch pipe in *Yue Lv Quan Shu* (《乐律全书》) *The Complete Book of the Tuning System* for the pipe makers so they could just follow the parameters, namely, the lengths and the internal and external diameters, and the sample drawing to make the pitch pipes. (*Lv Lv Jing Yi* the Internal Chapter (律吕精义》) *the Essence of Pitch-pipe Temperament Vol. 2).* Since not everyone is proficient in arithmetic, Zhu Zaiyu also carefully calculated the data of the equal temperament with different precisions according to different academic requirements for later generations to study. When calculating, Zhu Zaiyu generally counted to 9 digits (Without Taking the Internal Diameters No. 5 Part I. Lv Lv Jing Yi (《律吕精义》) the Essence of Pitch-pipe Temperament. (Vol. 2)) or 18 digits ("Without Taking the Internal Diameters No. 5 Part 2. Lv Lv Jing Yi (《律吕精义》) the Essence of Pitch-pipe Temperament. (Vol. 3)). When making the sample drawings of the musical instruments for the craftsmen, he only counted to 18 digits. (Musical Instrument Sample Drawing No.10 Part 1 (Lv Xue Xin Shuo (《律学新说》) A New Account of the Science of the Pitch-pipes (Vol.8)). However, due to the obstacles in mathematics and the limitation of social history, Zhu's theories had been ignorant for a long time. Even until the present day, some scholars only used $\sqrt[12]{2}$ as the proportion for the pipe length change and $\sqrt[24]{2}$ as the proportion for the pipe diameter change. They might have randomly assumed some other data used by Zhu Zaiyu concerning the experiments of the pitch pipes and string instruments as the data for analysis and experiments. They started from the wrong data and obtained wrong conclusions regarding Zhu Zaiyu's equal temperament law. (Chen Zhengsheng: Analysis of Zhu Zaiyu's Different-diameter Pitch Pipes (朱载堉异径管律分析). "Chinese Musicology (《中国音乐学》)", 1987, No.1).

In fact, Zhu Zaiyu not only strictly regulated the size of each part of pitch pipes, but also had strict requirements on the materials, methods, tools, and operating techniques for making the pitch pipes and gave detailed instructions. According to Zhu Zaiyu's requirements, the materials used to make pitch pipes "should be in line with the ancient musical laws. Copper would be better than bamboo." (Musical Instrument Manufacturing No. 7, Lv Xue Xin Shuo (《律学新说》) A New Account of the Science of the Pitch-pipes (Vol.1)).

In order to ensure the accuracy of the external diameters of the pitch pipes, Zhu Zaiyu specially made 36 special steel files. The basic method of making pitch pipes was to repair the inside and outside of the cast; the external square steel files and the internal circle steel files were used. As for the method of filing, there were also clear regulations. (The New and Old Pitch Pipes Experiment No. 7 (新旧律实验第七). *Lv Lv Jing Yi* the Internal Chapter (律吕精义》) *the Essence of Pitch-pipe Temperament Vol. 5).* Zhu Zaiyu also had requirements for the workers who made laws. (Musical Instrument Manufacturing No. 7, Lv Xue Xin Shuo (《律学新说》) A New Account of the Science of the Pitch-pipes (Vol.1)). Especially worth mentioning is the special mouthpiece of Zhu Zaiyu's pitch pipes with the shape of mouthpieces of horizontal flute and the cave entrance." (*Lv Lv Jing Yi* the Internal Chapter (律吕精义》) *the Essence of Pitch-pipe Temperament Vol.8).* This special mouthpiece is the finishing touch of Zhu Zaiyu's pitch pipes. Previously, we have proved that if Zhu Zaiyu's pitch pipes were not considered, based on the data of the open-ended pipes, there was the difference of about 1 cent values between different pitch pipes. The accumulated error of the 36 pitch pipes would be relatively obvious. Therefore, Zhu Zaiyu used a uniform mouthpiece to correct the above errors at one time, making the different-diameter pitch pipes fully compliant. However, it is regrettable that when the recent people tried to restore Zhu Zaiyu's pitch pipes, they almost did not consider the influence of this mouthpiece. Thus, the error of the conclusions was obvious.

The last point which could not be ignored was the requirements regarding the pipe players. (Wind Instruments. No. 8. *Lv Lv Jing Yi the Internal Chapter* (律吕精义》) *the Essence of Pitch-pipe Temperament Vol. 1).* Zhu gave a detailed account of the quality of the pipe players, the ends of the pipes, the breathing control, the angle of blowing the pipe, the position of the mouth, and the volumes as well as the texture of the pipes. However, when modern scholars experimented with pipes, they mostly used closed-end pipes, which violated Zhu Zaiyu's original intention in terms of pitch frequencies. An important reason for this misrecognition may be that the closed-tube pipe playing was easier, and the open-pipe playing was, as Zhu Zaiyu said, you can hardly blow out more than 2 of the pitches. Besides, there was a misconception among the academic circles which mistakenly think that the pitches of the open-ended pipes and closed pipes can be only one octave away. So most scholars extrapolated Zhu Zaiyu's open-ended pipe with the pitches of closed pipes, which has led to the inevitable errors of research conclusions.

In fact, if you want to carry out restoration and acoustic measurement research based on Zhu Zaiyu's original records, you must strictly comply with Zhu Zaiyu's technical regulations. In particular, the existence of mouthpieces cannot be ignored, nor can they be replaced by closed-ended pipes. Since Zhu Zaiyu also quoted the History of the Han Dynasty (《汉书》) as a historical basis, emphasizing the correctness of the pitches of the open-ended pipes and the difficulty of pitching based on the open-ended pipes, therefore Zhu Zaiyu's different-diameter pipes were made through a rigorous experiment. If the closed-end pitch pipes can be used to obtain a verifiable result, Zhu Zaiyu had not obviously bothered to spend so much time choosing the good work and open-ended pipes which could not be properly played by nearly eight out ten people.

Therefore, when we restore this set of pitch pipes, we should strictly abide by the norms formulated by Zhu Zaiyu before we could reach conclusions consistent with historical facts.

Upon the systematic investigation of Zhu Zaiyu's achievements in the equal temperament theories and experiments, it has already been fully evidenced: Zhu Zaiyu's different-diameter pitch pipes were the same as his new law for density rate, constituting a major contribution to the world civilization. Through rigorous mathematical calculations, many experimental explorations, and ingenious process designs, it solved the physical problem of pitch pipes with the mouth correction in line with the equal temperament law once and for all. Zhu also established the basic specifications for making the standard pitch pipes in conformity with the equal temperament and provided the first practical manufacturing scheme, physical model, and sound measurement results.

Due to Zhu Zaiyu's ingenious design of different-diameter pipes, it still took future generations special efforts and even several centuries before they could fully understand the subtlety of Zhu's theories. This once again proves the profoundness of ancient Chinese scientific and technological achievements and even the ancient Chinese civilization. No wonder Zhu Zaiyu also stated in his writing:

This is something that hasn't been in existence for more than two thousand years. Since the beginning of our current dynasty, all the musicians and scholars have been trying their best Pictures (12.10 and 12.11).

The research and publicity of Zhu Zaiyu's great achievements by modern Chinese scholars have been initiated since Mr. Liu Fu published the article "Zhu Zaiyu the Inventor of the Equal Temperament." Then through the joint efforts of Wang Guangqi, Yang Yinliu, Dai Nianzu, Chen Wannai, and other scholars as well as the specialized studies of musicology, it finally has been made clear both at home and abroad Zhu Zaiyu's theory and practice of the equal temperament had been a great contribution of ancient China to the world civilization. When we study Zhu Zaiyu's works in depth with modern physics theory and analyze the physical pitch pipes of different diameters, it is not difficult to find that Zhu Zaiyu's achievements are indeed convincing. Its equal temperament theoretical calculation results have been recognized as the world's first of its kind. At the same time, Zhu Zaiyu's unique Chinese traditional cultural style of analytical thinking and arithmetic method also showed that ancient Chinese scholars had theoretical thinking that was also

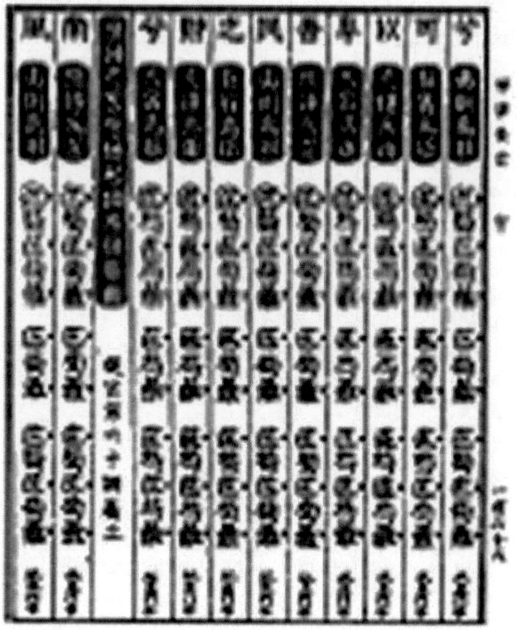

Picture 12.10 Zhu Zaiyu's scores of transition and modulation of equal temperament

comparable to other civilizations. What is more, Zhu Zaiyu also created the world's first of its kind in the discovery and adjustment of the mouth correction of the pitch pipes. He was the first in the history of human civilization who, under the guidance of the equal temperament law, made physical pitch pipes and corresponding musical instruments with strict compliance with the equal temperament requirements. This series of important achievements in physical acoustics showed that there had been indeed wonderful chapters in ancient Chinese acoustics.

In the West, it is generally believed that the earliest complete mathematical calculations for the equal temperament were that of Dutch scholar Simon Stevin (1548–1620) and French mathematician Marson Mersenne (1588–1648). Their calculation results were a few years or even decades later than Zhu Zaiyu. Most of them were theoretical deductions that had not been described. They could not be compared with the integrated solution of Zhu Zaiyu ranging from the algorithm calculation to the tuning system and production of pitch pipes and even the performance of transition and modulation. It was not until the eighteenth century that the musicians started to use the equal temperament law to compose tunes. In 1722, the works of Bach's Well-Tempered Clavier created by German composer J. Sebastian Bach (1685–1750) really made the equal temperament from theory into reality. Piano, the first equal tempered musical instrument in the West, was also created by Christopheri an Italian in around 1711. In contrast, Zhu Zaiyu's bamboo musical instruments capable of transition and modulation and the tuned music scores that had

Picture 12.11 Photocopy of Bach's piano manuscript of equal temperament

recorded various possible methods of modulation had been implemented and published for more than a century than the west.

Zhu Zaiyu's simple and easy-to-understand sample drawings of string and pipe standards make it easy for instrumental music craftsmen to produce pitch standard instruments and corresponding instruments that meet the equal temperament requirements. In the West, the work done by many scholars over a hundred years had been done by Zhu Zaiyu in ancient China, which was hardly achieved by many scholars at that time and even today. Behind the historical wonder of Zhu Zaiyu's formulation of the equal temperament law, there are still many mysteries of the history of science waiting for people to explore and discover. But in any case, the research results of many scholars in theory, music theory, and experiments so far have irrefutably proven that the theory and practice of Zhu Zaiyu's formulation of the equal temperament had been a great contribution of ancient China to world civilization!